# ADVANCES IN CHEMICAL PHYSICS

VOLUME 121

# ADVANCES IN
# CHEMICAL PHYSICS

*Edited by*

## I. PRIGOGINE

Center for Studies in Statistical Mechanics and Complex Systems
The University of Texas
Austin, Texas
and
International Solvay Institutes
Université Libre de Bruxelles
Brussels, Belgium

*and*

## STUART A. RICE

Department of Chemistry
and
The James Franck Institute
The University of Chicago
Chicago, Illinois

**VOLUME 121**

AN INTERSCIENCE PUBLICATION
**JOHN WILEY & SONS, INC.**

For ordering and customer service, call 1-800-CALL-WILEY

Library of Congress Catalog Number: 58-9935

ISBN 0-471-20504-4

Printed in the United States of America.

10 9 8 7 6 5 4 3 2 1

# CONTRIBUTORS TO VOLUME 121

ALEKSIJ AKSIMENTIEV, Computer Science Department, Material Science Laboratory, Mitsui Chemicals, Inc., Sodegaura-City, Chiba, Japan

MICHAL BEN-NUN, Department of Chemistry and the Beckman Institute, University of Illinois, Urbana, Illionis, U.S.A.

PAUL BLAISE, Centre d'Etudes Fondamentales, Université de Perpignan, Perpignan, France

DIDIER CHAMMA, Centre d'Etudes Fondamentales, Université de Perpignan, Perpignan, France

C. H. CHANG, Institute of Atomic and Molecular Sciences, Academia Sinica, Taipei, Taiwan, R.O.C.

R. CHANG, Institute of Atomic and Molecular Sciences, Academia Sinica, Taipei, Taiwan

D. S. F. CROTHERS, Theoretical and Computational Physics Research Division, Department of Applied Mathematics and Theoretical Physics, Queen's University Belfast, Belfast, Northern Ireland

MARCIN FIAŁKOWSKI, Institute of Physical Chemistry, Polish Academy of Science and College of Science, Department III, Warsaw, Poland

M. HAYASHI, Center for Condensed Matter Sciences, National Taiwan University, Taipei, Taiwan

OLIVIER HENRI-ROUSSEAU, Centre d'Etudes Fondamentales, Université de Perpignan, Perpignan, France

ROBERT HOŁYST, Institute of Physical Chemistry, Polish Academy of Science and College of Science, Department III, Warsaw, Poland; and Labo de Physique, Ecole Normale Superieure de Lyon, Lyon, France

F. C. HSU, Department of Chemistry, National Taiwan University, Taipei, Taiwan

K. K. LIANG, Institute of Atomic and Molecular Sciences, Academia Sinica, Taipei, Taiwan

S. H. LIN, Department of Chemistry, National Taiwan University, Taipei, Taiwan, R.O.C.; Institute of Atomic and Molecular Sciences, Academia Sinica, Taipei, Taiwan, R.O.C.

ASKOLD N. MALAKHOV (deceased), Radiophysical Department, Nizhny Novgorod State University, Nizhny Novgorod; Russia

TODD J. MARTÍNEZ, Department of Chemistry and the Beckman Institute, University of Illinois, Urbana, Illinois, U.S.A.

D. M. McSHERRY, Theoretical and Computational Physics Research Division, Department of Applied Mathematics and Theoretical Physics, Queen's University Belfast, Belfast, Northern Ireland

S. F. C. O'ROURKE, Theoretical and Computational Physics Research Division, Department of Applied Mathematics and Theoretical Physics, Queen's University Belfast, Belfast, Northern Ireland

ANDREY L. PANKRATOV, Institute for Physics of Microstructures of RAS, Nizhny Novgorod, Russia

Y. J. SHIU, Institute of Atomic and Molecular Sciences, Academia Sinica, Taipei, Taiwan

T.-S. YANG, Institute of Atomic and Molecular Sciences, Academia Sinica, Taipei, Taiwan, R.O.C.

ARUN YETHIRAJ, Department of Chemistry, University of Wisconsin, Madison, Wisconsin, U.S.A.

J. M. ZHANG, Institute of Atomic and Molecular Sciences, Academia Sinica, Taipei, Taiwan

# INTRODUCTION

Few of us can any longer keep up with the flood of scientific literature, even in specialized subfields. Any attempt to do more and be broadly educated with respect to a large domain of science has the appearance of tilting at windmills. Yet the synthesis of ideas drawn from different subjects into new, powerful, general concepts is as valuable as ever, and the desire to remain educated persists in all scientists. This series, *Advances in Chemical Physics*, is devoted to helping the reader obtain general information about a wide variety of topics in chemical physics, a field that we interpret very broadly. Our intent is to have experts present comprehensive analyses of subjects of interest and to encourage the expression of individual points of view. We hope that this approach to the presentation of an overview of a subject will both stimulate new research and serve as a personalized learning text for beginners in a field.

I. Prigogine
Stuart A. Rice

# CONTENTS

# ULTRAFAST DYNAMICS AND SPECTROSCOPY OF BACTERIAL PHOTOSYNTHETIC REACTION CENTERS

S. H. LIN

*Department of Chemistry, National Taiwan University, Taipei, Taiwan, R.O.C.;
Institute of Atomic and Molecular Sciences, Academia Sinica, Taipei,
Taiwan, R.O.C.*

C. H. CHANG, K. K. LIANG, R. CHANG, Y. J. SHIU, J. M. ZHANG,
and T.-S. YANG

*Institute of Atomic and Molecular Sciences, Academia Sinica,
Taipei, Taiwan, R.O.C.*

M. HAYASHI

*Center for Condensed Matter Sciences, National Taiwan University,
Taipei, Taiwan, R.O.C.*

F. C. HSU

*Department of Chemistry, National Taiwan University, Taipei, Taiwan, R.O.C.*

## CONTENTS

*Advances in Chemical Physics, Volume 121*, Edited by I. Prigogine and Stuart A. Rice.
ISBN 0-471-20504-4   © 2002 John Wiley & Sons, Inc.

1

# I.  INTRODUCTION

The photosynthetic reaction center (RC) of purple nonsulfur bacteria is the core molecular assembly, located in a membrane of the bacteria, that initiates a series of electron transfer reactions subsequent to energy transfer events. The bacterial photosynthetic RCs have been characterized in more detail, both structurally and functionally, than have other transmembrane protein complexes [1–52].

For convenience of discussion, a schematic diagram of bacterial photosynthetic RC is shown in Fig. 1 [29]. Conventionally, P is used to represent the special pair, which consists of two bacterial chlorophylls separated by $\sim 3$ Å, and B and H are used to denote the bacteriochlorophyll and bacteriopheophytin, respectively. The RC is embedded in a protein environment that comprise L and M branches. The initial electron transfer (ET) usually occurs from P to $H_L$ along the L branch in 1–4 picoseconds (ps) and exhibits the inverse temperature dependence; that is, the lower the temperature, the faster the ET. It should be noted that the distance between P and $H_L$ is about 15 Å [53–55].

The spectroscopy and dynamics of photosynthetic bacterial reaction centers have attracted considerable experimental attention [1–52]. In particular, application of spectroscopic techniques to RCs has revealed the optical features of the molecular systems. For example, the absorption spectra of *Rb. Sphaeroides* R26 RCs at 77 K and room temperature are shown in Fig. 2 [42]. One can see from Fig. 2 that the absorption spectra present three broad bands in the region of 714–952 nm. These bands have conventionally been assigned to the $Q_y$ electronic transitions of the P (870 nm), B (800 nm), and H (870 nm) components of RCs. By considering that the special pair P can be regarded as a dimer of two

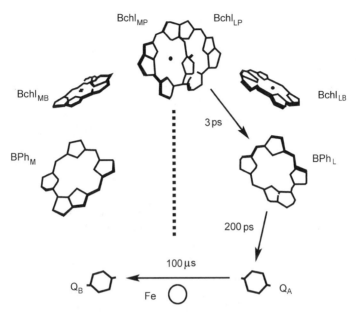

**Figure 1.** Structure of bacterial photosynthetic reaction center. Part of this figure is adapted from Ref. 29.

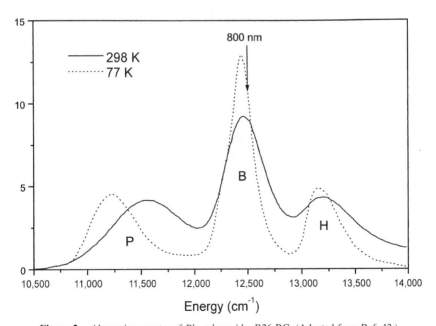

**Figure 2.** Absorption spectra of *Rb. sphaeroides* R26 RC. (Adapted from Ref. 42.)

bacteriochlorophylls, the P band can be assigned to the lower excitonic band of P. By taking into account some difference in the protein environment in the L and M branches [2,3,6–10,17–21,23,30–34,41,42,46,47,52], the 800-nm band can be attributed to $B_L$ and $B_M$ [30–42,46,47].

The vibrational frequency of the special pair P and the bacteriochlorophyll monomer B have also been extracted from the analysis of the Raman profiles [39,40,42,44,51]. Small's group has extensively performed hole-burning (HB) measurements on mutant and chemically altered RCs of *Rb. Sphaeroides* [44,45,48–50]. Their results have revealed low-frequency modes that make important contribution to optical features such as the bandwidth of absorption line-shape, as well as to the rate constant of the ET of the RCs.

The temperature effect on the absorption spectra is also shown in Fig. 2. One can see that the peak position and bandwidth of the P band increase with temperature, while in other bands (like the B and H bands) only its bandwidths show a positive temperature effect. It is important to note that even though the RC is a complicated system, its spectra are relatively simple and its bandwidth is not particularly broad. The above features of absorption spectra of RCs need to be taken into account when analyzing the observed absorption spectra.

In discussing the initial ET mechanism in RCs, the role of $B_L$ needs to be considered. Historically the initial ET was regarded as one-step process $P^*BH \rightarrow P^+BH^-$ with $B_L$ playing the role of bridge group [13]. That is, from the viewpoint of the superexchange ET, the $B_L$ group will provide the virtual states [32]. In Figs. 3 and 4, femtosecond (fs) time-resolved spectra of

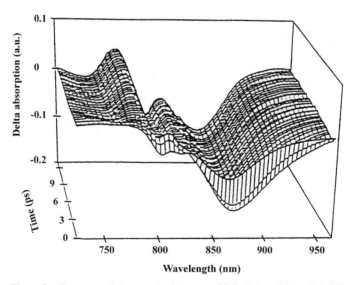

**Figure 3.**   Femtosecond time-resolved spectra of RC. (Adapted from Ref. 86.)

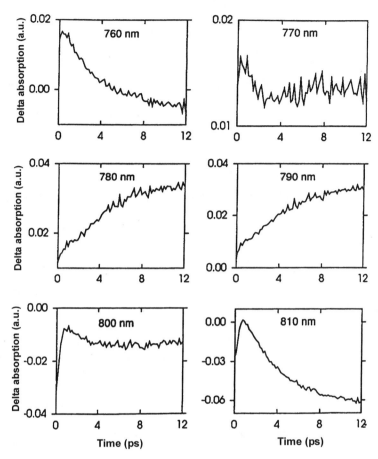

**Figure 4.** Dynamical behaviors of RC observed at various probing wavelengths. (Adapted from Ref. 86.)

R26 are shown. From Fig. 4, we can see that the dynamics of the initial ET depends on the probing wavelengths. This phenomenon is usually called the wavelength-dependent kinetics. Obviously, the one-step model cannot account for this wavelength-dependent kinetics phenomenon because in this case only one rate constant is involved. In order to understand the origin of wavelength-dependent kinetics, the detailed mechanism of the initial ET in RCs should be constructed.

The temperature dependence of the initial ET in RCs raises the interesting and important issue. Considering the fact that the initial ET in RCs takes place in the 1- to 4-ps range, one can ask the question as to whether or not one can use the conventional theory—for example, the Arrhenius-type rate constant

$$k_r = Ae^{-\Delta E_a/kT} \tag{1.1}$$

In conventional theories of rate processes, the temperature $T$ is usually involved. The involvement of $T$ implicitly assumes that vibrational relaxation is much faster than the process under consideration so that vibrational equilibrium is established before the system undergoes the rate process. For example, let us consider the photoinduced ET (see Fig. 5). From Fig. 5 we can see that for the case in which vibrational relaxation is much faster than the ET, vibrational equilibrium is established before the rate process takes place; in this case the ET rate is independent of the excitation wavelength and a thermal average ET constant can be used. On the other hand, for the case in which the ET is much faster than vibrational relaxation, the ET takes place from the pumped vibronic level (or levels) and thus the ET rate depends on the excitation wavelength and often quantum beat will be observed.

The events taking place in the RCs within the timescale of ps and sub-ps ranges usually involve vibrational relaxation, internal conversion, and photoinduced electron and energy transfers. It is important to note that in order to observe such ultrafast processes, ultrashort pulse laser spectroscopic techniques are often employed. In such cases, from the uncertainty principle $\Delta E \Delta t \sim \hbar/2$, one can see that a number of states can be coherently (or simultaneously) excited. In this case, the observed time-resolved spectra contain the information of the dynamics of both populations and coherences (or phases) of the system. Due to the dynamical contribution of coherences, the quantum beat is often observed in the fs time-resolved experiments.

Recently, rapid kinetics have been observed in anisotropy measurements applied to the P and B bands [10,21,31,39,51]. It is found that prior to ET, there exist a few steps in the P and B bands. These measurements provide detailed information on the P and B bands, especially for the nature of the electronic

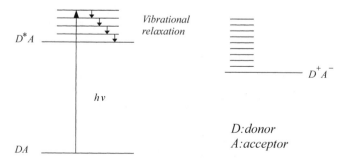

**Figure 5.** Effect of vibrational relaxation on photo-induced electron transfer.

states of these bands and the kinetic scheme of the rapid processes occurring among such electronic states.

Quite recently, Scherer and co-workers [39] have used a 10-fs laser pulse to excite $B^*$ at room temperature; and based on their detailed experimental results, they have proposed the following kinetic scheme:

$$P^*BH \rightarrow P^*_+BH \qquad 0.120 \text{ ps}$$
$$P^*_+BH \rightarrow P^*_-BH \qquad 0.065 \text{ ps}$$

Note that the usage of 10-fs laser pulse leads to rich oscillatory components as well as these rapid kinetics in their pump-probe time-resolved profiles. Obviously in this timescale, the temperature $T$ will have no meaning except for the initial condition before the pumping process. In addition, such oscillatory components may be due not only to vibrational coherence but also to electronic coherence. A challenging theoretical question may arise, for such a case, as to how one can describe these ultrafast processes theoretically.

Identification of electronic states of large molecular systems or assemblies such as reaction centers and light-harvesting complexes of photosynthetic bacteria is always a difficult and challenging task. One of the major reasons may be due to the fact that molecular orbital calculations for such ultralarge systems have not been quite developed yet in terms of efficiency, accuracy, and affordability, although several extensive works have been carried out quite recently [56–65]. We believe that even in the era of the high-performance computer, construction of a model Hamiltonian through analyzing spectroscopic results should provide an alternative way to understand the spectroscopy and dynamics of such ultralarge systems. One of the possible requirements for a successful model Hamiltonian is that it should explain not only the observed optical spectroscopic results but also the dynamics data without modifying any part of the Hamiltonian.

From the above discussion, we can see that the purpose of this paper is to present a microscopic model that can analyze the absorption spectra, describe internal conversion, photoinduced ET, and energy transfer in the ps and sub-ps range, and construct the fs time-resolved profiles or spectra, as well as other fs time-resolved experiments. We shall show that in the sub-ps range, the system is best described by the Hamiltonian with various electronic interactions, because when the timescale is ultrashort, all the rate constants lose their meaning. Needless to say, the microscopic approach presented in this paper can be used for other ultrafast phenomena of complicated systems. In particular, we will show how one can prepare a vibronic model based on the adiabatic approximation and show how the spectroscopic properties are mapped onto the resulting model Hamiltonian. We will also show how the resulting model Hamiltonian can be used, with time-resolved spectroscopic data, to obtain internal conversion rate constant and the electronic coupling. We hope this type of analysis will provide not only experimentalists with a molecular description for their observed results, but also molecular orbital (MO) calculation developers with useful molecular properties for improvements of their MO methods.

## II.  ABSORPTION SPECTROSCOPY OF MOLECULES

### A.  General Theory

Using the time-dependent perturbation method and the dipole approximation [53,66]

$$\hat{H}' = -\vec{\mu} \cdot \vec{E}_0 \cos \omega t \tag{2.1}$$

where $\vec{\mu}$ denotes the dipole operator, we obtain the optical absorption rate constant for the transition $k \to m$ as

$$W_{k \to m} = \frac{\pi}{2\hbar^2} |\vec{\mu}_{mk} \cdot \vec{E}_0|^2 \delta(\omega_{mk} - \omega) \tag{2.2}$$

where $\vec{\mu}_{mk}$ denotes the transition moment and $\delta(\omega_{mk} - \omega)$ represents the delta function. The single-level absorption rate constant $W_k$ is given by

$$W_k = \sum_m W_{k \to m} = \frac{\pi}{2\hbar^2} \sum_m |\vec{\mu}_{mk} \cdot \vec{E}_0|^2 \delta(\omega_{mk} - \omega) \tag{2.3}$$

while the thermal average rate constant can be expressed as

$$W = \sum_k P_k W_k = \frac{\pi}{2\hbar^2} \sum_k \sum_m P_k |\vec{\mu}_{mk} \cdot \vec{E}_0|^2 \delta(\omega_{mk} - \omega) \tag{2.4}$$

where $P_k$ denotes the Boltzmann distribution

$$P_k = \frac{\exp\left(-\frac{E_k}{kT}\right)}{Q(T)} \tag{2.5}$$

Here $Q(T)$ represents the partition function

$$Q(T) = \sum_k \exp\left(-\frac{E_k}{kT}\right) \tag{2.6}$$

If the dephasing (or damping) effect is included, the Lorentzian

$$D(\omega_{mk} - \omega) = \frac{1}{\pi} \cdot \frac{\gamma_{mk}}{(\omega_{mk} - \omega)^2 + \gamma_{mk}^2}$$

instead of the delta function, will be involved in $W$, that is,

$$W = \frac{\pi}{2\hbar^2} \sum_k \sum_m P_k |\vec{\mu}_{mk} \cdot \vec{E}_0|^2 D(\omega_{mk} - \omega) \tag{2.7}$$

If the systems are randomly oriented, then

$$W = \frac{\pi}{6\hbar^2} |\vec{E}_0|^2 \sum_k \sum_m P_k |\vec{\mu}_{mk}|^2 D(\omega_{mk} - \omega) \tag{2.8}$$

Here the following spatial average has been used:

$$\langle |\vec{\mu}_{mk} \cdot \vec{E}_0|^2 \rangle = |\vec{E}_0|^2 |\vec{\mu}_{mk}|^2 \langle \cos^2\theta \rangle = \frac{1}{3} |\vec{E}_0|^2 |\vec{\mu}_{mk}|^2 \tag{2.9}$$

Using the relation between $|\vec{E}_0|^2$ and light intensity I,

$$I = \frac{c}{8\pi} |\vec{E}_0|^2 \tag{2.10}$$

we obtain

$$W = \frac{4\pi^2}{3\hbar^2 c} I \sum_k \sum_m P_k |\vec{\mu}_{mk}|^2 D(\omega_{mk} - \omega) \tag{2.11}$$

To determine the absorption coefficient $\alpha(\omega)$, we can use the Beer–Lambert law and obtain

$$\alpha(\omega) = \frac{4\pi^2\omega}{3\hbar c} \sum_k \sum_m P_k |\vec{\mu}_{mk}|^2 D(\omega_{mk} - \omega) \tag{2.12}$$

Using the relations

$$D(\omega_{mk} - \omega) = \frac{1}{\pi} \cdot \frac{\gamma_{mk}}{(\omega_{mk} - \omega) + \gamma_{mk}^2} = \frac{1}{2\pi} \int_{-\infty}^{\infty} dt e^{it(\omega_{mk}-\omega)-\gamma_{mk}|t|} \tag{2.13}$$

and

$$\vec{\mu}_{mk} = \langle \psi_m |\vec{\mu}| \psi_k \rangle \tag{2.14}$$

we obtain

$$\alpha(\omega) = \frac{2\pi\omega}{3\hbar c} \int_{-\infty}^{\infty} dt e^{-it\omega-\gamma|t|} \langle \langle \vec{\mu} \cdot \vec{\mu}(t) \rangle \rangle = \frac{2\pi\omega}{3\hbar c} \int_{-\infty}^{\infty} dt e^{-it\omega-\gamma|t|} \sum_k P_k$$
$$\times \langle \psi_k |\vec{\mu} \cdot \vec{\mu}(t)| \psi_k \rangle \tag{2.15}$$

where

$$\vec{\mu}(t) = e^{it\hat{H}/\hbar}\vec{\mu}e^{-it\hat{H}/\hbar} \tag{2.16}$$

That is, $\alpha(\omega)$ can be expressed in terms of the dipole correlation function form.

For molecular system, the Born–Oppenheimer (B–O) adiabatic approximation is often used. In this case the molecular wave function can be expressed as

$$\psi_{av}(q, Q) = \Phi_a(q, Q)\Theta_{av}(Q) \tag{2.17}$$

that is, as a product of electronic wave function $\Phi_a(q, Q)$ and the wave function for nuclear motion $\Theta_{av}(Q)$. Here $q$'s and $Q$'s represent the electronic and nuclear coordinates, respectively. In the B–O approximation, we denote $k \rightarrow av$ and $m \rightarrow bv'$, and Fig. 6 shows electronic and vibrational energy levels of a molecular system. The absorption coefficient for molecular systems can be expressed as

$$\alpha(\omega) = \frac{4\pi^2\omega}{3a\hbar c}\sum_{v}\sum_{v'} P_{av}|\vec{\mu}_{bv',av}|^2 D(\omega_{bv',av} - \omega) \tag{2.18}$$

where $a$ is a function of refractive index introduced for solvent effect and

$$\vec{\mu}_{bv',av} = \langle\psi_{bv'}|\vec{\mu}|\psi_{av}\rangle = \langle\Phi_b\Theta_{bv'}|\vec{\mu}|\Phi_a\Theta_{av}\rangle = \langle\Theta_{bv'}|\vec{\mu}_{ba}|\Theta_{av}\rangle \tag{2.19}$$

In the Condon approximation, we have

$$\vec{\mu}_{bv',av} \cong \vec{\mu}_{ba}\langle\Theta_{bv'}|\Theta_{bv'}\rangle \tag{2.20}$$

where $\langle\Theta_{bv'}|\Theta_{av}\rangle$ denotes that vibrational overlap integral. Equation (2.20) can be applied only to the cases of allowed transitions. Substituting Eq. (2.20)

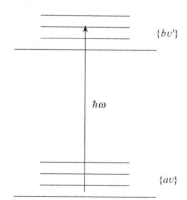

**Figure 6.** Schematic representation of a molecular system.

into Eq. (2.18) yields

$$\alpha(\omega) = \frac{4\pi^2\omega}{3a\hbar c}|\vec{\mu}_{ba}|^2 \sum_v \sum_{v'} P_{av}|\langle\Theta_{bv'}|\Theta_{av}\rangle|^2 D(\omega_{bv',av} - \omega) \qquad (2.21)$$

where $|\langle\Theta_{bv'}|\Theta_{av}\rangle|^2$ represents the Franck–Condon factor. Equation (2.21) shows that each transition is weighted by the Boltzmann distribution and Franck–Condon factor.

Next we shall express $\alpha(\omega)$ in terms of the correlation function. Notice that

$$\begin{aligned}
\alpha(\omega) &= \frac{2\pi\omega}{3a\hbar c}|\vec{\mu}_{ba}|^2 \sum_v \sum_{v'} P_{av} \int_{-\infty}^{\infty} dt\, e^{[it(\omega_{bv',av}-\omega)-\gamma_{ba}|t|]}|\langle\Theta_{bv'}|\Theta_{av}\rangle|^2 \\
&= \frac{2\pi\omega}{3a\hbar c}|\vec{\mu}_{ba}|^2 \int_{-\infty}^{\infty} dt\, e^{[it(\omega_{ba}-\omega)-\gamma_{ba}|t|]} \sum_v P_{av}\langle\Theta_{av}|e^{\frac{it}{\hbar}\hat{H}_b}e^{-\frac{it}{\hbar}\hat{H}_a}|\Theta_{av}\rangle \\
&= \frac{2\pi\omega}{3a\hbar c}|\vec{\mu}_{ba}|^2 \int_{-\infty}^{\infty} dt\, e^{[it(\omega_{ba}-\omega)-\gamma_{ba}|t|]}\langle\langle e^{\frac{it}{\hbar}\hat{H}_b}e^{-\frac{it}{\hbar}\hat{H}_a}\rangle\rangle \qquad (2.22)
\end{aligned}$$

The cumulant expansion method can be applied to Eq. (2.22).

In order to obtain an explicit expression for $\alpha(\omega)$, we shall consider a system consisting of a collection of harmonic oscillators, that is,

$$\Theta_{av} = \prod_i \chi_{av_i}(Q_i), \qquad \Theta_{bv'} = \prod_i \chi_{bv_i'}(Q_i') \qquad (2.23)$$

where $\chi_{av_i}(Q_i)$ and $\chi_{bv_i'}(Q_i')$ denote the vibrational wave functions belonging to electronic states $a$ and $b$. For displaced oscillators (see Fig. 7), at $T = 0$, Eq. (2.21) reduces to

$$\alpha(\omega) = \frac{4\pi^2\omega}{3a\hbar c}|\vec{\mu}_{ba}|^2 \sum_{v'} \prod_i |\langle\chi_{bv_i'}|\chi_{a0_i}\rangle|^2 D(\omega_{bv',a0} - \omega) \qquad (2.24)$$

Using the relations

$$\omega_{bv',a0} = \omega_{ba} + \sum_i (v_i'\omega_i) \qquad (2.25)$$

and

$$|\langle\chi_{bv_i'}|\chi_{a0_i}\rangle|^2 = \frac{S_i^{v_i'}}{v_i'!}e^{-S_i} \qquad (2.26)$$

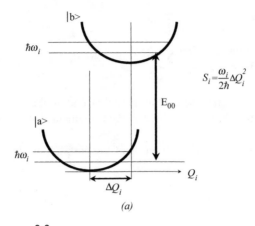

$$S_i = \frac{\omega_i}{2\hbar}\Delta Q_i^2$$

(a)

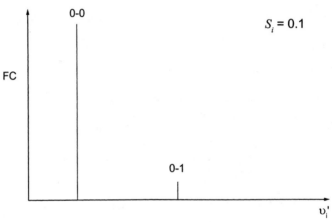

$S_i = 0.1$

(b)

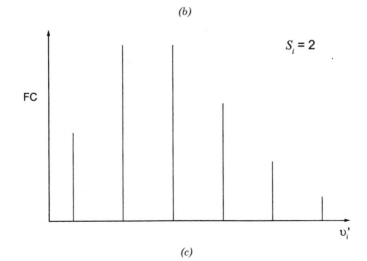

$S_i = 2$

(c)

we obtain

$$\alpha(\omega) = \frac{2\pi\omega}{3a\hbar c} |\vec{\mu}_{ba}|^2 \int_{-\infty}^{\infty} dt \exp[it(\omega_{bv',a0} - \omega) - \gamma_{ba}|t| - \sum_i S_i(1 - e^{it\omega_i})]$$

(2.27)

$S_i$ is often called the Huang–Rhys factor or coupling constant. The plot of the Franck–Condon factor versus $v_i'$ is also shown in Fig. 7.

Similarly for $T \neq 0$ we shall rewrite Eq. (2.24) as

$$\alpha(\omega) = \frac{2\pi\omega}{3a\hbar c} |\vec{\mu}_{ba}|^2 \int_{-\infty}^{\infty} dt \exp[it(\omega_{ba} - \omega) - \gamma_{ba}|t|] \prod_i G_i(t)$$

(2.28)

where

$$G_i(t) = \sum_{v_i} \sum_{v_i'} P_{av_i} |\langle \chi_{bv_i'} | \chi_{av_i} \rangle|^2 \exp\left[it\omega_i\left\{\left(v_i' + \frac{1}{2}\right) - \left(v_i + \frac{1}{2}\right)\right\}\right]$$

(2.29)

To simplify Eq. (2.29) we shall use the relation

$$|\langle \chi_{bv_i'} | \chi_{av_i} \rangle|^2 = \int_{-\infty}^{\infty} dQ_i \int_{-\infty}^{\infty} d\bar{Q}_i \chi_{bv_i'}(Q_i')\chi_{bv_i'}(\bar{Q}_i')\chi_{av_i}(Q_i)\chi_{av_i}(\bar{Q}_i)$$

(2.30)

and the Slater sum (or Mehler formula)

$$\sum_{n=0}^{\infty} \frac{e^{-(n+1/2)t}}{\sqrt{\pi}2^n n!} H_n(x)H_n(x')e^{-\frac{1}{2}(x^2+x'^2)}$$

$$= \frac{1}{(2\pi \sinh t)^{1/2}} \exp\left[-\frac{1}{4}\left\{(x+x')^2\tanh\left(\frac{t}{2}\right) + (x-x')^2\coth\left(\frac{t}{2}\right)\right\}\right]$$

(2.31)

to obtain

$$G_i(t) = \exp\left[-S_i\left\{\coth\frac{\hbar\omega_i}{2kT} - \operatorname{csch}\frac{\hbar\omega_i}{2kT}\cosh\left(it\omega_i + \frac{\hbar\omega_i}{2kT}\right)\right\}\right]$$

$$= \exp[-S_i\{(2\bar{n}_i + 1) - (\bar{n}_i + 1)e^{it\omega_i} - \bar{n}_i e^{-it\omega_i}\}]$$

(2.32)

**Figure 7.** (a) Schematic descriptions of the Huang–Rhys factor and the Franck–Condon factors for displaced oscillators. (b) Franck–Condon factor values for different vibrational transition for $S_i = 0.1$. (c) Franck–Condon factor values for different vibrational transition for $S_i = 2$.

and

$$\alpha(\omega) = \frac{2\pi\omega}{3a\hbar c} |\vec{\mu}_{ba}|^2 \int_{-\infty}^{\infty} dt \exp\left[ it(\omega_{ba} - \omega) - \gamma_{ba}|t| \right.$$
$$\left. - \sum_i S_i\{(2\bar{n}_i + 1) - (\bar{n}_i + 1)e^{it\omega_i} - \bar{n}_i e^{-it\omega_i}\} \right] \quad (2.33)$$

where $\bar{n}_i = (e^{\hbar\omega_i/kT} - 1)^{-1}$, the phonon distribution.

For the strong coupling case $\sum_j S_j \gg 1$, we can use the short-time approximation

$$e^{\pm it\omega_i} = 1 + (\pm it\omega_i) - \frac{\omega_i^2 t^2}{2} + \cdots \quad (2.34)$$

in Eq. (2.33) to carry out the integral

$$\alpha(\omega) = \frac{2\pi\omega}{3a\hbar c} |\vec{\mu}_{ba}|^2 \sqrt{\frac{2\pi}{\sum_i S_i\omega_i^2(2\bar{n}_i + 1)}} \exp\left[ -\frac{(\omega_{ba} + \sum_i S_i\omega_i - \omega)^2}{2\sum_i S_i\omega_i^2(2\bar{n}_i + 1)} \right] \quad (2.35)$$

In this case, the band shape of $\alpha(\omega)/\omega$ versus $\omega$ exhibits a Gaussian form with maximum at $\omega_{ba} + \sum_i S_i\omega_i$ (independent of $T$) and width $2\sum_i S_i\omega_i^2(2\bar{n}_i + 1)$ (dependent on $T$).

In summary, for displaced oscillators, absorption and emission spectra show a mirror image relation and for the strong coupling case, $\alpha(\omega)$ will exhibit a Gaussian band shape, absorption maximum independent of temperature, and bandwidth increasing with temperature. It should be noted that the distortion effect and Duschinsky effect have not been considered in this chapter, but these effects can be treated similarly.

For practical applications, the inhomogeneity effect needs to be taken into account. For the effect of inhomogeneity associated with $\omega_{ba}$, we can use

$$N(\omega_{ba}) = \sqrt{\frac{1}{\pi D_{ba}^2}} \exp\left( -\frac{(\omega_{ba} - \bar{\omega}_{ba})^2}{D_{ba}^2} \right) \quad (2.36)$$

to obtain

$$\langle e^{it\omega_{ba}} \rangle = \exp\left( it\bar{\omega}_{ba} - \frac{D_{ba}^2 t^2}{4} \right) \quad (2.37)$$

and

$$\langle \alpha(\omega) \rangle = \frac{2\pi\omega}{3a\hbar c} |\vec{\mu}_{ba}|^2 \int_{-\infty}^{\infty} dt \exp\left[ it(\bar{\omega}_{ba} - \omega) - \frac{D_{ba}^2 t^2}{4} - \gamma_{ba}|t| \right.$$
$$\left. - \sum_i S_i\{(2\bar{n}_i + 1) - (\bar{n}_i + 1)e^{it\omega_i} - \bar{n}_i e^{-it\omega_i}\} \right] \quad (2.38)$$

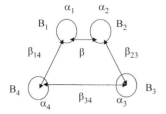

**Figure 8.** Model of excitonic interactions for the special pair (P) and accessory bacteriochlorophylls (B).

## B.  Applications to Photosynthetic RCs

Because we are concerned only with the analysis of the absorption spectra of P band and B band, we consider the excitonic interactions among P, $B_L$, and $B_M$ shown in Fig. 8. Here $(\alpha_1, \alpha_2, \alpha_3, \alpha_4)$ represent the diagonal matrix elements, while $(\beta, \beta_{14}, \beta_{23}, \beta_{34})$ represent the off-diagonal matrix elements [67]. As shown in Introduction, a main feature of the P band is that its absorption maximum shows a pronounced temperature shift [42,52]. According to the displaced oscillator model, the absorption maximum is independent of $T$. Although the distortion effect of potential surfaces will introduce some temperature shift, the effect cannot be as large as that shown in Fig. 2.

From Eq. (2.38) we can see that in order to analyze the absorption spectra it is necessary to know the vibrational modes involved and their Huang–Rhys factors. Fortunately for the B band, these data have been obtained by Mathies and Boxer (see Table I) by analyzing the resonance Raman spectra [42,46,47]. But they only analyzed the B band. The data shown in Table I for the P band are from the literature [42,46,47] and from our fitting of absorption spectra [68].

Here we shall assume that the temperature shift is due to the thermal expansion effect of proteins on the inter chromophore distances among P, $B_L$, and $B_M$. For simplicity we shall assume that the temperature dependence of interchromophore interactions are of the dipole–dipole interaction type. In this case, notice that the excitonic energy splitting of P is $\Delta E = E_{P+} - E_{P-}$,

TABLE I
Selected Vibrational Frequencies, Huang–Rhys Factors, and Electronic Energy Levels of the Involved States of RCs of *Rb. sphaeroides*

| $\omega_i/S_i$ | Vibrational Modes | | | | | | | |
|---|---|---|---|---|---|---|---|---|
| | 34 | 128 | 224 | 560 | 730 | 1160 | 1400 | 1520 |
| $P_-^*$ | 2.200 | 1.600 | 0.045 | 0.017 | 0.033 | 0.030 | 0.020 | 0.040 |
| $P_+^*$ | 0.160 | 0.570 | 0.045 | 0.017 | 0.033 | 0.030 | 0.020 | 0.040 |
| $B_1^*$ | 0.800 | 0.079 | 0.083 | 0.032 | 0.070 | 0.001 | 0.064 | 0.040 |
| $B_2^*$ | 0.800 | 0.079 | 0.083 | 0.032 | 0.070 | 0.001 | 0.064 | 0.040 |

TABLE II
Expansion Coefficients $\xi$ of Some Solvents

| Solvent | $\xi(10^{-4}\,K^{-1})$ |
|---------|------------------------|
| CCl$_4$ | 12.4 |
| EtOH | 11.2 |
| H$_2$O | 2.1 |
| C$_6$H$_6$ | 12.4 |

yielding

$$\Delta E = K'/R^3 = \frac{K'}{R_0^3} \times \frac{1}{\{1 + \xi(T - T_0)\}^3} \approx K_0\{1 - 3\xi(T - T_0)\} \qquad (2.39)$$

Using this model we found

$$\xi = 4.67 \times 10^{-4}/K \qquad (2.40)$$

This value is comparable with the thermal expansion coefficients of other solvents (see Table II).

With the localized basis set—that is, $|\chi_\ell\rangle = |e_\ell\rangle \prod_{m\neq\ell} |g_m\rangle$ where $|e_\ell\rangle$ ($|g_m\rangle$) is the electronically excited (ground) state of the Bchl $B_\ell$ ($B_m$)—the total electronic Hamiltonian for the model system is given by [56]

$$H = \begin{pmatrix} \bar{\alpha} + \Delta & \beta & 0 & \beta_{14} \\ \beta & \bar{\alpha} - \Delta & \beta_{23} & 0 \\ 0 & \beta_{23} & \alpha_3 & \beta_{34} \\ \beta_{14} & 0 & \beta_{34} & \alpha_4 \end{pmatrix} \qquad (2.41)$$

To diagonalize Eq. (2.41), one can adopt a delocalized basis set defined by

$$|e_k\rangle = \sum_{\ell=1}^{4} C_k^\ell |\chi_\ell\rangle \qquad (2.42)$$

The electronic transition moment between $|e_k\rangle$ and $|g\rangle = |g_1\rangle|g_2\rangle|g_3\rangle|g_4\rangle$ is given by

$$\vec{\mu}_{e_k g} = \langle e_k| \sum_{\ell=1}^{4} \vec{\mu}(B_\ell)|g\rangle = \sum_{\ell=1}^{4} C_k^\ell \langle\chi_\ell|\vec{\mu}_{B_\ell}|g_\ell\rangle \qquad (2.43)$$

The absorption intensity can then be expressed as the square of the transition moment. For example, the normalized intensity ratio of the second lowest to the lowest transition is given by

$$\frac{|\vec{\mu}_{e_1g}|^2}{|\vec{\mu}_{e_2g}|^2} = \frac{\sum_{\ell=1}^{4}\sum_{m=1}^{4} C_1^{\ell}C_1^{m}\cos\theta_{B_\ell B_m}}{\sum_{\ell=1}^{4}\sum_{m=1}^{4} C_2^{\ell}C_2^{m}\cos\theta_{B_\ell B_m}} \qquad (2.44)$$

where $\theta_{B_\ell B_m}$ is the angle between the electronic transition dipole moments of the two Bchls $B_\ell$ and $B_m$. In Eq. (2.44), we have assumed $|\vec{\mu}_{B_1}| = |\vec{\mu}_{B_2}| = |\vec{\mu}_{B_3}| = |\vec{\mu}_{B_4}| = |\vec{\mu}_B|$, although the protein environment is different for each Bchl.

For the case in which the magnitude of $\beta$ is larger than that of any $\beta_{\ell m}$ in Eq. (2.41), dimer basis set can be constructed. Notice that in our model, the dimer does not consist of two identical Bchls. By taking into account $\Delta$, our model considers a dimer comprising two electronic asymmetry Bchls.

We first define a unitary matrix as

$$U = \begin{pmatrix} \cos\theta & -\sin\theta & 0 & 0 \\ \sin\theta & \cos\theta & 0 & 0 \\ 0 & 0 & 1 & 0 \\ 0 & 0 & 0 & 1 \end{pmatrix} \qquad (2.45)$$

where

$$\tan\theta = \frac{\beta}{\alpha} \qquad (2.46a)$$

and

$$\alpha = \Delta + \sqrt{\Delta^2 + \beta^2} \equiv \Delta + \Omega \qquad (2.46b)$$

The unitary transformation applying Eq. (2.45) to Eq. (2.41) yields

$$\tilde{H} \equiv U^+HU = \tilde{H}_0 + \tilde{V} \qquad (2.47)$$

$$\tilde{H}_0 = \begin{pmatrix} \bar{\alpha} + \Omega & 0 & 0 & 0 \\ 0 & \bar{\alpha} - \Omega & 0 & 0 \\ 0 & 0 & \alpha_3 & 0 \\ 0 & 0 & 0 & \alpha_4 \end{pmatrix} \qquad (2.48a)$$

and

$$\tilde{V} = \begin{pmatrix} 0 & 0 & \beta_{23}\sin\theta & \beta_{14}\cos\theta \\ 0 & 0 & \beta_{23}\cos\theta & -\beta_{14}\sin\theta \\ \beta_{23}\sin\theta & \beta_{23}\cos\theta & 0 & \beta_{34} \\ \beta_{14}\cos\theta & -\beta_{14}\sin\theta & \beta_{34} & 0 \end{pmatrix} \qquad (2.48b)$$

The dimer basis set can be expressed as

$$|P_+\rangle = \cos\theta|\chi_1\rangle + \sin\theta|\chi_2\rangle \tag{2.49a}$$

$$|P_-\rangle = \sin\theta|\chi_1\rangle - \cos\theta|\chi_2\rangle \tag{2.49b}$$

$$|B_3\rangle = |\chi_3\rangle \tag{2.49c}$$

and

$$|B_4\rangle = |\chi_4\rangle \tag{2.49d}$$

In this case, the intensity ratio of the excitonic transitions becomes

$$\frac{|\vec{\mu}_{P_+g}|^2}{|\vec{\mu}_{P_-g}|^2} = \frac{1 + 2\cos\theta\sin\theta\cos\theta_{B_1B_2}}{1 - 2\cos\theta\sin\theta\cos\theta_{B_1B_2}} \tag{2.50}$$

Note that in the dimer model the absorption intensity depends on $\Delta$ and $\beta$.

There are several possible mechanisms that associate $\Delta$ and $\beta$ with temperature; we shall propose

$$\Delta(T) = \Delta_0[1 - 3\xi(T - T_0)] \tag{2.51}$$

and

$$\beta(T) = \beta_0[1 - 3\xi(T - T_0)] \tag{2.52}$$

as the first step to understand the temperature dependent absorption spectra of RCs. Of course, a more detailed mechanism should be investigated to clarify how structural rearrangements induced due to $T$ affect these quantities.

The energy gaps of R26.*Phe-a* mutant RCs at 50, 100, 150, 200, 250, and 295 K are available as listed in Table III. We apply Eqs. (2.51) and (2.52) and the obtained values $\Delta_0 = 223.7\,\text{cm}^{-1}$, $\beta_0 = 616.7\,\text{cm}^{-1}$, and $\xi = 4.67 \times 10^{-4}\,\text{K}$ to Eq. (2.41). The values of $\bar{\alpha}$, $\beta_{13}$, $\beta_{24}$, and $\beta_{34}$ in Eq. (2.41) still need to be determined. $\bar{\alpha}$ should be the middle value of the 0–0 transitions of the $P_+$ and $P_-$ bands, thus we set the $\bar{\alpha}$ value as $11{,}625\,\text{cm}^{-1}$. Moreover, we set $\alpha_3$ and $\alpha_4$ values as $12{,}403\,\text{cm}^{-1}$ and $12{,}500\,\text{cm}^{-1}$. The values of $\beta_{13}$, $\beta_{24}$, and $\beta_{34}$ can be adopted from the literature, and they are given by $20\,\text{cm}^{-1}$, $30\,\text{cm}^{-1}$, and $10\,\text{cm}^{-1}$, respectively. Note that we assume that $\beta_{13}$, $\beta_{24}$, and $\beta_{34}$ values are independent of temperature. This assumption is reasonable because the distances of $B_1 - B_3$, $B_2 - B_4$ and $B_3 - B_4$ are much larger than that of $B_1 - B_2$ so that distance changes due to thermal expansion should be ignored. Diagonalizing Eq. (2.41), the eigenenergies and the corresponding wave functions can be calculated.

TABLE III

The 0–0 Transitions of the $P_-$, $P_+$ and B Band and the Peak Intensity Ratios of $B/P_-$ and $P_+/P_-$ [a]

| | 0–0 Transition | | | Peak Ratio | |
| | $P_-$ | $P_+$ | B $S=0.8$ (34 cm$^{-1}$) | $B/P_-$ | $P_+/P_-$ |
|---|---|---|---|---|---|
| R26.Phe-a | 10,975.6 | 12,275.6 | 12,404 (806.2 nm) | 1.34 | 0.267 |
| 50 K | | | 12,504 (799.7 nm) | | |
| R26.Phe-a 100 K | 11,010.6 | 12,240.6 | 12,404 12,504 | 1.265 | 0.237 |
| R26.Phe-a 150 K | 11,070.6 | 12,215.6 | 12,404 12,504 | 1.180 | 0.210 |
| R26.Phe-a 200 K | 11,130.6 | 12,180.6 | 12,404 12,504 | 1.128 | 0.209 |
| R26.Phe-a 250 K | 11,190.6 | 12,150.6 | 12,404 12,504 | 1.010 | 0.130 |
| R26.Phe-a 295 K | 11,265.6 | 12,110.6 | 12,404 12,504 | 0.988 | 0.05 |
| R26 77 K | 10,982.6 | 12,250.6 | 12,400 12,475 | 1.27 | 0.368 |
| R26 85 K | 10,975.6 | 12,220.6 | 12,440 12,500 | 1.42 | 0.168 |
| R26 298 K | 11,257.6 | 12,150.6 | 12,400 12,475 | 0.974 | 0.082 |
| WT 1 K | 10,920.6 | 12,270.6 | 12,440 12,440 | 1.091 | 0.517 |
| WT 12 K | 10,940.6 | 12,275.6 | 12,440 12,440 | 0.958 | 0.517 |

[a] These data are obtained by the simulations of the absorption curves of the RCs of wild type, R26 and R26.*Phe-a* mutants of *Rb. sphaeroides* at different temperatures.

To calculate absorption spectra of multielectronic states, the transition dipole moment squares are needed for all the electronic states involved in the relevant optical transition. For this purpose, we next discuss the configurations of the transition dipole moments of the chromophores.

Equation (2.44) indicates that for the delocalized model, the transition dipole intensity ratio is related to $\theta_{B_\ell B_m}$. To obtain $\theta_{B_\ell B_m}$ at various temperatures, we utilize Breton's data regarding the angles between the electronic transition moments of the four BChls and the normal axis of the membrane. The calculated $\theta_{B_\ell B_m}$ are listed in Table IV.

By using Eq. (2.44) and applying $\theta_{B_1 B_2}$ (Table IV) and the wave function coefficients of the delocalized model at various temperatures to calculate the

TABLE IV

The Temperature Dependence of the Angle Relation, Transition Dipole Strength Ratio, and
Transition Dipole Orientation of the Chromophores in the R26.*Phe-a* RCs

| | $T/K$ | | | | | | |
|---|---|---|---|---|---|---|---|
| | 295 | 250 | 200 | 150 | 100 | 50 | 15 |
| $\theta_{B1B2}{}^a$ | 146.54 | 141 | 131.26 | 130 | 127.17 | 124.22 | 123.1 |
| $\theta_{B3}{}^b$ | 68 | 68 | 68 | 68 | 68 | 68 | 68 |
| $\theta_{B4}{}^b$ | 66 | 66 | 66 | 66 | 66 | 66 | 66 |
| *Transition Dipole Strength Ratio* | | | | | | | |
| (1) $\lvert\vec{\mu}_{e1g}\rvert^2/\lvert\vec{\mu}_{e2g}\rvert^2$ | 0.096 | 0.129 | 0.204 | 0.213 | 0.239 | 0.270 | 0.287 |
| (2) $\lvert\vec{\mu}_{e3g}\rvert^2/\lvert\vec{\mu}_{e2g}\rvert^2$ | 0.610 | 0.624 | 0.656 | 0.657 | 0.661 | 0.658 | 0.643 |
| (3) $\lvert\vec{\mu}_{e4g}\rvert^2/\lvert\vec{\mu}_{e2g}\rvert^2$ | 0.497 | 0.517 | 0.559 | 0.570 | 0.592 | 0.619 | 0.638 |
| (2) + (3) | 1.107 | 1.141 | 1.215 | 1.227 | 1.253 | 1.277 | 1.281 |
| *Transition Dipole Orientation* | | | | | | | |
| $\vec{\mu}_{e1g} \cdot \vec{\mu}_{e2g}$ | −0.230 | −0.2053 | −0.1566 | −0.1412 | −0.1136 | −0.0764 | −0.0438 |

$^a$ Angles (degrees) between the transition moments of the special pair dimer.
$^b$ Angles of the transition moments with the membrane normal.

peak ratios $\lvert\vec{\mu}_{e1g}\rvert^2/\lvert\vec{\mu}_{e2g}\rvert^2$, $\lvert\vec{\mu}_{e3g}\rvert^2/\lvert\vec{\mu}_{e2g}\rvert^2$, and $\lvert\vec{\mu}_{e4g}\rvert^2/\lvert\vec{\mu}_{e2g}\rvert^2$, they are found to simultaneously agree with the values presented in Table III. Table IV lists the calculation results of the delocalized model, including the angles between any two Bchls and the transition dipole moment intensity ratios $\lvert\vec{\mu}_{e1g}\rvert^2/\lvert\vec{\mu}_{e2g}\rvert^2$, $\lvert\vec{\mu}_{e3g}\rvert^2/\lvert\vec{\mu}_{e2g}\rvert^2$, and $\lvert\vec{\mu}_{e4g}\rvert^2/\lvert\vec{\mu}_{e2g}\rvert^2$ at various temperatures. Additionally, the inner products of $\vec{\mu}_{e1g}$ and $\vec{\mu}_{e2g}$—namely, $(\vec{\mu}_{e1g} \cdot \vec{\mu}_{e2g})$—are also presented in this Table IV. The inner product $(\vec{\mu}_{e1g} \cdot \vec{\mu}_{e2g})$ can be given by

$$(\vec{\mu}_{e1g} \cdot \vec{\mu}_{e2g}) = \sum_{\ell=1}^{4} \sum_{m=1}^{4} C_1^\ell C_2^m (\vec{\mu}_{B_\ell} \cdot \vec{\mu}_{B_m}) \cong \lvert\vec{\mu}_B\rvert^2 \sum_{\ell=1}^{4} \sum_{m=1}^{4} C_1^\ell C_2^m \cos\theta_{B_\ell B_m}$$

(2.53)

Because $(\vec{\mu}_{e1g} \cdot \vec{\mu}_{e2g})$ is not equal to zero in Table III, $\vec{\mu}_{e1g}$ is not perpendicular to $\vec{\mu}_{e2g}$. This situation is also reported by Hochstrasser and Scherer.

Note that both the delocalized and dimer models can be used to explain temperature dependence of the absorption spectra of R26.*Phe-a* RCs. This is due to the fact that the interactions [Eq. (2.48b)] among Bchls except for that between $B_1$ and $B_2$ are small compared with the diagonal element Hamiltonian in Eq. (2.48a).

Now we are in a position to calculate the absorption coefficient with displaced harmonic potential surfaces using Eq. (2.38). Substituting the

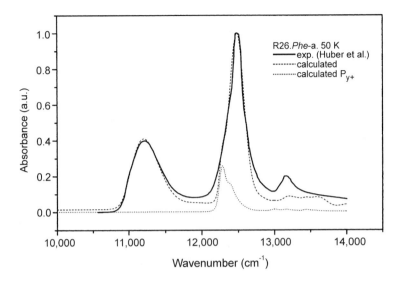

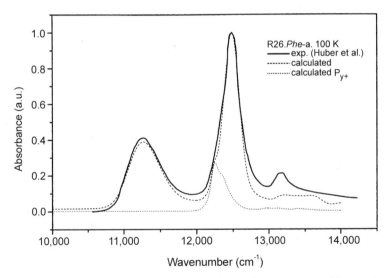

**Figure 9.** Absorption spectra of the R26.*Phe-a* RCs at various temperatures. The experimental data are taken from Ref. 52, and the theoretical absorption spectra are calculated by using the electronic structures and interactions presented in the text and the vibrational properties adopted from the previous work.

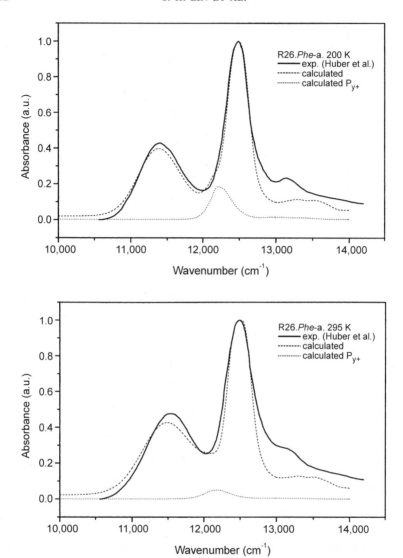

**Figure 9**   (*Continued*)

electronic and vibrational properties listed in Tables I and III into Eq. (2.38), we simulate absorption spectra of the R26.Phe-a RC at various temperatures. Figure 9 compares the experimental absorption spectra at various temperatures with the corresponding calculated absorption spectra using the delocalized model with the thermal expansion model. One can see good agreement between the experimental and theoretical spectra at all temperatures.

To study the difference between these two models, we calculate the anisotropy of several pairs of two electronic transitions by using

$$r(e_k, e_{k'}) = \frac{3(\vec{\mu}_{e_k g} \cdot \vec{\mu}_{e_{k'} g})^2 - (\vec{\mu}_{e_k g} \cdot \vec{\mu}_{e_k g})(\vec{\mu}_{e_{k'} g} \cdot \vec{\mu}_{e_{k'} g})}{(\vec{\mu}_{e_k g} \cdot \vec{\mu}_{e_k g})(\vec{\mu}_{e_{k'} g} \cdot \vec{\mu}_{e_{k'} g})} \qquad (2.54)$$

We calculate three models: (1) the dimer model presented in this work; (2) the dimer model employed by Scherer et al.; and (3) the delocalized model. Table V lists the calculated results as a function of temperature. The anisotropy values for $r(e_1, e_2)$ at 295 K are found to be quite different between the dimer and delocalized models. The difference is about 53%. Meanwhile, the differences at other temperatures are within about 13–14%. For $r(e_2, e_3)$, the differences are within 24–28%. Table V also lists the angle between $B_1$ and $B_2$ in the special pair of R26.*Phe-a* RCs as a function of temperature.

It should be noted that the calculated anisotropy may not be applied to fs time-resolved anisotropy measurements because fs time-resolved experiments involve pumping and probing conditions and may involve overlapping between the vibronic structures of several electronic states due to the use of fs laser pulses. Nevertheless, we think the calculated anisotropy using Eq. (2.54) can provide a reference in comparing models.

As mentioned in the previous section, the temperature-dependent absorption spectra of RCs are very important for the understanding of the molecular properties such as the electronic configurations, vibrational contributions, and transition moment relations of the Bchls in RCs. However, only in the R26. *Phe-a* mutant case have absorption spectra at various temperatures so far been available. Although the absorption spectra of the WT and R26 mutant RCs are available at a few temperatures like 1 K, 4 K, 77 K, and 298 K, the analyzed results are not so consistent (see Table III). It may be because the preparation

TABLE V
The Anisotropy Between the Transition Dipole Moments of the Chromophores in the Dimer Model and Delocalized Model[a]

| | Model | 295 K | 200 K | 100 K | 50 K |
|---|---|---|---|---|---|
| $r_{B1B2}$ | Dimer | −0.040 | −0.141 | −0.152 | −0.160 |
| | Dimer[b] | −0.039 | | | |
| | delocalized | −0.085 | −0.163 | −0.176 | −0.185 |
| $r_{B2B3}$ | Dimer | −0.096 | −0.083 | −0.080 | −0.077 |
| | delocalized | −0.127 | −0.115 | −0.110 | −0.106 |
| $\theta_{B1B2}$ | Dimer | 150.50 | 134.11 | 131.01 | 127.98 |
| | delocalized | 146.54 | 131.26 | 127.17 | 124.22 |

[a] The angle between $\vec{\mu}_{e_1 g}$ and $\vec{\mu}_{e_2 g}$ in different model also shown for comparison.
[b] From Ref. 39.

methods of RCs and the experimental conditions are somewhat different for different groups. Thus, to examine validity of the electronic state model and the thermal expansion model presented in this work, absorption spectra experiments of each RC at various temperatures should be carefully performed.

## III.  ELECTRON TRANSFER, ENERGY TRANSFER, AND RADIATIONLESS TRANSITIONS

### A.  Time-Dependent Perturbation Method

We shall start with the time-dependent Schrödinger equation [53,69–71]

$$\hat{H}\Psi = i\hbar \frac{\partial \Psi}{\partial t} \tag{3.1}$$

where the total Hamiltonian is divided into the noninteracting and the interacting parts:

$$\hat{H} = \hat{H}_0 + \hat{H}' \tag{3.2}$$

The Schrödinger equation of the noninteracting part

$$\hat{H}_0 \Psi_n^0 = i\hbar \frac{\partial \Psi_n^0}{\partial t} \tag{3.3}$$

is assumed to be solvable, and we can find the set of eigenfunctions $\psi_n$ with eigenenergies $E_n$ satisfying the time-independent Schrödinger equation

$$\hat{H}_0 \psi_n = E_n \psi_n \tag{3.4}$$

and the time-dependent solution is expressed as

$$\Psi_n^0 = \psi_n \exp\left(-\frac{itE_n}{\hbar}\right) \tag{3.5}$$

By using the expansion theorem, the total wave function can be expanded in the noninteracting basis set:

$$\Psi = \sum_n C_n(t)\Psi_n^0 \tag{3.6}$$

and by substituting Eq. (3.6) into the complete Schrödinger equation, we obtain the equation of motion of the expansion coefficients:

$$i\hbar \frac{dC_m}{dt} = \sum_n C_n \langle \Psi_m^0 | \hat{H}' | \Psi_n^0 \rangle \tag{3.7}$$

According to the perturbation method, we introduce the perturbation parameter $\lambda$ that labels the order of interaction,

$$i\hbar \frac{dC_m}{dt} = \lambda \sum_n C_n \langle \Psi_m^0 | \hat{H}' | \Psi_n^0 \rangle \tag{3.8}$$

and expand the coefficients in orders of $\lambda$, for example,

$$C_m(t) = C_m^{(0)}(t) + \lambda C_m^{(1)}(t) + \lambda^2 C_m^{(2)}(t) + \cdots \tag{3.9}$$

Suppose that the initial conditions are given by

$$C_k(0) = 1, \qquad C_n(0) = 0 \qquad (n \neq k) \tag{3.10}$$

at $t = 0$, which means that the system initially occupies only one state. For the case in which $\hat{H}'$ is time-independent, we find

$$i\hbar \frac{dC_m^{(1)}}{dt} = \langle \Psi_m^0 | \hat{H}' | \Psi_k^0 \rangle \tag{3.11}$$

or

$$C_m^{(1)}(t) = \frac{H'_{mk}}{\hbar \omega_{mk}} \left(1 - e^{it\omega_{mk}}\right) \tag{3.12}$$

Similarly we have, for the second order,

$$C_m^{(2)} = \sum_n \frac{H'_{mn} H'_{nk}}{\hbar \omega_{nk}} \left( \frac{1 - e^{it\omega_{mn}}}{\hbar \omega_{mn}} - \frac{1 - e^{it\omega_{mk}}}{\hbar \omega_{mk}} \right) \tag{3.13}$$

To the first-order approximation, we find

$$|C_m(t)|^2 = |C_m^{(1)}(t)|^2 = \frac{2}{\hbar^2} |H'_{mk}|^2 \frac{(1 - \cos \omega_{mk} t)}{\omega_{mk}^2} \tag{3.14}$$

As $t \to \infty$, we obtain

$$|C_m^{(1)}(t)|^2 = \frac{2\pi t}{\hbar^2} |H'_{mk}|^2 \delta(\omega_{mk}) \tag{3.15}$$

or

$$W_{k \to m} = \frac{2\pi}{\hbar^2} |H'_{mk}|^2 \delta(\omega_{mk}) \tag{3.16a}$$

Here the following expression for the delta function has been used:

$$\delta(\omega_{mk}) = \frac{1}{\pi} \lim_{t \to \infty} \frac{1}{t} \frac{(1 - \cos\omega_{mk}t)}{\omega_{mk}^2} \tag{3.16b}$$

Similarly, to the second-order approximation, we obtain

$$C_m(t) = C_m^{(1)} + C_m^{(2)} = \left( H'_{mk} + \sum_n \frac{H'_{mn}H'_{nk}}{\hbar\omega_{kn}} \right) \left( \frac{1 - e^{it\omega_{mk}}}{\hbar\omega_{mk}} \right) \tag{3.17}$$

As $t \to \infty$, Eq. (3.17) becomes

$$|C_m(t)|^2 = \frac{2\pi t}{\hbar^2} \left| H'_{mk} + \sum_n \frac{H'_{mn}H'_{nk}}{\hbar\omega_{kn}} \right|^2 \delta(\omega_{mk}) \tag{3.18}$$

or

$$W_{k \to m} = \frac{2\pi}{\hbar^2} \left| H'_{mk} + \sum_n \frac{H'_{mn}H'_{nk}}{\hbar\omega_{kn}} \right|^2 \delta(\omega_{mk}) \tag{3.19}$$

Thus, the thermal average constant is given by

$$W = \sum_k \sum_m P_k W_{k \to m} = \sum_k P_k W_k \tag{3.20}$$

where $P_k$ denotes the Boltzmann distribution function, and $W_k$ represents the single-level rate constant:

$$W_k = \sum_m W_{k \to m} = \frac{2\pi}{\hbar^2} \sum_m \left| H'_{mk} + \sum_n \frac{H'_{mn}H'_{nk}}{\hbar\omega_{kn}} \right|^2 \delta(\omega_{mk}) \tag{3.21}$$

## B. Electron and Energy Transfers

In the study of the ultrafast dynamics of photosynthetic bacterial reaction centers, we are concerned with the photoinduced electron transfer [72]

$$DA \xrightarrow{\hbar\omega} D^*A \to D^+A^- \tag{3.22}$$

and energy transfer

$$DA \xrightarrow{\hbar\omega} D^*A \to DA^* \tag{3.23}$$

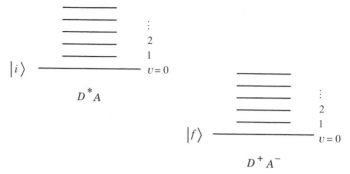

**Figure 10.** Vibronic levels of an electron transfer system. Here the initial electronic state is the $D^*A$ state and the final electronic state is the $D^+A^-$ charge separation state. The vibrational levels within the $D^*A$ manifold are labeled by $v$ and those within $D^+A^-$ by $v'$.

These two processes can be treated on equal footing. A schematic representation of vibronic levels of an electron transfer system is shown in Fig. 10. In other words, we are dealing with two vibronic manifolds, $\{iv\}$ for $D^*A$ and $\{fv'\}$ for $D^+A^-$ or $DA^*$. For convenience, hereafter we shall use the following notation:

$$T_{mk} = H'_{mk} + \sum_n \frac{H'_{mn}H'_{nk}}{\hbar\omega_{kn}} \qquad (3.24)$$

Using the Born–Oppenheimer adiabatic approximation, we obtain

$$T_{fv',iv} = H'_{fv',iv} + \sum_{n\mu} \frac{H'_{fv',n\mu}H'_{n\mu,iv}}{E_{iv} - E_{n\mu}} \qquad (3.25)$$

where, for example,

$$H'_{fv',iv} = \langle \Theta_{fv'} | H'_{fi} | \Theta_{iv} \rangle \qquad (3.26)$$

Here $\Theta_{fv'}$ and $\Theta_{iv}$ denote the vibrational wave function, and

$$H'_{fi} = \langle \Phi_f | \hat{H}' | \Phi_i \rangle \qquad (3.27)$$

where $\Phi_f$ and $\Phi_i$ represent the electronic wave functions. Using the Condon approximation, where the nuclear-coordinate dependence of the electronic interaction matrix elements are ignored, we find

$$T_{fv',iv} = H'_{fi}\langle \Theta_{fv'} | \Theta_{iv} \rangle + \sum_{n\mu} \frac{H'_{fn}H'_{ni}\langle \Theta_{fv'} | \Theta_{n\mu} \rangle \langle \Theta_{n\mu} | \Theta_{iv} \rangle}{E_{iv} - E_{n\mu}} \qquad (3.28)$$

For the case in which the electronic energy gap is much larger than the difference in vibrational energies, we can apply the Plazcek approximation $E_{iv} - E_{n\mu} \approx$

$E_i - E_n$ to Eq. (3.28) to obtain

$$T_{fv',iv} = \langle \Theta_{fv'} | \Theta_{iv} \rangle T_{fi} \tag{3.29}$$

where

$$T_{fi} = H'_{fi} + \sum_n \frac{H'_{fn} H'_{ni}}{E_i - E_n} \tag{3.30}$$

Here the closure relation

$$\sum_\mu \langle \Theta_{fv'} | \Theta_{n\mu} \rangle \langle \Theta_{n\mu} | \Theta_{iv} \rangle = \langle \Theta_{fv'} | \Theta_{iv} \rangle \tag{3.31}$$

has been used to obtain Eq. (3.30).

In this case, the thermal average rate constant given by Eq. (3.20) becomes

$$W_{i \to f} = \frac{2\pi}{\hbar} |T_{fi}|^2 \sum_v \sum_{v'} P_{iv} |\langle \Theta_{fv'} | \Theta_{iv} \rangle|^2 \delta(E_{fv'} - E_{iv}) \tag{3.32}$$

where $|\langle \Theta_{fv'} | \Theta_{iv} \rangle|^2$ is the Franck–Condon factor. It should be noted that in $T_{fi}$ given by Eq. (3.30), the first term $H'_{fi}$ represents the direct transfer process (or the transfer process through space), while the second term $\sum_n \frac{H'_{fn} H'_{ni}}{E_i - E_n}$ denotes the superexchange process (or the transfer through the bond process).

To simplify the expression of the thermal average rate $W_{i \to f}$ given by Eq. (3.32), we shall assume that both potential energy surfaces of the two excited vibronic manifolds of the *DA* system consist of a collection of harmonic oscillators, that is,

$$\Theta_{iv} = \prod_j \chi_{iv_j}(Q_j) \tag{3.33}$$

and

$$\Theta_{fv'} = \prod_j \chi_{fv'_j}(Q'_j) \tag{3.34}$$

Using the relation

$$\delta(\omega_{fv',iv}) = \frac{1}{2\pi} \int_{-\infty}^{\infty} dt e^{it\omega_{fv',iv}} \tag{3.35}$$

for the delta function, Eq. (3.32) becomes

$$W_{i \to f} = \frac{1}{\hbar^2} |T_{fi}|^2 \int_{-\infty}^{\infty} dt \, e^{it\omega_{fi}} \prod_j G_j(t) \tag{3.36}$$

where the so-called nuclear correlation function $G_j(t)$ is expressed as

$$G_j(t) = \sum_{v_j} \sum_{v_j'} P_{iv_j} |\langle \chi_{fv_j'} | \chi_{iv_j} \rangle|^2 \exp\left( it \left\{ \left( v_j' + \frac{1}{2} \right) \omega_j' - \left( v_j + \frac{1}{2} \right) \omega_j \right\} \right) \tag{3.37}$$

We consider the displaced oscillator case, that is, $\omega_j = \omega_j'$ and $\Delta Q_j = Q_j' - Q_j$. In this case, $G_j(t)$ has been evaluated [73]:

$$G_j(t) = \exp\left\{ -S_j \left[ \coth \frac{\hbar \omega_j}{2kT} - \csc h \frac{\hbar \omega_j}{2kT} \cosh\left( it\omega_j + \frac{\hbar \omega_j}{2kT} \right) \right] \right\} \tag{3.38}$$

or

$$G_j(t) = \exp[-S_j \{(2\bar{n}_j + 1) - \bar{n}_j e^{-it\omega_j} - (\bar{n}_j + 1) e^{it\omega_j} \}] \tag{3.39}$$

Substituting Eq. (3.39) into Eq. (3.36) yields

$$W_{i \to f} = \frac{1}{\hbar^2} |T_{fi}|^2 \int_{-\infty}^{\infty} dt \exp\left\{ it\omega_{fi} - \sum_j S_j [(2\bar{n}_j + 1) - \bar{n}_j e^{-it\omega_j} - (\bar{n}_j + 1) e^{it\omega_j}] \right\} \tag{3.40}$$

Various forms can be obtained from Eq. (3.40) for different limiting cases. For the case of strong coupling (i.e., $\sum_j S_j \gg 1$), we can use the short-time approximation—that is, make use of the expansion $\exp(it\omega_j) = 1 + it\omega_j + (it\omega_j)^2/2! + \cdots$ and keep the three leading terms—to obtain

$$W_{i \to f} = \frac{1}{\hbar^2} |T_{fi}|^2 \int_{-\infty}^{\infty} dt \exp\left[ it(\omega_{fi} + \sum_j S_j \omega_j) - \frac{t^2}{2} \sum_j S_j \omega_j^2 (2\bar{n}_j + 1) \right] \tag{3.41}$$

or

$$W_{i \to f} = \frac{1}{\hbar^2} |T_{fi}|^2 \sqrt{\frac{2\pi}{\sum_j S_j \omega_j^2 (2\bar{n}_j + 1)}} \exp\left[ -\frac{(\omega_{fi} + \sum_j S_j \omega_j)^2}{2 \sum_j S_j \omega_j^2 (2\bar{n}_j + 1)} \right] \tag{3.42}$$

This equation is referred to as the Marcus–Levich equation in which the tunneling effect is included.

In the classical regime—that is, $\hbar\omega_j/kT \ll 1$—we shall have $\bar{n}_j = kT/\hbar\omega_j \gg 1$, and Eq. (3.42) reduces to

$$W_{i \to f} = \frac{1}{\hbar^2}|T_{fi}|^2 \sqrt{\frac{\pi\hbar}{kT\sum_j S_j\omega_j}} \exp\left[-\frac{\hbar(\omega_{fi} + \sum_j S_j\omega_j)^2}{4kT\sum_j S_j\omega_j}\right] \qquad (3.43)$$

or

$$W_{i \to f} = \frac{1}{\hbar^2}|T_{fi}|^2 \sqrt{\frac{\pi\hbar^2}{\lambda kT}} \exp\left[-\frac{(\Delta G_{fi} + \lambda)^2}{4\lambda kT}\right] = A\exp\left[-\frac{\Delta E_a}{kT}\right] \qquad (3.44)$$

where the prefactor is

$$A = \frac{1}{\hbar^2}|T_{fi}|^2 \sqrt{\frac{\pi\hbar^2}{\lambda kT}} \qquad (3.45)$$

and the activation energy is

$$\Delta E_a = \frac{(\Delta G_{fi} + \lambda)^2}{4\lambda} \qquad (3.46)$$

Here the free energy change is

$$\Delta G_{fi} = \hbar\omega_{fi} = E_f - E_i \qquad (3.47)$$

and $\lambda$ is called the reorganization energy.

Equation (3.44) (in the Arrhenius form) is usually called the Marcus equation [74,75]. A special feature of the Marcus equation is that it predicts the parabolic dependence of the activation energy $\Delta E_a$ on the free energy change $\Delta G_{fi}$; that is, $\Delta E_a$ is related to the free energy change $\Delta G_{fi}$ in a parabolic form.

The reason that $\Delta G_{fi} = E_f - E_i$ is due to the fact that our system consists of a collection of harmonic oscillators with displaced surfaces between the initial and final electronic states. In this case, the vibrational partition functions are the same between the initial and final electronic states.

It can be shown that for the single-mode case $\Delta E_a$ represents the crossing point energy of the two potential curves belonging to the initial and final electronic states. For the multimode case, there are numerous crossing points and $\Delta E_a$ in this case represents the minimal crossing-point energy of these two potential surfaces (see Appendix I). In other words, for the multimode case of electron transfer the reaction coordinate is meaningful (or well-defined) only when the Marcus equation is valid—that is, in the strong coupling case (short-time approximation) and also in the classical regime (i.e., $\hbar\omega_j/kT \ll 1$). For the general case of electron transfer [see Eq. (3.40)], the activation energy and reaction coordinate are not defined.

In summary, to apply the Marcus theory of electron transfer, it is necessary to see if the temperature dependence of the electron transfer rate constant can be described by a function of the Arrhenius form. When this is valid, one can then determine the activation energy $\Delta E_a$; only under this condition can we use $\Delta E_a$ to determine if the parabolic dependence on $\Delta G_{fi}$ is valid and if the reaction coordinate is defined.

## C. Various Forms of Electron and Energy Transfer Rate Constants

In the previous section we have shown that the Marcus equation can be derived from Eq. (3.40). In this section, other forms of rate constants used in literatures will be derived. Notice that at $T = 0$, Eq. (3.40) reduces to

$$W_{i\to f} = \frac{1}{\hbar^2}|T_{fi}|^2 \int_{-\infty}^{\infty} dt \exp\left[it\omega_{fi} - \sum_j S_j(1 - e^{it\omega_j})\right] \qquad (3.48)$$

Introducing the average frequency $\bar{\omega}$ for $\omega_j$ and $S = \sum_j S_j$ yields

$$W_{i\to f} = \frac{1}{\hbar^2}|T_{fi}|^2 \int_{-\infty}^{\infty} dt \exp[it\omega_{fi} - S(1 - e^{it\bar{\omega}})] \qquad (3.49)$$

or

$$W_{i\to f} = \frac{1}{\hbar^2}|T_{fi}|^2 e^{-S} \int_{-\infty}^{\infty} dt e^{it\omega_{fi}} \sum_{n=0}^{\infty} \frac{(Se^{it\bar{\omega}})^n}{n!} = \frac{1}{\hbar^2}|T_{fi}|^2 e^{-S} \sum_{n=0}^{\infty} \frac{S^n}{n!} \delta(\omega_{fi} + n\bar{\omega})$$

$$= \frac{2\pi}{\hbar^2\bar{\omega}}|T_{fi}|^2 e^{-S} \frac{S^n}{n!} \qquad (3.50)$$

where $n = \omega_{if}/\bar{\omega}$.

Applying the method of steepest descent (or saddle-point method) to Eq. (3.49) yields

$$W_{i\to f} = \frac{1}{\hbar^2}|T_{fi}|^2 \sqrt{\frac{2\pi}{\bar{\omega}\omega_{if}}} e^{-S} \exp\left[-\frac{\omega_{if}}{\bar{\omega}}\left(\ln\frac{\omega_{if}}{S\bar{\omega}} - 1\right)\right] \qquad (3.51)$$

This is usually called the energy gap law expression and can be obtained from Eq. (3.50) by using the Stirling formula for $n!$. Equation (3.51) is often used by Mataga and his co-workers. See, for example, Ref. 76 and the citations therein.

Similarly for the $T \neq 0$ case, we have

$$W_{i \to f} = \frac{1}{\hbar^2} |T_{fi}|^2 \int_{-\infty}^{\infty} dt \exp[it\omega_{fi} - S\{(2\bar{n} + 1) - \bar{n}e^{-it\bar{\omega}} - (\bar{n} + 1)e^{it\bar{\omega}}\}] \quad (3.52)$$

or

$$W_{i \to f} = \frac{2\pi}{\hbar^2} |T_{fi}|^2 \sum_{m=0}^{\infty} \frac{(S\bar{n})^m}{m!} \frac{[S(\bar{n} + 1)]^{m + \frac{\omega_{if}}{\bar{\omega}}}}{(m + \frac{\omega_{if}}{\bar{\omega}})!} \quad (3.53)$$

This expression can be put in terms of modified Bessel functions. Applying the method of steepest descent to Eq. (3.52) yields

$$W_{i \to f} = \frac{1}{\hbar^2} |T_{fi}|^2 \sqrt{\frac{2\pi}{S\bar{\omega}^2 [\bar{n}e^{-it^*\bar{\omega}} + (\bar{n} + 1)e^{it^*\bar{\omega}}]}}$$
$$\times \exp[it^*\omega_{fi} - S\{(2\bar{n} + 1) - \bar{n}e^{-it^*\bar{\omega}} - (\bar{n} + 1)e^{it^*\bar{\omega}}] \quad (3.54)$$

where

$$e^{it^*\bar{\omega}} = \frac{1}{2(\bar{n} + 1)} \left[ \frac{\omega_{if}}{S\bar{\omega}} + \sqrt{\left(\frac{\omega_{if}}{S\bar{\omega}}\right)^2 + 4\bar{n}(\bar{n} + 1)} \right] \quad (3.55)$$

Another way to evaluate Eq. (3.40) by the method of steepest descent is to rewrite Eq. (3.40) as

$$W_{i \to f} = \frac{1}{\hbar^2} |T_{fi}|^2 \int_{-\infty}^{\infty} dt \exp\left[it\omega_{fi} - \sum_{j} S_j(1 - e^{it\omega_j})\right] G(t) \quad (3.56)$$

where

$$G(t) = \exp\left[\sum_{j} S_j \bar{n}_j (2 - e^{-it\omega_j} - e^{it\omega_j})\right] \quad (3.57)$$

It should be noted that at $T = 0$, $G(t) = 1$. In other words, the temperature effect of $W_{i \to f}$ is included in $G(t)$. Applying the method of steepest descent to Eq. (3.56) yields

$$W_{i \to f} = W_{i \to f}(0)G(t^*)$$
$$= W_{i \to f}(0)\exp\left[\sum_{j} S_j \bar{n}_j (2 - e^{-it\omega_j} - e^{it\omega_j})\right] \quad (3.58)$$

where the saddle-point value of $t^*$ is given by

$$it^* = \frac{1}{\omega} \ln \frac{\omega_{if}}{S\bar{\omega}} \tag{3.59}$$

which is the same as that used in obtaining Eq. (3.51).

Finally we shall derive the equation used by Bixon and Jortner. Suppose that an intramolecular vibrational mode, say $Q_\ell$, plays a very important role in electron transfer. To this mode, we can apply the strong-coupling approximation (or the short-time approximation). From Eq. (3.40), we have

$$W_{i \to f} = \frac{1}{\hbar^2} |T_{fi}|^2 \int_{-\infty}^{\infty} dt \exp\left[ it\omega_{fi} - \sum_j{}' S_j\{(2\bar{n}_j + 1) - \bar{n}_j e^{-it\omega_j} - (\bar{n}_j + 1)e^{it\omega_j}\} \right]$$
$$\times \exp[-S_\ell\{(2\bar{n}_\ell + 1) - \bar{n}_\ell e^{-it\omega_\ell} - (\bar{n}_\ell + 1)e^{it\omega_\ell}\}] \tag{3.60}$$

or

$$W_{i \to f} = \frac{1}{\hbar^2} |T_{fi}|^2 e^{-S_\ell(2\bar{n}_\ell+1)} \int_{-\infty}^{\infty} dt \exp\left[ it\omega_{fi} - \sum_j{}' S_j\{(2\bar{n}_j + 1) - \bar{n}_j e^{-it\omega_j} \right.$$
$$\left. - (\bar{n}_j + 1)e^{it\omega_j}\} \right] \times \sum_{m_\ell=0}^{\infty} \sum_{n_\ell=0}^{\infty} \frac{(S_\ell \bar{n}_\ell e^{-it\omega_\ell})^{m_\ell}}{m_\ell!} \times \frac{[S_\ell(\bar{n}_\ell + 1)e^{it\omega_\ell}]^{n_\ell}}{n_\ell!} \tag{3.61}$$

We then apply the short-time approximation (i.e., the strong coupling) to Eq. (3-61) to obtain

$$W_{i \to f} = \frac{1}{\hbar^2} |T_{fi}|^2 e^{-S_\ell(2\bar{n}_\ell+1)} \sum_{m_\ell=0}^{\infty} \sum_{n_\ell=0}^{\infty} \frac{(S_\ell \bar{n}_\ell)^{m_\ell}}{m_\ell!} \times \frac{[S_\ell(\bar{n}_\ell + 1)]^{n_\ell}}{n_\ell!}$$
$$\times \sqrt{\frac{2\pi}{\sum_j{}' S_j \omega_j^2 (2\bar{n}_j + 1)}} \exp\left[ -\frac{(\omega_{fi}(n_\ell m_\ell) + \sum_j{}' S_j \omega_j)^2}{2 \sum_j{}' S_j \omega_j^2 (2\bar{n}_j + 1)} \right] \tag{3.62}$$

where

$$\omega_{fi}(n_\ell m_\ell) = \omega_{fi} + n_\ell \omega_\ell - m_\ell \omega_\ell \tag{3.63}$$

In the classical regime for $\omega_j$ modes we obtain

$$W_{i \to f} = \frac{1}{\hbar^2} |T_{fi}|^2 e^{-S_\ell(2\bar{n}_\ell+1)} \sum_{m_\ell=0}^{\infty} \sum_{n_\ell=0}^{\infty} \frac{(S_\ell \bar{n}_\ell)^{m_\ell}}{m_\ell!} \times \frac{[S_\ell(\bar{n}_\ell + 1)]^{n_\ell}}{n_\ell!}$$
$$\times \sqrt{\frac{\pi \hbar^2}{\lambda' kT}} \exp\left[ -\frac{\Delta E_a(n_\ell m_\ell)}{kT} \right] \tag{3.64}$$

where

$$\lambda' = \sum_{j}^{j \neq \ell} {}' S_j \hbar \omega_j \tag{3.65}$$

and

$$\Delta E_a(n_\ell m_\ell) = \frac{(\hbar \omega_{fi}(n_\ell m_\ell) + \lambda')^2}{4\lambda'} \tag{3.66}$$

In the case where $\hbar \omega_\ell / kT \gg 1$, Eq. (3.64) becomes

$$W_{i \to f} = \frac{1}{\hbar^2} |T_{fi}|^2 e^{-S_\ell} \sum_{n_\ell=0}^{\infty} \frac{S_\ell^{n_\ell}}{n_\ell!} \sqrt{\frac{\pi \hbar^2}{\lambda' kT}} \exp\left[-\frac{\Delta E_a(n_\ell 0)}{kT}\right] \tag{3.67}$$

This equation is often used by Bixon and Jortner [77].

### D. Applications to Photosynthetic RCs

In 1991, Fleming and his co-workers have reported the temperature effect on the ET rate constants of *Rb. capsulatus* and its mutants (see Fig. 11 and Ref. 6). From Fig. 11, we can see that several species exhibit a pronounced inverse temperature

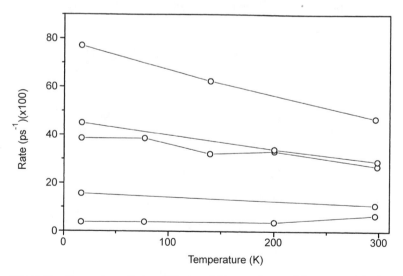

**Figure 11.** Temperature effect on ET rates of *Rb. capsulatas* and its mutants. The data are reproduced from Ref. 6.

TABLE VI
Vibrational Modes and Coupling Constants

| Transitions | Vibrational Modes | | | | | |
|---|---|---|---|---|---|---|
| | 100 | 224 | 750 | 1200 | 1400 | 1520 |
| | *Coupling Constants* | | | | | |
| $P \to P^*$ | $2.50^a$ | 0.03 | 0.02 | 0.03 | 0.02 | 0.04 |
| $P \to P^+$ | 0.3 | 0.1 | 0.07 | 0.01 | 0.09 | 0.1 |
| $H \to H^{-b}$ | 0.4 | 0.25 | 0.12 | 0.08 | 0.18 | 0.12 |

[a] Our result determined for absorption spectroscopy.
[b] We have assumed that the coupling constants for $B \to B^-$ are the same as those for $H \to H^-$, since the electronic structures of H and B are similar.

effect, while others show no temperature dependence. From Eq. (3.40) we can see that to determine the temperature dependence of ET rates, we need to know the electronic energy gap $\omega_{if}$ and the vibrational modes ($\omega_i$ and $S_i$) involved in ET. Using the $S_i$ and $\omega_i$ values given in Table VI, we calculated the temperature dependence of ET rates as a function of energy gaps $\omega_{if}$. The results are shown in Fig. 12 [54]. From Fig. 12 we can see that we can indeed theoretically reproduce the temperature-dependent behaviors of ET rates observed experimentally. The

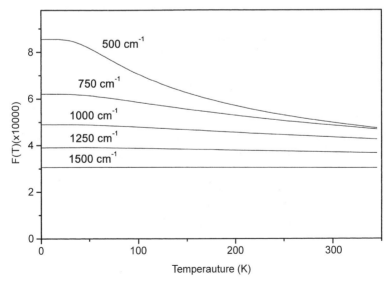

**Figure 12.** Calculated temperature dependence of ET rates. The value next to each curve is the corresponding energy gap.

reason that we have not attempted to fit the experimental results completely is because $\omega_{if}$ might depend on temperature due to the thermal expansion of proteins; furthermore, it is now believed that the ET from P to H is not a one-step process [71].

## E.  Theory of Radiationless Transitions

In the previous section we have shown that

$$W = \frac{2\pi}{\hbar} \sum_k \sum_m P_k \left| H'_{mk} + \sum_n \frac{H'_{mn}H'_{nk}}{\hbar\omega_{kn}} \right|^2 \delta(E_m - E_k) \qquad (3.68)$$

In the lower electronic manifolds, we can use the B–O adiabatic approximation as the basis set, that is,

$$W = \frac{2\pi}{\hbar} \sum_v \sum_{v'} P_{iv} \left| H'_{fv',iv} + \sum_{n\mu} \frac{H'_{fv',n\mu}H'_{n\mu,iv}}{E_{iv} - E_{n\mu}} \right|^2 \delta(E_{fv'} - E_{iv}) \qquad (3.69)$$

where for radiationless transitions, the interaction Hamiltonians can be considered as

$$\hat{H}'\psi_{iv} = \hat{H}'\Phi_i(q,Q)\Theta_{iv}(Q)$$
$$= \hat{H}'_{BO}\Phi_i(q,Q)\Theta_{iv}(Q) + \hat{H}'_{SO}\Phi_i(q,Q)\Theta_{iv}(Q) \qquad (3.70)$$

Here $\hat{H}'_{SO}$ denotes the spin–orbit coupling and $\hat{H}'_{BO}$ represents the B–O coupling due to the breakdown of the adiabatic approximation. Precisely,

$$\hat{H}'_{BO}\psi_{iv} = -\hbar^2 \sum_\ell \frac{\partial\Phi_i}{\partial Q_\ell}\frac{\partial\Theta_{iv}}{\partial Q_\ell} - \frac{\hbar^2}{2} \sum_\ell \frac{\partial^2\Phi_i}{\partial Q_\ell^2}\Theta_{iv} \qquad (3.71)$$

The second term in most cases is much smaller then the first term.

We first consider internal conversion [78]. Using Eq. (3.71) we obtain

$$W_{i \to f} = \frac{2\pi}{\hbar} \sum_v \sum_{v'} P_{iv} \left| \left\langle \Phi_f\Theta_{fv'} \left| -\hbar^2 \sum_\ell \frac{\partial\Phi_i}{\partial Q_\ell}\frac{\partial\Theta_{iv}}{\partial Q_\ell} \right\rangle \right|^2 \delta(E_{fv'} - E_{iv}) \qquad (3.72)$$

Using the Condon approximation, Eq. (3.72) becomes

$$W_{i \to f} = \frac{2\pi}{\hbar^2} |R_\ell(fi)|^2 \sum_v \sum_{v'} P_{iv} \left| \left\langle \Theta_{fv'} \left| \frac{\partial\Theta_{iv}}{\partial Q_\ell} \right\rangle \right|^2 \delta(E_{fv'} - E_{iv}) \qquad (3.73)$$

where

$$R_\ell(fi) = -\hbar^2 \langle \Phi_f | \frac{\partial}{\partial Q_\ell} | \Phi_i \rangle \tag{3.74}$$

Here for simplicity we assume that there is only one promoting mode $Q_\ell$. Equation (3.73) can be written as

$$W_{i \to f} = \frac{2\pi}{\hbar^2} |R_\ell(fi)|^2 \sum_v \sum_{v'} P_{iv} \left| \langle \chi_{fv'_\ell} | \frac{\partial}{\partial Q_\ell} | \chi_{iv_\ell} \rangle \right|^2 \prod_j{}' |\langle \chi_{fv'_j} | \chi_{iv_j} \rangle|^2 \delta(E_{fv'} - E_{iv}) \tag{3.75}$$

where $|\langle \chi_{fv'_j} | \chi_{iv_j} \rangle|^2$ denotes the Franck–Condon factor of the accepting mode $Q_j$. In other words, vibrational modes can be classified into promoting and accepting modes.

For simplicity, we shall assume that the promoting mode is neither distorted nor displaced; in this case we find

$$W_{i \to f} = \frac{2\pi}{\hbar} \left( \frac{\omega_\ell}{2\hbar} |R_\ell(fi)|^2 \right) \sum_v{}' \sum_{v'}{}' P'_{iv} \prod_j{}' |\langle \chi_{fv'_j} | \chi_{iv_j} \rangle|^2$$
$$\times [\bar{v}_\ell \delta(-\hbar\omega_\ell + E'_{fv'} - E'_{iv}) + (\bar{v}_\ell + 1)\delta(\hbar\omega_\ell + E'_{fv'} - E'_{iv})] \tag{3.76}$$

For the displaced oscillator model, Eq. (3.76) takes the following form:

$$W_{i \to f} = \frac{1}{\hbar^2} \left( \frac{\omega_\ell}{2\hbar} |R_\ell(fi)|^2 \right) \int_{-\infty}^{\infty} dt e^{it\omega_{fi}} [\bar{v}_\ell e^{-it\omega_\ell} + (\bar{v}_\ell + 1)e^{it\omega_\ell}]$$
$$\times \exp\left[ -\sum_j{}' S_j\{(2\bar{n}_j + 1) - \bar{n}_j e^{-it\omega_j} - (\bar{n}_j + 1)e^{it\omega_j}\} \right] \tag{3.77}$$

It should be noted that usually $\hbar\omega_\ell/kT \gg 1$; at this limit, Eq. (3.76) or Eq. (3.77) reduces to

$$W_{i \to f} = \frac{2\pi}{\hbar^2} \left( \frac{\omega_\ell}{2\hbar} |R_\ell(fi)|^2 \right) \sum_v{}' \sum_{v'}{}' P'_{iv} \prod_j{}' |\langle \chi_{fv'_j} | \chi_{iv_j} \rangle|^2 \delta(\hbar\omega_\ell + E'_{fv'} - E'_{iv})$$
$$= \frac{1}{\hbar^2} \left( \frac{\omega_\ell}{2\hbar} |R_\ell(fi)|^2 \right) \int_{-\infty}^{\infty} dt \exp\left[ it(\omega_{fi} + \omega_\ell) \right.$$
$$\left. - \sum_j{}' S_j\{(2\bar{n}_j + 1) - \bar{n}_j e^{-it\omega_j} - (\bar{n}_j + 1)e^{it\omega_j}\} \right] \tag{3.78}$$

In this case we can see that the promoting mode can accept at least one vibrational quantum (due to the term $\omega_{fi} + \omega_\ell$). This is usually called the propensity rule.

Next we consider intersystem crossing. In this case, due to the spin-forbiddenness, $\hat{H}'_{BO}$ does not make any contribution and we have

$$W_{i\to f} = \frac{2\pi}{\hbar} \sum_v \sum_{v'} P_{iv} |\langle \Theta_{fv'} \Phi_f | \hat{H}'_{SO} | \Phi_i \Theta_{iv} \rangle|^2 \delta(E_{fv'} - E_{iv}) \tag{3.79}$$

When we use the Condon approximation, Eq. (3.79) reduces to

$$W_{i\to f} = \frac{2\pi}{\hbar} |\langle \Phi_f | \hat{H}'_{SO} | \Phi_i \rangle|^2 \sum_v \sum_{v'} P_{iv} |\langle \Theta_{fv'} | \Theta_{iv} \rangle|^2 \delta(E_{fv'} - E_{iv}) \tag{3.80}$$

For the displaced oscillator case, Eq. (3.80) takes the form

$$W_{i\to f} = \frac{1}{\hbar^2} |\langle \Phi_f | \hat{H}'_{SO} | \Phi_i \rangle|^2 \int_{-\infty}^{\infty} dt \exp\left[ it\omega_{fi} - \sum_j S_j \{ (2\bar{n}_j + 1) \right.$$
$$\left. - \bar{n}_j e^{-it\omega_j} - (\bar{n}_j + 1)e^{it\omega_j} \} \right] \tag{3.81}$$

Equation (3.81) is similar to that in electron and energy transfer processes.

For the intersystem crossing $T_1 \leftrightarrow S_1$, the electronic coupling matrix element $\langle T_1 | \hat{H}'_{SO} | S_1 \rangle$ often vanishes. In this case, we have to take into account the higher-order terms in Eq. (3.69), that is,

$$\sum_{n\mu} \frac{\langle fv' | \hat{H}' | n\mu \rangle \langle n\mu | \hat{H}' | iv \rangle}{E_{iv} - E_{n\mu}}$$
$$= \sum_{n\mu} \frac{\langle fv' | \hat{H}'_{SO} | n\mu \rangle \langle n\mu | \hat{H}'_{BO} | iv \rangle}{E_{iv} - E_{n\mu}} + \sum_{n\mu} \frac{\langle fv' | \hat{H}'_{BO} | n\mu \rangle \langle n\mu | \hat{H}'_{SO} | iv \rangle}{E_{iv} - E_{n\mu}} \tag{3.82}$$

When we use the Plazcek and Condon approximations, Eq. (3.82) becomes

$$\sum_{n\mu} \frac{\langle fv' | \hat{H}' | n\mu \rangle \langle n\mu | \hat{H}' | iv \rangle}{E_{iv} - E_{n\mu}}$$
$$= \langle \Theta_{fv'} | Q_\ell | \Theta_{iv} \rangle \left[ \sum_n \frac{\langle \Phi_f | \hat{H}'_{SO} | \Phi_n \rangle \langle \Phi_n | - \hbar^2 \frac{\partial}{\partial Q_\ell} | \Phi_i \rangle}{E_i - E_n} \right.$$
$$\left. + \sum_{n'} \frac{\langle \Phi_f | - \hbar^2 \frac{\partial}{\partial Q_\ell} | \Phi_{n'} \rangle \langle \Phi_{n'} | \hat{H}'_{SO} | \Phi_i \rangle}{E_i - E_{n'}} \right] = \langle \Theta_{fv'} | Q_\ell | \Theta_{iv} \rangle T_{fi} \tag{3.83}$$

where

$$T_{fi} = \left[ \sum_n \frac{\langle \Phi_f | \hat{H}'_{SO} | \Phi_n \rangle \langle \Phi_n | - \hbar^2 \frac{\partial}{\partial Q_\ell} | \Phi_i \rangle}{E_i - E_n} + \sum_{n'} \frac{\langle \Phi_f | - \hbar^2 \frac{\partial}{\partial Q_\ell} | \Phi_{n'} \rangle \langle \Phi_{n'} | \hat{H}'_{SO} | \Phi_i \rangle}{E_i - E_{n'}} \right]$$

(3.84)

From Eq. (3.84) we can see that the summation over $n$ covers the singlet electronic states, while the summation over $n'$ covers the triplet electronic states. It follows that

$$W_{i \to f} = \frac{2\pi}{\hbar} |T_{fi}|^2 \sum_v \sum_{v'} P_{iv} \left| \langle \Theta_{fv'} | \frac{\partial}{\partial Q_\ell} | \Theta_{iv} \rangle \right|^2 \delta(E_{fv'} - E_{iv})$$

(3.85)

Here for simplicity it is assumed that there is only one promoting mode. Equation (3.85) should be compared with Eq. (3.73). In Eq. (3.85) both vibronic and spin–orbit coupling are involved in $W_{i \to f}$. Due to the rapid progress in *ab initio* calculations, it has now become possible to evaluate $W_{i \to f}$ by using potential energy surfaces information obtained from *ab initio* calculations.

### F. Förster–Dexter Theory

In the original Förster–Dexter theory [79–81] the superexchange terms in $T_{fi}$ have been ignored, that is, $T_{fi} = H'_{fi}$. We shall present the derivation of their theory in our framework. Consider the energy transfer

$$D^*A \to DA^*$$

(3.86)

For triplet–triplet transfer, we have

$$\Phi_i = \left| \overset{+}{\chi}_{D^*} \ \ \overset{+}{\chi}_D \ \ \overset{+}{\chi}_A \ \ \overset{-}{\chi}_A \right|$$

(3.87)

and

$$\Phi_f = \left| \overset{+}{\chi}_D \ \ \overset{-}{\chi}_D \ \ \overset{+}{\chi}_{A^*} \ \ \overset{+}{\chi}_A \right|$$

(3.88)

where $\chi_{D^*}$ and $\chi_{A^*}$ represent the molecular orbitals for excited electronic states. It follows that

$$\langle \Phi_f | \hat{H}' | \Phi_i \rangle = \langle \Phi_f | \sum_{i<j} \frac{e^2}{r_{ij}} | \Phi_i \rangle = -\langle \chi_D \chi_{A^*} | \frac{e^2}{r_{ij}} | \chi_A \chi_{D^*} \rangle$$

(3.89)

In other words, in this approximation for electronic states given by Eqs. (3.87) and (3.88) the triplet–triplet transition is due to the exchange interaction.

Next we consider the singlet–singlet transfer. In this case we have

$$\Phi_i = \frac{1}{\sqrt{2}}\left(\left|\chi_{D^*}^{-}\ \chi_{D}^{+}\ \chi_{A}^{+}\ \chi_{A}^{-}\right| - \left|\chi_{D^*}^{+}\ \chi_{D}^{+}\ \chi_{A}^{+}\ \chi_{A}^{-}\right|\right) \tag{3.90}$$

and

$$\Phi_f = \frac{1}{\sqrt{2}}\left(\left|\chi_{D}^{+}\ \chi_{D}^{-}\ \chi_{A^*}^{-}\ \chi_{A}^{+}\right| - \left|\chi_{D}^{+}\ \chi_{D}^{-}\ \chi_{A^*}^{+}\ \chi_{A}^{-}\right|\right) \tag{3.91}$$

$$\langle\Phi_f|\hat{H}'|\Phi_i\rangle = 2\langle\chi_D\chi_{A^*}|\frac{e^2}{r_{ij}}|\chi_{D^*}\chi_A\rangle - \langle\chi_D\chi_{A^*}|\frac{e^2}{r_{ij}}|\chi_A\chi_{D^*}\rangle \tag{3.92}$$

That is, in the singlet–singlet transition both Coulomb interaction and exchange interaction are involved. However, when the distance between $D$ and $A$ is large, the exchange term can be ignored, and we can use the multipole expansion for $e^2/r_{ij}$, that is,

$$\langle\Phi_f|\hat{H}'|\Phi_i\rangle = 2\langle\chi_D\chi_{A^*}|\frac{1}{R_{DA}^3}\left[(\vec{\mu}_D\cdot\vec{\mu}_A) - \frac{3(\vec{R}_{DA}\cdot\vec{\mu}_D)(\vec{R}_{DA}\cdot\vec{\mu}_A)}{R_{DA}^2}\right]|\chi_{D^*}\chi_A\rangle \tag{3.93}$$

where $\vec{\mu}_D$ and $\vec{\mu}_A$ denote the one-electronic dipole operator for $D$ and $A$, respectively. Equation (3.93) can be written as

$$\langle\Phi_f|\hat{H}'|\Phi_i\rangle = \frac{1}{R_{DA}^3}\left[(\vec{\mu}_{DD^*}\cdot\vec{\mu}_{AA^*}) - \frac{3(\vec{R}_{DA}\cdot\vec{\mu}_{DD^*})(\vec{R}_{DA}\cdot\vec{\mu}_{AA^*})}{R_{DA}^2}\right] \tag{3.94}$$

where

$$\vec{\mu}_{DD^*} = -\sqrt{2}\langle\chi_D|\vec{\mu}_D|\chi_{D^*}\rangle, \qquad \vec{\mu}_{AA^*} = -\sqrt{2}\langle\chi_A|\vec{\mu}_A|\chi_{A^*}\rangle \tag{3.95}$$

This is due to the fact that if

$$\Phi_g = \left|\chi_{D}^{+}\ \chi_{D}^{-}\ \chi_{A}^{+}\ \chi_{A}^{-}\right| \tag{3.96}$$

then

$$\vec{\mu}_{DD^*} = \langle\Phi_i|\sum_i e\vec{r}_i|\Phi_g\rangle = -\sqrt{2}\langle\chi_{D^*}|e\vec{r}_i|\chi_D\rangle \tag{3.97}$$

and

$$\vec{\mu}_{AA^*} = \langle\Phi_f|\sum_i e\vec{r}_i|\Phi_g\rangle = -\sqrt{2}\langle\chi_{A^*}|e\vec{r}_i|\chi_A\rangle \tag{3.98}$$

Next we shall show that the electronic energy transfer rate can be put into the spectral overlap form. Notice that by ignoring the super-exchange term we have

$$W_{i \to f} = \frac{2\pi}{\hbar} \sum_v \sum_{v'} P_{iv} |\langle fv' | \hat{H}' | iv \rangle|^2 \delta(E_{fv'} - E_{iv}) \tag{3.99}$$

For singlet–singlet transfer, the electronic matrix element is given by

$$
\begin{aligned}
\langle \Phi_f | \hat{H}' | \Phi_i \rangle &= \frac{1}{\varepsilon R_{AD}^3} \left[ (\vec{\mu}_{DD^*} \cdot \vec{\mu}_{AA^*}) - \frac{3(\vec{\mu}_{DD^*} \cdot \vec{R}_{AD})(\vec{\mu}_{AA^*} \cdot \vec{R}_{AD})}{R_{AD}^2} \right] \\
&= \frac{|\vec{\mu}_{DD^*}||\vec{\mu}_{AA^*}|}{\varepsilon R_{AD}^3} \Omega_{AD}
\end{aligned}
\tag{3.100}
$$

where $\varepsilon$ denotes the dielectric constant and $\Omega_{AD}$ describes the relative orientation between $A$ and $D$. Substituting Eq. (3.100) into Eq. (3.99) yields

$$
\begin{aligned}
W_{i \to f} = \frac{2\pi}{\hbar} \frac{|\Omega_{AD}|^2}{\varepsilon^2 R_{AD}^6} &\left[ \sum_v \sum_{v'} P_{iv} |\langle \Theta_{fv'} | \vec{\mu}_{DD^*} | \Theta_{iv} \rangle|^2 \right]_D \\
&\times \left[ \sum_v \sum_{v'} P_{iv} |\langle \Theta_{fv'} | \vec{\mu}_{AA^*} | \Theta_{iv} \rangle|^2 \right]_A \delta(E_{fv'} - E_{iv})
\end{aligned}
\tag{3.101}
$$

The absorption coefficient $\alpha_{if}(\omega)$ can be expressed as

$$\alpha_{if}(\omega) = \frac{4\pi^2 \omega}{3\hbar \alpha_a c} \sum_v \sum_{v'} P_{iv} |\langle \Theta_{fv'} | \vec{\mu}_{fi} | \Theta_{iv} \rangle|^2 \delta(\omega_{fv',iv} - \omega) \tag{3.102}$$

where $\alpha_a$ is a function of refraction index introduced to correct the medium effect. Equation (3.102) can be rewritten as

$$\alpha_{if}(\omega) = \frac{2\pi \omega}{3\hbar \alpha_a c} \int_{-\infty}^{\infty} dt \, e^{-it\omega} \sum_v \sum_{v'} P_{iv} |\langle \Theta_{fv'} | \vec{\mu}_{fi} | \Theta_{iv} \rangle|^2 e^{it\omega_{fv',iv}} \tag{3.103}$$

or

$$\int_{-\infty}^{\infty} e^{it'\omega} \alpha_{if}(\omega) \frac{d\omega}{\omega} = \frac{4\pi^2}{3\hbar \alpha_a c} \sum_v \sum_{v'} P_{iv} |\langle \Theta_{fv'} | \vec{\mu}_{fi} | \Theta_{iv} \rangle|^2 e^{it'\omega_{fv',iv}} = \frac{4\pi^2}{3\hbar \alpha_a c} F(t) \tag{3.104}$$

Notice that in terms of $F(t)$ we find

$$W_{i\to f} = \frac{1}{\hbar}\frac{|\Omega_{AD}|^2}{\varepsilon^2 R_{AD}^6}\int_{-\infty}^{\infty} dt F_D(t)F_A(t) \tag{3.105}$$

Using the relation

$$F_A(t) = \frac{3\hbar\alpha_a c}{4\pi^2}\int_{-\infty}^{\infty}\frac{d\omega}{\omega}e^{it\omega}\alpha_{if}^{(A)}(\omega) \tag{3.106}$$

we obtain

$$W_{i\to f} = \frac{1}{\hbar}\frac{|\Omega_{AD}|^2}{\varepsilon^2 R_{AD}^6}\frac{3\hbar\alpha_a c}{4\pi^2}\int_{-\infty}^{\infty} dt F_D(t)\int_{-\infty}^{\infty}\frac{d\omega}{\omega}e^{it\omega}\alpha_{if}^{(A)}(\omega) \tag{3.107}$$

Carrying out the integration with $t$ in Eq. (3.107) yields

$$W_{i\to f} = \frac{1}{\hbar}\frac{|\Omega_{AD}|^2}{\varepsilon^2 R_{AD}^6}\frac{3\hbar\alpha_a c}{2\pi}\int_{-\infty}^{\infty}\frac{d\omega}{\omega}\alpha_{if}^{(A)}(\omega)$$
$$\times\left[\sum_v\sum_{v'} P_{iv}|\langle\Theta_{fv'}|\vec{\mu}_{fi}|\Theta_{iv}\rangle|^2\delta(\omega_{fv',iv}+\omega)\right]_D \tag{3.108}$$

It has been shown that the normalized intensity distribution function of emission for $i\to f$ can be written as

$$I_{if}(\omega) = \frac{4\alpha_e\omega^3}{3\hbar c^3 A_{if}}\sum_v\sum_{v'} P_{iv}|\langle\Theta_{fv'}|\vec{\mu}_{fi}|\Theta_{iv}\rangle|^2\delta(\omega_{fv',iv}+\omega) \tag{3.109}$$

where $A_{if}$ denotes the radiative rate constant. It follows that

$$W_{i\to f} = \frac{9|\Omega_{AD}|^2\alpha_a c^4 A_{if}^{(D)}}{8\pi\varepsilon^2 R_{AD}^6\alpha_e}\int_{-\infty}^{\infty}\frac{d\omega}{\omega^4}\alpha_{if}^{(A)}(\omega)I_{if}^{(D)}(\omega) \tag{3.110}$$

This is the famous Förster relation that expresses the singlet–singlet transition rate constant in terms of the spectral overlap of the emission spectra of $D$ and absorption spectra of $A$.

Next we consider the triplet–triplet transition. In the Condon approximation we have

$$W_{i\to f} = \frac{2\pi}{\hbar}|H'_{fi}|^2\sum_v\sum_{v'} P_{iv}|\langle\Theta_{fv'}|\Theta_{iv}\rangle|^2\delta(E_{fv'} - E_{iv}) \tag{3.111}$$

or

$$W_{i \to f} = \frac{1}{\hbar} |H'_{fi}|^2 \int_{-\infty}^{\infty} dt \left( \sum_v \sum_{v'} e^{it\omega_{fv',iv}} P_{iv} |\langle \Theta_{fv'} | \Theta_{iv} \rangle|^2 \right)_D$$

$$\times \left( \sum_v \sum_{v'} e^{it\omega_{fv',iv}} P_{iv} |\langle \Theta_{fv'} | \Theta_{iv} \rangle|^2 \right)_A \qquad (3.112)$$

or

$$W_{i \to f} = \frac{1}{\hbar} |H'_{fi}|^2 \int_{-\infty}^{\infty} dt\, f_A(t) f_D(t) \qquad (3.113)$$

where for example,

$$f_A(t) = \left( \sum_v \sum_{v'} P_{iv} e^{it\omega_{fv',iv}} |\langle \Theta_{fv'} | \Theta_{iv} \rangle|^2 \right)_A \qquad (3.114)$$

For allowed transitions, $f_A(t)$ can be expressed as

$$f_A(t) = \frac{\int_{-\infty}^{\infty} \frac{d\omega}{\omega} e^{it\omega} \alpha_{if}^{(A)}(\omega)}{\int_{-\infty}^{\infty} \frac{d\omega}{\omega} \alpha_{if}^{(A)}(\omega)} = \int_{-\infty}^{\infty} d\omega\, e^{it\omega} \sigma_{if}^{(A)}(\omega) \qquad (3.115)$$

where $\sigma_{if}^{(A)}(\omega)$ represents the normalized distribution function of absorption,

$$\sigma_{if}^{(A)}(\omega) = \frac{\frac{\alpha_{if}^{(A)}(\omega)}{\omega}}{\int_{-\infty}^{\infty} d\omega \frac{\alpha_{if}^{(A)}(\omega)}{\omega}} \qquad (3.116)$$

It follows that

$$W_{i \to f} = \frac{2\pi}{\hbar^2} |H'_{fi}|^2 \int_{-\infty}^{\infty} d\omega \sigma_{if}^{(A)}(\omega) \left( \sum_v \sum_{v'} P_{iv} |\langle \Theta_{fv'} | \Theta_{iv} \rangle|^2 \delta(\omega_{fv',iv} + \omega) \right)_D$$

$$(3.117)$$

or

$$W_{i \to f} = \frac{2\pi}{\hbar^2} |H'_{fi}|^2 \int_{-\infty}^{\infty} d\omega \sigma_{if}^{(A)}(\omega) \eta_{if}^{(D)}(\omega) \qquad (3.118)$$

where

$$\eta_{if}^{(D)}(\omega) = \left( \sum_{v} \sum_{v'} P_{iv} |\langle \Theta_{fv'} | \Theta_{iv} \rangle|^2 \delta(\omega_{fv', iv} + \omega) \right)_D \tag{3.119}$$

which is the normalized function of emission spectra. Equation (3.118) is similar to the equation derived by Dexter for triplet–triplet transfer.

## IV. RELAXATION DYNAMICS OF A SYSTEM IN A HEAT BATH

### A. Density Matrix Method

We shall start with the definition of density matrix [82–84]. For this purpose, we consider a two-state system. According to the expansion theorem we have

$$\Psi = C_a(t)u_a + C_b(t)u_b = C_a^0(t)\Psi_a^0(t) + C_b^0(t)\Psi_b^0(t) \tag{4.1}$$

where

$$\Psi_a^0(t) = \exp\left(-\frac{it}{\hbar}E_a\right)u_a, \qquad \Psi_b^0(t) = \exp\left(-\frac{it}{\hbar}E_b\right)u_b \tag{4.2}$$

The matrix elements of the density matrix $\hat{\rho}$ of the system are defined as follows:

$$\rho_{aa} = C_a(t)C_a^*(t) = |C_a(t)|^2 = |C_a^0(t)|^2 \tag{4.3}$$

$$\rho_{bb} = C_b(t)C_b^*(t) = |C_b(t)|^2 = |C_b^0(t)|^2 \tag{4.4}$$

and

$$\rho_{ab} = C_a^0(t)C_b^0(t)^* e^{-\frac{it}{\hbar}(E_b - E_a)} = \rho_{ba}^* \tag{4.5}$$

It follows that

$$\hat{\rho} = \begin{pmatrix} \rho_{aa} & \rho_{ab} \\ \rho_{aa} & \rho_{bb} \end{pmatrix} \tag{4.6}$$

From Eqs. (4.3)-(4.6), we can see that the diagonal matrix elements $\rho_{aa}$ and $\rho_{bb}$ of $\hat{\rho}$ represent the populations (or probabilities) of the system, while the off-diagonal matrix elements $\rho_{ab}$ and $\rho_{ba}$ represent the phase or coherences of the system. A main feature of $\rho_{ab}$ (or $\rho_{ba}$) is that for $\rho_{ab}$ to be nonzero, both $C_a(t)$ and $C_b(t)$ have to be nonzero. That is, to create a coherence $\rho_{ab}$, both $a$ and $b$ have to be simultaneously pumped. Another feature of $\rho_{ab}$ is that it is oscillatory with the oscillatory frequency related to the energy difference $(E_b - E_a)/\hbar$. From the uncertainty principle $\Delta E \Delta t \sim \hbar$, we can see that when an ultrashort pulse is used ($\Delta t$ small), $\Delta E$ is very large; this means that a great number of states are coherently (or simultaneously) populated. That is, coherences are created and because of these coherences, in fs experiments quantum beat is often observed.

Next we discuss the properties of $\hat{\rho}$. Notice that

$$\text{Tr}(\hat{\rho}) = \rho_{aa} + \rho_{bb} = 1 \tag{4.7}$$

because of the normalization of the wave function $\Psi$. The average value of a dynamical variable M can be calculated in terms of $\hat{\rho}$ as follows:

$$\langle M \rangle = \langle \Psi | M | \Psi \rangle = \int d\tau \Psi^* M \Psi = \text{Tr}(\hat{M}\hat{\rho}) = \text{Tr}(\hat{\rho}\hat{M}) \tag{4.8}$$

To obtain the equation of motion for $\hat{\rho}$, we consider the time-dependent Schrödinger equation

$$\hat{H}\Psi = i\hbar \frac{\partial \Psi}{\partial t} \tag{4.9}$$

where $\hat{H} = \hat{H}_0 + \hat{H}'$ and

$$\hat{H}_0 u_a = E_a u_a, \qquad \hat{H}_0 u_b = E_b u_b \tag{4.10}$$

Substituting Eq. (4.1) into Eq. (4.9) yields

$$i\hbar \frac{\partial C_a}{\partial t} = E_a C_a + C_b H'_{ab}, \qquad i\hbar \frac{\partial C_b}{\partial t} = E_b C_b + C_a H'_{ba} \tag{4.11}$$

It follows that

$$\frac{d\rho_{aa}}{dt} = \frac{d}{dt}(C_a C_a^*) = -\frac{i}{\hbar}[\hat{H}, \hat{\rho}]_{aa}, \qquad \frac{d\rho_{ab}}{dt} = \frac{d}{dt}(C_a C_b^*) = -\frac{i}{\hbar}[\hat{H}, \hat{\rho}]_{ab} \tag{4.12}$$

and

$$\frac{d\hat{\rho}}{dt} = -\frac{i}{\hbar}[\hat{H}, \hat{\rho}] = -i\hat{L}\hat{\rho} \tag{4.13}$$

where $\hat{L}$ is usually called the Liouville operator (a superoperator). Equation (4.13) is called the Liouville equation, which is the equation of motion for $\hat{\rho}$. In general we have

$$\frac{d\rho_{nm}}{dt} = -i \sum_{n'} \sum_{m'} L_{nm}^{n'm'} \rho_{n'm'}, \qquad L_{nm}^{n'm'} = \frac{1}{\hbar}(H_{nn'}\delta_{mm'} - H_{m'm}\delta_{nn'}) \qquad (4.14)$$

To show an application we introduce the effective Hamitonian method. For this purpose we first set $H' = 0$. In this case, Eq. (4.11) becomes

$$i\hbar \frac{\partial C_a}{\partial t} = E_a C_a \qquad (4.15)$$

which can be integrated

$$C_a(t) = C_a(0)e^{-\frac{it}{\hbar}E_a} \qquad (4.16)$$

According to the effective Hamiltonian method, the energy $E_a$ is complex, that is,

$$E_a = E_a^0 - \frac{i}{2}\hbar\gamma_a \qquad (4.17)$$

and

$$C_a(t) = C_a(0)\exp\left(-\frac{it}{\hbar}E_a^0 - \frac{\gamma_a t}{2}\right) \qquad (4.18)$$

The density matrix element $\rho_{aa}$ takes the form

$$\rho_{aa}(t) = C_a(t)C_a^*(t) = C_a(0)C_a^*(0)e^{-\gamma_a t} = \rho_{aa}(0)e^{-\gamma_a t} \qquad (4.19)$$

Equation (4.19) shows that $\gamma_a$ is related to the lifetime of an $a$ state. Similarly, we have

$$\rho_{ab}(t) = C_a(t)C_b^*(t) = C_a(0)C_b^*(0)e^{-\frac{it}{\hbar}(E_a - E_b)}e^{-\frac{t}{2}(\gamma_a + \gamma_b)} = \rho_{ab}(0)e^{it\omega_{ba}}e^{-\gamma_{ab}t} \qquad (4.20)$$

where $\gamma_{ab}$ is called the dephasing constant and is related to $\gamma_a$ and $\gamma_b$ as

$$\gamma_{ab} = \frac{1}{2}(\gamma_a + \gamma_b) = \frac{1}{2}\left(\frac{1}{\tau_a} + \frac{1}{\tau_b}\right) \qquad (4.21)$$

where $\tau_a$ and $\tau_b$ represent the lifetimes.

In summary, we obtain

$$\frac{d\hat{\rho}}{dt} = -\frac{i}{\hbar}[\hat{H}_0, \hat{\rho}] - \hat{\Gamma}\hat{\rho} \qquad (4.22)$$

where $\hat{\Gamma}$ is damping operator, and

$$\frac{d\rho_{aa}}{dt} = -\gamma_a\rho_{aa}, \qquad \frac{d\rho_{bb}}{dt} = -\gamma_b\rho_{bb} \qquad (4.23)$$

$$\frac{d\rho_{ab}}{dt} = -(i\omega_{ab} + \gamma_{ab})\rho_{ab}, \qquad \gamma_{ab} = \frac{1}{2}(\gamma_a + \gamma_b) \qquad (4.24)$$

To show the importance of the damping operator, we apply the effective Hamiltonian to optical absorption. In this case, we have

$$\hat{H} = \hat{H}_0 + \hat{V}(t) \qquad (4.25)$$

where $\hat{V}(t)$ denotes the interaction between the system and radiation field:

$$\hat{V}(t) = -\vec{\mu} \cdot \vec{E}(t) \qquad (4.26)$$

Here $\vec{\mu}$ is dipole operator and $\vec{E}(t)$ is electric field in optical radiation. Due to the application of $\hat{V}(t)$, the system becomes nonstationary and the rate of energy absorption can be calculated:

$$Q = \left\langle \frac{\partial \hat{H}}{\partial t} \right\rangle = \left\langle \frac{\partial \hat{V}}{\partial t} \right\rangle = -\langle \vec{\mu} \rangle \cdot \frac{d\vec{E}}{dt} = -\vec{P} \cdot \frac{d\vec{E}}{dt} \qquad (4.27)$$

where

$$\vec{P} = \langle \vec{\mu} \rangle = \mathrm{Tr}(\hat{\rho}\vec{\mu}) = \mathrm{Tr}(\vec{\mu}\hat{\rho}) \qquad (4.28)$$

and $\vec{P}$ is commonly called polarization. Notice that

$$\vec{E}(t) = \vec{E}_0 \cos\omega t = \vec{E}(\omega)e^{-it\omega} + \vec{E}(-\omega)e^{it\omega} \qquad (4.29)$$

and that $\vec{P}$, the induced dipole moment, can be written as

$$\vec{P}(t) = \vec{P}(\omega)e^{-it\omega} + \vec{P}(-\omega)e^{it\omega} \qquad (4.30)$$

Substituting Eqs. (4.29) and (4.30) into Eq. (4.27) yields

$$Q = -i\omega[\vec{P}(\omega)e^{-it\omega} + \vec{P}(-\omega)e^{it\omega}] \cdot [\vec{E}(\omega)e^{-it\omega} + \vec{E}(-\omega)e^{it\omega}]$$
$$= -i\omega[\vec{P}(\omega) \cdot \vec{E}(-\omega) - \vec{P}(-\omega) \cdot \vec{E}(\omega)] \qquad (4.31)$$

Here the oscillatory terms like $e^{2it\omega}$ and $e^{-2it\omega}$ have been ignored, which is called the rotating wave approximation (RWA). Linear susceptibility $\chi(\omega)$ is defined by

$$\vec{P}(\omega) = \chi(\omega)\vec{E}(\omega), \qquad \vec{P}(-\omega) = \chi(-\omega)\vec{E}(-\omega) \qquad (4.32)$$

where $\vec{P}(\omega) = \vec{P}^*(-\omega)$. Writing $\chi(\omega)$ as

$$\chi(\omega) = \chi'(\omega) + i\chi''(\omega) \tag{4.33}$$

we obtain

$$Q = 2\omega i\chi''(\omega)|\vec{E}(\omega)|^2 \tag{4.34}$$

That is, $Q$ is related only to the imaginary part of $\chi(\omega)$. From the above discussion we can see that the central problem here is the calculation of $\vec{P}$, which is related to $\hat{\rho}$ by

$$\vec{P} = \vec{\mu}_{ab}\rho_{ba} + \rho_{ab}\vec{\mu}_{ba} = \vec{\mu}_{ab}\rho_{ba} + cc \tag{4.35}$$

$\hat{\rho}$ is determined by solving the stochastic Liouville equation

$$\frac{d\hat{\rho}}{dt} = -\frac{i}{\hbar}[\hat{H}_0, \hat{\rho}] - \frac{i}{\hbar}[\hat{V}, \hat{\rho}] - \hat{\Gamma}\hat{\rho} \tag{4.36}$$

or

$$\frac{d\rho_{ba}}{dt} = -(i\omega_{ba} + \gamma_{ba})\rho_{ba} + \frac{i}{\hbar}V_{ba}(\rho_{bb} - \rho_{aa}) \tag{4.37}$$

where

$$V_{ba} = -\vec{\mu}_{ba} \cdot [\vec{E}(\omega)e^{-it\omega} + \vec{E}(-\omega)e^{it\omega}] \tag{4.38}$$

In RWA, we set

$$\rho_{ba}(t) = \rho_{ba}(\omega)e^{-it\omega} \tag{4.39}$$

that is, the term like $\rho_{ba}(-\omega)e^{it\omega}$ has been neglected, to obtain

$$\frac{d\rho_{ba}(\omega)}{dt} = -[i(\omega_{ba} - \omega) + \gamma_{ba}]\rho_{ba}(\omega) + \frac{i}{\hbar}\vec{\mu}_{ba} \cdot \vec{E}(\omega)(\rho_{aa} - \rho_{bb}) \tag{4.40}$$

Applying the steady-state approximation to Eq. (4.40) yields

$$\rho_{ba}(\omega) = \frac{\frac{i}{\hbar}\vec{\mu}_{ba} \cdot \vec{E}(\omega)}{i(\omega_{ba} - \omega) + \gamma_{ba}}(\rho_{aa} - \rho_{bb}) \tag{4.41}$$

$$\vec{P}(t) = \rho_{ba}(\omega)\vec{\mu}_{ab} + \rho_{ba}^*(\omega)\vec{\mu}_{ba} = \vec{P}(\omega)e^{-it\omega} + \vec{P}(-\omega)e^{it\omega} \tag{4.42}$$

$$\vec{P}(\omega) = \frac{\frac{i}{\hbar}\vec{\mu}_{ab}(\vec{\mu}_{ba} \cdot \vec{E}(\omega))}{i(\omega_{ba} - \omega) + \gamma_{ba}}(\rho_{aa} - \rho_{bb}), \quad \chi(\omega) = \frac{\frac{i}{\hbar}\vec{\mu}_{ab}\vec{\mu}_{ba}}{i(\omega_{ba} - \omega) + \gamma_{ba}}(\rho_{aa} - \rho_{bb}) \tag{4.43}$$

and

$$\chi''(\omega) = \frac{1}{\hbar}(\vec{\mu}_{ab}\vec{\mu}_{ba})\frac{\gamma_{ba}}{(\omega_{ba} - \omega)^2 + \gamma_{ba}^2}(\rho_{aa} - \rho_{bb}) \qquad (4.44)$$

That is, $\chi''(\omega)$ has the Lorentzian shape function. The bandwidth is determined by the damping (or dephasing) constant.

## B. Reduced Density Matrix and Damping Operator

In the previous section we discussed the effective Hamiltonian method; a main feature of this method is that it results in the appearance of damping operator $\hat{\Gamma}$ in the Liouville equation. However, the damping operator is introduced in an *ad hoc* manner. In this section we shall show that the damping operator results from the interaction between the system and heat bath.

Next we shall consider the case

$$\hat{H} = \hat{H}_s + \hat{H}_b + \hat{H}' \qquad (4.45)$$

where $\hat{H}'$ denotes the interaction between the system and heat bath. We shall attempt to derive the equation of motion for the density matrix of the system $\hat{\sigma}$, reduced density matrix, by starting from

$$\frac{d\hat{\rho}}{dt} = -\frac{i}{\hbar}[\hat{H}, \hat{\rho}] \qquad (4.46)$$

For this purpose we introduce

$$\sigma_{mn} = \sum_{\alpha} \rho_{m\alpha, n\alpha} \qquad (4.47)$$

where $(m, n)$ describes the system while $(\alpha, \beta)$ describes the bath. Equation (4.47) can be written as

$$\hat{\sigma} = \mathrm{Tr}_b(\hat{\rho}) \qquad (4.48)$$

We shall use the projection operator $\hat{D}$ to pick up the matrix element $\rho_{m\alpha, n\alpha}$, that is,

$$\hat{\rho}_1 = \hat{D}\hat{\rho}, \qquad \hat{\rho}_2 = (1 - \hat{D})\hat{\rho} \qquad (4.49)$$

and

$$\hat{\rho}_1 + \hat{\rho}_2 = \hat{\rho} \qquad (4.50)$$

where

$$D_{m\alpha,n\beta}^{m'\alpha',n'\beta'} = \delta_{\alpha\alpha'}\delta_{mm'}\delta_{nn'}\delta_{\beta\beta'}\delta_{\alpha\beta} \tag{4.51}$$

Notice that

$$(\hat{D}\hat{\rho})_{m\alpha,n\beta} = \sum_{m'\alpha'}\sum_{n'\beta'} D_{m\alpha,n\beta}^{m'\alpha',n'\beta'} \rho_{m'\alpha',n'\beta'} = \delta_{\alpha\beta}\rho_{m\alpha,n\beta} \tag{4.52}$$

Applying the Laplace transformation

$$\hat{\rho}(p) = \int_0^\infty e^{-pt}\hat{\rho}(t)\,dt \tag{4.53}$$

we obtain

$$p\hat{\rho}(p) - \hat{\rho}(0) = -i\hat{L}\hat{\rho}(p) \tag{4.54}$$

$$p\hat{\rho}_1(p) - \hat{\rho}_1(0) = -i\hat{D}\hat{L}\hat{\rho}_1(p) - i\hat{D}\hat{L}\hat{\rho}_2(p) \tag{4.55}$$

and

$$p\hat{\rho}_2(p) - \hat{\rho}_2(0) = -i(1 - \hat{D})\hat{L}\hat{\rho}_1(p) - i(1 - \hat{D})\hat{L}\hat{\rho}_2(p) \tag{4.56}$$

Eliminating $\hat{\rho}_2(p)$ yields

$$p\hat{\rho}_1(p) - \hat{\rho}_1(0) = -i\hat{D}\hat{L}\hat{\rho}_1(p) - i\hat{D}\hat{L}\frac{1}{p + i(1 - \hat{D})\hat{L}}\hat{\rho}_2(0) - \hat{M}(p)\hat{\rho}_1(p) \tag{4.57}$$

where $\hat{M}(p)$ denotes the memory kernel

$$\hat{M}(p) = \hat{D}\hat{L}\frac{1}{p + i(1 - \hat{D})\hat{L}}(1 - \hat{D})\hat{L} \tag{4.58}$$

It follows that

$$\frac{d\hat{\rho}_1}{dt} = -i\hat{D}\hat{L}\hat{\rho}_1 - i\hat{D}\hat{L}e^{-it(1-\hat{D})\hat{L}}\hat{\rho}_2(0) - \int_0^t d\tau\hat{M}(\tau)\hat{\rho}_1(t - \tau) \tag{4.59}$$

where

$$\hat{M}(t) = \hat{D}\hat{L}e^{-it(1-\hat{D})\hat{L}}(1 - \hat{D})\hat{L} \tag{4.60}$$

An important case is

$$\rho_{m\alpha,n\alpha} = \sigma_{mn}\rho_{\alpha\alpha}^{(b)} \tag{4.61}$$

where $\rho_{\alpha\alpha}^{(b)}$ represents the equilibrium distribution. In this case we obtain

$$\frac{d\hat{\sigma}}{dt} = -i\bar{L}\hat{\sigma} - \int_0^t d\tau \bar{M}(\tau)\hat{\sigma}(t-\tau) \tag{4.62}$$

where

$$\bar{L} = Tr_b[\hat{D}\hat{L}\hat{\rho}^{(b)}], \qquad \bar{M}(\tau) = Tr_b[\hat{M}(\tau)\hat{\rho}^{(b)}] \tag{4.63}$$

It follows that

$$\frac{d\sigma_{mn}}{dt} = -i\sum_{m'n'} \bar{L}_{mn}^{m'n'} \sigma_{m'n'} - \int_0^t d\tau \sum_{m'n'} \bar{M}(\tau)_{mn}^{m'n'} \sigma_{m'n'}(t-\tau) \tag{4.64}$$

where for $m \neq n$

$$\bar{L}_{mn}^{m'n'} = (\hat{L}_s)_{mn}^{m'n'} + \frac{1}{\hbar}(\bar{H}'_{mm'}\delta_{nn'} - \bar{H}'_{n'n}\delta_{mm'}) = (\hat{L}_s)_{mn}^{m'n'} + (\bar{L}')_{mn}^{m'n'} = (\hat{L}_s + \bar{L}')_{mn}^{m'n'} \tag{4.65}$$

Here, for example,

$$\bar{H}'_{mm'} = \sum_\beta \rho_{\beta\beta}^{(b)} H'_{m\beta,m'\beta} \tag{4.66}$$

Next we consider the calculation of

$$M(\tau)_{m\alpha,n\alpha}^{m'\beta,n'\beta} = \frac{1}{\hbar^2} \sum_{m''} \sum_{\alpha'}{}^{\stackrel{\alpha'\neq\alpha}{\prime}} [H_{m\alpha,m''\alpha'} e^{-i\tau\omega_{m''\alpha',n\alpha}}(H_{m''\alpha',m'\beta}\delta_{n\alpha,n'\beta} - H_{n'\beta,n\alpha}\delta_{m''\alpha',m'\beta})$$
$$- H_{m''\alpha',n\alpha} e^{-i\tau\omega_{m\alpha,m''\alpha'}}(H_{m\alpha,m'\beta}\delta_{m''\alpha',n'\beta} - H_{n'\beta,m''\alpha'}\delta_{m\alpha,m'\beta})] \tag{4.67}$$

It follows that

$$\bar{M}(\tau)_{mn}^{m'n'} = \sum_{\alpha\beta} \rho_{\beta\beta}^{(b)} M(\tau)_{m\alpha,n\alpha}^{m'\beta,n'\beta} \tag{4.68}$$

or

$$\bar{M}(\tau)_{mn}^{m'n'} = \frac{\delta_{nn'}}{\hbar^2} \sum_{m''} \sum_{\alpha'} \overset{\alpha' \neq \alpha}{\sum_{\alpha}} \rho_{\alpha\alpha}^{(b)} H_{m\alpha, m''\alpha'} H_{m''\alpha', m'\alpha} e^{-i\tau\omega_{m''\alpha', n\alpha}}$$

$$+ \frac{\delta_{mm'}}{\hbar^2} \sum_{m''} \sum_{\alpha} \rho_{\alpha\alpha}^{(b)} H_{m''\alpha', n\alpha} H_{n'\alpha, m''\alpha'} e^{-i\tau\omega_{m\alpha, m''\alpha'}}$$

$$- \frac{1}{\hbar^2} \sum_{\alpha'} \overset{\alpha' \neq \beta}{\sum_{\beta}} {}' \rho_{\beta\beta}^{(b)} H_{m\alpha, m'\beta} H_{n'\beta, n\alpha} \left( e^{-i\tau\omega_{m'\beta, n\alpha}} + e^{-i\tau\omega_{m\alpha, n'\beta}} \right) \quad (4.69)$$

and

$$\sum_{m'n'} \bar{L}_{mn}^{m'n'} \sigma_{m'n'} = \frac{1}{\hbar} (\bar{H}_{mm} - \bar{H}_{nn}) \sigma_{mn} + \frac{1}{\hbar} \bar{H}_{mn} (\sigma_{nn} - \sigma_{mm})$$

$$+ \frac{1}{\hbar} \sum_{m'} {}'' (\bar{H}_{mm'} \sigma_{m'n} - \sigma_{mm'} \bar{H}_{m'n}) \quad (4.70)$$

In Eq. (4.62), $\hat{\sigma}(t - \tau)$ can be expanded as

$$\hat{\sigma}(t - \tau) = \hat{\sigma}(t) + \frac{d\hat{\sigma}}{dt}(-\tau) + \frac{1}{2!} \frac{d^2\hat{\sigma}}{dt^2}(-\tau)^2 + \cdots + \frac{1}{n!} \frac{d^n\hat{\sigma}}{dt^n}(-\tau)^n + \cdots$$

$$= \left[ \sum_{n=0}^{\infty} \frac{1}{n!} \frac{d^n}{dt^n} (-\tau)^n \right] \hat{\sigma}(t) \quad (4.71)$$

or approximately

$$\hat{\sigma}(t - \tau) = e^{-i\tau\bar{L}} \hat{\sigma}(t) = e^{-i\tau\bar{L}_0} \hat{\sigma}(t) \quad (4.72)$$

Equation (4.62) can then be written as

$$\frac{d\hat{\sigma}}{dt} = -i\bar{L}\hat{\sigma} - \hat{R}(t)\hat{\sigma}(t) \quad (4.73)$$

where

$$\hat{R}(t) = \int_0^t d\tau \, \bar{M}(\tau) e^{-i\tau\bar{L}_0} \quad (4.74)$$

or

$$R(t)^{m'n'}_{mn} = \frac{\delta_{nn'}}{\hbar^2} \sum_{m''} \sum_{\alpha'}^{\alpha' \neq \alpha} {}' \sum_{\alpha} \rho^{(b)}_{\alpha\alpha} H_{m\alpha,m''\alpha'} H_{m''\alpha',m'\alpha} \varsigma(\omega_{m''\alpha',m\alpha})$$

$$+ \frac{\delta_{mm'}}{\hbar^2} \sum_{m''} \sum_{\alpha'}^{\alpha' \neq \alpha} {}' \sum_{\alpha} \rho^{(b)}_{\alpha\alpha} H_{m''\alpha',n\alpha} H_{n'\alpha,m''\alpha'} \varsigma(\omega_{n'\alpha,m''\alpha'})$$

$$- \frac{1}{\hbar^2} \sum_{\alpha}^{\alpha \neq \beta} \sum_{\beta} {}' \rho^{(b)}_{\beta\beta} H_{m\alpha,m'\beta} H_{n'\beta,n\alpha} [\varsigma(\omega_{n'\beta,n\alpha}) + \varsigma(\omega_{m\alpha,m'\beta})] \quad (4.75)$$

Here, for example,

$$\varsigma(\omega_{m\alpha,m'\beta}) = \int_0^t d\tau e^{-i\tau\omega_{m\alpha,m'\beta}} = \frac{\sin\omega_{m\alpha,m'\beta}t}{\omega_{m\alpha,m'\beta}} - i\frac{(1 - \cos\omega_{m\alpha,m'\beta}t)}{\omega_{m\alpha,m'\beta}} \quad (4.76)$$

and as $t \to \infty$

$$\varsigma(\omega_{m\alpha,m'\beta}) = \pi\delta(\omega_{m\alpha,m'\beta}) - i\frac{P}{\omega_{m\alpha,m'\beta}} \quad (4.77)$$

Therefore

$$\frac{d\hat{\sigma}}{dt} = -i\hat{L}\hat{\sigma} - \hat{R}(t)\hat{\sigma}(t) \quad (4.78)$$

$$\frac{d\sigma_{nn}}{dt} = -\frac{i}{\hbar} \sum_{m'} {}' (\bar{H}'_{nm'}\sigma_{m'n} - \sigma_{nm'}\bar{H}'_{m'n})$$

$$- \sum_m R(t)^{mm}_{nn}\sigma_{mm} - \sum_{m'} \sum_{n'}^{m' \neq n'} R(t)^{m'n'}_{nn}\sigma_{m'n'} \quad (4.79)$$

and for $m \neq n$

$$\frac{d\sigma_{mn}}{dt} = -(i\bar{\omega}_{mn} + R(t)^{mn}_{mn})\sigma_{mn} - \frac{i}{\hbar}\bar{H}'_{mn}(\sigma_{nn} - \sigma_{mm})$$

$$- \frac{i}{\hbar} \sum_{m'} {}'' (\bar{H}'_{mm'}\sigma_{m'n} - \sigma_{mm'}\bar{H}'_{m'n}) - \sum_{m'} \sum_{n'} R(t)^{m'n'}_{mn}\sigma_{m'n'} \quad (4.80)$$

where

$$\bar{\omega}_{mn} = \frac{1}{\hbar}(\bar{H}_{mm} - \bar{H}_{nn}) \quad (4.81)$$

Equations (4.79) and (4.80) are usually refered to as the Redfield equation. If $\bar{H}' = 0$, then

$$\frac{d\hat{\sigma}}{dt} = -i\hat{L}_s\hat{\sigma} - \hat{R}(t)\hat{\sigma} \tag{4.82}$$

$$\frac{d\sigma_{nn}}{dt} = -\sum_m R(t)_{nn}^{mm}\sigma_{mm} - \sum_{m'}\sum_{n'}{}'R(t)_{nn}^{m'n'}\sigma_{m'n'} \tag{4.83}$$

and

$$\frac{d\sigma_{mn}}{dt} = -(i\omega_{mn}^{(s)} + R_{mn}^{mn})\sigma_{mn} - \sum_{m'}\sum_{n'}R(t)_{mn}^{m'n'}\sigma_{m'n'} \tag{4.84}$$

where for $m' \neq m$ and in the Markoff approximation

$$R(t)_{mm}^{m'm'} = -\frac{2\pi}{\hbar^2}\sum_\alpha\sum_\beta{}'\rho_{\beta\beta}^{(b)}|H'_{m\alpha,m'\beta}|^2\delta(\omega_{m\alpha,m'\beta}) \tag{4.85}$$

$$R(t)_{mm}^{mm} = \frac{2\pi}{\hbar^2}\sum_{m''}{}'\sum_\alpha\sum_\beta{}'\rho_{\alpha\alpha}^{(b)}|H'_{m\alpha,m''\beta}|^2\delta(\omega_{m\alpha,m''\beta}) \tag{4.86}$$

or

$$R(t)_{mm}^{mm} = -\sum_{m''}{}'R(t)_{m''m''}^{mm} \tag{4.87}$$

and

$$\begin{aligned}
R(t)_{mn}^{mn} &= \frac{1}{\hbar^2}\sum_{m''}{}'\sum_\alpha\sum_\beta{}'\rho_{\alpha\alpha}^{(b)}|H'_{m\alpha,m''\beta}|^2\varsigma(\omega_{m''\beta,m\alpha}) \\
&+ \frac{1}{\hbar^2}\sum_{m''}{}'\sum_\alpha\sum_\beta{}'\rho_{\alpha\alpha}^{(b)}|H'_{n\alpha,m''\beta}|^2\varsigma(\omega_{n\alpha,m''\beta}) + \Gamma_{mn}^{mn}(d) \\
&= \mathrm{Re}\{R(t)_{mn}^{mn}\} + i\,\mathrm{Im}\{R(t)_{mn}^{mn}\} + \Gamma_{mn}^{mn}(d) \tag{4.88}
\end{aligned}$$

$\Gamma_{mn}^{mn}(d)$ denotes the pure dephasing

$$\Gamma_{mn}^{mn}(d) = \frac{\pi}{\hbar^2}\sum_\alpha\sum_\beta\rho_{\alpha\alpha}^{(b)}(H'_{m\alpha,m\beta} - H'_{n\alpha,n\beta})^2\delta(\omega_{\alpha\beta}) \tag{4.89}$$

Here we have

$$
\begin{aligned}
\operatorname{Re}\{R(t)_{mn}^{mn}\} &= \frac{\pi}{\hbar^2} \sum_{m''}^{m''\neq m}{}' \sum_{\alpha}^{\alpha\neq\beta} \sum_{\beta} {}' \rho_{\alpha\alpha}^{(b)} |H'_{m\alpha,m''\beta}|^2 \delta(\omega_{m''\beta,m\alpha}) \\
&\quad + \frac{\pi}{\hbar^2} \sum_{n''}^{n''\neq n}{}' \sum_{\alpha}^{\alpha\neq\beta} \sum_{\beta} {}' \rho_{\alpha\alpha}^{(b)} |H'_{n\alpha,n''\beta}|^2 \delta(\omega_{n''\beta,n\alpha}) \\
&= \frac{1}{2}\{R(t)_{mm}^{mm} + R(t)_{nn}^{nn}\}
\end{aligned}
\tag{4.90}
$$

and

$$
\begin{aligned}
\operatorname{Im}\{R(t)_{mn}^{mn}\} &= -\frac{1}{\hbar^2} \sum_{m''}^{m''\neq m}{}' \sum_{\alpha}^{\alpha\neq\beta} \sum_{\beta} {}' \rho_{\alpha\alpha}^{(b)} |H'_{m\alpha,m''\beta}|^2 \frac{P}{\omega_{m''\beta,m\alpha}} \\
&\quad - \frac{1}{\hbar^2} \sum_{n''}^{n''\neq n}{}' \sum_{\alpha}^{\alpha\neq\beta} \sum_{\beta} {}' \rho_{\alpha\alpha}^{(b)} |H'_{n\alpha,n''\beta}|^2 \frac{P}{\omega_{n\alpha,n''\beta}}
\end{aligned}
\tag{4.91}
$$

In summary, we obtain

$$
\begin{aligned}
\frac{d\sigma_{nn}}{dt} &= -\frac{i}{\hbar} \sum_{m}^{m\neq n}{}' (\bar{H}'_{nm}\sigma_{mn} - \sigma_{nm}\bar{H}'_{mn}) \\
&\quad - \sum_{m} R(t)_{nn}^{mm}\sigma_{mm} - \sum_{m'}^{m'\neq n'} \sum_{n'} {}'R(t)_{nn}^{m'n'}\sigma_{m'n'}
\end{aligned}
\tag{4.92}
$$

and

$$
\begin{aligned}
\frac{d\sigma_{mn}}{dt} &= -(i\bar{\omega}_{mn} + R(t)_{mn}^{mn})\sigma_{mn} - \frac{i}{\hbar}\bar{H}'_{mn}(\sigma_{nn} - \sigma_{mm}) \\
&\quad - \frac{i}{\hbar}\sum_{m'}{}''(\bar{H}'_{mm'}\sigma_{m'n} - \sigma_{mm'}\bar{H}'_{m'n}) - \sum_{m'}\sum_{n'} R(t)_{mn}^{m'n'}\sigma_{m'n'}
\end{aligned}
\tag{4.93}
$$

For the case in which the Markoff approximation is not used, we find

$$
\begin{aligned}
\frac{d\sigma_{nn}}{dt} &= \frac{2}{\hbar} \sum_{m}^{m\neq n}{}' \operatorname{Im}\left[\bar{H}'_{nm}\sigma_{mn}(0)e^{-t(i\bar{\omega}_{mn}+R_{mn}^{mn})} \right. \\
&\quad \left. - \frac{i}{\hbar}|\bar{H}'_{mn}|^2 \int_0^t d\tau e^{-\tau(i\bar{\omega}_{mn}+R_{mn}^{mn})}\{\sigma_{nn}(t-\tau) - \sigma_{mm}(t-\tau)\}\right] \\
&\quad - \sum_{m} R(t)_{nn}^{mm}\sigma_{mm} - \sum_{m'}^{m'\neq n'} \sum_{n'} {}'R(t)_{nn}^{m'n'}\sigma_{m'n'}
\end{aligned}
\tag{4.94}
$$

and

$$\frac{d\sigma_{mn}}{dt} = -(i\bar{\omega}_{mn} + R(t)^{mn}_{mn})\sigma_{mn} + \frac{i}{\hbar}\bar{H}'_{mn}(\sigma_{mm} - \sigma_{nn})$$

$$-\frac{i}{\hbar}\sum_{m'}{}''\left[\bar{H}'_{mm'}\sigma_{m'n}(0)e^{-it(i\bar{\omega}_{m'n}+R^{m'n}_{m'n})} - \sigma_{mm'}(0)\bar{H}'_{m'n}e^{-it(i\bar{\omega}_{mm'}+R^{mm'}_{mm'})}\right.$$

$$+\frac{i}{\hbar}\bar{H}'_{mm'}\bar{H}'_{m'n}\int_0^t d\tau\{e^{-i\tau(i\bar{\omega}_{m'n}+R^{m'n}_{m'n})}(\sigma_{m'm'}(t-\tau) - \sigma_{nn}(t-\tau))$$

$$\left.+ e^{-i\tau(i\bar{\omega}_{mm'}+R^{mm'}_{mm'})}(\sigma_{m'm'}(t-\tau) - \sigma_{mm}(t-\tau))\}\right] \qquad (4.95)$$

## V.  ULTRAFAST DYNAMICS

### A.  Dynamics of Populations and Coherences

In the previous section the equation of motion for the reduced matrix $\hat{\sigma}$ has been derived. In this section we shall study the dynamical processes taking place in this system. For this purpose we set $\hat{H}_s = \hat{H}_0 + \hat{H}'$, where $\hat{H}_0$ determines the basis set and $\hat{H}'$ describes the nature of rate process, that is,

$$\frac{d\hat{\sigma}}{dt} = -\frac{i}{\hbar}[\hat{H}_0, \hat{\sigma}] - \frac{i}{\hbar}[\hat{H}', \hat{\sigma}] - \hat{\Gamma}\hat{\sigma} \qquad (5.1)$$

Figure 13 shows a schematic representation of $\hat{H}'$ and manifolds $\{n\}$ and $\{m\}$. Notice that

$$\frac{d\sigma_{nn}}{dt} = -\frac{i}{\hbar}\sum_m(\hat{H}'_{nm}\sigma_{mn} - \sigma_{nm}\hat{H}'_{mn}) - \sum_\ell \Gamma^{\ell\ell}_{nn}\sigma_{\ell\ell} - \sum_k^{k\neq n}\sum_\ell{}'\Gamma^{k\ell}_{nn}\sigma_{k\ell} \qquad (5.2)$$

and

$$\frac{d\sigma_{mn}}{dt} = -(i\omega_{mn} + \gamma_{mn})\sigma_{mn} - \frac{i}{\hbar}\sum_\ell(\hat{H}'_{m\ell}\sigma_{\ell n} - \sigma_{m\ell}\hat{H}'_{\ell n}) - \sum_{m'}^{m'\neq m}\sum_{n'}^{n'\neq n}{}'\Gamma^{m'n'}_{mn}\sigma_{m'n'} \qquad (5.3)$$

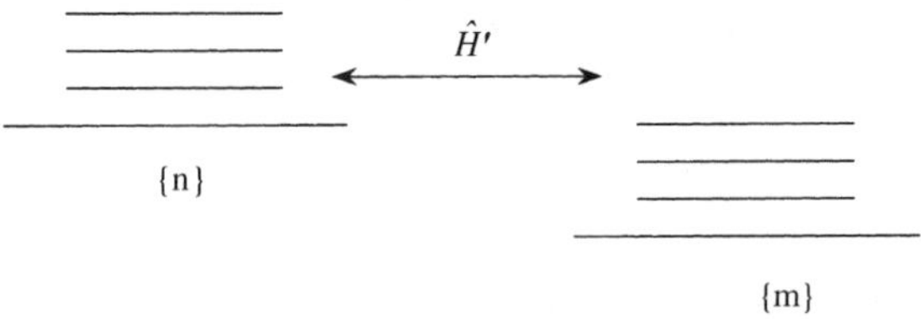

**Figure 13.**   Schematic representation of $\hat{H}'$ and manifolds $\{n\}$ and $\{m\}$.

The question is how to solve Eqs. (5.2) and (5.3). To be consistent with the approximations involved in $\Gamma_{nn}^{\ell\ell}$, $\Gamma_{nn}^{k\ell}$, $\gamma_{mn}$ (i.e., the real part of $\Gamma_{mn}^{mn}$), and $\Gamma_{mn}^{m'n'}$, we have $\sigma_{k\ell} = e^{-t(i\omega_{k\ell}+\gamma_{k\ell})}\sigma_{k\ell}(0)$ and $\sigma_{m'n'} = e^{-t(i\omega_{m'n'}+\gamma_{m'n'})}\sigma_{m'n'}(0)$.

We shall use the projection operator method to derive the Pauli master equation. With the Liouville equation, we separate the Liouville operator into two parts:

$$\frac{d\hat{\sigma}}{dt} = -i\hat{L}_s\hat{\sigma} - \hat{\Gamma}\hat{\sigma}, \qquad \hat{L}_s = \hat{L}_0 + \hat{L}' \tag{5.4}$$

First we perform the Laplace transformation,

$$p\hat{\sigma}(p) - \hat{\sigma}(0) = -i\hat{L}_s\hat{\sigma}(p) - \hat{\Gamma}\hat{\sigma}(p) \tag{5.5}$$

and then introduce the projection operator $\hat{D}$

$$\hat{\sigma}_1(p) = \hat{D}\hat{\sigma}(p), \qquad D_{mn}^{m'n'} = \delta_{mn}\delta_{mm'}\delta_{nn'} \tag{5.6}$$

which contains the diagonal matrix elements. It follows that if we let

$$i\bar{L}_s = i\hat{L}_s + \hat{\Gamma} \qquad \text{or} \qquad \bar{L}_s = \hat{L}_s - i\hat{\Gamma} \tag{5.7}$$

then we find

$$\frac{d\hat{\sigma}}{dt} = -i\bar{L}_s\hat{\sigma} \tag{5.8}$$

and if we let $\hat{M}(p)$ denote the memory kernel

$$\hat{M}(p) = \hat{D}\bar{L}_s\frac{1}{p + i(1-\hat{D})\bar{L}_s}(1-\hat{D})\bar{L}_s, \qquad \hat{M}(t) = \hat{D}\bar{L}_s e^{-it(1-\hat{D})\bar{L}_s}(1-\hat{D})\bar{L}_s \tag{5.9}$$

then we obtain

$$\frac{d\hat{\sigma}_1}{dt} = -i\hat{D}\bar{L}_s\hat{\sigma}_1 - i\hat{D}\bar{L}_s e^{-it(1-\hat{D})\bar{L}_s}\hat{\sigma}_2(0) - \int_0^t d\tau\hat{M}(\tau)\hat{\sigma}_1(t-\tau) \tag{5.10}$$

and

$$\frac{d\hat{\sigma}_2}{dt} = e^{-it(1-\hat{D})\bar{L}_s}\hat{\sigma}_2(0) - \int_0^t d\tau e^{-i\tau(1-\hat{D})\bar{L}_s}[i(1-\hat{D})\bar{L}_s]\hat{\sigma}_1(t-\tau) \tag{5.11}$$

In the random phase approximation (RPA), that is, when the dephasing is rapid in the timescale under consideration, we have

$$\hat{\sigma}_2(0) = 0 \tag{5.12}$$

$$\frac{d\hat{\sigma}_1}{dt} = -i\hat{D}\bar{L}_s\hat{\sigma}_1 - \int_0^t d\tau \hat{M}(\tau)\hat{\sigma}_1(t-\tau) \tag{5.13}$$

or

$$\hat{\sigma}_1(p) = \frac{1}{p + i\hat{D}\bar{L}_s + \hat{M}(p)}\hat{\sigma}_1(0) \tag{5.14}$$

Notice that

$$\frac{d\sigma_{nn}}{dt} = -\sum_\ell \Gamma_{nn}^{\ell\ell}\sigma_{\ell\ell} - \sum_\ell \int_0^t d\tau M(\tau)_{nn}^{\ell\ell}\sigma_{\ell\ell}(t-\tau) \tag{5.15}$$

and in the lowest approximation

$$M(t)_{nn}^{\ell\ell} = \frac{2}{\hbar^2}\delta_{\ell n}\sum_k{}^{k\neq\ell}{}'|H_{nk}'|^2 e^{-\gamma_{nk}t}\cos\omega_{nk}t - \frac{2}{\hbar^2}|H_{n\ell}'|^2 e^{-\gamma_{n\ell}t}\cos\omega_{n\ell}t \tag{5.16}$$

That is, for $n \neq \ell$

$$M(t)_{nn}^{\ell\ell} = -\frac{2}{\hbar^2}|H_{n\ell}'|^2 e^{-\gamma_{n\ell}t}\cos\omega_{n\ell}t \tag{5.17}$$

and for $\ell = n$

$$M(t)_{nn}^{nn} = \frac{2}{\hbar^2}\sum_k{}^{k\neq n}{}'|H_{nk}'|^2 e^{-\gamma_{nk}t}\cos\omega_{nk}t = -\sum_k M(t)_{nn}^{kk} = -\sum_k M(t)_{kk}^{nn} \tag{5.18}$$

Therefore

$$\frac{d\sigma_{nn}}{dt} = -\sum_\ell \Gamma_{nn}^{\ell\ell}\sigma_{\ell\ell} - \int_0^t d\tau \left[ M(\tau)_{nn}^{nn}\sigma_{nn}(t-\tau) + \sum_{\ell\neq n}{}'M(\tau)_{nn}^{\ell\ell}\sigma_{\ell\ell}(t-\tau) \right] \tag{5.19}$$

or

$$\frac{d\sigma_{nn}}{dt} = -\sum_\ell \Gamma_{nn}^{\ell\ell}\sigma_{\ell\ell} + \sum_\ell{}' \int_0^t d\tau M(\tau)_{nn}^{\ell\ell}[\sigma_{\ell\ell}(t-\tau) - \sigma_{nn}(t-\tau)] \tag{5.20}$$

According to the Markov approximation, for example,

$$M(t)_{nn}^{\ell\ell} = -W_{n\ell}\delta(t) \tag{5.21}$$

and

$$W_{n\ell} = -\int_0^\infty dt M(t)_{nn}^{\ell\ell} \tag{5.22}$$

Eq. (5.20) becomes

$$\frac{d\sigma_{nn}}{dt} = -\sum_\ell \Gamma_{nn}^{\ell\ell}\sigma_{\ell\ell} + \sum_\ell{}' W_{n\ell}(\sigma_{\ell\ell} - \sigma_{nn}) \tag{5.23}$$

where $W_{n\ell}$ represents the rate constant for the transition $n \leftrightarrow \ell$:

$$W_{n\ell} = \frac{2\pi}{\hbar^2}|H'_{n\ell}|^2 \frac{1}{\pi}\frac{\gamma_{n\ell}}{\gamma_{n\ell}^2 + \omega_{n\ell}^2} \tag{5.24}$$

## B. Applications to Nonadiabatic Processes

Using Eq. (5.23), that is,

$$\frac{d\sigma_{nn}}{dt} = -\sum_\ell \Gamma_{nn}^{\ell\ell}\sigma_{\ell\ell} + \sum_\ell W_{n\ell}(\sigma_{\ell\ell} - \sigma_{nn}) \tag{5.25}$$

various cases can be considered. For example, we are interested in the dynamical processes taking place between the vibronic manifolds, $\{av\}$ and $\{bu\}$ (see Fig. 14). In this case we have

$$\frac{d\sigma_{av,av}}{dt} = -\sum_{v'} \Gamma_{av,av}^{av',av'}\sigma_{av',av'} + \sum_u W_{av,bu}(\sigma_{bu,bu} - \sigma_{av,av}) \tag{5.26}$$

Here it is assumed that the interaction between the system and the heat bath does not induce the nonadiabatic transition. Similarly we have

$$\frac{d\sigma_{bu,bu}}{dt} = -\sum_{v'} \Gamma_{bu,bu}^{bu',bu'}\sigma_{bu',bu'} + \sum_u W_{av,bu}(\sigma_{av,av} - \sigma_{bu,bu}) \tag{5.27}$$

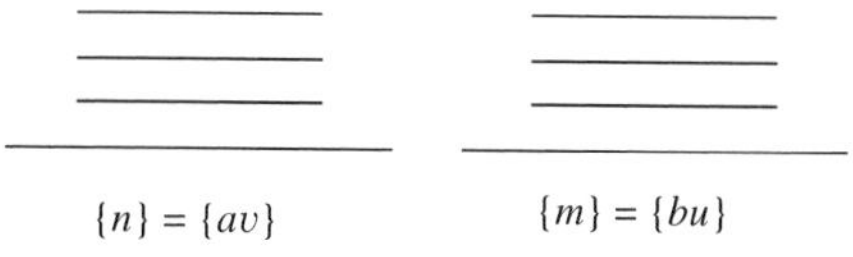

$\{n\} = \{av\}$ $\qquad$ $\{m\} = \{bu\}$

**Figure 14.** Schematic representation of manifolds $\{av\}$ and $\{bu\}$.

**Case 1:** *Fast Vibrational Relaxation. If the vibrational relaxation is much faster than nonadiabatic transitions, then*

$$\sigma_{av,av} = \sigma_{aa}P_{av}, \qquad \sigma_{bu,bu} = \sigma_{bb}P_{bu} \qquad (5.28)$$

*where $P_{av}$ and $P_{bu}$ denote the Boltzmann distribution factors. It follows that*

$$\frac{d\sigma_{aa}}{dt} = W_{b\to a}\sigma_{bb} - W_{a\to b}\sigma_{aa} \qquad (5.29)$$

*and*

$$\frac{d\sigma_{bb}}{dt} = -W_{b\to a}\sigma_{bb} + W_{a\to b}\sigma_{aa} \qquad (5.30)$$

*where, for example,*

$$W_{a\to b} = \frac{2\pi}{\hbar}\sum_u\sum_v P_{av}|H'_{bu,av}|^2\delta(E_{bu} - E_{av}) \qquad (5.31)$$

*is a thermal average rate constant.*

**Case 2:** *Slow Vibrational Relaxation. In this case we have*

$$\frac{d\sigma_{nn}}{dt} = \sum_m W_{nm}(\sigma_{mm} - \sigma_{nn}) \qquad (5.32)$$

*where $m \equiv bu$ and $n \equiv av$. For simplicity we shall assume that only one $av$ state is initially populated, that is,*

$$\frac{d\sigma_{mm}}{dt} = W_{nm}(\sigma_{nn} - \sigma_{mm}) \qquad (5.33)$$

*Applying the Laplace transformation, we obtain*

$$p\bar{\sigma}_{mm}(p) = W_{nm}(\bar{\sigma}_{nn} - \bar{\sigma}_{mm}), \qquad \bar{\sigma}_{mm} = \frac{\bar{\sigma}_{nn}}{p + W_{nm}} \qquad (5.34)$$

*and*

$$p\bar{\sigma}_{nn} - \bar{\sigma}_{nn}(0) = \sum_m W_{nm}(\bar{\sigma}_{mm} - \bar{\sigma}_{nn}) \qquad (5.35)$$

*or*

$$\bar{\sigma}_{nn} = \frac{\sigma_{nn}(0)}{p\left(1 + \sum_m \frac{W_{nm}}{p+W_{nm}}\right)} \tag{5.36}$$

*In the long-time range, $p \ll W_{nm}$ and*

$$\bar{\sigma}_{nn} = \frac{\sigma_{nn}(0)}{p(1+M)}, \qquad \sigma_{nn}(t) = \frac{\sigma_{nn}(0)}{1+M} \tag{5.37}$$

*where $M = \sum_m 1$. Similarly, in short-time range, $p \gg W_{nm}$ and*

$$\bar{\sigma}_{nn} = \frac{\sigma_{nn}(0)}{p + \sum_m W_{nm}}, \qquad \sigma_{nn}(t) = \sigma_{nn}(0)e^{-tW_n} \tag{5.38}$$

*where*

$$W_n = \sum_m W_{nm} \tag{5.39}$$

*$W_n$ denotes the single-level rate constant.*

## VI.   PUMP-PROBE AND TRANSIENT ABSORPTION EXPERIMENT

We start with the Hamiltonian

$$\hat{H} = \hat{H}_0 + \hat{H}' \tag{6.1}$$

where

$$\hat{H}' = -\vec{\mu} \cdot \vec{E}(t) = -\vec{\mu} \cdot \vec{E}_0 \cos \omega t \tag{6.2}$$

and $\vec{E}$ represents the electric field of optical radiation. That is, the interaction between the system and radiation field is assumed to be described by the dipole approximation. In Section IV we showed that

$$Q = 2\omega\chi''(\omega)|\vec{E}_0(\omega)|^2 \tag{6.3}$$

That is, the rate of energy absorption $Q$ is linearly related to the imaginary part of linear susceptibility $\chi(\omega)$.

To calculate $\chi(\omega)$ we have to calculate the polarizability $\vec{P}(t)$, which is related to the reduced density matrix $\hat{\rho}(t)$. [Here, for convenience, $\hat{\rho}$ is used instead of $\hat{\sigma}(t)$.] The reduced density matrix satisfies the Liouville equation:

$$\frac{d\hat{\rho}}{dt} = -i\hat{L}_0\hat{\rho} - i\hat{L}'\hat{\rho} - \hat{\Gamma}\hat{\rho} \tag{6.4}$$

We shall let

$$\hat{\rho}(t) = e^{-it\hat{L}_0'}\hat{\sigma}(t), \qquad \hat{L}_0' = \hat{L}_0 - i\hat{\Gamma} \tag{6.5}$$

to obtain

$$\frac{d\hat{\sigma}}{dt} = -ie^{it\hat{L}_0'}\hat{L}'e^{-it\hat{L}_0'}\hat{\sigma}(t) = -i\bar{L}'(t)\hat{\sigma}(t) \tag{6.6}$$

where

$$\bar{L}'(t) = e^{it\hat{L}_0'}\hat{L}'e^{-it\hat{L}_0'} \tag{6.7}$$

It follows that

$$\hat{\sigma}(t) = \hat{\sigma}_i - i\int_{t_i}^{t} d\tau \bar{L}'(\tau)\hat{\sigma}(\tau) \tag{6.8}$$

where $\hat{\sigma}_i$ denotes the $\hat{\sigma}(t)$ at $t = t_i$. To the first-order approximation, Eq. (6.8) yields

$$\hat{\sigma}^{(1)}(t) = -i\int_{t_i}^{t} d\tau \bar{L}'(\tau)\hat{\sigma}_i, \qquad \hat{\rho}^{(1)}(t) = e^{-it\hat{L}_0'}\hat{\sigma}^{(1)}(t) \tag{6.9}$$

or

$$\hat{\rho}^{(1)}(t) = -i\int_{t_i}^{t} d\tau e^{-i(t-\tau)\hat{L}_0'}\hat{L}'(\tau)e^{-i(\tau-t_i)\hat{L}_0'}\hat{\rho}(t_i) \tag{6.10}$$

and the polarizability can be evaluated up to the first order:

$$\vec{P}(t) = \vec{P}^{(1)}(t) = \text{Tr}(\hat{\rho}^{(1)}(t)\vec{\mu}) = \text{Tr}(\vec{\mu}\hat{\rho}^{(1)}(t)) \tag{6.11}$$

For the manifolds $\{n\}$, $\{m\}$, and $\{k\}$ shown in Fig. 15, we find

$$\vec{P}(t) = \sum_{\ell}\sum_{m} \rho_{\ell m}^{(1)}(t)\vec{\mu}_{m\ell} + \text{c.c.} \tag{6.12}$$

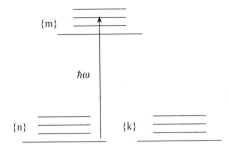

**Figure 15.** Schematic representation of manifolds $\{n\}$, $\{m\}$ and $\{k\}$.

and

$$\rho_{\ell m}^{(1)}(t) = -\frac{i}{\hbar} \int_{t_i}^{t} d\tau e^{-i(t-\tau)\omega'_{\ell m}} \sum_{\ell'} [H'_{\ell\ell'}(\tau) e^{-i(\tau-t_i)\omega'_{\ell'm}} \rho(t_i)_{\ell'm}$$
$$- H'_{\ell'm}(\tau) e^{-i(\tau-t_i)\omega'_{\ell\ell'}} \rho(t_i)_{\ell\ell'}] \tag{6.13}$$

Because the coherence like $\rho(t_i)_{\ell'm}$, which involves the higher electronic state, is difficult to create, we shall remove this contribution. Thus

$$\rho_{\ell m}^{(1)}(t) = \frac{i}{\hbar} e^{-i(t-\tau)\omega'_{\ell m}} \sum_{\ell'} (e^{-it_i\omega'_{\ell\ell'}} \rho(t_i)_{\ell\ell'}) \int_{t_i}^{t} d\tau e^{-i\tau\omega'_{\ell'm}} H'(\tau)_{\ell'm} \tag{6.14}$$

and

$$\vec{P}(t) = -\frac{1}{\hbar} \sum_{\ell} \sum_{\ell''} \sum_{m} \rho(\Delta t)_{\ell\ell''} (\vec{\mu}_{m\ell} \vec{\mu}_{\ell''m} \cdot \vec{E}(-\omega) e^{it\omega}) \frac{[1 - e^{-i\Delta t(\omega+\omega'_{\ell''m})}]}{\omega + \omega_{\ell''m} - i\gamma_{\ell''m}} + \text{c.c.} \tag{6.15}$$

where $\Delta t = t - t_i$ and $\rho(\Delta t)_{\ell\ell''}$ represents the population (when $\ell'' = \ell$) or coherence (when $\ell'' \neq \ell$) of the system at $t$.

It follows that in the long-time region,

$$\chi(-\omega) = -\frac{1}{\hbar} \sum_{\ell} \sum_{\ell''} \sum_{m} \rho(\Delta t)_{\ell\ell''} \frac{(\vec{\mu}_{\ell''m} \vec{\mu}_{m\ell})}{\omega + \omega_{\ell''m} - i\gamma_{\ell''m}} \tag{6.16}$$

and

$$\chi(\omega) = -\frac{1}{\hbar} \sum_{\ell} \sum_{\ell''} \sum_{m} \rho(\Delta t)_{\ell\ell''} \frac{(\vec{\mu}_{\ell''m} \vec{\mu}_{m\ell})}{\omega + \omega_{\ell m} + i\gamma_{\ell m}} \tag{6.17}$$

or

$$\chi''(\omega) = \frac{i}{2\hbar} \sum_\ell \sum_{\ell''} \sum_m \rho(\Delta t)_{\ell\ell''} (\vec{\mu}_{\ell''m} \vec{\mu}_{m\ell}) \left( \frac{1}{\omega + \omega_{\ell m} + i\gamma_{\ell m}} - \frac{1}{\omega + \omega_{\ell''m} - i\gamma_{\ell''m}} \right)$$

(6.18)

Equation (6.18) shows that $\chi''(\omega)$ consists of the incoherent contribution $\chi''(\omega)_i$ and coherent contribution $\chi''(\omega)_c$, that is,

$$\chi''(\omega) = \chi''(\omega)_i + \chi''(\omega)_c$$

(6.19)

where

$$\chi''(\omega)_i = \frac{1}{\hbar} \sum_\ell \sum_m \rho(\Delta t)_{\ell\ell} (\vec{\mu}_{\ell m} \vec{\mu}_{m\ell}) \frac{\gamma_{\ell m}}{\gamma_{\ell m}^2 + (\omega + \omega_{\ell m})^2}$$

(6.20)

and

$$\chi''(\omega)_c = \frac{i}{2\hbar} \sum_\ell \sum_{\ell''}^{\ell \neq \ell''} {}' \sum_m \rho(\Delta t)_{\ell\ell''} (\vec{\mu}_{\ell''m} \vec{\mu}_{m\ell}) \left( \frac{1}{\omega + \omega_{\ell m} + i\gamma_{\ell m}} - \frac{1}{\omega + \omega_{\ell''m} - i\gamma_{\ell''m}} \right)$$

(6.21)

An interesting case will be the contribution to $\chi''(\omega)$ due to the electronic coherence, that is,

$$\chi''(\omega)_{c(\text{electronic})} = \frac{i}{2\hbar} \sum_n \sum_k \sum_m \rho(\Delta t)_{nk} (\vec{\mu}_{km} \vec{\mu}_{mn})$$

$$\times \left( \frac{1}{\omega + \omega_{nm} + i\gamma_{nm}} - \frac{1}{\omega + \omega_{km} - i\gamma_{km}} \right)$$

(6.22)

The above theory is usually called the generalized linear response theory because the linear optical absorption initiates from the nonstationary states prepared by the pumping process [85–87]. This method is valid when pumping pulse and probing pulse do not overlap. When they overlap, third-order $\vec{P}^{(3)}$ or $\chi^{(3)}(\omega)$ should be used. In other words, Eq. (6.4) should be solved perturbatively to the third-order approximation. From Eqs. (6.19)–(6.22) we can see that in the time-resolved spectra described by $\chi''(\omega)$, the dynamics information of the system is contained in $\hat{\rho}(\Delta t)$, which can be obtained by solving the reduced Liouville equations. Application of Eq. (6.19) to stimulated emission monitoring vibrational relaxation is given in Appendix III.

## VII. EARLY EVENTS IN PHOTOSYNTHETIC RCs

From the discussion presented in previous sections, vibrational relaxation (Appendix II) plays a very important role in the initial ET in photosynthetic RCs. This problem was first studied by Martin and co-workers [4] using *Rb. capsulatas* $D_{LL}$. In this mutant, the ultrafast initial ET is suppressed and the ultrafast process taking place in the ps range is mainly due to vibrational relaxation. They have used the pumping laser at $\lambda_{pump} = 870$ nm and probed at $\lambda_{probe} = 812$ nm at 10 K. The laser pulse duration in this case is 80 fs. Their experimental results are shown in Fig. 16, where one can observe that the fs time-resolved spectra exhibit an oscillatory build-up. To analyze these results, we use the relation

$$\chi''(\omega) = \sum_{v} \rho(\Delta t)_{av,av}\alpha_{av,av}(\omega) + \sum_{v}^{v \neq v'}\sum_{v'} \text{Re}[\rho(\Delta t)_{av,av'}\alpha_{av,av'}(\omega)] \quad (7.1)$$

which shows that the fs time-resolved spectra consist of the contributions from population $\rho(\Delta t)_{av,av}$ and from coherences (or phases) $\rho(\Delta t)_{av,av'}$. In Eq. (7.1) $\alpha_{av,av}(\omega)$ and $\alpha_{av,av'}(\omega)$ represent the band-shape functions associated with populations and coherences, respectively.

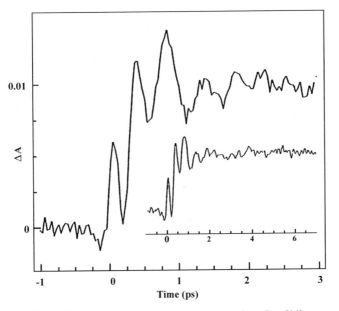

**Figure 16.** Quantum beat observed in *Rb. capsulatus* $D_{LL}$ [14].

The dynamical behaviors of $\rho(\Delta t)_{av,av}$ and $\rho(\Delta t)_{av,av'}$ have to be determined by solving the stochastic Liouville equation for the reduced density matrix; the initial conditions are determined by the pumping process. For the purpose of qualitative discussion, we assume that the 80-fs pulse can only pump two vibrational states, say $v = 0$ and $v = 1$ states. In this case we obtain

$$\rho(\Delta t)_{a1,a1} = \rho(0)_{a1,a1} e^{-\gamma_{10}\Delta t} \tag{7.2}$$

$$\rho(\Delta t)_{a0,a0} = \rho(0)_{a0,a0} + \rho(0)_{a1,a1}(1 - e^{-\gamma_{10}\Delta t}) \tag{7.3}$$

and

$$\rho(\Delta t)_{a1,a0} = \rho(0)_{a1,a0} e^{-\Delta t(i\omega_v + \Gamma_{10})} \tag{7.4}$$

where $\gamma_{10}$, $\Gamma_{10}$, and $\omega_v$ represent the vibrational relaxation rate constant, dephasing constant, and vibrational frequency, respectively. Using Eqs. (7.2)–(7.4) to analyze the fs time-resolved spectra shown in Fig. 16, we can obtain the information of $\gamma_{10}$, $\Gamma_{10}$, and $\omega_v$. This type of analysis has been carried out, and the calculated fs time-resolved spectra are shown in Fig. 17. Comparing Figs. 16 and 17, we can see that the agreement between experiment and theory is very reasonable and we obtain

$$\gamma_{10} = 2\,\text{ps}^{-1} \tag{7.5}$$

$$\Gamma_{10} = 1.2\,\text{ps}^{-1} \tag{7.6}$$

$$\omega_v = 100\,\text{cm}^{-1} \tag{7.7}$$

From Eqs (7.5) and (7.6) we can deduce that the pure dephasing rate is $\gamma_{10}(d) = 0.2\,\text{ps}^{-1}$ and the vibrational relaxation takes place in the timescale of 0.5 ps for the 100-cm$^{-1}$ mode. More results related to vibrational relaxation have been reported by Martin group [1–5]. In this chapter we choose 0.3 ps for the 100-cm$^{-1}$ mode, and the vibrational relaxation rates for other modes are scaled with their vibrational frequencies.

Next we shall discuss how to theoretically construct the three-dimensional fs time-resolved spectra. For this purpose, recent fs time-resolved spectra reported by Scherer's group for *Rb. sphaeroides* R26 with $\lambda_{\text{excitation}} = 800$ nm at room temperature are shown in Fig. 18 is considered [39]. They have used a laser pulse of 30 fs. To theoretically construct the fs time-resolved spectra, we need the potential surfaces; for displaced surfaces we need the vibrational frequencies $\omega_i$ and their Huang–Rhys factors $S_i$. For bacterial photosynthetic RCs, these physical constants are given in Table I. In addition to the potential surfaces, we need interactions between different electronic states which are shown in

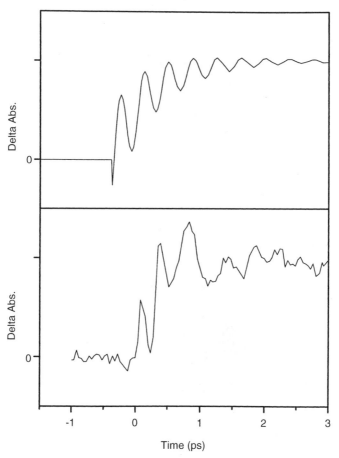

**Figure 17.** Calculated fs time-resolved spectra for *Rb. capsulatus* $D_{LL}$ [71]. $\omega_v = 100$ cm$^{-1}$, $\lambda_1 = 2$ ps$^{-1} = \gamma_{10}$, $\gamma_2 = 1.2$ ps$^{-1} = \Gamma_{10}$.

Fig. 19 for the wild type and R26 of *Rb. Sphaeroides*. In Fig. 19 we also show the calculated rate constants of various processes. We regard $PB^*H \rightarrow P^*_{Y+}BH$ and $P^*_{Y+}BH \rightarrow P^*_{Y-}BH$ as internal conversion. The timescales shown in Fig. 19 are compatible with those discussed in the literature [2,10,16,21,23,25,31, 36,39,51].

For the case in which the electronic transition is much faster than vibrational relaxation, one has to use the single-vibronic level rate constant; and in analyzing the transient absorption or stimulated emission spectra, the single-vibronic level absorption or stimulated emission coefficient should be used. For

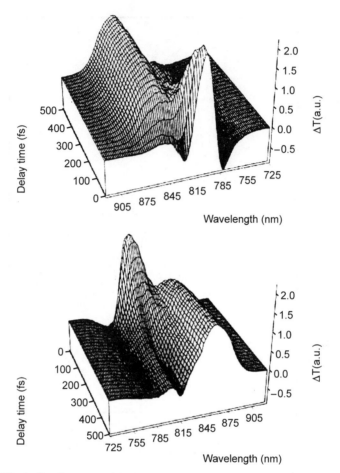

**Figure 18.** Femtosecond time-resolved spectra for *Rb. sphaeroides* R26 [39].

example, for single-vibronic level stimulated emission spectra, we have

$$\alpha_{bv}(\omega) = \frac{4\pi^2\omega}{3\hbar c n_e(\omega)} \sum_u |\langle \Psi_{au}|\vec{\mu}|\Psi_{bv}\rangle|^2 D(\omega_{au,bv} + \omega) \tag{7.8}$$

In the case of displaced harmonic surfaces with inhomogeneity in the electronic energy gap, $\langle \alpha_{bv}(\omega) \rangle$ becomes

$$\langle \alpha_{bv}(\omega) \rangle = \frac{2\pi^2\omega}{3\hbar c n_e(\omega)} \text{Re} \int_0^\infty dt \exp[(\bar{\omega}_{ba} + \omega) - (tD_{ba}/2)^2 \prod_\ell G_{v\ell}(t) \tag{7.9}$$

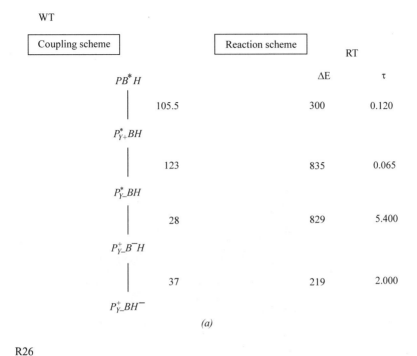

**Figure 19.** Electronic states coupling schemes for WT [Fig. 19(a)] and R26 [Fig. 19(b)] of *Rb. sphaeroides*.

Here, $G_{v\ell}(t)$ is given by

$$G_{v_\ell}(t) = \exp[-S_\ell(1 - e^{it\omega_\ell})] \sum_{m_\ell=0}^{v_\ell} \frac{v_\ell![S_\ell(e^{it\omega_\ell/2} - e^{-it\omega_\ell/2})^2]^{v_\ell - m_\ell}}{m_\ell![(v_\ell - m_\ell)!]^2} \qquad (7.10)$$

Using Table I and Fig. 19, we theoretically construct the fs time-resolved spectra for R26 of *Rb. Sphaeroides* at room temperature; the calculated spectra are

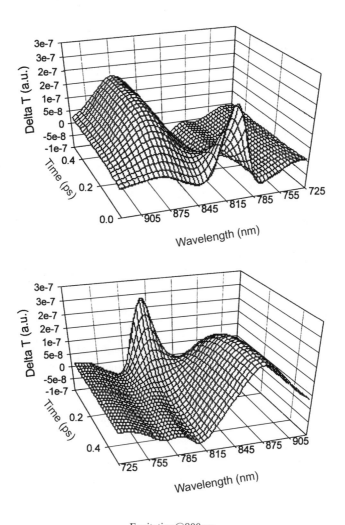

Excitation@800nm

**Figure 20.** Calculated fs time-resolved spectra for *Rb. sphaeroides* R26.

shown in Fig. 20. Figure 21 shows the calculated single vibronic emission spectra used in the calculation of Fig. 20. Comparing Fig. 18 and 20, we can see that the agreement is quite satisfactory.

Recently, Scherer et al. have used the 10-fs laser pulse with $\lambda_{excitation} = 860$ nm to study the dynamical behavior of *Rb. Sphaeroides* R26 at room temperatures. In this case, due to the use of the 10-fs pulse both P band and B band are coherently excited. Thus the quantum beat behaviors are much more complicated. We have used the data given in Table I and Fig. 19 to simulate the quantum beat behaviors (see also Fig. 22). Without including the electronic coherence, the agreement between experiment and theory can not be accomplished.

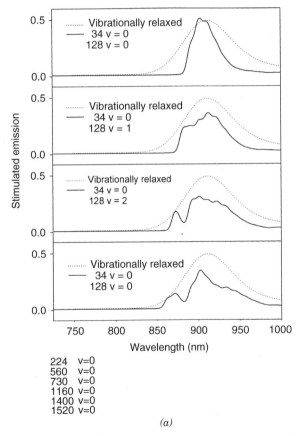

*(a)*

**Figure 21.** Calculated single-vibronic level spectra. The spectra are calculated as a function of the quantum numbers of the protein mode $\omega_{34}$ and the marker mode $\omega_{128}$: (a) $v_{34} = 0$, (b) $v_{34} = 1$, (c) $v_{34} = 2$.

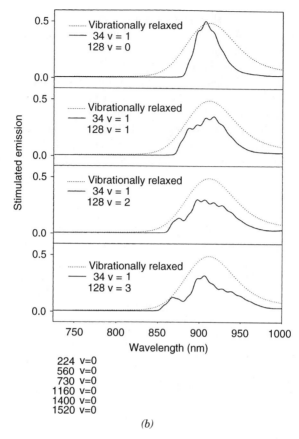

*(b)*

**Figure 21** *(Continued)*

# VIII.   CONCLUDING REMARKS

In this chapter we have shown that the dynamics and spectroscopy of the initial events taking place in bacterial photosynthetic RCs can be described by the model shown in Table I and Fig. 19. Using these physical constants we can calculate the absorption spectra, ET rate constants, and fs time-resolved spectra. It should be noted that for processes taking place in sub-ps range, it is more reasonable not to use "rate constant" because the concept of "rate constant" requires the validity of the Markoff approximation [82,88]. Instead the

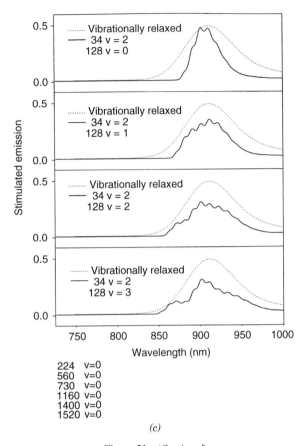

*(c)*

**Figure 21** *(Continued)*

Hamiltonian that involves different potential surfaces and their interactions should be used. The nature of vibrational relaxation is very important in fs processes; it requires much more experimental and theoretical investigations.

In order to determine the physical mechanism of initial ET including other rapid kinetics in photosynthetic RCs, it is necessary to construct a vibronic model that comprises the electronic and vibrational states of the system. It is also important to take into account temperature effect in both experiments and theories. In particular, we should stress that most of MO calculations carried out for RCs are based on the crystallographic structures. However, the structure at room temperature may be different from that obtained from the X-ray analysis,

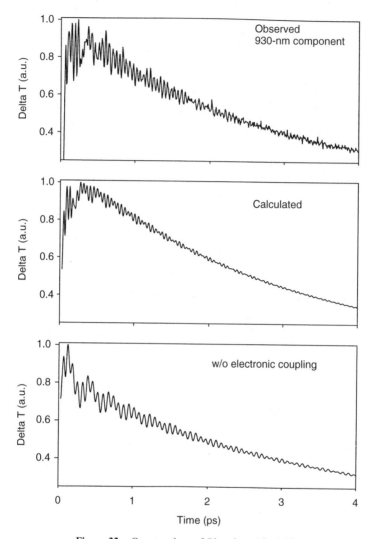

**Figure 22.** Quantum beat of *Rb. sphaeroides* R26.

which, in turn, leads to differences in the electronic states of RCs at room temperature. Our theoretical analysis based on the thermal expansion model indicates that the temperature significantly affects the electronic states of RCs. Thus, we would like to emphasize that extreme caution should be needed especially when anisotropy data at room temperature are analyzed based on the MO calculations.

## APPENDIX I

In this appendix we shall discuss the relation between $\Delta E_a$ and the potential energy surface crossing. First we consider the single-mode case; in this case the potential curves for the initial and final electronic states are given by

$$U_i(Q) = E_i + \frac{1}{2}\omega^2 Q^2 \tag{I.1}$$

and

$$U_f(Q') = E_f + \frac{1}{2}\omega^2 Q'^2 \tag{I.2}$$

where $Q' = Q + \Delta Q$ for the displaced oscillator case (Fig. A1). At the crossing point we have

$$U_i(Q_c) = U_f(Q'_c) \tag{I.3}$$

or

$$E_i + \frac{1}{2}\omega^2 Q_c^2 = E_f + \frac{1}{2}\omega^2 (Q_c + \Delta Q)^2 \tag{I.4}$$

It follows that

$$\Delta Q_c = \frac{(E_i - E_f - \frac{1}{2}\omega^2 \Delta Q^2)}{\omega^2 \Delta Q} \tag{I.5}$$

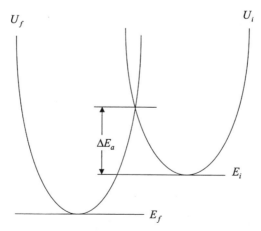

**Figure A1.**    Schematic representation of potential energy surfaces $U_i$ and $U_f$.

and

$$\Delta E_a = U_i(Q_c) - E_i = \frac{1}{2}\omega^2 Q_c^2 = \frac{(E_f - E_i + \frac{1}{2}\omega^2 \Delta Q^2)^2}{2\omega^2 \Delta Q_c^2} \tag{I.6}$$

Using $S = \frac{\omega}{2\hbar}\Delta Q^2$, we obtain

$$\Delta E_a = \frac{(\Delta G_{fi} + \lambda)^2}{4\lambda} \tag{I.7}$$

where $\lambda = S\hbar\omega$.

Next we consider the multimode case. In this case we have

$$U_i(\vec{Q}) = E_i + \sum_j \frac{1}{2}\omega_j^2 Q_j^2 \tag{I.8}$$

and

$$U_f(\vec{Q}') = E_f + \sum_j \frac{1}{2}\omega_j^2 Q_j'^2 \tag{I.9}$$

At the crossing point,

$$U_i(\vec{Q}) = U_f(\vec{Q}') \tag{I.10}$$

To find the minimum crossing point we shall use the Lagrange multiplier method. That is, we consider

$$U_i(\vec{Q}) = E_i + \sum_j \frac{1}{2}\omega_j^2 Q_j^2 + \lambda(U_i - U_f) \tag{I.11}$$

or

$$U_i(\vec{Q}) = E_i + \sum_j \frac{1}{2}\omega_j^2 Q_j^2 \lambda\left[E_i + \sum_j \frac{1}{2}\omega_j^2 Q_j^2 - E_f - \sum_j \frac{1}{2}\omega_j^2 (Q_j + \Delta Q_j)^2\right] \tag{I.12}$$

where $\lambda$ denotes the Lagrange multiplier. Applying the condition for finding a minimum (or maximum), we find

$$\frac{\partial U_i}{\partial Q_j} = \omega_j^2 Q_j + \lambda[\omega_j^2 Q_j - \omega_j^2(Q_j + \Delta Q_j)] \tag{I.13}$$

Setting $\left(\frac{\partial U_i}{\partial Q_j}\right)_c = 0$ yields

$$Q_{jc} = \lambda \Delta Q_j \tag{I.14}$$

$$E_i + \frac{1}{2}\sum_j \omega_j^2 Q_{jc}^2 = E_f + \frac{1}{2}\sum_j \omega_j^2 (Q_{jc} + \Delta Q_j)^2 \tag{I.15}$$

and

$$\lambda = \frac{E_i - E_f - \frac{1}{2} \sum_j \omega_j^2 \Delta Q_j^2}{\sum_j \omega_j^2 \Delta Q_j^2} \tag{I.16}$$

It follows that

$$\Delta E_a = U_i(Q_c) - E_i = \frac{1}{2} \sum_j \omega_j^2 Q_{jc}^2 = \frac{\lambda^2}{2} \sum_j \omega_j^2 \Delta Q_j^2 \tag{I.17}$$

or

$$\Delta E_a = \frac{(E_f - E_i + \frac{1}{2} \sum_j \omega_j^2 \Delta Q_j^2)^2}{2 \sum_j \omega_j^2 \Delta Q_j^2} = \frac{(\Delta G_{fi} + \lambda)^2}{4\lambda} \tag{I.18}$$

where $\lambda = \sum_j S_j \hbar \omega_j$.

## APPENDIX II. VIBRATIONAL RELAXATION AND DEPHASING

From the above discussion, we can see that vibrational relaxation plays a very important role in ultrafast phenomena. In this appendix we shall present a model that can describe vibrational relaxation and dephasing (especially pure dephasing). Suppose that

$$\hat{H} = \hat{H}_s + \hat{H}_b + \hat{H}' \tag{II.1}$$

where

$$\hat{H}_s = \frac{1}{2} P^2 + \frac{1}{2} \omega^2 Q^2 \tag{II.2}$$

$$\hat{H}_b = \sum_i \left( \frac{1}{2} p_i^2 + \frac{1}{2} \omega_i^2 q_i^2 \right) \tag{II.3}$$

and

$$\hat{H}' = \frac{1}{3!} \sum_i \sum_j \left( \frac{\partial^3 V}{\partial Q \partial q_i \partial q_j} \right)_0 Q q_i q_j + \frac{1}{3!} \sum_i \left( \frac{\partial^3 V}{\partial Q^2 \partial q_i} \right)_0 Q^2 q_i$$

$$+ \frac{1}{4!} \sum_i \sum_j \sum_k \left( \frac{\partial^4 V}{\partial Q \partial q_i \partial q_j \partial q_k} \right)_0 Q q_i q_j q_k$$

$$+ \frac{1}{4!} \sum_i \sum_j \left( \frac{\partial^4 V}{\partial Q^2 \partial q_i \partial q_j} \right)_0 Q^2 q_i q_j + \cdots \tag{II.4}$$

That is, we consider a system oscillator embedded in a bath of a collection of harmonic oscillators. The interaction between the system and heat bath is described by anharmonic couplings. For convenience of discussion, we shall use the following notations for the values of different derivatives of the potential:

$$\alpha_{ij} = \frac{1}{3!}\left(\frac{\partial^3 V}{\partial Q \partial q_i \partial q_j}\right)_0, \qquad \alpha_i = \frac{1}{3!}\left(\frac{\partial^3 V}{\partial Q^2 \partial q_i}\right)_0$$

$$\beta_{ijk} = \frac{1}{4!}\left(\frac{\partial^4 V}{\partial Q \partial q_i \partial q_j \partial q_k}\right)_0, \qquad \beta_{ij} = \frac{1}{4!}\left(\frac{\partial^4 V}{\partial Q^2 \partial q_i \partial q_j}\right)_0, \quad \text{etc.} \tag{II.5}$$

We first consider the vibrational relaxation that can be induced by $\alpha_{ij}Qq_iq_j$ (three-phonon processes) or $\beta_{ijk}Qq_iq_jq_k$ (four-phonon processes). In the three-phonon processes there are two accepting modes, while in the four-phonon processes there are three accepting modes. To calculate the rate of vibrational relaxation, we use

$$R_{mm}^{nn} = -\frac{2\pi}{\hbar^2} \sum_{\alpha}^{\alpha \neq \beta} \sum_{\beta}' \rho_{\beta\beta}^{(b)} |H'_{m\alpha,n\beta}|^2 \delta(\omega_{m\alpha,n\beta}) \tag{II.6}$$

for the $n \rightarrow m$ transition. We shall use the three-phonon process as an example. In this case we have

$$H'_{m\alpha,n\beta} = \langle m\alpha | Q \sum_i \sum_j \alpha_{ij} q_i q_j | n\beta \rangle = \langle m|Q|n \rangle \langle \alpha | \sum_i \sum_j \alpha_{ij} q_i q_j | \beta \rangle \tag{II.7}$$

where

$$\langle m|Q|n \rangle = \sqrt{\frac{(n+1)\hbar}{2\omega}} \langle m|n+1 \rangle + \sqrt{\frac{n\hbar}{2\omega}} \langle m|n-1 \rangle \tag{II.8}$$

and $\omega$ denotes the vibrational frequency of the system oscillator. For the vibrational relaxation we have $m = n - 1$, and

$$R_{n-1n-1}^{nn} = -\frac{2\pi}{\hbar^2}\left(\frac{n\hbar}{2\omega}\right) \sum_{\alpha}^{\alpha \neq \beta} \sum_{\beta}' \rho_{\beta\beta}^{(b)} \left|\langle \alpha | \sum_i \sum_j \alpha_{ij} q_i q_j | \beta \rangle\right|^2 \delta(\omega_{\alpha\beta} - \omega) \tag{II.9}$$

or

$$R_{n-1n-1}^{nn} = nR_{00}^{11} \tag{II.10}$$

and

$$R_{00}^{11} = -\frac{\pi}{\hbar\omega} \sum_{\alpha}^{\alpha \neq \beta} \sum_{\beta}' \rho_{\beta\beta}^{(b)} \left|\langle \alpha | \sum_i \sum_j \alpha_{ij} q_i q_j | \beta \rangle\right|^2 \delta(\omega_{\alpha\beta} - \omega) \tag{II.11}$$

representing the vibrational relaxation rate constant for the $n = 1 \rightarrow m = 0$ transition. Using the relations

$$|\alpha\rangle = \prod_i |\alpha_i\rangle, \qquad |\beta\rangle = \prod_i |\beta_i\rangle \tag{II.12}$$

we obtain

$$R_{00}^{11} = -\frac{\pi}{\hbar\omega} \sum_\beta \rho_{\beta\beta}^{(b)} \sum_i \sum_j \alpha_{ij}^2 \left( \frac{\hbar^2}{4\omega_i\omega_j} (\beta_i + 1)(\beta_j + 1) \right) \delta(\omega_i + \omega_j - \omega) \tag{II.13}$$

or

$$R_{00}^{11} = -\frac{\pi}{\hbar\omega} \sum_i \sum_j \left( \frac{\hbar^2 \alpha_{ij}^2}{4\omega_i\omega_j} \right) (\bar{\beta}_i + 1)(\bar{\beta}_j + 1)\delta(\omega_i + \omega_j - \omega) \tag{II.14}$$

where $\bar{\beta}_i$ denotes the phonon distribution,

$$\bar{\beta}_i = \frac{1}{e^{\hbar\omega_i/kT} - 1} \tag{II.15}$$

Using the relation

$$\sum_j \rightarrow \int \rho(\omega_j)\, d\omega_j \tag{II.16}$$

to transform the sum over states into integration over density of states, Eq. (II.14) becomes

$$R_{00}^{11} = -\frac{\pi}{\hbar\omega} \sum_i \left( \frac{\hbar^2 \alpha_{ij}^2}{4\omega_i\omega_j} \right) (\bar{\beta}_i + 1)(\bar{\beta}_j + 1)\rho(\omega_j) \tag{II.17}$$

where $\omega_j = \omega - \omega_i$.

Next we consider the pure dephasing,

$$\Gamma_{mn}^{mn}(d) = \frac{\pi}{\hbar^2} \sum_\alpha \sum_\beta \rho_{\alpha\alpha}^{(b)} (H'_{m\alpha,m\beta} - H'_{n\alpha,n\beta})^2 \delta(\omega_{\alpha\beta}) \tag{II.18}$$

For $H'_{m\alpha,m\beta}$ and $H'_{n\alpha,n\beta}$ to be nonzero, from Eq. (II.4) we can see that up to the fourth-order approximation only the term $\sum_i \sum_j \beta_{ij} Q^2 q_i q_j$ can play the role of pure dephasing, that is,

$$\hat{H}' = Q^2 \sum_i \sum_j \beta_{ij} q_i q_j \tag{II.19}$$

$$H'_{m\alpha,m\beta} = \langle m|Q^2|m\rangle \langle \alpha| \sum_i \sum_j \beta_{ij} q_i q_j |\beta\rangle$$

$$= \left(m + \frac{1}{2}\right) \frac{\hbar}{\omega} \langle \alpha| \sum_i \sum_j \beta_{ij} q_i q_j |\beta\rangle \tag{II.20a}$$

and

$$\Gamma_{mn}^{mn}(d) = \frac{\pi}{\hbar^2} (m-n)^2 \left(\frac{\hbar^2}{\omega^2}\right) \sum_\alpha \sum_\beta \rho_{\alpha\alpha}^{(b)} \left| \langle \alpha| \sum_i \sum_j \beta_{ij} q_i q_j |\beta\rangle \right|^2 \delta(\omega_{\alpha\beta}) \tag{II.20b}$$

or

$$\Gamma_{mn}^{mn}(d) = \frac{\pi}{\hbar^2} (m-n)^2 \left(\frac{\hbar^2}{\omega^2}\right) \sum_i \sum_j \left(\frac{\hbar^2 \beta_{ij}^2}{4\omega_i \omega_j}\right) (\bar{\alpha}_i + 1)\bar{\alpha}_j \delta(\omega_i - \omega_j) \tag{II.20c}$$

or

$$\Gamma_{mn}^{mn}(d) = \frac{\pi}{\omega^2} (m-n)^2 \sum_i \left(\frac{\hbar^2 \beta_{ij}^2}{4\omega_i \omega_j}\right) (\bar{\alpha}_i + 1)\bar{\alpha}_j \rho(\omega_j) \tag{II.20d}$$

where $\omega_i = \omega_j$. From Eq. (II.20d) we can see that at $T = 0$, $\Gamma_{mn}^{mn}(d) = 0$.

Finally we consider the temperature effect on the spectral shift. To the lowest-order approximation we have

$$\bar{H}'_{mm} = \sum_\beta \rho_{\beta\beta}^{(b)} H'_{m\beta,m\beta} \tag{II.21}$$

From Eq. (II.4) we can see that for Eq. (II.21) we use

$$\hat{H}' = Q^2 \sum_i \beta_{ii} q_i^2 \tag{II.22}$$

to obtain

$$\bar{H}'_{mm} = \sum_\beta \rho_{\beta\beta}^{(b)} \langle m|Q^2|m\rangle \langle \beta| \sum_i \beta_{ii} q_i^2 |\beta\rangle \tag{II.23}$$

or

$$\bar{H}'_{mm} = \left(m + \frac{1}{2}\right)\frac{\hbar}{\omega}\sum_i \left(\bar{\beta}_i + \frac{1}{2}\right)\frac{\hbar\beta_{ii}}{\omega_i} \tag{II.24}$$

or

$$\bar{H}'_{nn} - \bar{H}'_{mm} = (n - m)\frac{\hbar}{\omega}\sum_i \left(\bar{\beta}_i + \frac{1}{2}\right)\frac{\hbar\beta_{ii}}{\omega_i} \tag{II.25}$$

This describes the temperature dependence of spectral shifts.

From the above discussion, we can see that different terms in $\hat{H}'$ given by Eq. (II.4) play different roles in vibrational relaxation and dephasing and in spectral shifts. The above treatment for vibrational relaxation can only be applied to the case where the system oscillator frequency $\omega$ is not too much larger than the bath mode frequencies $\omega_i$. For the case where $\omega \gg \omega_i$, multi-phonon processes are involved in vibrational relaxation. In this case the adiabatic approximation basis set can be used for calculating the rates of vibrational relaxation.

In Eqs. (II.1)–(II.4) we have assumed that there is only one system oscillator. In the case where there exists more than one oscillator mode, in addition to the processes of vibrational relaxation directly into the heat bath, there are the so-called cascade processes in which the highest-frequency system mode relaxes into the lower-frequency system modes with the excess energy relaxed into the heat bath. These cascade processes can often be very fast. The master equations of these complicated vibrational relaxation processes can be derived in a straightforward manner.

## APPENDIX III

We consider a model for the pump-probe stimulated emission measurement in which a pumping laser pulse excites molecules in a ground vibronic manifold $\{g\}$ to an excited vibronic manifold $\{n\}$ and a probing pulse applied to the system after the excitation. The probing laser induces stimulated emission in which transitions from the manifold $\{n\}$ to the ground-state manifold $\{m\}$ take place. We assume that there is no overlap between the two optical processes and that they are separated by a time interval $\tau$. On the basis of the perturbative density operator method, we can derive an expression for the time-resolved profiles, which are associated with the imaginary part of the transient linear susceptibility, that is,

$$\bar{\bar{\chi}}''(\omega_{pr}, \tau) = -\frac{1}{\hbar}\sum_{n,n'}\sum_m \mathrm{Im}\left\{\rho_{n,n'}(\tau)\frac{i(\vec{\mu}_{n',m}\otimes\vec{\mu}_{m,n})}{i(\omega_{n',m} - \omega_{pr}) + 1/T_{pr}}\right\} \tag{III.1}$$

where $\rho_{n,n'}(\tau)$ denotes the density matrix element and $\omega_{pr}$, $1/T_{pr}$, $\omega_{n',m}$, and $\vec{\mu}_{n',m}$ are, respectively, the central frequency of the probing laser pulse, its bandwidth, the transition frequency, and the transition dipole moment.

Based on Eq. (III.1), we shall derive a single-mode representation of $\vec{\chi}''(\omega_{pr}, \tau)$. In the adiabatic approximation, one can find $|n\rangle = |b\rangle|bv\rangle$, $|n'\rangle = |b\rangle|bv'\rangle$, $|m\rangle = |a\rangle|au\rangle$, and $|m'\rangle = |a\rangle|au'\rangle$ where $b$ and $a$, respectively, represent the electronically excited and ground states and $v$ or $v'$ ($u$ or $u'$) is the vibrational state in the electronic state $b$ ($a$). We assume that the probing pulse is so short that molecular dephasing processes during the optical process can be ignored. Using the above basis set, Eq. (III.1) becomes

$$\vec{\chi}''(\omega_{pr}, \tau) = -\frac{1}{\hbar} \sum_{v,v'} \sum_{u} \text{Im}\left\{ \rho_{bv,bv'}(\tau) \frac{i(\vec{\mu}_{bv',au} \otimes \vec{\mu}_{au,bv})}{i(\omega_{bv',au} - \omega_{pr}) + 1/T_{pr}} \right\} \quad \text{(III.2)}$$

One can easily find in Eq. (III.2) that $\vec{\chi}''(\omega_{pr}, \tau)$ consists of the coherent part $\vec{\chi}''_{co}(\omega_{pr}, \tau)$ and the incoherent part $\vec{\chi}''_{in}(\omega_{pr}, \tau)$, that is,

$$\vec{\chi}''(\omega_{pr}, \tau) = \vec{\chi}''_{co}(\omega_{pr}, \tau) + \vec{\chi}''_{in}(\omega_{pr}, \tau) \quad \text{(III.3)}$$

where

$$\vec{\chi}''_{in}(\omega_{pr}, \tau) = -\frac{1}{\hbar} \sum_{v} \sum_{u} \text{Im}\left\{ \rho_{bv,bv}(\tau) \frac{i(\vec{\mu}_{bv,au} \otimes \vec{\mu}_{au,bv})}{i(\omega_{bv,au} - \omega_{pr}) + 1/T_{pr}} \right\} \quad \text{(III.4)}$$

and

$$\vec{\chi}''_{co}(\omega_{pr}, \tau) = -\frac{1}{\hbar} \sum_{v \neq v'} \sum_{u} \text{Im}\left\{ \rho_{bv,bv'}(\tau) \frac{i(\vec{\mu}_{bv',au} \otimes \vec{\mu}_{au,bv})}{i(\omega_{bv',au} - \omega_{pr}) + 1/T_{pr}} \right\} \quad \text{(III.5)}$$

where $\omega_{bv',au} = \omega_{b,a} + (v' - u)\omega$ and $\omega_{b,a}$ is the electronic transition frequency and $\omega$ denotes the vibrational frequency. The first and second terms on the right-hand side of Eq. (III.3) are associated with the dynamics of vibrational population and coherence. We shall first consider the population dynamics. In the Condon approximation, Eq. (III.4) becomes

$$\vec{\chi}''_{in}(\omega_{pr}, \tau) = -\frac{(\vec{\mu}_{b,a} \otimes \vec{\mu}_{u,b})}{\hbar} \sum_{v} \text{Re}\{\rho_{bv,bv}(\tau) F_{bv,bv}(\omega_{pr})\} \quad \text{(III.6)}$$

Here $F_{bv,bv}(\omega_{pr})$ denotes the band-shape function of $\vec{\chi}''_{in}(\omega_{pr}, \tau)$ and is given by

$$F_{bv,bv}(\omega_{pr}) = \sum_{v} \frac{|\langle au|bv\rangle|^2}{i(\omega_{bv,au} - \omega_{pr}) + 1/T_{pr}} \quad \text{(III.7)}$$

where $\langle au|bv \rangle$ denotes the overlap integral of the vibrational wave functions. The vibrational overlap is expressed, in terms of the displaced harmonic oscillator basis, as

$$\langle au|bv \rangle = e^{-S/2} \sum_{i=0}^{v} \sum_{j=0}^{u} \delta_{u-j,v-i} S^{(i+j)/2} \frac{v!}{i!(v-i)!} \left(\frac{u!}{v!}\right)^{1/2} \frac{(-1)^j}{j!} \quad (\text{III.8})$$

where $S$ is the coupling constant (or Huang–Rhys factor) associated with the displacement in the normal coordinate between the equilibrium positions of each potential. Hereafter we ignore the suffix $b$ in $\rho_{bv,bv'}(\tau)$ for simplicity. The temporal development of the population in the vibrational state $v$, $\rho_{v,v}(\tau)$, satisfies the master equation [89]:

$$\frac{\partial \rho_{v,v}(\tau)}{\partial \tau} = -\Gamma_{v,v}^{v,v} \rho_{v,v}(\tau) + \sum_{\substack{v' \neq v \\ \text{and/or} \\ v'' \neq v}} \Gamma_{v,v}^{v',v''} \rho_{v',v''}(\tau) \quad (\text{III.9})$$

where $\Gamma_{v,v}^{v,v}$ denotes the population decay constant of the vibrational state $v$ and the diagonal element of $\Gamma_{v,v}^{v',v''}$; for example, $\Gamma_{v,v}^{v',v'}$ represents the rate constant of the vibrational population transfer from $v'$ to $v$, while the off-diagonal element (i.e., $v \neq v'$) stands for the vibrational coherence transfer from $v \leftrightarrow v'$ to $v \leftrightarrow v$. Those decay and transfer processes are induced by an interaction with the heat-bath modes. Having focused on the vibrational relaxation, we have ignored in the derivation of Eq. (III.9) the vibrational couplings that deplete the electronically excited state via nonradiative transitions.

In the secular approximation [89], we can eliminate the coherence terms [e.g., $\rho_{v',v''}(\tau)(v' \neq v'')$] in Eq. (III.9) such that the only diagonal terms contribute to the vibrational transitions through which the vibrational populations in various states are coupled. By applying the ladder model [89] to the interaction between the vibrational and heat-bath modes, the vibrational population decay constant $\Gamma_{v,v}^{v,v}$ is expressed as

$$\Gamma_{v,v}^{v,v} = \{v + (1+v)\exp(-\hbar\omega/k_B T)\}\gamma_{1 \to 0} \quad (\text{III.10})$$

The first term in Eq. (III.10) is originated from the decay from $v$ to $v-1$, while the second one, with the Boltzmann factor, is due to the thermally activated transition from $v$ to $v+1$. The vibrational population transfer, on the other hand, is only allowed to undergo the $v \to v \pm 1$ transitions in the ladder model. Thus, the rate constant of the transfer is given by

$$\Gamma_{v,v}^{v',v''} = \Gamma_{v,v}^{v-1,v-1} \delta_{v',v-1} \delta_{v'',v-1} + \Gamma_{v,v}^{v+1,v+1} \delta_{v',v+1} \delta_{v'',v+1} \quad (\text{III.11})$$

where $\Gamma_{v,v}^{v-1,v-1}$ and $\Gamma_{v,v}^{v+1,v+1}$ pertain to, respectively, the vibrational population transfer from $v$ to $v+1$ and that from $v+1$ to $v$, and they are given by

$$\Gamma_{v,v}^{v-1,v-1} = \{v\exp(-\hbar\omega/k_BT)\}\gamma_{1\to0} \tag{III.12}$$

and

$$\Gamma_{v,v}^{v+1,v+1} = (1+v)\gamma_{1\to0} \tag{III.13}$$

One can easily see from Eq. (III.12) that the population transfer from $v-1$ to the above state takes place via accepting energy from the heat-bath modes. Notice that $\gamma_{1\to0}$ in Eq. (III.11) represents the rate constant of the vibrational transition $v=1 \to v=0$ and is given by [89]

$$\gamma_{1\to0} = \frac{2\pi}{\hbar}\frac{1}{2C^2}\sum_{\alpha,\beta}P_\alpha|\langle\beta|\hat{F}_B|\alpha\rangle^2\delta(E_\beta - E_\alpha - \hbar\omega) \tag{III.14}$$

where $C^2 = \omega/\hbar$, and $\alpha$ and $\beta$ ($E_\alpha$ and $E_\beta$) denotes the states of the heat-bath modes (their energies), $\hat{F}_B$ is the heat-bath operator, and $P_\alpha$ represents the thermal distribution function in the heat-bath modes. Substituting Eqs. (III.10) and (III.11) into Eq. (III.9), one can obtain

$$\frac{\partial\rho_{v,v}(\tau)}{\partial\tau} = -\{v + (1+v)\exp(-\hbar\omega/k_BT)\}\gamma_{1\to0}\rho_{v,v}(\tau)$$
$$+ (1+v)\gamma_{1\to0}\rho_{v+1,v+1}(\tau)$$
$$\times v\exp(-\hbar\omega/k_BT)\}\gamma_{1\to0}\rho_{v-1,v-1}(\tau) \tag{III.15}$$

where the terms with the Boltzmann factor are associated with the thermally induced vibrational transitions as mentioned above; they vanish at low temperature, when $\hbar\omega \gg k_BT$ holds.

With the ultrashort pulse excitation, the initial condition of Eq. (III.15) is given by [90]

$$\rho_{v,v}(0) = \left|\frac{\vec{\mu}_{b,a}\cdot\vec{E}_{pu}(\omega_{pu})}{\hbar}\right|^2\sum_{u'}P_{u'}|\langle bv|au\rangle|^2 \tag{III.16}$$

where $P_{u'} = \exp(-u'\hbar\omega/k_BT)/\{1 - \exp(-\hbar\omega/k_BT)\}$. The initial condition is a function of $S$ as seen in Eq. (III.8).

The coherent component [i.e., Eq. (III.5)] can be treated in a similar fashion as well and one can find

$$\vec{\vec{\chi}}_{co}''(\omega_{pr},\tau) = -\frac{(\vec{\mu}_{b,a}\otimes\vec{\mu}_{u,b})}{\hbar}\sum_{v,v'}^{v\neq v'}\mathrm{Re}\{\rho_{bv,bv'}(\tau)F_{bv,bv'}(\omega_{pr})\} \tag{III.17}$$

where

$$F_{bv,bv'}(\omega_{pr}) = \sum_u \frac{\langle bv'|au\rangle\langle au|bv\rangle}{i(\omega_{bv',au} - \omega_{pr}) + 1/T_{pr}}$$ (III.18)

In the secular approximation and with the ladder model, the time evolution of the vibrational coherence, $\rho_{v,v'}(\tau)$, is determined by

$$P_{v,v'}(\tau) = \exp\{-(i\omega_{v,v'} + \Gamma_{v,v'}^{v,v'})\tau\}\rho_{v,v'}(0)$$ (III.19)

where $\omega_{v,v'} = (v - v')\omega$ and

$$\rho_{v,v'}(0) = \left|\frac{\vec{\mu}_{b,a} \cdot \vec{E}_{pu}(\omega_{pu})}{\hbar}\right|^2 \sum_{u'} P_{u'}\langle bv'|au'\rangle\langle au'|bv\rangle$$ (III.20)

By omitting the pure dephasing processes, which is warranted at low temperatures, the dephasing constant $\Gamma_{v,v'}^{v,v'}$ in Eq. (III.19) can be expressed, in terms of the population decay constants of the states $v$ and $v'$, as

$$\Gamma_{v,v'}^{v,v'} = \frac{1}{2}\{v + v' + (2 + v + v')\exp(-\hbar\omega/k_B T)\}\gamma_{1\to0}$$ (III.21)

Having considered only the single mode case so far, we can also derive an expression of $\vec{\chi}''(\omega_{pr}, \tau)$ for a multimode system in a similar fashion. In the two-mode case, for instance, $\vec{\chi}''(\omega_{pr}, \tau)$ can be divided into three terms, each of which corresponds to interference between the vibrational processes of the two modes. It should be noted here that within the same approximations as used above, the density matrix of the two modes during the time interval $\tau$ can be expressed as a product of each mode's matrix.

## Acknowledgments

This work was supported by the National Science Council of R.O.C. and Academia Sinica. The authors wish to thank Professors Fleming, Sundström, van Grondelle, and Yoshihara for helpful discussions.

## References

1. J.-L. Martin, J. Breton, A. J. Hoff, A. Migus, and A. Antonetti, *Proc. Natl. Acad. Sci. USA* **83**, 857 (1986).

2. G. R. Fleming, J.-L. Martin, and J. Breton, *Nature* **333**, 190 (1988).

3. J. Breton, J.-L. Martin, G. R. Fleming, and J.-C. Lambry, *Biochemistry* **27**, 8276 (1988).

4. M. H. Vos, J.-C. Lambry, S. J. Robles, J. Breton, and J.-L. Martin, *Proc. Natl. Acad. Sci. USA* **89**, 613 (1992).

5.  M. H. Vos, M. R. Jones, C. N. Hunter, J. Breton, J.-C. Lambry, and J.-L. Martin, *Biochemistry* **33**, 6750 (1994).

6.  C.-K. Chan, L. X.-Q. Chen, T. J. DiMagno, D. K. Hanson, S. L. Nance, M. Schiffer, J. R. Norris, and G. R. Fleming, *Chem. Phys. Lett.* **176**, 366 (1991).

7.  M. Du, S. J. Rosenthal, X. Xie, T. J. DiMagno, M. Schmidt, D. K. Hanson, M. Schiffer, J. R. Norris, and G. R. Fleming, *Proc. Natl. Acad. Sci. USA* **89**, 8517 (1992).

8.  Y. Jia, D. M. Jones, T. Joo, Y. Nagasawa, M. J. Lang, and G. R. Fleming, *J. Phys. Chem.* **99**, 6263 (1995).

9.  M.-L. Groot, J.-Y. Yu, R. Agarwal, J. R. Norris, and G. R. Fleming, *J. Phys. Chem.* **102**, 5923 (1998).

10. D. M. Jonas, M. J. Lang, Y. Nagasawa, T. Joo, and G. R. Fleming, *J. Phys. Chem.* **100**, 12660 (1996).

11. J. Breton, J.-L. Martin, A. Migus, A. Antonetti, and A. Orszag, *Proc. Natl. Acad. Sci. USA* **83**, 5121 (1986).

12. N. W. Woodbury, M. Becker, D. Middendorf, and W. W. Person, *Biochemistry* **24**, 7516 (1985).

13. C. Kirmaier and D. Holten, *Proc. Natl. Acad. Sci. USA* **87**, 3552 (1990).

14. M. H. Vos, J.-C. Lambry, S. J. Robles, J. Breton, and J.-L. Martin, *Proc. Natl. Acad. Sci. USA* **88**, 8885 (1991).

15. M. H. Vos, F. Rappaport, J.-C. Lambry, J. Breton, and J.-L. Martin, *Nature* **363**, 320 (1993).

16. M. H. Vos, J. Breton, and J.-L. Martin, *J. Phys. Chem. B* **101**, 9820 (1997).

17. C.-K. Chan, T. J. DiMagno, L. X.-Q. Chen, J. R. Norris, and G. R. Fleming, *Proc. Natl. Acad. Sci. USA* **88**, 11202(1991).

18. Z. Wang, R. M. Pearlstein, Y. Jia, G. R. Fleming, and J. R. Norris, *Chem. Phys.* **176**, 421 (1993).

19. D. M. Jones, M. J. Lang, Y. Nagasawa, T. Joo, and G. R. Fleming, *J. Chem. Phys.* **100**, 12660 (1996).

20. S. Maiti, G. C. Walker, B. R. Cowen, R. Pippenger, C. C. Moser, P. L. Dutton, and R. M. Hochstrasser, *Proc. Natl. Acad. Sci. USA* **91**, 10360 (1994).

21. G. C. Walker, S. Maiti, B. R. Cowen, C. C. Moser, P. L. Dutton, and R. M. Hochstrasser, *J. Phys. Chem.* **98**, 5778 (1994).

22. K. Dressler, E. Umlauf, S. Schmidt, P. Hamm, W. Zinth, S. Buchanan, and H. Michel, *Chem. Phys. Lett.* **183**, 270 (1991).

23. S. Schmidt, T. Arlt, P. Hamm, H. Huber, T. Nägele, J. Wachtveil, M. Meyer, H. Scheer, and W. Zinth, *Chem. Phys. Lett.* **223**, 116 (1994).

24. P. Hamm and W. Zinth, *J. Phys. Chem.* **99**, 13537 (1995).

25. I. H. M. van Stokkun, L. M. Beekman, M. R. Jones, M. E. van Brederode, and R. van Grondelle, *Biochemistry* **36**, 11360 (1997).

26. M. E. van Brederode, M. R. Jones, and R. van Grondelle, *Chem. Phys. Lett.* **268**, 143 (1997).

27. A. M. Streltsov, T. J. Aartsma, A. J. Hoff, and V. A. Shuvalov, *Chem. Phys. Lett.* **266**, 347 (1997).

28. S. Spörlein, W. Zinth, and J. Wachtveitl, *J. Phys. Chem. B* **102**, 7492 (1998).

29. J. W. Lewis, R. A. Goldbeak, D. S. Kliger, X. Xie, R. C. Dunn, and J. D. Simon, *J. Phys. Chem.* **96**, 5243 (1992).

30. N. J. Cherepy, A. P. Shreve, L. J. Moore, S. G. Boxer, and R. A. Mathies, *Proc. Natl. Acad. Sci. USA* **88**, 11207 (1991).

31. K. Wynne, G. Haran, G. D. Reid, C. C. Moser, P. L. Dutton, and R. M. Hochstrasser, *J. Phys. Chem.* **100**, 5140 (1996).

32. W. Holzapfel, U. Finkele, W. Kaiser, D. Oesterhelt, H. Scheer, H. U. Stilz, and W. Zinth, *Proc. Natl. Acad. Sci. USA* **87**, 5168 (1990).

33. T. Arlt, S. Schmidt, W. Kaiser, C. Lauterwasser, M. Meyer, H. Scheer, and W. Zinth, *Proc. Natl. Acad. Sci. USA* **90**, 11757 (1993).

34. P. Hamm, M. Zurek, W. Mäntele, M. Meyer, H. Scheer, and W. Zinth, *Proc. Natl. Acad. Sci. USA* **92**, 1826 (1995).

35. W. Zinth, T. Arlt, and J. Wachtveitl, *Ber. Bunsenges. Phys. Chem.* **100**, 1962 (1996).

36. M. E. van Brederode, M. R. Jones, F. van Mourik, I. H. M. van Stokkun, and R. van Grondelle, *Biochemistry* **36**, 6856 (1997).

37. R. J. Stanley, B. King, and S. G. Boxer, *J. Phys. Chem.* **100**, 12052 (1996).

38. A. M. Streltsov, S. I. E. Vulto, A. Ya. Shkuropatov, A. J. Hoff, T. J. Aartsma, and V. A. Shuvalov, *J. Phys. Chem. B* **102**, 7293 (1998).

39. D. C. Arnett, C. C. Moser, P. L. Dutton, and N. F. Scherer, *J. Phys. Chem. B* **103**, 2014 (1999).

40. G. Hartwich, H. Lossau, A. Ogrodnik, and M. E. Michel-Beyerle, in *The Reaction Center of Photosynthetic Bacteria*, M. E. Michel-Beyerle, ed., Springer-Verlag, Berlin, 1996, pp. 199.

41. N. J. Cherepy, A. R. Holzwarth, and R. A. Mathies, *Biochemistry* **34**, 5288 (1995).

42. N. J. Cherepy, A. P. Shreve, L. J. Moore, S. G. Boxer, and R. A. Mathies, *J. Phys. Chem. B* **101**, 3250 (1997).

43. N. R. S. Reddy, P. A. Lyle, and G. J. Small, *Photosyn. Res.* **31**, 167 (1992).

44. N. R. S. Reddy, S. V. Kolaczkowski, and G. J. Small, *J. Phys. Chem.* **97**, 6934 (1993).

45. N. R. S. Reddy, S. V. Kolaczkowski, and G. J. Small, *Science* **260**, 68 (1993).

46. N. J. Cherepy, A. P. Shreve, L. J. Moore, S. G. Boxer, and R. A. Mathies, *J. Phys. Chem.* **98**, 6023 (1994).

47. N. J. Cherepy, A. P. Shreve, L. J. Moore, S. G. Boxer, and R. A. Mathies, *Biochemistry* **36**, 8559 (1997).

48. S. G. Johnson, D. Tang, R. Jankowiak, J. M. Hayes, and G. J. Small, *J. Phys. Chem.* **94**, 5849 (1990).

49. G. J. Small, J. M. Hayes, and R. J. Sillbey, *J. Phys. Chem.* **96**, 7499 (1992).

50. P. A. Lyle, S. V. Kolaczkowski, and G. J. Small, *J. Phys. Chem.* **97**, 6924 (1993).

51. G. Haran, K. Wynne, C. C. Moser, P. L. Dutton, and R. M. Hochstrasser, *J. Phys. Chem.* **100**, 5562 (1996).

52. H. Huber, M. Meyer, H. Scheer, W. Zinth, and J. Wachveitl, *Photosyn. Res.* **55**, 153 (1998).

53. J. Deisenhofer and M. Michel, *Science* **245**, 1463 (1989).

54. S. H. Lin, M. Hayashi, R. G. Alden, S. Suzuki, X. Z. Gu, and Y. Y. Lin, in *Femtosecond Chemistry*, Manz and Wöste, eds., VCH Publishers, New York, 1995, pp. 633–667.

55. M. Hayashi, T.-S. Yang, C. H. Chang, K. K. Liang, R. Chang, and S. H. Lin, *Int. J. Quantum Chem.* **80**, 1043 (2000).

56. J. M. Zhang, Y. J. Shiu, M. Hayashi, K. K. Liang, C. H. Chang, V. Gulbinas, C. M. Yang, F. C. Hsu, T.-S. Yang, H. Z. Wang, and S. H. Lin, *J. Phys. Chem. A* (submitted).

57. G. D. Scholes, R. D. Harcourt, and G. R. Fleming, *J. Phys. Chem. B* **101**, 7302 (1997).

58. M. A. Thompson and M. C. Zerner, *J. Am. Chem. Soc.* **113**, 8210 (1991).

59. M. A. Thompson, M. C. Zerner, and J. Fajer, *J. Phys. Chem.* **95**, 5693 (1991).

60. W. W. Parson and A. Warshel, *J. Am. Chem. Soc.* **109**, 6143 (1987).

61. P. O. J. Scherer and S. F. Fischer, *Chem. Phys.* **131**, 115 (1989).

62. P. O. J. Scherer, S. F. Fischer, C. R. D. Lancaster, G. Fritzsch, S. Schmidt, T. Arlt, K. Dressler, and W. Zinth, *Chem. Phys. Lett.* **223**, 110 (1994).

63. P. O. J. Scherer, C. Scharnagl, and S. F. Fischer, *Chem. Phys.* **197**, 333 (1995).

64. P. O. J. Scherer and S. F. Fischer, *Chem. Phys. Lett.* **268**, 133 (1997).

65. P. O. J. Scherer and S. F. Fischer, *Spectrochim. Acta Part A* **54**, 1191 (1998).

66. H. Eyring, S. H. Lin, and S. M. Lin, *Basic Chemical Kinetics*, Wiley-Interscience, New York, 1980, Chapters 7 and 8.

67. C. H. Chang, M. Hayashi, K. K. Liang, R. Chang, and S. H. Lin, *J. Phys. Chem. B* **105**, 1216 (2001).

68. C. H. Chang, M. Hayashi, R. Chang, K. K. Liang, T.-S. Yang, and S. H. Lin, *J. Chin. Chem. Soc.* **47**, 785 (2000), UPS-99 Special Issue.

69. S. H. Lin, *J. Chem. Phys.* **90**, 7103 (1989).

70. M. Hayashi, T.-S. Yang, K. K. Liang, C. H. Chang, and S. H. Lin, *J. Chin. Chem. Soc.* **47**, 741 (2000), UPS-99 Special Issue.

71. R. G. Alden, W. D. Cheng, and S. H. Lin, *Chem. Phys. Lett.* **194**, 318 (1992).

72. S. H. Lin, W. Z. Xiao, and W. Dietz, *Phys. Rev.* **47E**, 3698 (1993).

73. S. H. Lin, *J. Chem. Phys.* **44**, 3768 (1966).

74. R. A. Marcus, *J. Chem. Phys.* **24**, 966 (1956).

75. V. G. Levich, *J. Phys. Chem.* **78**, 249 (1966).

76. N. Mataga and T. Kubota, *Molecular Interaction and Electronic Spectra*, Marcel Dekker, New York, 1970.

77. M. Bixon and J. Jortner, *J. Phys. Chem.* **95**, 1941 (1991).

78. S. H. Lin, *J. Chem. Phys.* **44**, 3759 (1966).

79. T. Förster, *Ann. Phys.* **2**, 55 (1948).

80. T. Förster, *Disc. Faraday Trans.* **27**, 7 (1965).

81. D. L. Dexter, *J. Chem. Phys.* **21**, 836 (1953).

82. S. H. Lin, R. G. Alden, R. Islampour, H. Ma, and A. A. Villaeys, *Density Matrix Method and Femtosecond Processes*, World Scientific, Singapore, 1991.

83. B. Fain, S. H. Lin, and N. Hamer, *J. Chem. Phys.* **91**, 4485 (1989).

84. S. H. Lin, B. Fain, and N. Hamer, *Adv. Chem. Phys.* **79**, 133 (1990).

85. R. G. Alden, M. Hayashi, J. P. Allen, N. W. Woodbury, H. Murchison, and S. H. Lin, *Chem. Phys. Lett.* **208**, 350 (1993).

86. S. H. Lin, R. G. Alden, M. Hayashi, S. Suzuki, and H. A. Murchison, *J. Phys. Chem.* **97**, 12566 (1993).

87. S. H. Lin, M. Hayashi, W. Z. Xiao, S. Suzuki, X. Z. Gu, and M. Sugawara, *Chem. Phys.* **197**, 435 (1995).

88. M. Hayashi, T.-S. Yang, K. K. Liang, C. H. Chang, R. Chang, and S. H. Lin, *J. Chin. Chem. Soc.* **47**, 117 (2000), UPS-99 Special Issue.

89. S. H. Lin, *J. Chem. Phys.* **61**, 3810 (1974).

90. X. Z. Gu, M. Hayashi, S. Suzuki, and S. H. Lin, *Biochim, Biophys Acta*, **1229**, 215 (1995)

# POLYMER MELTS AT SOLID SURFACES

ARUN YETHIRAJ

*Department of Chemistry, University of Wisconsin, Madison, Wisconsin, U.S.A.*

## CONTENTS

## I. INTRODUCTION

The behavior of polymers confined between surfaces is of importance in a number of problems of practical interest. These include polymer processing,

*Advances in Chemical Physics, Volume 121*, Edited by I. Prigogine and Stuart A. Rice.
ISBN 0-471-20504-4 © 2002 John Wiley & Sons, Inc.

lubrication, the permeation of natural gases through porous materials [1], membrane separation processes [2], and adsorption processes [3]. Adsorption processes alone are numerous and wide-ranging in their applications [4]: They are used in fields as diverse as catalysis, where physical adsorption is used extensively for the characterization of catalyst properties [5]; adhesives, where it is believed that the adsorption of the adhesives on the solid determines the strength of the adhesion [6]; and surface coating and paint technology, where poor adsorption of the paint or coating on the surface could cause blistering or peeling of the coating [7]. In colloid and surface science, the addition of polymers to colloidal suspensions can either enhance or decrease the tendency of the particles to flocculate depending on the degree of polymer adsorption and the nature of the polymer-induced forces between the colloidal particles [8]. The selectivity of adsorbents to certain adsorbates makes adsorption important in purification and separation processes. In many of these applications the properties of the liquid within a few angstroms of the surface are crucial. Advances in experimental techniques have allowed one to probe these systems on molecular lengthscales, and this has spurred theoretical research in this area. This chapter describes recent progress on liquid state methods for the theoretical study of polymers at or in between solid surfaces.

Bulk polymer melts are characterized by (a) local liquid-like structure on monomeric lengthscales caused by packing effects and (b) universal structure on a lengthscale of the order of the size of the molecules caused by chain connectivity. The local structure is influenced strongly by the chemical composition of the monomers and intramolecular constraints such as bond angle and torsional potentials, whereas the long-range structure is largely a function of the average size of the molecules. In most applications of polymer melts at surfaces the local structure is more important than the long-range structure, and a quantity of central importance is the density profile of the liquid at a surface. In polymer melts, the density profiles show interesting structure in the immediate vicinity of the surface, and this structure persists for only a few nanometers in most cases.

This chapter is concerned with the application of liquid state methods to the behavior of polymers at surfaces. The focus is on computer simulation and liquid state theories for the structure of continuous-space or off-lattice models of polymers near surfaces. The first computer simulations of off-lattice models of polymers at surfaces appeared in the late 1980s, and the first theory was reported in 1991. Since then there have been many theoretical and simulation studies on a number of polymer models using a variety of techniques. This chapter does not address or discuss the considerable body of literature on the adsorption of a single chain to a surface, the scaling behavior of polymers confined to narrow spaces, or self-consistent field theories and simulations of lattice models of polymers. The interested reader is instead guided to review articles [9–11] and books [12–15] that cover these topics.

Computer simulations have the advantage of being conceptually simple and exact (for a given model) but, for polymers, have the disadvantage of being computationally intensive. Therefore, although research in this area was ignited by computer simulations, simulations of confined polymers have been restricted to relatively simple models and to short chains. Theoretical development in this area has paralleled what happened earlier in the theory of simple liquids at surfaces; that is, the earliest theory for nonuniform polymers was an integral equation theory, and since then several density functional theories have been developed. Integral equation theories start with the set of integral equations for a mixture of polymer molecules and a spherical adsorbent and then take the limit as the adsorbent becomes infinitely dilute and infinitely large. Density functional theories start with an expression for the free energy of the fluid as a functional of the density profile. The equilibrium density profile is obtained by minimizing this functional with respect to variations in the density profile. An exact free energy functional is not known, and several approximate schemes have been proposed. Density functional theories are found to be in very good agreement with the computer simulations for density profiles of hard chains at hard walls, generally better than the integral equation theories. One limitation of density functional theory is related to the fact that in order to obtain predictions for a polymer melt at a surface, it is necessary to obtain the density profile of a single-polymer molecule in an arbitrary external field. An exact solution to this problem is not available; consequently, density functional theories often use rather unrealistic ideal systems in which all intramolecular interactions other than the bonding constraints are neglected, or resort to computer simulations of a single chain in an external field. These latter theories, though accurate, can therefore be computationally demanding.

The rest of this chapter is organized as follows: The next three sections describe computer simulations, integral equations, and density functional theories, respectively, and some conclusions are presented in the final section.

## II. MONTE CARLO SIMULATIONS

Computer simulations of confined polymers have been popular for several reasons. For one, they provide exact results for the given model. In addition, computer simulations provide molecular information that is not available from either theory or experiment. Finally, advances in computers and simulation algorithms have made reasonably large-scale simulations of polymers possible in the last decade. In this section I describe computer simulations of polymers at surfaces with an emphasis on the density profiles and conformational properties of polymers at single flat surfaces.

## A. Molecular Models

The starting point in any simulation, of course, is the choice of a molecular model. In polymers several levels of coarse-grained modeling are possible. An atomistic model of polyethylene would incorporate all the carbon and hydrogen atoms on the molecule and specify interatomic potentials between all atoms as well as the vibrational, bending, and torsional potentials for each bond. With appropriate parameterization of the intramolecular and intermolecular potentials, such a model is expected to faithfully represent real polyethylene, and the simulation results may be compared directly to experiment. The problem is that simulation of such a model can be prohibitively expensive, thus restricting the timescales, lengthscales, and degrees of polymerization that can be investigated. In addition, it is not always possible to relate cause and effect. As a result, very few studies of atomistic models of polymers at surfaces have been reported in the literature. A first level of coarse-graining is to group the $CH_2$ and $CH_3$ groups into united atoms and specify interaction potentials between these groups. These interactions include bond bending, torsional rotation, and bond stretching potentials in addition to site–site interaction potentials of, for example, the Lennard-Jones form. United atom models represent experimental systems rather well, and excellent parameterizations of the potential parameters are available [16,17]. However, even these models require significant computational investment, and united atom studies of polymers at surfaces have been restricted to fairly short chains. The next level of coarse-graining is when the model incorporates just the essential features of polymer behavior—that is, chain connectivity and intramolecular and intermolecular interactions. Such off-lattice models have been very popular, and the bulk of our understanding of polymer behavior in confined geometry comes from these models. Finally, the simplest class of models is the lattice model, where each polymer site is restricted to the site of a regular lattice. There have been many lattice simulations of polymers at surfaces [9,10,18–23]. Although these simulations provide valuable insight into the effect of confinement on the configurational entropy of polymers, with modest computational effort, they are less useful for polymer melts where the local ordering at the surface is of interest.

The focus of this chapter is on an intermediate class of models, a picture of which is shown in Fig. 1. The polymer molecule is a string of beads that interact via simple site–site interaction potentials. The simplest model is the "freely jointed hard-sphere chain" model where each molecule consists of a pearl necklace of tangent hard spheres of diameter $\sigma$. There are no additional bending or torsional potentials. The next level of complexity is when a stiffness is introduced that is a function of the bond angle. In the "semiflexible chain" model, each molecule consists of a string of hard spheres with an additional bending potential, $E_B = k_B T \epsilon (1 + \cos \theta)$, where $k_B$ is Boltzmann's constant, $T$ is

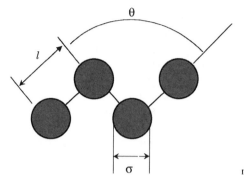

**Figure 1.**   Geometry of model polymer molecules.

the temperature, $\epsilon$ is the dimensionless stiffness parameter, and $\theta$ is the bond angle (see Fig. 1). For $\epsilon = 0$ this model reduces to the freely jointed chain and for $\epsilon = \infty$ this model reduces to the tangent-sphere "shish kabob" rod model. The next step higher is the "fused-hard-sphere chain" model where each molecule consists of a string of hard spheres with fixed bond lengths $l$ and bond angles $\theta$. In order to mimic alkanes the bond length is fixed at $l = 0.4\ \sigma$, and the bond angle is fixed at $\theta = 109°$. Intramolecular excluded volume is incorporated for beads separated by four or more bonds. Incorporating torsional rotations and dispersion interactions into this model brings one to the united atom model. The similarities and differences in the behavior of these model systems provides insight into the properties of polymers at surfaces. The vast majority of work in this area model the surface as smooth and flat. There have been a few studies of polymers at spherical surfaces [24–26] and in spherical cavities [27], and there are no significant differences between polymer behavior in these geometries when compared to flat surfaces. Therefore, in this chapter I will restrict the discussion to model surfaces that contain a hard component that makes them impenetrable to the centers of sites of the molecules and that, in some cases, contain a slowly varying attractive component.

## B.   Density Profiles of Hard Chains at Hard Walls

The quantity of primary interest in the study of nonuniform fluids is the density profile of the fluid at a surface. Dickman and Hall [28] reported the density profiles of freely jointed hard-sphere chains at hard walls. Their focus was on the equation of state of melts of hard-chain polymers, and they performed simulations of polymers at hard walls because the bulk pressure, $P$, can be calculated from the value of the density profile at the surface using the wall sum rule:

$$P = k_B T \rho(0) \tag{1}$$

where $\rho(z)$ is the site density profile at a single hard wall (i.e., the interaction potential between the sites and the wall is zero for $z > 0$ and infinity for $z < 0$). In practice the density profile at a single wall is obtained by simulating two walls sufficiently far apart so that there is a bulk-like region with a flat density profile in the middle region of the simulation box.

Dickman and Hall [28] found that at low densities the polymer sites were depleted near the surface, and at high densities the polymer sites were enhanced near the surface. This behavior has been seen in several other studies as well [29–36]. Figure 2 depicts the density profile of freely jointed tangent sphere chains (with $N = 20$ beads per chain) between hard walls [36]. The wall separation is large enough that there is a bulk-like region in the middle region, and these density profiles therefore are similar to those of a fluid at a single wall. In the figure, $\eta$ is the packing fraction, defined as $\eta \equiv \pi \rho_{av} \sigma^3 / 6$ where $\rho_{av}$ is the number of sites per unit volume. The density profiles are normalized by the "bulk" density, $\rho$, in the middle region between the surfaces. For low densities, $\eta \leq 0.2$, the density of polymer sites is lower at the surface than in the bulk, whereas for high densities, $\eta \geq 0.3$, the density of polymer sites at the surface is higher than that in the bulk. At the highest density the profile shows oscillations that arise from a liquid-like packing or layering of sites at the surface. Experimentally, these packing effects result in an oscillatory force between surfaces (as a function of their separation) when they are immersed in a polymer melt. Bitsanis and Hadziioannou [32] and Yethiraj and Woodward [37]

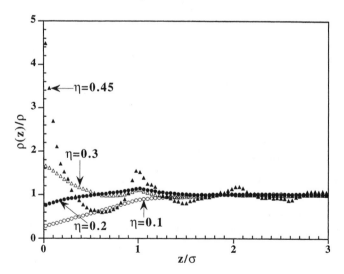

**Figure 2.** Normalized density profiles of freely jointed hard chains at a hard wall for $N = 20$ and various packing fractions (as marked). The surface is located at $z = 0$ and is impenetrable to the centers of the sites.

investigated the effect of adsorbing surfaces and found that a strongly adsorbing surface resulted in a higher density at the surface as well as more pronounced layering.

The qualitative features observed in the density profiles can be explained using entropic arguments. A single-polymer molecule experiences a loss in configurational entropy as it approaches the surface because the number of accessible conformations are smaller in the vicinity of the surface than they are in the bulk (the presence of the surface prohibits some configurations that would otherwise have been allowed). On the other hand, packing entropy favors a high density of polymers at the surface because when a bead is placed near the surface, it creates more free volume in the rest of the fluid. The actual density profile is therefore a result of a competition between packing and configurational entropic effects. One can estimate the density at which packing effects win over the configurational entropic effects by noting that this happens when $\rho(0) = \rho$ (and therefore $P = k_B T \rho$), where $\rho$ is the bulk density. Because the bulk pressure of a polymer solution scales with the correlation length, $\xi$, as $P \sim k_B T / \xi^3$, the density at which packing effects become important is when $\xi \sim \rho^{-1/3}$. Since $\rho \sigma^3 \sim 1$ in polymer melts, this occurs when the correlation length is of the order of the size of the polymer beads (or segments). In other words, configurational entropic effects should dominate in dilute and semidilute solutions, and packing entropic effects should dominate in concentrated solutions and melts. This is an important point because the majority of applications of polymers at surface (e.g., lubrication, coatings, and adhesives) involve polymer melts. In this case, because packing effects dominate, the density profile—or, alternatively, wall–fluid correlations—will decay on a lengthscale of the size of the polymer beads rather than decaying on a lengthscale of the size of the polymer chains.

The competition between packing and configurational entropy is also observed when the chains are stiffer, and in this case the qualitative shape of the density profiles can be a function stiffness at melt-like conditions. At low densities the density profiles of semiflexible chains look similar for all values of $\epsilon$ (although there are small quantitative differences) and therefore these density profiles are not shown. Figure 3 depicts the density profiles of hard semiflexible chains ($N = 20$, $\eta = 0.3$) at a hard wall [36]. At higher densities the effect of molecular stiffness on the shape of the density profiles is difficult to predict from the simple physical arguments used for fully flexible molecules. This is because as the chains become stiffer, several entropic effects come into play, and the actual density profile reflects a balance between these. On the segmental level, stiffer molecules are locally more open (the persistence length is larger), and this allows the segments to approach each other and the wall more closely. The average bond angles of the molecules also changes, however, and this places some geometric restrictions on packing. Fine structure due to these

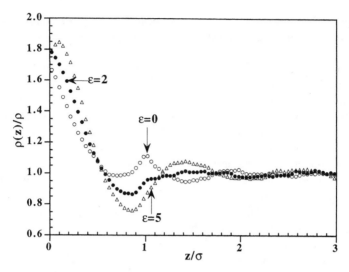

**Figure 3.** Normalized density profiles of semiflexible chains at a hard wall for $N = 20$ and $\eta = 0.3$.

effects are observed in simulations of semiflexible triatomic fluids [38]. On the molecular level, the restrictions on configurational entropy on stiffer chains are also higher because the molecules are larger. The resulting density profiles have features that are quite different from the $\epsilon = 0$ freely jointed case. For example, for stiff chains the peak in the density profile for $\eta = 0.3$ does not occur at $z = 0$ but instead occurs a small distance away from the wall, and the cusp at $z = \sigma$ is not present.

For more realistic polymer models, configurational entropic effects play a larger role. One consequence is that the high value of the density at the surface is only observed for short chains. Figures 4(a) and 4(b) depict density profiles for freely rotating fused-sphere chains (with fixed bond angles and bond lengths as described earlier) with $N = 4$ and 16, respectively [39]. To compare with the hard-chain model, note that a reduced density of $\rho\sigma^3 = 1.8$ for 16-mer fused-sphere chains corresponds (roughly) to a packing fraction of about 0.45 [40]. As is the case for the other hard-chain models, the competition between packing and configurational entropic effects is observed. However, for $N = 16$ the peak in the density profile at the surface is not as high at the largest densities, although the oscillations in the density profile are quite pronounced at larger distances. The former is a manifestation of the fact that the fused-sphere chains do not pack as efficiently amongst each other as do the freely jointed hard chains. An interesting feature is the many cusps and shoulders in the density profiles of fused-sphere chains at high densities. These are a consequence of the fixed bond angles in the model, and the location of these cusps can be predicted

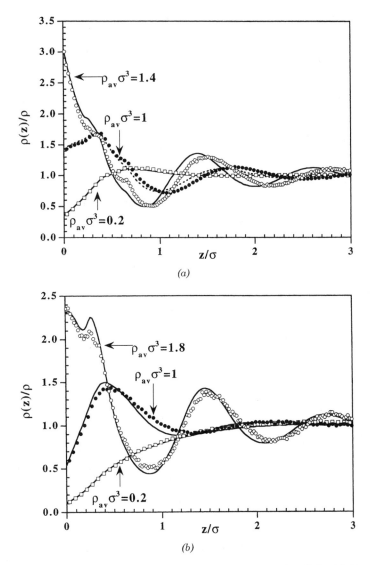

**Figure 4.** Density profiles of fused-hard-sphere chains at a hard wall for (a) $N = 4$ and (b) $N = 16$. Also shown are predictions of the density functional theory of Yethiraj [39].

from the geometry of the molecular model. The lines are theoretical predictions that will be discussed in a later section.

The entropic restrictions are less severe on the end sites of the chain than they are on the middle sites along the chain backbone. Consequently, the density

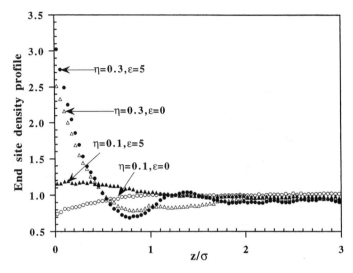

**Figure 5.** End-site density profiles for semiflexible chains normalized to the value in the middle region.

of the end sites at the surface is always higher than that of the middle sites. Figure 5 depicts end-site density profiles for semiflexible chains, and it shows that increasing either density or stiffness causes an increase in the value of the end-site density at the surface. (The density of the middle sites in long chains are almost identical to the total site density profiles and are therefore not shown.) If the end enhancement is defined by the ratio of the end density at the wall to the average value in the cell, it is found that this enhancement is independent of $N$ over the range $8 < N < 100$. Attempts have been made to measure these end enhancement effects via neutron reflectivity [41]. No significant end enhancement is seen in the experiments, which could be because the density profiles are assumed to be a monotonic function of distance from the surface in the analysis of the experiments whereas in reality the density profiles are oscillatory.

The center of mass density profiles shed light on the competition between configurational and packing entropy at the molecular level. There is much less structure in the center-of-mass profile $\rho_{CM}(z)$ than there is in the site density profiles. Figure 6 depicts center-of-mass profiles for several cases. For fully flexible chains $\rho_{CM}(z)$ is zero at the surface, has a peak at a distance corresponding roughly to the size of the molecule, and then decays to its "bulk" value. As the density is increased (at fixed stiffness), the peak in the profile is pushed closer to the wall. This is a manifestation of the packing of the molecules at a molecular level [31,33] and is also observed for stiffer chains. Increasing

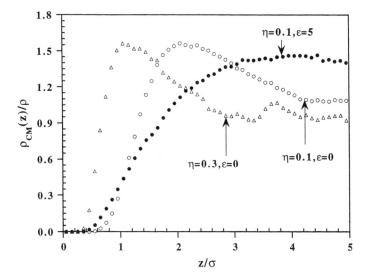

**Figure 6.** Centre-of-mass density profiles for semiflexible chains.

the stiffness (at fixed density) causes the molecular centers to prefer the region away from the wall because the stiffer molecules are larger, and entropic restrictions near a surface are greater.

## C. Solvation Forces

In order to determine the force between plates as a function of their separation, one would have to perform a series of simulations with different wall separations and with the chemical potential of the fluid fixed at the bulk value. This is technically feasible, but very computationally intensive [42]. The qualitative behavior of the force law can, however, be estimated from the density profile of a fluid at a single wall using the wall sum rule and a superposition approximation [31,43]. The basic idea is that the density profile [denoted $\rho_H(z)$] of a fluid between two walls at a separation $H$ can be obtained from the density profile [denoted $\rho_1(z)$] of the same fluid at a single wall using

$$\rho_H(z) = \rho_1(z) + \rho_1(H - z) - \rho \qquad (2)$$

The forces per unit area on the wall is therefore given by $F/k_B T = \rho_H(0) = \rho_1(0) + \rho_1(H) - \rho$. Subtracting the force the fluid exerts on a single wall gives the solvation force:

$$\frac{F_s}{k_B T} = \rho_1(H) - \rho \qquad (3)$$

where $F_s$ is the solvation force per unit area of the walls and $\rho_1(H)$ is the value of the density profile at a distance $H$ from a *single* wall. The superposition approximation has been shown to be accurate for the force between walls in flexible hard-chain fluids [42]. Because the density profiles are relatively insensitive to $H$ for $H \geq 10$, it is reasonable to assume that the density profile of a fluid at a single wall is given by the total density profiles monitored in the simulations described above.

The qualitative behavior of the solvation force can thus be estimated from the density profile. At low densities, the solvation force will be a primarily attractive function of wall separation, as has been also reported in other simulations [44]. This is the origin of the so-called depletion force between surfaces immersed in polymer solutions. At high densities the solvation forces will be oscillatory, and this behavior is explained by noting that under such conditions the fluid layers at the surface. Consider two surfaces immersed in a reservoir of fluid. As the surfaces are brought closer together, the fluid in between them gets compressed, thus increasing the force between the surfaces. At some separation, however, it is more efficient to push out a layer of the fluid rather than compress the fluid further, and the force becomes attractive. The physical origin of oscillatory surface forces is therefore identical to the origin of the oscillatory density profiles.

### D.  Chain Conformations

One measure of the conformational properties of the chains are the components of the mean-square end-to-end distance, defined as

$$\langle R_\alpha^2 \rangle = \langle (r_{N\alpha} - r_{1\alpha})^2 \rangle \tag{4}$$

where $r_{j\alpha}$ is the $\alpha$th ($\alpha = x$, $y$, or $z$) component of the position of bead $j$ of the chain. In a bulk fluid, these components are equal, and near the surface there is a distortion caused by the presence of the surface. A typical example is shown in Fig. 7, which depicts $\langle R_\alpha^2 \rangle$ for freely jointed chains as a function of the center-of mass position for $\eta = 0.1$ and $N = 20$. ($\langle R_{xy}^2 \rangle$ is the average over values the $x$ and $y$ directions.) For center-of-mass positions near the surface, $\langle R_z^2 \rangle$ is very small and $\langle R_{xy}^2 \rangle$ is quite large. As one moves away from the surface, the parallel component decreases in magnitude and the perpendicular component increases in magnitude. The components have similar values in the middle of the pore where the fluid is isotropic. The trends seen above are consistent with the conclusion that the chains are quite flattened near the surface. However, the flattening of the chains is not necessary for these trends to be observed. It has been pointed out [23] that a single-polymer molecule (or a random walk) is a highly anisotropic cigar-shaped object, and merely reorienting these cigars near a surface could account for the apparent flattening seen in the $\langle R_\alpha^2 \rangle$ profiles.

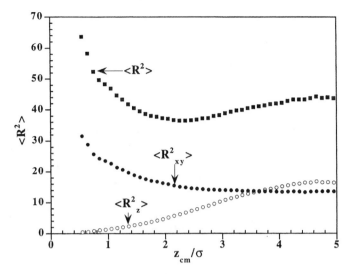

**Figure 7.** Components of the mean-square end-to-end distance as a function of center-of-mass position for freely jointed hard chains for $N = 20$ and $\eta = 00.1$.

In other words, the loss in configurational entropy is due to a loss in molecular orientational entropy rather than due to a loss of allowed chain conformations or shapes. Also shown in the figure is the total $\langle R^2 \rangle$ ($\equiv \langle R_x^2 \rangle + \langle R_y^2 \rangle + \langle R_z^2 \rangle$) as a function of the distance of the center of mass from the surface. The figure shows that $\langle R^2 \rangle$ does increase as the polymer centers of mass approach the surface, and by quite a significant amount (about 50%), although the effect is felt only for a distance of about 2 $\sigma$. This suggests that both a reorienting of the molecules and a flattening occur in off-lattice systems.

A better estimate of the shape of the polymer molecules, since they are highly anisotropic, is a representation of each molecule in terms of an equivalent spheroid with the same moment of inertia [45,46]. This is achieved by diagonalizing the moment of inertia tensor to obtain the eigenvectors **a**, **b**, and **c** and the principal moments $I_{aa}$, $I_{bb}$, and $I_{cc}$. The moment of inertia tensor of molecule $j$ is given by

$$I_{j\alpha\beta} = \sum_i (x_i^2 \delta_{\alpha\beta} - x_{i\alpha}x_{i\beta}) \tag{5}$$

where sum is over all the sites of molecule $j$, $\alpha$ and $\beta$ are the Cartesian coordinates, $\delta_{\alpha\beta}$ is the Kronecker delta, $x_{i\alpha}$ is the distance in the $\alpha$ direction of site $i$ from the center of mass, and $x_i^2 = \sum_\alpha x_{i\alpha}^2$. The eigenvector **a** is the one that corresponds to the smallest eigenvalue, and I refer to it as the molecular axis vector. The lengths of the semi-axis vectors are given by $a^2 = 5(I_{bb} + I_{cc} - I_{aa})/2N$, $b^2 = 5(I_{aa} + I_{cc} - I_{bb})/2N$, and $c^2 = 5(I_{bb} + I_{aa} - I_{cc})/2N$. The variation of the root-mean-square axis lengths (defined as $a \equiv \sqrt{\langle a^2 \rangle}$, etc.) with center-of-mass position contains more information about the molecular conformations than do the components of $\langle R^2 \rangle$.

Figure 8 depicts $a$, $b$, and $c$ as a function of the center-of-mass position for freely jointed chains with $N = 20$ and $\eta = 0.1$. For center-of-mass positions greater than about 2 $\sigma$, there is very little dependence of the size on position. As the chain centers come closer, however, the molecules become both longer and thinner. The distance at which the conformations become perturbed corresponds roughly to the larger of the two minor semi-axis lengths (in this case, $b$). These results are consistent with a physical picture where the molecules (which can be thought of as cigar-shaped objects) prefer to reorient rather than change their conformation as they approach the surface. When the centers of mass reach distances that are of the order of half the thickness of the cigars, a reorientation is no longer possible, and the only way the molecular centers can approach closer are by distorting the chain conformations. This is entropically quite expensive, as evidenced by the low (but nonzero) center-of-mass density for distances smaller than about 2 $\sigma$. The decrease in size with increasing density (for $\epsilon = 0$) is due to the molecules becoming both thinner and shorter. The shape of the $a$ profile is very similar to that of the $\langle R^2 \rangle$ profile in Figure 7, because the increase in $\langle R^2 \rangle$ is accompanied by an increase in the lengths of the

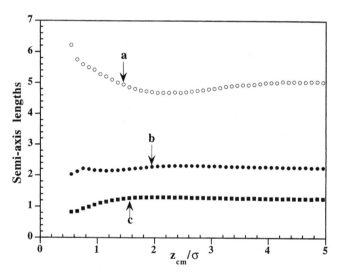

**Figure 8.**   Principal semi-axes dimensions as a function of center-of-mass position for freely jointed hard chains for $N = 20$ and $\eta = 0.1$.

molecules. These conclusions have been recently verified by experiments on thin polymer films by Jones et al. [47], who found that the chain conformations were not perturbed from the bulk value in thin films, but the chains tended to be swollen in the thinnest films that they studied.

In order to estimate the orientations of the molecules with respect to the surface, it is convenient to define a molecular axis orientational correlation function, $G_2(z)$, by

$$G_2(z) = \frac{3\langle \cos^2\gamma \rangle - 1}{2} \tag{6}$$

where $\gamma$ is the angle between the molecular axis vector $\mathbf{a}$ and the $z$ axis, and $z$ is the position of the center of mass. If the molecular axes align parallel to the surface, then $G_2 = -0.5$; if they align perpendicular to the surface, then $G_2 = 1$; and if the fluid is isotropic, then $G_2 = 0$. Figure 9 depicts $G_2(z)$ as a function of the center-of-mass position for freely jointed chains for $N = 20$ and $\eta = 0.1$. As expected from the earlier discussion, when the centers of mass are close to the wall, the molecular axes are aligned parallel to the surface. This alignment decreases as the centers of mass move away from the surface.

## E.   Effect of Attractive Interactions

In contrast to the considerable attention focused on the behavior of athermal polymers at surfaces, there are very few simulation studies that have investigated the

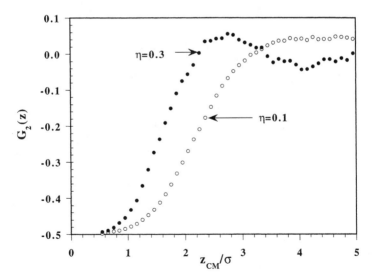

**Figure 9.** Orientational correlation function of the principal axis as a function of center-of-mass distance for freely jointed hard chains for $N = 20$.

effect of fluid–fluid and wall–fluid attractive forces on the polymer properties. Bitsanis and Hadziioannou [32] and Yethiraj and Woodward [37] investigated the effect of adsorbing surfaces and found that a strongly adsorbing surface resulted in a higher density at the surface as well as more pronounced layering. Binder and coworkers, in a series of papers [48–54], have investigated the adsorption behavior of polymers at surfaces, including the effect of the strength of the wall fluid attraction on the conformational and dynamic properties of off-lattice bead spring chains. The only systematic study of the effect of relative strengths of the fluid–fluid and wall–fluid attractions is the work of Patra and Yethiraj [55], who studied fused-sphere chains at surfaces. In their model, beads that are on different chains interact via a hard sphere plus Yukawa potential given by $\beta u(r) = \infty$ for $r < \sigma$, and $\beta u(r) = -\epsilon_{\mathrm{ff}} \frac{\exp[-\lambda(r/\sigma-1)]}{r/\sigma}$ for $r > \sigma$, where $\lambda$ is the inverse range of Yukawa potential (with $\lambda = 2.5$ to mimic a Lennard-Jones attractive tail), and $\beta = 1/k_B T$. The wall–fluid potential was $\phi(z) = u_{\mathrm{wf}}(z) + u_{\mathrm{wf}}(H - z)$, where $H$ is the separation of the walls and $z$ is the perpendicular distance from one of the walls, and $\beta u_{\mathrm{wf}}(z) = \infty$ for $z < 0$ or $z > H$, and $\beta u_{\mathrm{wf}}(z) = -\epsilon_{\mathrm{wf}} \exp(-\lambda z/\sigma)$ for $0 < z < H$.

The density profiles are now determined not only by a balance of various entropic effects but also from a balance between wall–fluid and fluid–fluid attractions. When a fluid–fluid attraction is introduced, the molecules prefer the region away from the surface because there now is an energetic preference for the molecules to be amongst each other. On the other hand, an attractive wall–fluid potential causes an enhancement of chain density at the surface relative to the case where the wall–fluid interaction is absent. The former effect can be explained as follows. Near the surface, the total attraction experienced by any bead due to the rest of the fluid is reduced roughly by a factor of 2 compared to the bulk because of the presence of the surface which excludes the other sites from half the volume (corresponding to $z < 0$). This causes the region near the surface to be energetically less favorable than the bulk region, and fluid–fluid attractions cause a depletion of sites from the surface. This "depletion" effect is counteracted by a wall–fluid attraction which provides an energetic incentive for the chain sites to be at the surface. These enhancement and depletion effects in the presence of attractions are shown in Fig. 10(a) for 8-mers at $\rho\sigma^3 = 0.6$ and various combinations of $\epsilon_{\mathrm{wf}}$ and $\epsilon_{\mathrm{ff}}$. When the wall–fluid and fluid–fluid attractions are of comparable strength (i.e., $\epsilon_{wf} = \epsilon_{ff} = 0.2$), the depletion mechanism due to the bulk attraction dominates over the wall–fluid attraction. For stronger bulk fluid attractions ($\epsilon_{ff} = 0.5$), the liquid completely dries off the surface. Figure 10(b) depicts density profiles of 8-mers at a density $\rho\sigma^3 = 1.4$. The presence of chains near the walls even when the bulk fluid–fluid attraction ($\epsilon_{ff}$) is present indicates the dominance of packing entropic effects near the surface over the enthalpic effects. The qualitative features are similar to what is seen at low densities except that there is no drying at the surface. The trends in

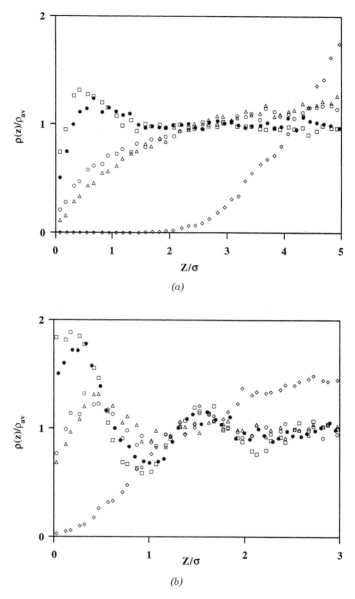

**Figure 10.** Effect of wall–fluid and fluid–fluid attractions on the density profiles of fused-sphere chains for $N = 8$ and (a) $\rho_{av}\sigma^3 = 0.6$ and (b) $\rho_{av}\sigma^3 = 1.4$. $\epsilon_{wf}$ and $\epsilon_{ff}$ represent dimensionless strengths of wall–fluid and fluid–fluid attractions, respectively. The symbols coorrespond to values of $(\epsilon_{wf}, \epsilon_{ff}) = (0,0)$ (●), $(0.2,0)$ (□), $(0.2,0.2)$ (○), $(0,0.2)$ (△), $(0,0.5)$ (◇).

the behavior of the density profiles in the presence of added attractions remain the same for longer chains, except that the depletion effects are stronger in the latter case.

In polymer blends, or mixtures, the primary question is whether one of the components segregates preferentially to the surface. One of the reasons this is of interest is because most commercial polymers contain more than one component and a surface segregation of one of the components from a miscible mixture during, for example, extrusion of the material, could affect the surface finish of the product. Because polymer blends are generally dense liquids, from the previous discussion it is clear that packing effects are expected to dominate the surface properties.

An interesting case is the effect of chain topology on the surface segregation from polymer blends. A series of experiments involving polyolefin materials (where the elementary chemical units are similar $CH_X$ groups) demonstrated [56,57] that from a blend of polymers with different degrees of branching, the more branched polymers always segregated to the surface, independent of the type of surface. For example, from a blend of linear polyethylene (PE) and branched polyethylethylene (PEE) polymers, the PEE segregated preferentially to the surface. From a physical standpoint, this is an interesting problem because while configurational entropic effects are expected to favor the branched molecules at the surface, packing entropic effects are expected to favor the linear molecules at the surface. This is because the branched molecules (for fixed number of monomers) are conformationally smaller and suffer a smaller loss in configurational entropy, but the linear molecules can pack more efficiently at the surface. The presence of wall–fluid and fluid–fluid attractions further complicates the issue.

A simple model for the investigation of this system is a blend of polymers with the two components identical in every respect except for the arrangement of the beads along the chain. For $N = 19$ one can construct a branched chain where every other backbone bead (of which there are 13 in all) has one additional bead attached to it. The surface segregation from a blend of linear 19-mers and these branched molecules, with fluid–fluid and wall–fluid attractive interactions present, has been investigated using Monte Carlo simulation [58]. The density profiles (at a hard wall) of the two components of this blend, with and without fluid–fluid attractions, are shown in Figs. 11(a) and 11(b), respectively. (The interaction potentials for the wall–fluid and fluid–fluid attractions are the same as those used for the fused-sphere chains depicted in Fig. 10.) In the absence of fluid–fluid attractions, the linear chains are favored at the surface because they can pack more efficiently. Similarly, in an athermal blend of stiff and flexible chains, the stiffer chains are found at the surface even though they are conformationally larger [59,60]. In the presence of fluid–fluid attractions the branched chains are favored at the surface. The latter segregation is caused by

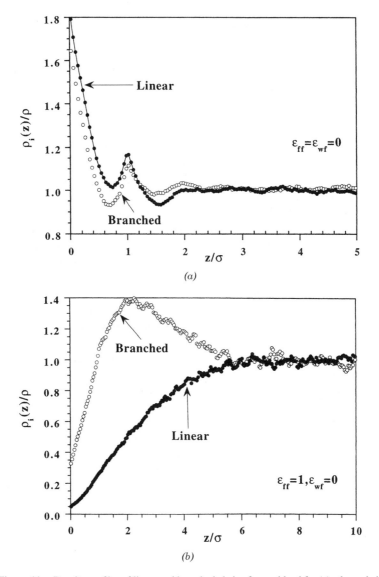

**Figure 11.** Density profiles of linear and branched chains from a blend for (a) athermal chains and (b) chains with fluid–fluid attractions present.

the fact that the linear polymers can pack amongst each other more efficiently than the branched polymers and there is therefore an enthalpic advantage for them to be in the bulk where the coordination number is larger than near the surface. The segregation of the branched polymers is accompanied by a depletion of polymer beads at the surface. A general conclusion of this work is that the fluid–fluid interactions can cause a surface segregation in polymer blends with the component with weaker interactions generally segregating to the surface. This is true of isotopic blends, where the weaker interacting deuterated species segregates to the surface, or blends with topological differences where the more branched chains are effectively weaker interacting due to more inefficient packing in the condensed phase.

## F.  Summary and Outlook

There have been many simulation studies on the behavior of polymers at surfaces. Athermal chains show depletion and enhancement effects where the chain sites are depleted from the surface at low densities (due to configurational entropic effects) and enhanced at the surface at high densities (due to packing entropic effects). Fluid–fluid and wall–fluid attractive potentials cause a depletion and enhancement of chains sites at the surface, respectively. In polymer blends, fluid–fluid attractive interactions cause the segregation of the weaker interacting component to the surface. These model studies have provided us with a good understanding of the basic phenomenology and physics of polymers at surfaces and have been useful in explaining experimental phenomena such as the segregation from blends of polymers.

One of the interests in confined polymers arises from adsorption behavior—that is, the intake or partitioning of polymers into porous media. Simulation of confined polymers in equilibrium with a bulk fluid requires simulations where the chemical potentials of the bulk and confined polymers are equal. This is a difficult task because simulations of polymers at constant chemical potential require the insertion of molecules into the fluid, which has poor statistics for long chains. Several methods for simulating polymers at constant chemical potential have been proposed. These include biased insertion methods [61,62], "novel" simulation ensembles [63,64], and simulations where the pore is physically connected to a large bulk reservoir [42]. Although these methods are promising, so far they have not been implemented in an extensive study of the partitioning of polymers into porous media. This is a fruitful avenue for future research.

There have also been a number of simulations of more realistic models of polymers at surfaces [65–77]. The behavior of these more realistic models of polymers is similar to that of the model systems discussed above with no real surprises. Of course, the use of realistic models allows a direct comparison with experiment. For example, surface forces apparatus measurements [78] show that in some branched alkanes the force is a monotonic rather than oscillatory function of the separation. This is a surprising result because these branched alkanes pack quite efficiently (in fact they crystallize under some conditions), and this would imply that the surface forces should be oscillatory. Several simulations [68,71,74,77] have addressed this issue and are consistent with oscillatory surface forces, in disagreement with experiment. The experimental results are normally explained as a nonequilibrium effect. Because most simulations of realistic models are performed with molecular dynamics simulations, they provide valuable information of the dynamics of the liquid near the interface. Generally, the dynamics are sensitive to the nature of the solid–fluid interaction and are different near the surface when compared to a bulk fluid. The differences can often be rationalized by noting that the local density at the surface is considerably higher than a bulk liquid.

Finally, a relatively new area in the computer simulation of confined polymers is the simulation of nonequilibrium phenomena [72,79–87]. An example is the behavior of fluids undergoing shear flow, which is studied by moving the confining surfaces parallel to each other. There have been some controversies regarding the use of thermostats and other technical issues in the simulations. If only the walls are maintained at a constant temperature and the fluid is allowed to heat up under shear [79–82], the results from these simulations can be analyzed using continuum mechanics, and excellent results can be obtained for the transport properties from molecular simulations of confined liquids. This avenue of research is interesting and could prove to be important in the future.

## III.   INTEGRAL EQUATION THEORIES

Integral equation theories are widely used in the theoretical study of liquids. There are two broad classes of integral equation theories: those based on the Born–Green–Yvon (BGY) hierarchy and those based on the Ornstein–Zernike (OZ) equation [88]. Although the formalism is exact in both classes, it is generally easier to fashion approximations in the case of the OZ-equation-based approach, and this type of theory has therefore been more popular. Surprisingly, the BGY approach has never been implemented for nonuniform polymers, and this section is therefore restricted to a discussion of the OZ-equation-based approach.

### A.   Uniform Fluids

A quantity of central importance in the study of uniform liquids is the pair correlation function, $g(r)$, which is the probability (relative to an ideal gas) of finding a particle at position $r$ given that there is a particle at the origin. All other structural and thermodynamic properties can be obtained from a knowledge of $g(r)$. The calculation of $g(r)$ for various fluids is one of the long-standing problems in liquid state theory, and several accurate approaches exist. These theories can also be used to obtain the density profile of a fluid at a surface.

The starting point for the calculation of $g(r)$ is the Ornstein–Zernike (OZ) equation, which, for a one-component system of liquids interacting via spherically symmetric potentials (e.g. Argon), is [89]

$$h(r) = c(r) + \rho \int d\mathbf{r}' c(\mathbf{r}') h(|\mathbf{r} - \mathbf{r}'|) \tag{7}$$

where $\rho$ is the number density of the fluid, $h(r) \equiv g(r) - 1$ is the total correlation function, and $c(r)$ is the direct correlation function, defined via Eq. (7). In Fourier space the OZ equation may be written as

$$\hat{h}(k) = \hat{c}(k) + \rho \hat{c}(k) \hat{h}(k) \tag{8}$$

where the carets denote three-dimensional Fourier transforms defined by

$$\hat{h}(k) = \frac{4\pi}{k} \int_0^\infty r \sin(kr) h(r) \, dr \tag{9}$$

When supplemented with a closure relation, Eq. (7) can be solved for $h(r)$ and $c(r)$. For example, the Percus–Yevick (PY) closure is given by [89]

$$c(r) = (1 - e^{\beta u(r)}) g(r) \tag{10}$$

where $u(r)$ is the interaction potential. For hard-sphere liquids the PY closure is particularly simple: $g(r) = 0$ for $r < \sigma$, and $c(r) = 0$ for $r > \sigma$. The PY closure is quantitatively very accurate for the pair correlation function and equation of state of hard-sphere liquids (when compared to simulations).

If each polymer is modeled as being composed of $N$ beads (or sites) and the interaction potential between polymers can be written as the sum of site–site interactions, then generalizations of the OZ equation to polymers are possible. One approach is the polymer reference interaction site model (PRISM) theory [90] (based on the RISM theory [91]) which results in a nonlinear integral equation given by

$$\hat{h}(k) = \hat{\omega}(k) \hat{c}(k) \hat{\omega}(k) + \rho \hat{\omega}(k) \hat{c}(k) \hat{h}(k) \tag{11}$$

$$= \hat{\omega}(k) \hat{c}(k) \hat{S}(k) \tag{12}$$

where $\rho$ here is the number density of polymer sites (not chains), $\hat{\omega}(k)$ is the single-chain structure factor, $h(r)$ and $c(r)$ are the total and direct correlation functions, respectively, averaged over all the beads on the polymers, and $\hat{S}(k) \equiv \hat{\omega}(k) + \rho \hat{h}(k)$ is the static structure factor. The single-chain structure factor may be obtained from the intramolecular correlations via

$$\omega(r) = \frac{1}{N} \sum_{i=1}^{N} \sum_{j=1}^{N} \omega_{ij}(r) \tag{13}$$

where $\omega_{ij}(r)$ is the probability that beads $i$ and $j$ on the same chain are a distance $r$ apart. In the simplest implementation of the PRISM theory, $\hat{\omega}(k)$ is assumed to be known and the standard closure approximations used for simple liquids (e.g., PY) are employed without change.

In the PRISM approach, information regarding chain conformations and local chemistry (e.g., bond-angle and bond-length constraints, stiffness, and branching) is input through the single-chain structure factor, $\hat{\omega}(k)$. In general this function depends on the pair correlations and must be calculated self-consistently with $g(r)$. In most implementations of the theory, however, $\hat{\omega}(k)$ is assumed to be known *a priori*. This can be justified by invoking the Flory ideality hypothesis, which states that chains in a melt behave essentially like ideal chains because intramolecular interactions exactly counteract intermolecular interactions.

The OZ (or PRISM) equation with closure relation can be solved using a Picard iteration procedure. One starts with a guess for the function $\gamma(r) = h(r) - c(r)$, either $\gamma(r) = 0$ or the value of $\gamma(r)$ from some condition close to the condition of interest. Using the closure relation, $c(r)$ is then obtained from $\gamma(r)$. With the PY closure, for example, we obtain

$$c(r) = (e^{-\beta u(r)} - 1)(1 + \gamma(r)) \tag{14}$$

$\hat{c}(k)$ is then evaluated numerically, and the next guess for $\gamma(r)$ is obtained from the OZ equation followed by an inverse Fourier transform of $\hat{\gamma}(k)$. For numerical convenience the functions are discretized and the Fourier transforms are performed using the fast Fourier transform (FFT) algorithm.

### B. Polymers at a Surface

Integral equations can also be used to treat nonuniform fluids, such as fluids at surfaces. One starts with a binary mixture of spheres and polymers and takes the limit as the spheres become infinitely dilute and infinitely large [92–94]. The sphere polymer pair correlation function is then simply related to the density profile of the fluid.

The PRISM equations for a bulk mixture of a sphere (species 1) and a polymer (species 2) can be written as

$$h_{11}(r) = c_{11}(r) + \rho_1 c_{11} * h_{11}(r) + \rho_2 c_{12} * h_{22}(r) \tag{15}$$

$$h_{12}(r) = c_{12} * \omega(r) + \rho_1 c_{11} * h_{12}(r) + \rho_2 c_{12} * h_{22}(r) \tag{16}$$

$$h_{22}(r) = \omega * c_{22} * \omega(r) + \rho_1 \omega * c_{12} * h_{12}(r) + \rho_2 \omega * c_{22} * h_{22}(r) \tag{17}$$

where $h_{ij}(r)$ and $c_{ij}(r)$ are, respectively, the total and direct correlation functions between species $i$ and $j$, and the asterisks denote convolution integrals. In the limit as $\rho_1 \to 0$, this set of equations reduces to

$$h_w(r) = c_w * \omega(r) + \rho c_w * h(r) = c_w * S(r) \tag{18}$$

$$h(r) = \omega * c * \omega(r) + \rho \omega * c * h(r) \tag{19}$$

where $\rho_2$ has been replaced by $\rho$, the polymer site density, the subscript 12 has been replaced by "$w$" for wall–fluid, and the subscript 22 has been dropped. Equation (19) is just the PRISM equation for a bulk polymer fluid, that is, Eq. (12). Equation (18) can be written in Fourier space as

$$\hat{h}_w(k) = \hat{S}(k)\hat{c}_w(k) \tag{20}$$

or in real space, using bipolar coordinates, as

$$h_w(r) = \frac{2\pi}{r} \int_0^\infty tS(t)\, dt \int_{|r-t|}^{r+t} sc_w(s)\, ds \tag{21}$$

To obtain the correlation functions for a fluid at a surface, the substitutions $r' = r - R$ and $s' = s - R$ are made in Eq. (21), where $R$ is the radius of the sphere, and the limit is taken as $R \to \infty$. With the above substitutions, Eq. (21) takes the form

$$h_w(r') = 2\pi \int_0^\infty tS(t)\, dt \int_{r'-t}^{r'+t} \frac{s'+R}{r'+R} c_w(s')\, ds' \tag{22}$$

which, in the $R \to \infty$ limit, becomes

$$h_w(z) = 2\pi \int_0^\infty tS(t)\, dt \int_{z-t}^{z+t} c_w(s')\, ds' \tag{23}$$

where the variable $z$ is used to denote the perpendicular distance from the surface of the (infinitely large) sphere. Equation (23) is the OZ equation for a polymer fluid at a surface. It can be written in a more compact form in Fourier space. Defining a one-dimensional Fourier transform by

$$\tilde{f}(k) = \frac{1}{2\pi} \int_{-\infty}^{+\infty} f(z)e^{ikz}\, dz \tag{24}$$

and performing this transform on Eq. (23) gives

$$\tilde{h}_w(k) = \hat{S}(k)\tilde{c}_w(k) \tag{25}$$

which looks just like Eq. (20) except that one-dimensional rather than three-dimensional Fourier transforms are used for $c_w$ and $h_w$. Most applications have considered fluids between two surfaces. For this geometry it is convenient to define new functions, $g_w(z) \equiv h_w(z) + 1$ and $C_w(z) = c_w(z) + 1/\hat{S}(0)$. With these substitutions, Eq. (25) can be written as

$$\tilde{g}_w(k) = \hat{S}(k)\tilde{C}_w(k) \tag{26}$$

which is referred to as the wall–PRISM equation.

Standard closure approximations may be used for the wall–PRISM equation. For example, the PY closure is

$$c_w(z) = (1 - e^{-\beta\phi(z)})g_w(z) \tag{27}$$

where $\phi(z)$ is the fluid-surface potential (or external field), and $g_w(z) = 1 + h_w(z)$. The density profile is given by $\rho(z) = \rho g_w(z)$. The wall-PRISM equation is solved by first calculation $\hat{S}(k)$ (e.g., using the PRISM theory) and then employing a Picard iteration procedure for the wall–PRISM equation, similar to the one described earlier for the PRISM equation.

## C. Multiple-Site Models and Multicomponent Systems

The wall–PRISM theory has been extended to multiple site models [95]. A simple example of a multiple-site model is a vinyl polymer (e.g., polypropylene) where there are three types of united-atom sites corresponding to $CH_2$, CH, and $CH_3$ groups. Ignoring end effects as before the PRISM equations take the form

$$\hat{\mathbf{H}}(k) = \hat{\mathbf{\Omega}}(k)\hat{\mathbf{C}}(k)[\hat{\mathbf{\Omega}}(k) + \hat{\mathbf{H}}(k)] \tag{28}$$

where $\hat{\mathbf{H}}$, $\hat{\mathbf{\Omega}}$, and $\hat{\mathbf{C}}$ are matrices of correlation functions and are defined below. The total correlation function matrix is composed of elements

$$H_{\alpha\beta}(r) = \rho_\alpha\rho_\beta h_{\alpha\beta}(r) = \rho_\alpha\rho_\beta(g_{\alpha\beta}(r) - 1) \tag{29}$$

where $g_{\alpha\beta}(r)$ is the pair correlation function between sites $\alpha$ and $\beta$ on different chains, and $\rho_\alpha$ is the density of sites of type $\alpha$. The intramolecular structure factor matrix is defined in reciprocal space according to

$$\hat{\Omega}_{\alpha\beta}(k) = \frac{\rho_\alpha}{N}\sum_{i\in\alpha}\sum_{j\in\beta}\left\langle\frac{\sin kr_{ij}}{kr_{ij}}\right\rangle \tag{30}$$

where sites $\alpha$ and $\beta$ are on the same chain and the average is over all conformations. The direct correlation functions $C_{\alpha\beta}(r)$ are defined by the PRISM equations.

The wall–PRISM equations take the same form as in the simple homopolymer case, that is,

$$\rho_\alpha \tilde{g}_{w,\alpha}(k) = \sum_\beta [\hat{\Omega}_{\alpha\beta} + \hat{H}_{\alpha\beta}] \tilde{C}_{w,\beta}(k) \tag{31}$$

where the tilde and carets have the same meaning as before. The density profile of site $\alpha$ is then given by $\rho_\alpha(z) = \rho_\alpha g_{w,\alpha}(z)$.

The equations that describe a blend or mixture of homopolymers is similar to that shown for multisite systems above. The only difference is that the matrix of intramolecular correlation functions is different and is given by

$$\hat{\Omega}_{\alpha\beta}(k) = \rho_\alpha \delta_{\alpha\beta} \hat{\omega}_\alpha(k) \tag{32}$$

where $\delta_{\alpha\beta}$ is the Kronecker delta, and $\hat{\omega}_\alpha(k)$ is the single-chain structure factor of chains of species $\alpha$.

## D. Predictions for Hard Chains

The wall–PRISM equation has been implemented for a number of hard-chain models including freely jointed [94] and semiflexible [96] tangent hard-sphere chains, freely rotating fused-hard-sphere chains [97], and united atom models of alkanes, isotactic polypropylene, polyisobutylene, and polydimethyl siloxane [95]. In all implementations to date, to my knowledge, the theory has been used exclusively for the structure of hard-sphere chains at smooth structureless hard walls.

Yethiraj and Hall [94] studied the density profiles, surface forces, and partition coefficient of freely jointed tangent hard-sphere chains between hard walls. The theory was able to capture the depletion of chain sites at the surface at low densities and the enhancement of chain sites at the surface at high densities. This theory is in qualitative agreement with simulations for the density profiles and partitioning of 4 and 20 bead chains, although several quantitative deficiencies are present. At low densities the theory overestimates the value of the density profile near the surface. Furthermore, it predicts a quadratic variation of density with distance near the surface, whereas in reality the density profile should be linear in distance, for long chains. At high densities the theory underestimates the value of the density near the surface. The theory is quite accurate, however, for the partition coefficient for hard chains in slit-like pores.

The accuracy of the theory diminishes somewhat for models where the packing is dissimilar from a hard-sphere fluid. Yethiraj and co-workers have compared theoretical predictions for the density profiles of semiflexible hard-sphere chains [96] and fused-hard-sphere freely rotating chains [97] at hard walls. In the former case, the theoretical predictions are not even in qualitative agreement with simulations for high values of stiffness. The theory misses the disappearance and reappearance of liquid-like packing with increasing stiffness, instead predicting a monotonic increase in the oscillations in the density profile as the stiffness is increased. In the rod limit, the wall–PRISM theory predicts peaks in the wrong positions altogether. For the fused-sphere chains the theory is in reasonable agreement with the simulations, although the quantitative agreement is poorer than for the tangent sphere chains. In addition, the predictions become very sensitive to the single-chain structure input into the theory, contrary to the case for freely jointed hard chains.

For the united atom models of realistic polymers the wall PRISM theory predicts interesting structure near the surface [95]. For example, the side chains are found preferentially in the immediate vicinity of the surface and shield the backbone from the surface. This behavior is expected from entropic considerations. Computer simulations of these systems would be of considerable interest.

Recently the wall–PRISM theory has been used to investigate the forces between hydrophobic surfaces immersed in polyelectrolyte solutions [98]. Polyelectrolyte solutions display strong peaks at low wavevectors in the static structure factor, which is a manifestation of liquid-like order on long lengthscales. Consequently, the force between surfaces confining polyelectrolyte solutions is an oscillatory function of their separation. The wall–PRISM theory predicts oscillatory forces in salt-free solutions with a period of oscillation that scales with concentration as $\rho^{-1/3}$ and $\rho^{-1/2}$ in dilute and semidilute solutions, respectively. This behavior is explained in terms of liquid-like ordering in the bulk solution which results in liquid-like layering when the solution is confined between surfaces. In the presence of added salt the theory predicts the possibility of a predominantly attractive force under some conditions. These predictions are in accord with available experiments [99,100].

The wall–PRISM theory has also been implemented for binary polymer blends. For blends of stiff and flexible chains the theory predicts that the stiffer chains are found preferentially in the immediate vicinity of the surface [60]. This prediction is in agreement with computer simulations for the same system [59,60]. For blends of linear and star polymers [101] the theory predicts that the linear polymers are in excess in the immediate vicinity of the surface, but the star polymers are in excess at other distances. Therefore, if one looks at the integral of the difference between the density profiles of the two components, the star polymers segregate to the surface in an integrated sense, from purely entropic effects.

The integral equation theory is a simple means of studying the density profiles of dense polymer melts at surfaces where the structure is dominated by repulsive interactions. Under these circumstances the theory is in qualitative agreement with simulations and experiments. It is therefore a convenient means of investigating the entropic effects of polymer confinement under melt-like conditions. The theory has never, to my knowledge, been used for models where attractive (dispersion) interactions are present. This is because even for simple liquids, the theory is known to be fairly diseased in its treatment of attractive forces. For example, it completely misses the drying transition when a Lennard-Jones at a hard wall is cooled toward the coexistence region [89].

## IV.  DENSITY FUNCTIONAL THEORY

There are many varieties of density functional theories depending on the choice of ideal systems and approximations for the excess free energy functional. In the study of non-uniform polymers, density functional theories have been more popular than integral equations for a variety of reasons. A survey of various theories can be found in the proceedings of a symposium on chemical applications of density functional methods [102]. This section reviews the basic concepts and tools in these theoretical methods including techniques for numerical implementation.

### A.  Simple Liquids

The basic idea in density functional theory is that the grand free energy of an inhomogeneous fluid is a minimum with respect to variations in the one-body or singlet density profile, $\rho(\mathbf{r})$. Therefore all the thermodynamic and structural properties can be determined by minimizing this functional with respect to the density profile. Although this functional is not known, in general, several good approximate schemes are available and are reviewed in recent articles [103,104]. For simple liquids, exact expressions are available for an ideal gas, and the crux of density functional theory is to describe the nonideal (or excess) properties. For polymeric liquids, the ideal gas is a many-body problem and cannot be solved exactly. However, the approximations for the excess free energy functional can be obtained from methods similar to those used for simple liquids.

The starting point is the definition of the partition function, $\Xi$, in the grand canonical ensemble:

$$\Xi = \sum_{n \geq 0} \frac{\lambda^n}{\Lambda^{3n} n!} \int e^{-\beta V_n} e^{-\beta \Phi_n} \, d\mathbf{r}^n \tag{33}$$

where $n$ is the number of atoms, $\lambda = \exp(\beta\mu)$ is the activity, $\mu$ is the chemical potential, $\Lambda$ is the thermal deBroglie wavelength, $V_n$ is the potential energy from interatomic interactions, and $\Phi_n = \sum_i \phi(\mathbf{r}_i)$ is the potential energy due to the external field, $\phi(\mathbf{r})$. The singlet density $\rho(\mathbf{r})$ is defined by

$$\rho(\mathbf{r}) = \frac{1}{\Xi} \sum_{n \geq 1} \frac{\lambda^n}{\Lambda^{3n}(n-1)!} \int e^{-\beta V_n} e^{-\beta \Phi_n} \, d\mathbf{r}_2 \, d\mathbf{r}_3 \dots d\mathbf{r}_n \tag{34}$$

and the grand free energy is given by

$$\Omega = -k_B T \ln \Xi \tag{35}$$

In density functional theory, one starts with $\Omega$ as a functional of the density profile. It can be shown that at equilibrium this functional is minimal with respect to variations in $\rho(\mathbf{r})$, that is,

$$\frac{\delta \Omega}{\delta \rho(\mathbf{r})} = 0 \tag{36}$$

and this condition is used to evaluate the equilibrium density profile as well as the free energy. This functional is related to the Helmholtz free energy functional $F[\rho(\mathbf{r})]$ via Legendre transform,

$$\Omega[\rho(\mathbf{r})] = F[\rho(\mathbf{r})] + \int d\mathbf{r}(\phi(\mathbf{r}) - \mu)\rho(\mathbf{r}) \tag{37}$$

For an ideal gas, the functional $F[\rho(\mathbf{r})]$ is known exactly. Because $V_n = 0$, the partition function and density profile are given, respectively, by

$$\Xi = \sum_{n \geq 0} \frac{\lambda^n}{\Lambda^{3n} n!} \left[ \int \exp\left(-\beta \phi(\mathbf{r})\right) d\mathbf{r} \right]^n \tag{38}$$

and

$$\begin{aligned}
\rho(\mathbf{r}) &= \frac{1}{\Xi} \sum_{n \geq 1} \frac{\lambda^n}{\Lambda^{3n}(n-1)!} e^{-\beta \phi(\mathbf{r})} \left[ \int \exp\left(-\beta \phi(\mathbf{r})\right) d\mathbf{r} \right]^{n-1} \\
&= \frac{\lambda}{\Lambda^3} e^{-\beta \phi(\mathbf{r})} \frac{1}{\Xi} \sum_{n \geq 1} \frac{\lambda^{n-1}}{\Lambda^{3(n-1)}(n-1)!} \left[ \int \exp\left(-\beta \phi(\mathbf{r})\right) d\mathbf{r} \right]^{n-1} \\
&= \frac{\lambda}{\Lambda^3} e^{-\beta \phi(\mathbf{r})}
\end{aligned} \tag{39}$$

Substituting the equation for $\rho(\mathbf{r})$ into that for $\Xi$, we have

$$\Xi = \sum_{n \geq 0} \frac{1}{n!} \left[ \int \rho(\mathbf{r}) \, d\mathbf{r} \right]^n \tag{40}$$

$$= \exp \left[ \int \rho(\mathbf{r}) \, d\mathbf{r} \right] \tag{41}$$

Using this result and

$$\ln \Lambda^3 \rho(\mathbf{r}) = \beta(\mu - \phi(\mathbf{r})) \tag{42}$$

in the definition of $F[\rho(\mathbf{r})]$, that is,

$$F = -k_B T \ln \Xi + \int (\mu - \phi) \rho(\mathbf{r}) \, d\mathbf{r} \tag{43}$$

gives

$$F = k_B T \int \rho(\mathbf{r})(\ln \Lambda^3 \rho(\mathbf{r}) - 1) \, d\mathbf{r} \tag{44}$$

There are several ways of obtaining functionals for nonideal systems. In most cases the free energy functional is expressed as the sum of an ideal gas term, a hard-sphere term, and a term due to attractive forces. Below, I present a scheme by which approximate expression for the free energy functional may be obtained. This approach relies on the relationship between the free energy functional and the direct correlation function. Because the direct correlation functions are defined through functional derivatives of the excess free energy functional, that is,

$$\frac{\delta^2 F_{\mathrm{EX}}[\rho(\mathbf{r})]}{\delta \rho(\mathbf{r}) \delta \rho(\mathbf{r}')} = -k_B T c(\mathbf{r}, \mathbf{r}') \tag{45}$$

where $F_{\mathrm{EX}}$ is the excess (over ideal gas) free energy of the fluid, approximations for the functional can be obtained from a functional integration of the above definition. This is advantageous because accurate theories for the direct correlation function exist for uniform simple liquids. Integrating Eq. (45) using the linear path

$$\rho_\alpha(\mathbf{r}) = \alpha \rho(\mathbf{r}) \tag{46}$$

gives

$$\beta F_{EX}[\rho(\mathbf{r})] = -k_B T \int_0^1 d\alpha \int_0^\alpha d\alpha' \int d\mathbf{r} \int d\mathbf{r}' \rho(\mathbf{r})\rho(\mathbf{r}')c(\mathbf{r}, \mathbf{r}'; \alpha) \qquad (47)$$

which is an exact relation.

The simplest choice for $F_{EX}$ is to approximate the nonuniform fluid direct correlation function in Eq. (47) with the uniform fluid direct correlation at some bulk density, that is, to set

$$c(\mathbf{r}, \mathbf{r}'; \alpha) \approx c(|\mathbf{r} - \mathbf{r}'|; \rho) \qquad (48)$$

where the right-hand side is the $c(r)$ for a uniform fluid. The resulting expression for $F_{EX}[\rho(\mathbf{r})]$ is

$$\beta F_{EX}[\rho(\mathbf{r})] = -\frac{k_B T}{2} \int d\mathbf{r} \int d\mathbf{r}' \rho(\mathbf{r})\rho(\mathbf{r}')c(|\mathbf{r} - \mathbf{r}'|; \rho) \qquad (49)$$

This approximation amounts to truncating the functional expansion of the excess free energy at second order in the density profile. This approach is accurate for Lennard-Jones fluids under some conditions, but has fallen out of favor because it is not capable of describing wetting transitions and coexisting liquid–vapor phases [105–107]. Incidentally, this approximation is identical to the hypernetted chain closure to the wall–OZ equation [103].

Without a loss of generality, the direct correlation function can be decomposed into three contributions by defining a new function $\Delta c(\mathbf{r}, \mathbf{r}')$ by

$$c(\mathbf{r}, \mathbf{r}') = c_{HC}(\mathbf{r}, \mathbf{r}') - \beta u(|\mathbf{r} - \mathbf{r}'|) + \Delta c(\mathbf{r}, \mathbf{r}') \qquad (50)$$

where $u(r)$ is the interatomic potential and $c_{HC}$ is the hard-core contribution. The integrals over $c_{HC}(\mathbf{r}, \mathbf{r}')$ and $u(|\mathbf{r} - \mathbf{r}'|)$ can now be performed to give

$$F[\rho(\mathbf{r})] = F_{ID}[\rho(\mathbf{r})] + F_{HC}[\rho(\mathbf{r})] + \frac{1}{2} \int d\mathbf{r} \int d\mathbf{r}' \rho(\mathbf{r})\rho(\mathbf{r}')u(|\mathbf{r} - \mathbf{r}'|)$$

$$- k_B T \int d\mathbf{r} \int d\mathbf{r}' \rho(\mathbf{r})\rho(\mathbf{r}') \int_0^1 d\alpha \int_0^\alpha d\alpha' \Delta c(\mathbf{r}, \mathbf{r}'; \alpha') \qquad (51)$$

In general, different approximations are invoked for the hard-core contribution and the attractive contribution to the free energy functional. For the hard-core contribution, two accurate approximations can be obtained from the fundamental measure theory [108] and the weighted density approximation

(WDA) [109,110]. I describe the latter because it is easily extended to flexible polymers. In the WDA,

$$F_{HC}[\rho(\mathbf{r})] = \int \rho(\mathbf{r}) f(\bar{\rho}) \, d\mathbf{r} \tag{52}$$

where $f(\bar{\rho})$ is the excess (over ideal gas) free energy per site of the bulk hard-sphere fluid evaluated at a site density $\bar{\rho}(\mathbf{r})$,

$$\bar{\rho}(\mathbf{r}) = \int \rho(\mathbf{r}') w(|\mathbf{r} - \mathbf{r}'|) \, d\mathbf{r}' \tag{53}$$

is the weighted density, and $w(r)$ is the weighting function, normalized so that $\int w(\mathbf{r}) \, d\mathbf{r} = 1$. The function $f(\rho)$ can be obtained from an equation of state for the bulk fluid, which is assumed to be known. The central approximation is the choice of the weighting function. The simplest choice for $w(r)$ is [109,111]

$$w(r) = \frac{3}{4\pi\sigma^3} \Theta(\sigma - r) \tag{54}$$

where $\Theta(r)$ is the Heaviside step function. A more accurate theory can be obtained from the Curtin–Ashcroft recipe [112], where $w(r)$ is obtained by forcing the free energy functional to satisfy Eq. (45) for a uniform fluid. This results in the nonlinear equation

$$-k_B T \hat{c}(k) = 2f'(\rho)\hat{w}(k) + \rho f''(\rho)\hat{w}^2(k) + 2\rho f'(\rho)\hat{w}(k)\frac{\partial \hat{w}(k)}{\partial \rho} \tag{55}$$

where primes denote derivatives with respect to density, and the bulk fluid direct correlation function is assumed to be known.

The simplest choice for the attractive contribution to the free energy functional is to set

$$\Delta c(\mathbf{r}, \mathbf{r}') \approx 0 \tag{56}$$

which is identical to a van der Waals approximation for the attractive forces. Although very simple, this approach gives good results for Lennard-Jones fluids at Lennard-Jones surfaces under some conditions [113]. A more elaborate approximation is to set the attractive part of the direct correlation function to be equal to that of a uniform fluid with bulk density $\rho$, that is,

$$\Delta c(\mathbf{r}, \mathbf{r}') \approx \Delta c(|\mathbf{r} - \mathbf{r}'|; \rho) \tag{57}$$

This approximation gives excellent results for the behavior of ionic fluids at charged surfaces [114,115].

## B.  Polymeric Liquids

The primary complication in going from simple liquids to polymers is the intramolecular interactions and constraints in the latter. Each molecule now has $N$ monomers, and there are bonding constraints between neighboring monomers along the backbone as well as other interactions such as excluded volume torsional rotation potentials, bond bending potentials, and so on. If $\mathbf{R}$ denotes the positions of all the $N$ monomers on a polymer molecule (i.e., $\mathbf{R} \equiv \{\mathbf{r}_1, \mathbf{r}_2, \ldots, \mathbf{r}_N\}$, where $\mathbf{r}_i$ is the position of bead $i$ on the chain), then the intramolecular interactions, denoted $V(\mathbf{R})$, in general, depend on all these positions. It can be shown that the grand free energy is minimal with respect to variations in either the molecular density, $\rho_M(\mathbf{R})$, or the individual sites densities, $\rho_\alpha(\mathbf{r})$, that is,

$$\frac{\delta\Omega}{\delta\rho_M(\mathbf{R})} = 0 \tag{58}$$

$$\frac{\delta\Omega}{\delta\rho_\alpha(\mathbf{r})} = 0 \tag{59}$$

and either of these conditions may be used to determine the density profile.

### 1.  Ideal Gas Functional

The central issue in the density functional theory of polymers is the choice of the ideal system, and earlier work focused on the problem of finding suitable approximations to the ideal gas free energy functional. In the limit that the density is slowly varying, one can obtain an exact expression for the ideal gas functional similar to what was derived for ideal gases of simple liquids. Following the methods used earlier, it can be shown that the ideal gas free energy functional for a fluid of noninteracting monomers (i.e., without either intramolecular or intermolecular interactions) is given by

$$F[\{\rho_\alpha(\mathbf{r})\}] = \sum_\alpha \int d\mathbf{r}\,\rho_\alpha(\mathbf{r})[\ln\lambda_\alpha\rho_\alpha(\mathbf{r}) - 1] \tag{60}$$

where $\lambda_\alpha$ is a constant with units of length and will be dropped hereafter because it cancels out of all equations that need to be solved. In principle, both intramolecular and intermolecular contributions could be incorporated as excess terms, as has been suggested by Chandler et al. [116,117]. The problem is that the intramolecular functional is highly nonlinear and difficult to approximate. Such an approach has never been implemented or tested, although some approximations have been suggested by Chandler et al. [116,117].

In the united atom limit—that is, when all bond lengths are equal to zero—
the ideal gas free energy functional is of course just given by

$$F_{ua} = \int d\mathbf{r}\rho(\mathbf{r})\left[\ln\frac{\rho(\mathbf{r})}{N} - 1\right] \tag{61}$$

where $\rho(\mathbf{r}) = \sum_{\alpha}\rho_{\alpha}(\mathbf{r})$. All the site density profiles are equal because the sites
are coincident. McMullen and Freed [118] have argued that this is a reasonable
approximation when the field is slowly varying and may be considered constant
over lengthscales of the order of the size of the molecules. Such situations exist,
for example, at interfaces between polymer solutions, especially near the critical
point. The first few corrections to this approximation have also been evaluated
[119]. Of course, approximations of this nature are not very useful for polymer
melts at surfaces. For example, for an ideal polymer in an external field the united
atom approximation gives

$$\rho(\mathbf{r}) = N\exp\left[\sum_{\alpha}\{\phi(\mathbf{r}) - \mu_{\alpha}\}\right] \tag{62}$$

For a fluid at a hard wall this implies $\rho(\mathbf{r}) = $ constant. As expected, because the
theory leaves out the entropic effects of confinement, namely chain connectivity,
it is in qualitative error for the "depletion effects" for a polymer at a surface
[118–120].

A different approach is to express the ideal gas free energy functional in
terms of the molecular density, $\rho_M(\mathbf{R})$, instead of the site densities. In this case
the ideal gas functional is known exactly:

$$F_{ID}[\rho_M(\mathbf{R})] = k_BT\int d\mathbf{R}\rho_M(\mathbf{R})[\ln\rho_M(\mathbf{R}) - 1] + \int d\mathbf{R}V(\mathbf{R})\rho_M(\mathbf{R}) \tag{63}$$

To see that this is exact, consider an ideal chain fluid in a external field $\Phi(\mathbf{R})$. For
this fluid,

$$\Omega = F_{ID} + \int d\mathbf{R}\rho_M(\mathbf{R})(\mu - \Phi) \tag{64}$$

and the variational principle, that is,

$$\frac{\delta\Omega}{\delta\rho_M(\mathbf{R})} = 0 \tag{65}$$

gives

$$\rho_M(\mathbf{R}) = \exp[-\beta V(\mathbf{R}) + \beta\mu - \beta\Phi(\mathbf{R})] \tag{66}$$

which is exact. The complication with this choice of free energy functional is that $V(\mathbf{R})$ is a many-body potential and further progress is difficult without computation.

For freely jointed chains, if all intramolecular interactions other than bonding constraints are neglected, such an ideal system is still tractable. In this case,

$$e^{-\beta V(\mathbf{R})} = \exp\left(-\sum_{\alpha} v_b(\mathbf{r}_\alpha - \mathbf{r}_{\alpha-1})\right) \tag{67}$$

where $v_b$ is the bonding potential. Another system for which this ideal gas functional can be written down in closed form is for short alkanes in the rotational isomeric state approximation. In this case $V(\mathbf{R})$ has discrete states corresponding to the three energy minima for each torsional rotation in the molecule. This simplicity has been exploited by Seok and Oxtoby [121] in their study of the freezing of alkanes.

In general, use of the ideal gas functional in terms of the molecular density requires computation. Despite the computational intensive nature of the resulting theory, this is probably the most widely used functional for polymers and is described greater detail below. As mentioned earlier, the approximations for the excess free energy functional are similar to those used for simple liquids. The exact expression for the ideal gas functional in this case is

$$F_{\mathrm{ID}}[\rho_M(\mathbf{R})] = k_B T \int d\mathbf{R} \rho_M(\mathbf{R})[\ln \rho_M(\mathbf{R}) - 1] + \int d\mathbf{R} V(\mathbf{R})\rho_M(\mathbf{R}) \tag{68}$$

The many-body nature of $\rho_M(\mathbf{R})$, however, complicates matters.

### 2. Excess Free Energy Functional

As before, $F[\rho_M]$ can be expressed as the sum of an (exactly known) ideal part and excess ($F_{\mathrm{EX}}$) part, that is

$$F[\rho_M(\mathbf{R})] = k_B T \int d\mathbf{R} \rho_M(\mathbf{R})[\ln \rho_M(\mathbf{R}) - 1] + \int d\mathbf{R} V(\mathbf{R})\rho_M(\mathbf{R}) + F_{\mathrm{EX}}[\rho_M(\mathbf{R})] \tag{69}$$

A simple and successful approximation for the excess free energy functional is to first assume that the excess free energy functional is only a functional of the average site density profile, denoted $\rho(\mathbf{r})$, and then invoke standard approximations, similar to those used for simple liquids, for $F_{\mathrm{EX}}$. With a judicious choice of $F_{\mathrm{EX}}[\rho_M]$, the free energy functional can be exactly decomposed as

$$F[\rho_M(\mathbf{R})] = F_{\mathrm{ID}}[\rho_M(\mathbf{R})] + F_{\mathrm{EX}}[\rho(\mathbf{r})] \tag{70}$$

where $F_{ID}$ takes the interpretation of being the exact free energy of the ideal chain constrained to have a site density $\rho(\mathbf{r})$. If one assumes that this functional is given by

$$F_{ID}[\rho_M(\mathbf{R})] = k_B T \int d\mathbf{R} \rho_M(\mathbf{R})[\ln \rho_M(\mathbf{R}) - 1] + \int d\mathbf{R} V(\mathbf{R}) \rho_M(\mathbf{R}) \tag{71}$$

then the task reduces to finding approximations for $F_{EX}[\rho(\mathbf{r})]$.

Three different approximations for $F_{EX}[\rho(\mathbf{r})]$ have been employed. The first approximation, due to McCoy, Curro and coworkers [122–125], is to truncate the functional expansion at second order, that is,

$$F_{EX}[\rho(\mathbf{r})] \approx -\frac{k_B T}{2} \int d\mathbf{r} d\mathbf{r}' \rho(\mathbf{r}) \rho(\mathbf{r}') c(|\mathbf{r} - \mathbf{r}'|; \rho) \tag{72}$$

where $c(r)$ is the direct correlation function of the bulk polymer melt evaluated at the bulk site density $\rho$ and can be obtained, for example, from the PRISM theory. This approximation is used for both the hard-core and the attractive contribution to the excess free energy.

The other two approaches divide the excess functional into a hard-core and an attractive part with different approximations for the two. Rosinberg and coworkers [126–129] have derived a functional from Wertheim's first-order perturbation theory of polymerization [130] in the limit of complete association. Woodward, Yethiraj, and coworkers [39,131–137] have used the weighted density approximation for the hard-core contribution to the excess free energy functional, that is,

$$F_{HC}[\rho(\mathbf{r})] = \int \rho(\mathbf{r}) f(\bar{\rho}) \, d\mathbf{r} \tag{73}$$

where $f(\rho)$ is the excess (over ideal gas) free energy per site of the bulk hard-chain fluid evaluated at a site density $\rho$, and the weighted density and weighting function are defined in the same way as for simple liquids [Eq. (53)]. Because the range of the direct correlations in the fluid are expected to be of the order of the bead diameter, the simplest approximation is the step function approximation of Eq. (54). Although more sophisticated choices for $w(r)$ are possible and have been implemented [39], for most cases the simplest choice of $w(r)$ is adequate. Implementation of these theories to systems that include attractions are few [129,137], and in all cases the attractions were treated in a mean-field approximation, that is, using Eq. (56).

### 3. Numerical Implementation

The density functional theory has the structure of a self-consistent field theory where the density profile is obtained from a simulation of a single chain in the field due to the rest of the fluid and the surface. A formal minimization of $\Omega$ gives

$$\frac{\delta \Omega}{\delta \rho_M(\mathbf{R})} = \frac{\delta F}{\delta \rho_M(\mathbf{R})} + \Phi(\mathbf{R}) - \mu = 0 \qquad (74)$$

which may be written as

$$\rho_M(\mathbf{R}) = \exp\left[-\beta V(\mathbf{R}) + \beta\mu - \beta\Phi(\mathbf{r}) + \beta \sum_{i=1}^{N} \lambda(\mathbf{r}_i)\right] \qquad (75)$$

where $\lambda(\mathbf{r})$ is the self-consistent field given by

$$\lambda(\mathbf{r}) = \frac{\delta F_{\text{EX}}}{\delta \rho(\mathbf{r})} \qquad (76)$$

If there were no intramolecular interactions (such as bonding or excluded volume), then $V(\mathbf{R}) = 0$, and the next guess for the density profile can be obtained directly from Eq. (75). The presence of $V(\mathbf{R})$ necessitates either a multidimensional integration or (more conveniently) a single-chain simulation.

The implementation of the density functional theory entails the iterative solution of one equation:

$$\rho(\mathbf{r}) = \int d\mathbf{R}\left[\sum_{i=1}^{N} \delta(\mathbf{r} - \mathbf{r}_i)\right] \exp\left[-\beta V(\mathbf{R}) + \beta\mu - \beta\Phi(\mathbf{r}) + \beta \sum_{i=1}^{N} \lambda(\mathbf{r}_i)\right] \qquad (77)$$

with $\lambda(\mathbf{r})$ given by Eq. (76). This equation is solved by discretizing the density profile, treating the value of the density at each point as an independent variable, and then solving the resulting nonlinear set of equations.

One way to obtain predictions for the density profile is to use a Picard iteration procedure. In this method one starts with an initial guess for the density profile. The field $\lambda(\mathbf{r})$ is then calculated using Eq. (76), and a new estimate for the density profile is obtained using Eq. (77). The latter requires the simulation of a single chain with intramolecular interaction $V(\mathbf{R})$ in an effective field $\chi(\mathbf{R}) = \Phi(\mathbf{R}) + \Lambda(\mathbf{R})$, where $\Lambda(\mathbf{R}) \equiv \sum_\alpha \lambda(\mathbf{r}_\alpha)$. The density profile is then calculated from

$$\rho(\mathbf{r}) = \left\langle \sum_{i=1}^{N} \delta(\mathbf{r} - \mathbf{r}_i) \right\rangle \qquad (78)$$

where the average is for a single chain in external field $\chi$ and is normalized so that $\int \rho(\mathbf{r}) \, d\mathbf{r} = \rho_{av}$ is the average density desired. The new density profile is compared to the old one; the procedure is stopped if convergence has been achieved, and it is continued otherwise. The Picard iteration scheme is beset with problems, however. Except at low densities, the method does not converge unless the new estimate for the density profile is mixed with the old one. As one nears the true solution, the method becomes unstable and very small mixing parameters (i.e., using very little of the *new* density profile) are required. This instability problem has been encountered before in solutions of the Ornstein–Zernike equation.

A more efficient way of solving the DFT equations is via a Newton–Raphson (NR) procedure as outlined here for a fluid between two surfaces. In this case one starts with an initial guess for the density profile. The self-consistent fields are then calculated and the next guess for density profile is obtained through a single-chain simulation. The difference from the Picard iteration method is that an NR procedure is used to estimate the new guess from the density profile from the old one and the one monitored in the single-chain simulation. This requires the computation of a Jacobian matrix in the course of the simulation, as described below.

The region between the walls is first divided into bins, and the density at the midpoint of each bin is treated as an independent variable. If the density is desired at $M$ discrete points, then the numerical method reduces to simultaneously solving $M$ equations in $M$ unknowns:

$$\psi_k \equiv \rho(z_k) - \left\langle \sum_{i=1}^{N} \delta(z_k - z_i) \right\rangle = 0, \qquad k = 1, 2, \ldots, M \tag{79}$$

where $\rho(z_k)$ is the value of the density at point $z_k$. (Since the effective field is a functional of the density profile, $\rho(z)$ also appears in the second term through the definition of the ensemble average.) Given a density profile $\rho_l(z)$, a new estimate for the density profile $\rho_{l+1}(z)$ is obtained from the NR recipe:

$$\rho_{l+1}(z_k) = \rho_l(z_k) - \sum_{j=1}^{M} J_{kj}^{-1} \psi_j \tag{80}$$

where $J_{ij} \equiv \partial \psi_i / \partial \rho(z_j)$ are the elements of the Jacobian matrix, $\mathbf{J}$, and $J_{ij}^{-1}$ are the elements of its inverse. Elements of $\mathbf{J}$ are obtained by differentiating Eq. (79):

$$J_{ij} = \frac{\partial \psi_i}{\partial \rho(z_j)} = \delta_{ij} + \left\langle \left[ \sum_{l=1}^{N} \delta(z_i - z_l) \right] \left[ \sum_{l=1}^{N} \frac{\partial \lambda(z_l)}{\partial \rho(z_j)} \right] \right\rangle \tag{81}$$

and must be monitored in the simulation along with the density profile. When the density profile is not very strongly varying it is convenient to approximate $\mathbf{J}$ by

$$J_{ij} \approx \delta_{ij} + \left\langle \left[ \sum_{l=1}^{N} \delta(z_i - z_l) \right] \right\rangle \left[ \sum_{l=1}^{N} \frac{\partial \lambda(z_l)}{\partial \rho(z_j)} \right] \tag{82}$$

(the last term has been pulled outside the ensemble average) without loss of performance and some savings in computer time.

### 4. Summary

The density functional approach has been very successful in describing the behavior of hard chains at hard walls. All the theoretical approaches that incorporate chemical bonding at some level into the ideal gas free energy functional are accurate for the depletion and enhancement effects and for the layering of the chains at the surface. For freely jointed hard chains at hard walls the theories of Woodward [131], Kierlik and Rosinberg [128], Sen et al. [122], and Yethiraj and Woodward [37] are about equally accurate for the density profiles. When other intramolecular interactions are present, such as torsional rotational potentials or bond bending potentials, a single-chain simulation is required in order to obtain accurate results. The most tested theory is that of Yethiraj and Woodward [37], as extended by Yethiraj [39] and by Patra and Yethiraj [137]. This approach is accurate (as discussed later) in most cases, although it is computationally demanding.

## V. CONCLUSIONS

### A. Comparison of Theories to Computer Simulations

The majority of comparisons of theory to computer simulations have been performed for freely jointed hard chains, and most of the theories are fairly accurate for this model. Typical examples are shown in Figs. 12(a) and 12(b), which compare predictions of the wall–PRISM theory of Yethiraj and Hall [94] (with the Percus–Yevick closure for the wall–fluid correlation function) and the density functional theory of Yethiraj and Woodward (YW), respectively, to simulations of freely jointed tangent sphere hard chains [36] with $N = 20$, $H = 10\sigma$, and for various packing fractions. Both theories are in qualitative agreement with the simulations and predict depletion and enhancement effects and the transition from one behavior to the other with increasing density. The integral equation theory tends to overestimate the value of the density at the surface for low densities and underestimate the value of the density at the surface for high densities. The density functional theory, on the other hand, is in excellent quantitative agreement with the simulations at all distances.

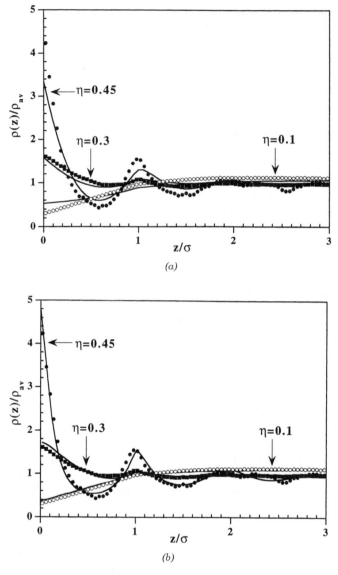

**Figure 12.** Comparison of predictions of the (a) wall–PRISM theory [94] and (b) density functional theory of Yethiraj and Woodward [37] to Monte Carol simulations for the density profile of 20-mers at hard walls for various packing fractions (as marked). The density profiles are normalized to the average value in the cell.

The accuracy of the theory does not diminish when the external field has a slowly varying component in addition to the hard-core repulsion. Figure 13 compares predictions of the wall–PRISM theory and the YW theory to computer simulations of freely jointed hard chains at surface that include a hard-core plus a soft component; that is, the external field is given by

$$\phi(z) = u_{wf}(z) + u_{wf}(H - z), \qquad 0 < z < H$$
$$= \infty \qquad\qquad\qquad \text{otherwise} \qquad (83)$$

$$\beta u_{wf}(z) = \epsilon_{wf}\left\{1 + c\left[\frac{1}{(1 + z/d)^9} - \frac{1}{(1 + z/d)^3}\right]\right\}\Theta(p\sigma - z) \qquad (84)$$

$c = 3^{3/2}/2$, $d = p\sigma/(3^{1/6} - 1)$, $p = 0.5$, and $\Theta(x) = 1$ for $x > 0$. The main figure depicts results for $\epsilon_{wf} = -2$, while the inset depicts those for $\epsilon_{wf} = 2$, both for $H = 10\,\sigma$, and the density profiles are normalized with respect to the bulk value. In this comparison the hypernetted chain closure is used in the wall–PRISM theory because it is more accurate than the Percus–Yevick closure. Incorporating an additional attraction causes a further enhancement of chain sites at the surface, and the additional repulsion causes a depletion of chains sites at the surface. Both theories are in excellent agreement with the computer simulations except that when the density profile is significantly oscillatory the

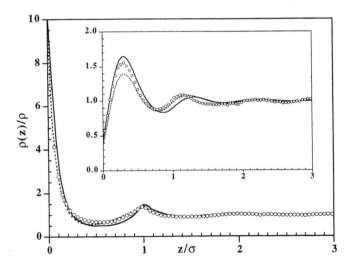

**Figure 13.** Comparison of prediction of the wall–PRISM theory with the hypernetted chain closure ($\cdots$) and the density functional theory of Yethiraj and Woodward [37] (— — —) to Monte Carlo simulations for the density profile of freely jointed hard 20-mers for $\eta = 0.35$. The wall–fluid potential consists of a hard core plus short ranged attractive and repulsive (*inset*) potential.

theoretical predictions tend to be slightly out of phase with the simulation results. For $\epsilon = -2$ the wall–PRISM theory is in slightly better agreement with the simulations than the YW theory, whereas for $\epsilon = 2$ the YW theory is slightly more accurate than the wall–PRISM theory.

The YW theory uses the weighted density approximation with a very simple step-function choice for the weighting function. Yethiraj [39] has investigated a more sophisticated choice for the weighting function using the Curtin–Ashcroft recipe with the direct correlation function obtained from the bulk fluid RISM theory. For freely jointed hard chains at low densities, as well as for fused-sphere chains, there is little difference between the predictions using the simple weighting function and the Curtin–Ashcroft weighting function, but when there is significant layering in the liquid, the latter is considerably more accurate. Figure 14 compares the density functional theory with the Curtin–Ashcroft weighting function (solid lines) and the simple weighting function (dashed lines) to Monte Carlo simulations of freely jointed 20-mers at high melt-like densities. One can see that the simple theory predicts density oscillations that are out of phase with the simulation results. The density functional theory with the Curtin–Ashcroft weighting function is in quantitative agreement with the simulation data on all lengthscales. In fact it is about as accurate for hard chains as the Curtin–Ashcroft theory is for hard spheres.

The density functional theory is at least as accurate for other models of polymers as it is for freely jointed hard chains, but this is not the case with the

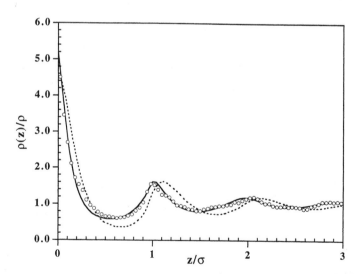

**Figure 14.** Comparison of predictions of the density functional theory of Yethiraj with the simple weighting function ($\cdots$) and the Curtin–Ashcroft weighting function (— — —) to Monte Carlo simulations for the density profile of 20-mers at hard walls for $\eta = 0.45$.

wall–PRISM theory. Figure 15 compares the density functional theories of Yethiraj and Woodward [37] (denoted YW) and Sen et al. [122] (denoted SCMC) to Monte Carlo simulations of semiflexible chains with $\epsilon = 5$ and $\eta = 0.3$ and 0.2 (inset). The differences between the theories is pronounced for this model. The wall-PRISM theory is not even in qualitative agreement with the simulations and predicts enhanced layering in the density profiles when in fact such layering is absent. In some cases (see inset) it predicts strong enhancement effects when in fact the chains are depleted from the surface. The density functional theories are quite accurate when compared to the simulations with the YW theory being clearly the more accurate of the two. Part of the accuracy of the YW theory arises from the fact that the theory satisfies the exact wall sum rule, and since it uses an accurate equation of state as input, the value at the surface is in good agreement with simulations. The SCMC theory, on the other hand, does not satisfy the wall sum rule which is in fact fortunate because the equation of state used by the theory is not very accurate [40].

The density functional theories are also accurate for the density profiles of fused-sphere chains. Figures 4(a) and 4(b) compare the theory of Yethiraj [39] (which is a DFT with the Curtin–Ashcroft weighting function) to Monte Carlo simulations of fused-hard-sphere chains at hard walls for $N = 4$ and 16, respectively. For both chain lengths the theory is in quantitative agreement with the simulation results and appears to get more accurate as the chain length is increased. Similarly good results were also found by SCMC who compared

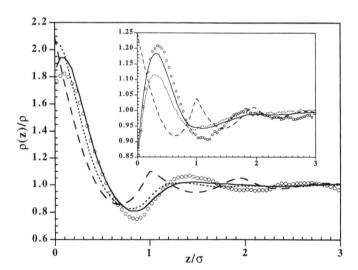

**Figure 15.** Comparison of predictions of the density functional theories of Yethiraj and Woodward [37] (— — —), Sen et al. [122] (· · · lines), and wall–PRISM theory [96] (– – –) to Monte Carlo simulations of semiflexible 20-mers for $\epsilon = 5$ and $\eta = 0.2$ (*inset*) and 0.3.

their theoretical predictions to molecular dynamics simulations of short alkanes [122]. In contrast, the wall–PRISM theory is not very accurate for this model. Figure 16 compares the wall–PRISM theory (with Percus–Yevick closure) to computer simulations of fused-hard-sphere chains at hard walls for $N = 16$. At low densities, the theory tends to overestimate the density at the surface, and at high densities it tends to underestimate the density at the surface. At the highest density the theory also significantly underestimates the degree of liquid-like ordering at the surface. The quantitative agreement with simulations is clearly not nearly as good as it was for hard chains.

All of the above tests were for hard chains at surfaces. The only comparison between theory and simulation for various values of fluid–fluid and bulk fluid attractions is that done by Patra and Yethiraj (PY) [137], who presented a simple van der Waals DFT for polymers and compared to simulations of fused-sphere chains. In their theory, PY used the Yethiraj functional [39] for the hard-chain contribution to the free energy and a simple mean-field term for the attractive contribution. Their excess free energy functional is given by

$$F[\rho(\mathbf{r})] = \int \rho(\mathbf{r})f(\bar{\rho})\,d\mathbf{r} + \frac{1}{2}\int d\mathbf{r}\int d\mathbf{r}'\rho(\mathbf{r})\rho(\mathbf{r}')u(|\mathbf{r}-\mathbf{r}'|) \qquad (85)$$

with the weighted density calculated using the Curtin–Ashcroft recipe. Their density functional theory is in qualitative agreement with the simulations for the density profiles in all cases, and it is in excellent agreement with the simulations

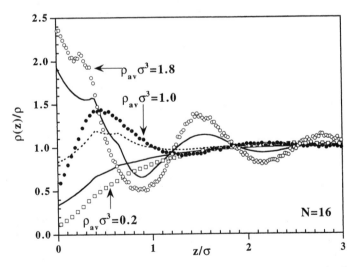

**Figure 16.** Comparison of the wall–PRISM theory to computer simulations for the density profiles of fused-hard-sphere chains at a hard wall for $N = 16$.

except when the density profiles are dominated by the bulk fluid interactions, which happens for low densities at high values of $\epsilon_{ff}$.

Figures 17(a) and 17(b) compare DFT predictions to Monte Carlo simulations [55] for the density profiles of 8-mers for parameter values of $(\epsilon_{wf}, \epsilon_{ff}) = (0.2, 0)$, $(0, 0.2)$, respectively, and for three densities in each case. The interaction potentials are the same as described earlier in connection with Fig. 10, $H = 10\sigma$, and the density profiles are normalized by the average density. When $\epsilon_{ff} = 0$ the DFT is in excellent agreement with the simulations and captures the enhancement/depletion entropic effects as well as the effect of the wall fluid attraction. When a fluid–fluid attraction is present, the performance of the DFT deteriorates somewhat, although it does capture the fact that the bulk attractions play a much more prominent role than the surface attractions. The theory underestimates the value of the density profile at the surface at high densities, and it significantly overestimates the drying off of the liquid from the surface at low densities. Overall, the theory is in good agreement with the simulations, except for $\rho = 0.2$ and $\epsilon_{ff} = 0.2$ where the agreement with simulation is poor. The performance of the theory for other chain lengths is similar to that seen in Fig. 17, except that at low densities and $\epsilon_{ff} = 0.2$ the theory progressively gets worse as the chain length is increased.

The inaccuracy of the theory for long chains and low densities can be attributed to the van der Waals approximation in the theory. For a bulk fluid, the excess internal energy, $U_{EX}$, per molecule is given by

$$U_{EX} = 2\pi N\rho \int_0^\infty r^2 u(r)g(r)\, dr \tag{86}$$

The simple van der Waals approximation assumes $g(r) = 1$ and for long-chain polymers at low densities this significantly *overestimates* the magnitude of the internal energy of the bulk fluid. This is because in polymers, intermolecular contacts are much reduced by the presence of beads on the same chain in the immediate vicinity of any given bead (the correlation hole effect). This results in $g(r) \ll 1$ at short distances at low densities. Because the theory overestimates the effect of bulk fluid attractions, it is expected to significantly overestimate the critical temperature, and it is not surprising that it significantly exaggerates the drying of the chains from the surface. The correlation hole effect becomes more significant for longer chains (the value of $g(r)$ at contact varies linearly with $1/N$ in hard chains [139]), and so the theory become less accurate as the chain length is increased.

## B. Summary

The last decade has seen active research in the implementation of liquid state methods to the behavior of polymers at surfaces or confined in small pores.

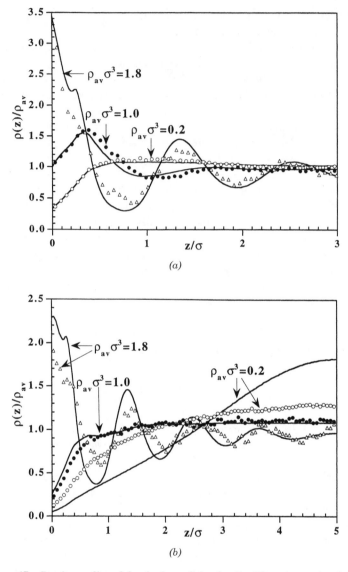

**Figure 17.** Density profiles of fused-sphere chains for $N = 18$, various reduced densities (as marked) and for strengths of wall–fluid and fluid–fluid attractions of $(\epsilon_{wf}, \epsilon_{ff}) =$ (a) (0.2, 0) and (b) (0, 0.2). Lines are theoretical predictions of Patra and Yethiraj [137].

There have been extensive computer simulations and liquid state theories, and a good understanding of these systems is now available. The majority of work has focused on simple hard-chain systems, and the depletion and enhancement effects in these systems are well understood and there are several theories that are in quantitative agreement with computer simulations. In contrast, there has been relatively little attention focused on the effect of wall–fluid and fluid–fluid attractions on the behavior of confined polymers. Only the simplest of DFT approaches has been attempted, and the results are promising although the quantitative performance leaves a lot to be desired.

Several avenues of future research suggest themselves. The first is the development of functionals to treat the contribution to the free energy of the attractive part of the intermolecular potential. Accurate functionals exist for simple liquids, and their implementation for polymers should be interesting. Similarly, the development of better closures (beyond the simple atomic closures) to the wall–PRISM approach should allow using this theory for realistic systems. Although the wall–PRISM theory is not as accurate as the density functional methods, it is very simple to implement and this is a distinct advantage. The addressing of these issues should allow the study of a number of interesting problems including systems such as confined polymer blends, where attractive forces play an crucial role, and polyelectrolyte solutions, where electrostatic forces play a crucial role. Finally, the BGY approach has never been implemented for polymers. This is surprising because, unlike the OZ-based approach, the BGY equation enforces a force balance and is ideally suited to the study of nonuniform liquids. It would appear that with a few small advances the theory of nonuniform polymers will be in a position to tackle a variety of challenging problems in colloid and polymer science that are inaccessible to any other theoretical method.

## Acknowledgments

I gratefully acknowledge support from the Alfred P. Sloan Foundation, the National Science Foundation (through grants CHE 9502320 and CHE 9732604 to A.Y. and CHE 9522057 to the Department of Chemistry), and the Alexander von Humboldt Foundation. Part of this work was completed when I was on sabbatical at the Indian Institute of Science in Bangalore. I thank Professor Biman Bagchi and other members of the S.S.C.U. for their hospitality.

## References

1. J. D. Walls, A. M. Nur, and T. Bourbie, *J. Pet. Technol* **April**, 930 (1982).

2. W. J. Schell, *Hydrocarbon Processing* **August**, 43 (1983).

3. D. M. Ruthven, *Principles of Adsorption and Adsorption Processes*, Wiley, New York, 1984.

4. D. Nicholson and N. G. Parsonage, *Computer Simulation and the Statistical Mechanics of Adsorption*, Academic, New York, 1980.

5. J. M. Thomas and R. M. Lambert, eds., *Characterization of Catalysts*, Wiley, New York, 1980.

6. D. M. Brewis and D. Briggs, eds., *Industrial Adhesion Problems*, Wiley, New York, 1985.

7. W. M. Krumbhaar, *The Chemistry of Synthetic Surface Coatings*, Reinhold, New York, 1937.

8. R. J. Hunter, *Foundations of Colloid Science*, Clarendon, Oxford, 1987.

9. I. A. Bitsanis and G. Ten Brinke, *J. Chem. Phys.* **99**, 3100 (1993).

10. G. D. Smith, D. Y. Yoon, and R. L. Jaffe, *Macromolecules* **25**, 7011 (1992).

11. F. Schmid, *J. Phys. Cond. Mater.* **10**, 8105 (1998).

12. P. -G. de Gennes, *Scaling Concepts in Polymer Physics*, Cornell University Press, Ithaca, 1979.

13. E. Eisenriegler, *Polymers at Surfaces*, World Scientific, New York, 1993.

14. K. Binder, ed., *The Monte Carlo Method in Condensed Matter Physics*, Springer, Berlin, 1995.

15. S. Granick, ed., *Polymers in Confined Environments, Advances in Polymer Science*, Vol. 138, Springer, Berlin, 1999.

16. C. D. Wick, M. G. Martin, and J. I. Siepmann, *J. Phys. Chem. B* **104**, 8001 (2000).

17. S. K. Nath, B. J. Banaszak, and J. J. de Pablo, *J. Chem. Phys.* **114**, 3612 (2001).

18. W. G. Madden, *J. Chem. Phys.* **87**, 1405 (1987).

19. W. G. Madden, *J. Chem. Phys.* **88**, 3934 (1988).

20. G. Ten Brinke, G. Hadziioannou, D. Ausserre, S. Hirz, and C. Frank, *Bull. Am. Phys. Soc.* **33**, 498 (1988).

21. G. Ten Brinke, D. Ausserre, and G. Hadziioannou, *J. Chem. Phys.* **89**, 4374 (1988).

22. K. F. Mansfield and D. N. Theodorou, *Macromolecules* **22**, 3143 (1989).

23. T. Pakula, *J. Chem. Phys.* **95**, 4685 (1991).

24. A. Yethiraj, C. K. Hall, and R. Dickman, *J. Colloid Interface Sci.* **151**, 102 (1992).

25. R. Dickman and A. Yethiraj, *J. Chem. Phys.* **100**, 4683 (1994).

26. B. C. Freasier and C. E. Woodward, *Comput. Theoret. Polym. Sci.* **9**, 141 (1999).

27. S. Jorge and A. Rey, *J. Chem. Phys.* **106**, 5720 (1997).

28. R. Dickman and C. K. Hall, *J. Chem. Phys.* **89**, 3168 (1988).

29. S. K. Kumar, M. Vacatello, and D. Y. Yoon, *J. Chem. Phys.* **89**, 5209 (1988).

30. A. Yethiraj and C. K. Hall, *J. Chem. Phys.* **91**, 4827 (1989).

31. A. Yethiraj and C. K. Hall, *Macromolecules* **23**, 1635 (1990).

32. I. Bitsanis and G. Hadziioannou, *J. Chem. Phys.* **92**, 3827 (1990).

33. S. K. Kumar, M. Vacatello, and D. Y. Yoon, *Macromolecules* **23**, 2189 (1990).

34. A. Yethiraj and C. K. Hall, *Macromolecules* **24**, 709 (1991).

35. A. Yethiraj and C. K. Hall, *J. Chem. Phys.* **95**, 1999 (1991).

36. A. Yethiraj, *J. Chem. Phys.* **101**, 2489 (1994).

37. A. Yethiraj and C. E. Woodward, *J. Chem. Phys.* **102**, 5499 (1995).

38. S. Phan, E. Kierlik, M. L. Rosinberg, A. Yethiraj, and R. Dickman, *J. Chem. Phys.* **102**, 2141 (1995).

39. A. Yethiraj, *J. Chem. Phys.* **109**, 3269 (1998).

40. A. Yethiraj, *J. Chem. Phys.* **102**, 6874 (1995).

41. W. Zhao, X. Zhao, M. H. Rafailovich, J. Sokolov, R. J. Composto, S. D. Smith, M. Satkowski, T. P. Russell, W. D. Dozier, and T. Mansfield, *Macromolecules* **26**, 561 (1993).

42. A. Yethiraj and C. K. Hall, *Mol. Phys.* **73**, 503 (1991).

43. I. K. Snook and W. van Megen, *J. Chem. Phys.* **72**, 2907 (1980).

44. M. J. Grimson, *Chem. Phys. Lett.* **180**, 129 (1991).

45. K. Solc and W. H. Stockmayer, *J. Chem. Phys.* **54**, 2756 (1971).

46. M. R. Wilson and M. P. Allen, *Mol. Phys.* **80**, 277 (1993).

47. R. L. Jones, S. K. Kumar, D. L. Ho, R. M. Briber, and T. P. Russell, *Nature* **400**, 146 (1999).

48. A. Milchev and K. Binder, *J. Phys. (Paris) II,* **6**, 21 (1996).

49. A. Milchev and K. Binder, *Eur. Phys. J. B* **3**, 477 (1998).

50. J. S. Wang and K. Binder, *J. Phys. I* **1**, 1583 (1991).

51. A. Milchev, W. Paul, and K. Binder, *Macromol. Theory Simul.* **3**, 305 (1994).

52. A. Milchev and K. Binder, *Macromolecules* **29**, 343 (1996).

53. K. Binder, A. Milchev, and J. Baschnagel, *Annu. Rev. Mater. Sci.* **26**, 107 (1996).

54. R. B. Pandey, A. Milchev, and K. Binder, *Macromolecules* **30**, 1194 (1997).

55. C. N. Patra, A. Yethiraj, and J. G. Curro, *J. Chem. Phys.* **111**, 1608 (1999).

56. U. Steiner, J. Klein, E. Eiser, A. Budkowski, and L. J. Fetters, *Science* **258**, 1126 (1992).

57. M. Sikka, N. Singh, A. Karim, and F. S. Bates, *Phys. Rev. Lett.* **70**, 307 (1993).

58. A. Yethiraj, *Phys. Rev. Lett.* **74**, 2018 (1995).

59. A. Yethiraj, S. K. Kumar, A. Hariharan, and K. S. Schweizer, *J. Chem. Phys.* **100**, 4691 (1994).

60. S. K. Kumar, A. Yethiraj, K. S. Schweizer, and F. A. M. Leermakers, *J. Chem. Phys.* **103**, 10332 (1995).

61. L. F. Vega, A. Z. Panagiotopoulos, and K. E. Gubbins, *Chem. Eng. Sci.* **49**, 2921 (1994).

62. H. Tang, I. Szleifer, and S. K. Kumar, *J. Chem. Phys.* **100**, 5367 (1994).

63. F. A. Escobedo and J. J. de Pablo, *J. Chem. Phys.* **105**, 4391 (1996).

64. B. Svensson and C. E. Woodward, *Mol. Phys.* **87**, 1363 (1996).

65. M. W. Ribarsky and U. Landman, *J. Chem. Phys.* **97**, 1937 (1992).

66. F. Elhebil and S. Premilat, *Biophys. Chem.* **42**, 195 (1992).

67. R. G. Winkler and R. Hentschke, *J. Chem. Phys.* **99**, 5405 (1993).

68. Y. Wang, K, Hill, and J. G. Harris, *Langmuir* **9**, 1983 (1993).

69. I. A. Bitsanis and C. M. Pan, *J. Chem. Phys.* **99**, 5520 (1993).

70. M. Vacatello, *Mol. Simul.* **13**, 245 (1994).

71. Y. T. Wang, K. Hill, and J. G. Harris, *J. Chem. Phys.* **100**, 3276 (1994).

72. P. Padilla and S. Toxvaerd, *J. Chem. Phys.* **101**, 1490 (1994).

73. T. Matsuda, G. D. Smith, R. G. Winkler, and D. Y. Yoon, *Macromolecules* **28**, 165 (1995).

74. J. Gao, W. D. Luedtke, and U. Landman, *J. Chem. Phys.* **106**, 4309 (1997) and references therein.

75. R. G. Winkler, R. H. Schmid, A. Gerstmair, and P. Reineker, *J. Chem. Phys.* **104**, 8103 (1996).

76. R. K. Ballamudi and I. A. Bitsanis, *J. Chem. Phys.* **105**, 7774 (1996).

77. J. C. Wang and K. A. Fichthorn, *J. Chem. Phys.* **108**, 1653 (1998).

78. J. N. Israelachvili and S. J. Kott, *J. Chem. Phys.* **88**, 7162 (1988).

79. R. S. Khare, J. J. de Pablo, and A. Yethiraj, *Macromolecules* **29**, 7910 (1996).

80. R. S. Khare, J. J. de Pablo, and A. Yethiraj, *J. Chem. Phys.* **107**, 2589 (1997).

81. A. Yethiraj, R. Khare, and J. J. de Pablo, *Polym. Mater. Sci. Eng.* **77**, 642 (1997).

82. R. S. Khare, J. J. de Pablo, and A. Yethiraj, *J. Chem. Phys.* (in press).

83. M. J. Stevens, M. Mondello, G. S. Grest, S. T. Cui, H. D. Cochran, and P. T. Cummings, *J. Chem. Phys.* **106**, 7303 (1997).

84. S. A. Gupta, H. D. Cochran, and P. T. Cummings, *J. Chem. Phys.* **107**, 10335 (1997).

85. A. Koike, *Macromolecules* **31**, 4605 (1998); A. Koike and M. Yoneya, *J. Phys. Chem. B* **102**, 3669 (1998).

86. A. Jabbarzadeh, J. D. Atkinson, and R. I. Tanner, *J. Non-Newtonian Fluid Mech.* **77**, 53 (1998).

87. A. Koike, *J. Phys. Chem. B* **103**, 4578 (1999).

88. D. A. McQuarrie, *Statistical Mechanics*, Harper and Row, New York, 1979.

89. J.-P. Hansen and I. R. McDonald, *Theory of Simple Liquids*, Academic, New York, 1986.

90. J. G. Curro and K. S. Schweizer, *J. Chem. Phys.* **87**, 1842 (1987). For a review, see K. S. Schweizer and J. G. Curro, *Adv. Chem. Phys.* **98**, 1 (1997).

91. D. Chandler and H. C. Andersen, *J. Chem. Phys.* **57**, 1930 (1972).

92. D. Henderson, F. F. Abraham, and J. A. Barker, *Mol. Phys.* **31**, 1291 (1975).

93. Y. Zhou and G. Stell, *Mol. Phys.* **66**, 767 (1989).

94. A. Yethiraj and C. K. Hall, *J. Chem. Phys.* **95**, 3749 (1991).

95. J. G. Curro, J. D. Weinhold, J. D. McCoy, and A. Yethiraj, *Comput. Theoret. Polym. Sci.* **8**, 159 (1998).

96. A. Yethiraj, *ACS Symp. Ser.* **629**, 274 (1996).

97. A. Yethiraj, J. G. Curro, K. S. Schweizer, and J. D. McCoy, *J. Chem. Phys.* **98**, 1635 (1993).

98. A. Yethiraj, *J. Chem. Phys.* **111**, 1797 (1999).

99. A. J. Milling, *J. Phys. Chem.* **100**, 9896 (1996).

100. B. Kolaric, W. Jaeger, and R. von Klitzing, *J. Phys. Chem. B* **104**, 5096 (2000).

101. A. Yethiraj, *Comput. Theoret. Polym. Sci.* **10**, 115 (2000).

102. B. B. Laird, R. B. Ross, and T. Ziegler, eds., *Chemical Applications of Density-Functional Theory*, ACS Symposium Series, Vol. 629, American Chemical Society, Washington, D.C., 1996.

103. R. Evans, in *Fundamentals of Inhomogeneous Fluids*, D. Henderson, ed., Dekker, New York, 1992.

104. J. K. Percus, *Acc. Chem. Res.* **27**, 224 (1994).

105. M. J. Grimson and G. Rickayzen, *Mol. Phys.* **42**, 47 (1981).

106. R. Evans, P. Tarazona, and U. M. B. Marconi, *Mol. Phys.* **50**, 993 (1983).

107. R. Evans and U. M. B. Marconi, *Phys. Rev. A* **34**, 3504 (1986).

108. Y. Rosenfeld, *J. Phys. Cond. Mater.* **8**, 9289 (1996); Y. Rosenfeld and P. Tarazona, *Mol. Phys.* **95**, 141 (1998).

109. S. Nordholm, M. Johnson, and B. C. Freasier, *Aust. J. Chem.* **33**, 2139 (1980).

110. P. Tarazona, *Phys. Rev. A* **31**, 2672 (1985).

111. G. Stell, in *The Equilibrium Theory of Classical Fluids*, H. L. Frisch and J. L. Lebowitz, eds., W. A. Benjamin, New York, 1964.

112. W. A. Curtin and N. W. Ashcroft, *Phys. Rev. A* **32**, 2909 (1985).

113. P. C. Ball and R. Evans, *Mol. Phys.* **63**, 159 (1988).

114. Z. Tang, L. Mier-y-Teran, H. T. Davis, L. E. Scriven, and H. S. White, *Mol. Phys.* **71**, 369 (1990); Z. Tang, L. E. Scriven, and H. T. Davis, *J. Chem. Phys.* **97**, 9258 (1992).

115. C. N. Patra and S. K. Ghosh, *Phys. Rev. E* **47**, 4088 (1993).

116. D. Chandler, J. D. McCoy, and S. J. Singer, *J. Chem. Phys.* **85**, 5971 (1986).

117. D. Chandler, J. D. McCoy, and S. J. Singer, *J. Chem. Phys.* **85**, 5977 (1986).

118. W. E. McMullen and K. F. Freed, *J. Chem. Phys.* **92**, 1413 (1990).

119. W. E. McMullen, *J. Chem. Phys.* **95**, 8507 (1991).

120. H. Tang and K. F. Freed, *J. Chem. Phys.* **94**, 1572 (1991).

121. C. Seok and D. W. Oxtoby, *J. Chem. Phys.* **109**, 7982 (1998).

122. S. Sen, J. M. Cohen, J. D. McCoy, and J. G. Curro, *J. Chem. Phys.* **101**, 9010 (1994).

123. S. K. Nath, J. D. McCoy, J. P. Donley, and J. G. Curro, *J. Chem. Phys.* **103**, 1635 (1995).

124. J. D. McCoy and S. K. Nath, *ACS Symp. Ser.* **629**, 246 (1996).

125. J. B. Hooper, J. D. McCoy, and J. G. Curro, *J. Chem. Phys.* **112**, 3090 (2000).

126. E. Kierlik and M. L. Rosinberg, *J. Chem. Phys.* **100**, 1716 (1994).

127. E. Kierlik and M. L. Rosinberg, *J. Chem. Phys.* **99**, 3950 (1993).

128. E. Kierlik and M. L. Rosinberg, *J. Chem. Phys.* **100**, 1716 (1994).

129. E. Kierlik, S. Phan, and M. L. Rosinberg, *ACS Symp. Ser.* **629**, 212 (1996).

130. M. S. Wertheim, *J. Stat. Phys.* **35**, 19 (1984).

131. C. E. Woodward, *J. Chem. Phys.* **94**, 3183 (1991).

132. C. E. Woodward, *J. Chem. Phys.* **97**, 695 (1992).

133. C. E. Woodward, *J. Chem. Phys.* **97**, 4525 (1992).

134. C. E. Woodward and A. Yethiraj, *J. Chem. Phys.* **100**, 3181 (1994).

135. B. C. Freasier and C. E. Woodward, *Comput. Theoret. Polym. Sci.* **9**, 141 (1999).

136. A. Yethiraj, *Chem. Eng. J* **74**, 109 (1999).

137. C. N. Patra and A. Yethiraj, *J. Chem. Phys.* **112**, 1579 (2000).

138. S. K. Nath, P. F. Nealey, and J. J. de Pablo, *J. Chem. Phys.* **110**, 7483 (1999).

139. A. Yethiraj, *J. Chem. Phys.* **101**, 9104 (1994).

# MORPHOLOGY OF SURFACES IN MESOSCOPIC POLYMERS, SURFACTANTS, ELECTRONS, OR REACTION–DIFFUSION SYSTEMS: METHODS, SIMULATIONS, AND MEASUREMENTS

ALEKSIJ AKSIMENTIEV

*Computer Science Department, Material Science Laboratory, Mitsui Chemicals, Inc., Sodegaura-City, Chiba, Japan*

MARCIN FIAŁKOWSKI

*Institute of Physical Chemistry, Polish Academy of Science and College of Science, Department III, Warsaw, Poland*

ROBERT HOŁYST

*Institute of Physical Chemistry, Polish Academy of Science and College of Science, Department III, Warsaw, Poland; and Labo de Physique, Ecole Normale Superieure de Lyon, Lyon, France*

## CONTENTS

*Advances in Chemical Physics, Volume 121*, Edited by I. Prigogine and Stuart A. Rice.
ISBN 0-471-20504-4  © 2002 John Wiley & Sons, Inc.

## I. INTRODUCTION

Surfaces are ubiquitous in nature because they accompany every phase transition [1–4]. For example, let us consider a symmetric A–B homopolymer blend. If we quench the mixture from the one-phase region (where both components are well-mixed) to the thermodynamically unstable region (spinodal region, where the spontaneous demixing occurs) by the temperature quench or a rapid evaporation of the solvent, we observe a formation of A-rich and B-rich domains separated by an interface. The system is bicontinuous; A-rich and B-rich domains percolate in the system, and the single interface separates them. The growth of the domains is strongly related to the geometrical properties of the interface, especially local mean curvature and global curvature distribution. The study of surfaces in such systems were in the past hampered by the lack of adequate experimental tools and theoretical methods. The breakthrough was achieved in 1997 when Jinnai et al. [5–7] from the Hashimoto group developed a technique for the direct visualization of the interface in the three-dimensional (3D) polymer system undergoing spinodal decomposition. From the 3D image obtained by the use of the laser scanning confocal microscopy (LSCM), the authors were able to obtain the curvature distribution of the interface and the scaling of the Gaussian and mean curvatures. The same authors combined the methods for the 3D reconstruction developed in LSCM experiments with spinodally decomposing blends and transmission electron microtomography (TEMT) to visualize the 3D structure of the gyroid phase in the triblock copolymer system and to measure its distribution of curvatures [8]. The theoretical studies followed soon [9]. In particular, application of methods of topology and geometry of surfaces to the asymmetric and symmetric polymer blends revealed the scaling of the Euler characteristic and allowed scientists to study the transition between bicontinuous morphology characteristic for symmetric system and droplet morphology characteristic for the asymmetric blends [9,10]. This chapter is devoted to the presentation of various theoretical tools for the study of surfaces in connection with specific examples—for example, in polymer and surfactant systems, electron liquid gas systems, or reaction–diffusion systems. The precise purpose is to present the methods for the very accurate calculation of the Gaussian and mean curvatures, Euler characteristic, surface area, and volume fractions for systems with internal surfaces such as diblock copolymer systems or polymer blends undergoing phase separating transition (spinodal decomposition) and discuss various implications of the present studies.

In what follows we will discuss systems with internal surfaces, ordered surfaces, topological transformations, and dynamical scaling. In Section II we shall show specific examples of mesoscopic systems with special attention devoted to the surfaces in the system—that is, periodic surfaces in surfactant systems, periodic surfaces in diblock copolymers, bicontinuous disordered interfaces in spinodally decomposing blends, ordered charge density wave patterns in electron liquids, and dissipative structures in reaction–diffusion systems. In Section III we will present the detailed theory of morphological measures: the Euler characteristic, the Gaussian and mean curvatures, and so on. In fact, Sections II and III can be read independently because Section II shows specific models while Section III is devoted to the numerical and analytical computations of the surface characteristics. In a sense, Section III is robust; that is, the methods presented in Section III apply to a variety of systems, not only the systems shown as examples in Section II. Brief conclusions are presented in Section IV.

## A.  Ordered, Periodic Surfaces

The periodic surface is the surface that moves onto itself under a unit translation in one, two, or three coordinate directions similarly as in the periodic arrangement of atoms in regular crystals. The most interesting are triply periodic surfaces that are periodic in all three directions forming structures that have various crystallographic symmetries. Here we mostly concentrate on smooth surfaces with cubic symmetry.

The paradigm structures for all periodic surfaces are minimal periodic surfaces. A patch of the minimal surface can be visualized in a simple experiment of Plateau (nineteenth-century Belgian physicist): Use a soap solution and dip a metal frame (not necessarily planar) in it. The film that forms on the frame will adopt the shape such as to minimize the surface free energy—that is, to minimize the area of the surface (hence the term *minimal surface*). Such surfaces have zero mean curvature at every point of the surface as was shown by Meusnier and Laplace in the eighteenth century [11]. Surfaces are characterized locally by the Gaussian, $K$, and mean, $H$, curvature. $K$ and $H$ are given by the following equations: $H = \frac{1}{2}(1/R_1 + 1/R_2)$, $K = 1/(R_1 R_2)$, where $R_1$ and $R_2$ are principal radii of curvature (Fig. 1). (the full presentation of curvatures is given in Section III). In the saddle part of the surface such as the one shown on Fig. 1, the principal radii of curvatures have different signs. The four typical surface motifs are as follows: spherical ($H > 0$, $K > 0$), cylindrical ($H > 0$, $K = 0$), planar ($H = 0$, $K = 0$) and saddle ($K < 0$, $H = 0$). The curvature of the surface is related to thermodynamic quantities by the Poisson–Laplace equation $H = \Delta P/2\sigma$, where $H$ is the mean curvature at the interface of two homogeneous media, $\Delta P$ is the pressure difference across the interface, and $\sigma$ is the surface tension.

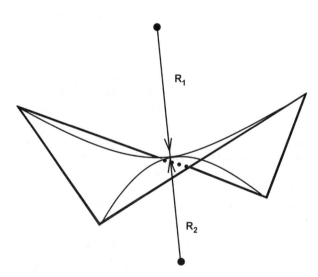

**Figure 1.** Piece of a hyperbolic surface. $R_1$ and $R_2$ are the principal radii.

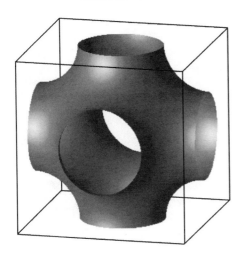

**Figure 2.** Minimal Schwartz cubic surface P.

The first periodic (in one direction only) minimal surface [12] discovered in 1776 was a helicoid: The surface was swept out by the horizontal line rotating at the constant rate as it moves at a constant speed up a vertical axis. The next example (periodic in two directions) was discovered in 1830 by Herman Scherk. The first triply periodic minimal surface was discovered by Herman Schwarz in 1865. The P and D Schwarz surfaces are shown in Figs. 2 and 3. The revival of interest in periodic surfaces was due to (a) the observation[13–16] that at suitable thermodynamic conditions, bilayers of lipids in water solutions form triply periodic surfaces and (b) the discovery of new triply periodic minimal

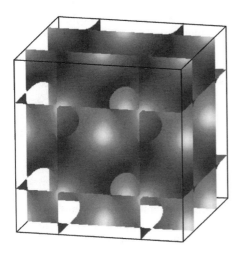

**Figure 3.** Minimal Schwartz diamond surface D.

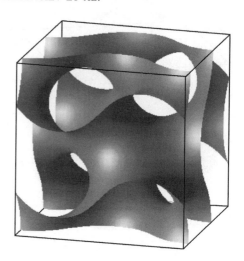

**Figure 4.** Minimal Schoen gyroid surface G.

surfaces (among them the gyroid phase, Fig. 4) by mathematician Schoen [17]. If we draw a surface through the middle of the triply periodic lipid bilayer, it follows from the geometrical constraints that it must be a triply periodic minimal surface [18].

## B.  Surfaces in Surfactant Systems

In the ordered ternary mixture of water, oil, and surfactant the surface covered with the surfactants divides volume into the water-rich and the oil-rich subvolumes. The diffusion measurements provide direct information about the bicontinuity of the structure and therefore about the existence of the periodic surface. The pulsed gradient nuclear magnetic resonance (NMR) self-diffusion technique gives the self-diffusion rates of all the components in the structure [19], providing a direct check on the continuity and the extension of a region occupied by the components. If any of the components of the system is closed in a finite (small) volume, then its effective diffusion constant (measured as a mean-squared displacement divided by time) approaches zero, whereas in a continuous structure the effective diffusion coefficient is nonzero. The X-ray scattering provides information about the symmetry of the structure and there-fore discerns between different periodic surfaces. Here we concentrate on the cubic symmetries. The set of the reflections for the P cubic structure ($Im\bar{3}m$ symmetry, Fig. 2) index to $\sqrt{2} : \sqrt{4} : \sqrt{6} : \sqrt{10} : \sqrt{12} : \sqrt{14} : \sqrt{16}$, that of the D structure ($Pn\bar{3}m$ symmetry Fig. 3) index to $\sqrt{2} : \sqrt{3} : \sqrt{4} : \sqrt{6} : \sqrt{8} : \sqrt{9} : \sqrt{10}$, and that of the G structure ($Ia\bar{3}d$ symmetry, Fig. 4) index to $\sqrt{6} : \sqrt{8} : \sqrt{14} : \sqrt{16} : \sqrt{20} : \sqrt{22} : \sqrt{24}$. Finally the measurements of the surface area inside the cubic cell can yield partial information about a topology of the structure. One can expect a small unit cell and the small surface area per side of the unit cell in structures of the simple topology [20] and the large values for the structures of complex topology. Typically the surface area per side of the cubic cell is between 2 and 4 for simple topology surfaces such as P, D, G, I-WP [17], O-CT0 [17], and so on, but it can be larger than 7 for the complex topology surfaces[21–23].

In binary mixtures of water, surfactants, or lipids the most common structure is the gyroid one, G, existing usually on the phase diagram between the hexagonal and lamellar mesophases. This structure has been observed in a very large number of surfactant systems [13–16,24–27] and in the computer simulations of surfactant systems [28]. The G phase is found at rather high surfactant concentrations, usually much above 50% by weight.

Other cubic phases can be found even at very low concentration of surfactants in ternary mixtures (with water and oil) [20,29,30]. For the didodecyldimethylammonium bromide–water–styrene system the periodic surfaces are found over the huge range of water fraction from 11% to over 80%. However, when the volume fractions of the oil and the water are equal, one finds the cubic phases in a narrow window of the surfactant concentration around 0.5 weight fraction [31,32]. Most of the studies concentrated on the cases with large amounts of water and surfactant and the influence of added oil on the phase transitions. It was shown that the emerging phases depend on the properties of the oil—that is, whether it penetrates the surfactant or swells the bilayer. In the highly asymmetric case of a large amount of water and surfactant and a small amount of oil, it has been found that with increasing the volume fraction of water, one finds the following progression of cubic phases: G → D → P. These studies are time-consuming due to the very long equilibration times of weeks or even months. Nevertheless, the experimental data are very rich in this case [31–35].

## C.  Surfaces in Copolymer Systems

An A–B diblock copolymer is a polymer consisting of a sequence of A-type monomers chemically joined to a sequence of B-type monomers. Even a small amount of incompatibility (difference in interactions) between monomers A and monomers B can induce phase transitions. However, A-homopolymer and B-homopolymer are chemically joined in a diblock; therefore a system of diblocks cannot undergo a macroscopic phase separation. Instead a number of order–disorder phase transitions take place in the system between the isotropic phase and spatially ordered phases in which A-rich and B-rich domains, of the size of a diblock copolymer, are periodically arranged in lamellar, hexagonal, body-centered cubic (bcc), and the double gyroid structures. The covalent bond joining the blocks rests at the interface between A-rich and B-rich domains.

At the mesoscopic level of description the Landau–Ginzburg model of the phase transitions in diblock copolymer system was formulated by Leibler [36]

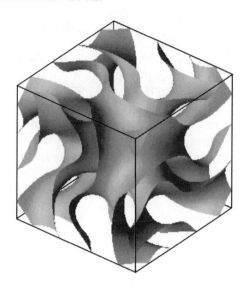

**Figure 5.**   The double gyroid surface DG.

and later refined by Fredrickson et al. [37]. Matsen et al. [38–40] applied the
self-consistent field theory to the phase transitions in this system. For $f = 0.5$
( $f$ is the volume fraction of A monomers in the system), only the lamellar phase
is stable with periodic stack of flat interfaces. As we decrease $f$, the double
gyroid phase appears (Fig. 5). This phase consists of two gyroid surfaces in a
unit cell (see Fig. 4). They form boundaries between A-rich and B-rich domains.
The phase has Ia$\bar{3}$d symmetry [41] and each surface has genus 5 per unit cell
[41,42]. Next a transition to the hexagonal phase and bcc phase is observed. In
the hexagonal phase the minority component forms cylinders arranged on the
hexagonal lattice and inserted in the matrix of the majority component. In the
bcc phase (symmetry Im$\bar{3}$m) the minority component forms spheres arranged on
the bcc lattice and surrounded by the majority component. The last phase that
appears in the system is the CPS phase, which consists of the spheres of the
minority phase arranged either on the face-centered cubic (fcc) or hexagonal
close-packed (hcp) lattice. The theoretical studies of the system were greatly
stimulated by experiments. The cylindrical, spherical, and lamellar phases have
been known for a long time [43], but only recently the novel bicontinuous
structures have been discovered in the diblock copolymer systems [44-46]. The
originally discovered bicontinuous ordered structure has been misidentified in
the transmission electron microscopy (TEM) experiments [46]; later it has been
shown that this structure is the double gyroid phase [47] (Fig. 5). The possible
number of different morphologies in block copolymers is infinite; many new
triblock copolymers have been found [48–50].

## D. Polymerization of Surfactant Surfaces

Apart from the same order found in surfactant systems and in the polymer systems, there is also a very deep connection between these systems at the technological level. The recent discovery that organic, ordered surfaces formed in water solutions by surfactants can be used as templates for organic or inorganic polymerization reactions where the final product is an ordered mesoporous material with well-defined pore sizes (ranging from 2 to 20 nm) and shapes [51] brought the subject of ordering in surfactant systems into the realm of modern polymer technology. The results are fascinating: Composite structures (organic silica) have been synthesized with lamellar [52–54], hexagonal [51,53,54], and cubic [51,53], hierarchical arrangement of tubules inside tubules [55] and other [56] topologies. Their possible applications range from contact lenses, slow drug release systems [56], and mesoporous sieves and catalysts [51–54] to mesoporous electrodes for batteries, fuel cells, electrochemical capacitors, sensors [57], and water purification systems [58]. The mesoporous silica structures are prepared in the water surfactant system [51]. After the polymerization of the silica oligomers the surfactant can be burned and the inorganic sieve emerges. The sieves are remarkable because they have the pore sizes ranging from 20 Å to hundreds of angstroms; the pore sizes have well-defined shape and size; the structure is ordered over micrometer length scale. The process is complicated and still not fully understood. It has been described [53] as follows: There is a multidentate binding of anionic silicate oligomers to the cationic surfactant, the polymerization takes place in the interfacial region, and there is a charge density matching between the surfactant and the silicate. The synthesis can be carried out under the conditions in which silicate alone does not condense (at pH from 12 to 14 and silicate concentration between 0.5% and 5%) and/or at the small concentration of surfactant [53,57]. It points out that at certain conditions the formation of the ordered structure results from the interplay between the silica and surfactants. The formation of the ordered structure at high temperature is fast in comparison to the polymerization reaction; the polymerized silica mesostructure obtained in this two-stage process is more stable than the one obtained at low temperature. The polymerization of silicate at the interface is favored because the surfactant partially screens the charges and the concentration of silicate species is high in this region. There is indication that the silicate and the polymerization process may take part in the ordering process especially when the polymerization is fast enough. Apart from the lamellar and hexagonal microstructures, the bicontinuous gyroid structure is formed. For a high fraction of both components in the $CTA–SiO_2$ system, the single infinite silicate sheet separates the surfactant into two equal and disconnected subvolumes. The midplane of the silicate sheet is the minimal gyroid periodic surface.

## E. Periodic Surfaces in Other Systems

The formation of periodic surfaces in etioplast in plants and the usage of periodic surfaces for the crystallization of high-molecular-weight membrane proteins [59] shows that these surfaces are related to biological problems. After the discovery of various structures formed by carbon (i.e., wires, fullerenes), it has been also shown theoretically that the carbon atoms can be arranged to form triply periodic surfaces [60,61]. In ionic crystals, the points at which the electrostatic potential is constant comprise the periodic surface; its symmetry can be determined by the symmetry adapted from the distribution of charges [62]. It is also shown that there is a very strong connection between chemical structures and periodic surfaces [63,64].

The chroloplasts are organelles involved in photosynthesis, presenting a complex system of parallel bilayers. When the plant is grown for several days in the absence of light, the lamellar structure of chloroplast transforms into the periodic surface of the same topology and symmetry as the P Schwarz surface (named also a prolamellar body or etioplast), although it is not a minimal surface [65–67]. The cell membranes in living organisms—for example, endoplasmic reticulum (where cell proteins and lipids are synthesized)—are known to have a complex three-dimensional morphology, which, as shown by Landh [68], is the same as for the triply periodic minimal surfaces. Three fundamental structures identified so far in cellular organelles are: gyroid G Schoen surface and D and P Schwarz surfaces. The partition of space by the surface in the cell organelles enables cells to control the concentration of various molecules and their transport across the bilayers. There are indications that multiply continuous structures (like, e.g., double gyroid phase; see Fig. 5) exists in mitochondria of the very large cone cells in tree shrews [69]. The skeleton elements of the sea stars, sea urchins, or sea cucumbers are the monocrystals of a magnesium-rich calcite, where each crystal is not limited by the planar surface but is, instead, limited by the surface of the topology of P surface [70].

The periodic zero potential surfaces (POPS) occur most naturally in the ionic crystals, where we have a periodic distribution of charges [62]. Although POPS are not minimal surfaces [64,71], they may resemble them closely [63]. The cesium chloride CsCl structure has the same topology and symmetry as the P Schwarz surface and is called $P^*$ surface to distinguish it from the P minimal surface. The aluminosilicate of the zeolite sodalite ($Na_4Al_3Si_{12}O_{26}Cl$) form a $P^*$ surface, similarly as the perovskite structure ($CaTiO_3$). The $D^*$ surface can be found in the zeolite structure of faujasite [72] ($M_{57}^IAl_{57}Si_{135}O_{384}^{[53]}$), similarly to the cubic Laves phase $MgCu_2$. Finally the $Y^{**}$ the analog of the G surface is found in the compounds with the garnet structure $Ca_3[Al_2Si_3O_{12}]$, materials important for their applications as optoelectric color displays and magnetic bubble memory. Also in the $Ta_6Cl_{15}$ compound the surface is found. A detailed description is given by von Schnering and Nesper [62].

Mackay and Terrones [73] predicted that a new kind of ordered graphite foam, related to fullerenes with topology similar to periodic surfaces, may be found. These new structures are constructed by introducing seven- or eight-membered ring of the carbon atoms into sheets of six-membered rings, thus giving saddle-shape surfaces. In fact, such structures should be more stable than fullerenes [60] $C_{60}$ mainly because the $120°$ bond angles in ordinary graphite are almost preserved in the seven-membered and eight-membered carbon rings. In this way P, D, G, and I-WP surfaces were constructed of energies lower than the $C_{60}$ form of carbon. The bending and saddle-splay modulus [74] for these structures [61] is 190 kT and $-65$ kT, making the graphite surfaces an order of magnitude stiffer than bilayers of amphiphilic system [75].

## F.  Topological Transformations

Many natural phenomena involve the topological changes of surfaces in a system. In order to describe such phenomena, one uses the Euler characteristic and/or genus, because they describe quantitatively the topology of the surface. The genus, $g$, of a closed surface is equal to the number of holes in it. Therefore the Euler characteristic $\chi = 2(1 - g)$ is 2 for a sphere (since $g = 0$), 0 for a torus ($g = 1$) and –2 for two tori joined by a handle (passage) ($g = 2$). We shall present more technical description of the computation of $\chi$ in Section III. The Euler characteristic for a system of disjoint surfaces is equal to the sum of the Euler characteristic of individual surfaces. If we join two surfaces by a passage, the Euler characteristic of a system will change by $-2$, which is easy to see. Let us take two spheres for which we have $\chi = 4$. If we join them by a passage, we will get a single closed surface with $\chi = 2$. Therefore a passage changed $\chi$ of the system by $-2$. If a droplet appears in a system, the Euler characteristic changes by $+2$. Therefore we have two typical topological elements: a droplet and a passage (Fig. 6).

For example, the process of spinodal decomposition is accompanied by topological transformation of the interface between the domains. The interface between the A-rich and B-rich domains has a large and negative Euler characteristic at the beginning of the spinodal decomposition, since the interface is highly interconnected and the initial structure is bicontinuous. At the end of the process the Euler characteristic is 0 because in the final equilibrium configuration, two phases are separated by a single flat interface (in an external field). When the system is quenched into the metastable region, the demixing proceeds via the Lifshitz–Slyozov–Wagner (LSW) mechanism— that is, growth of large droplets and disappearance of small droplets. In this case the Euler characteristic, initially large and positive, should reach 0 at the end of the process.

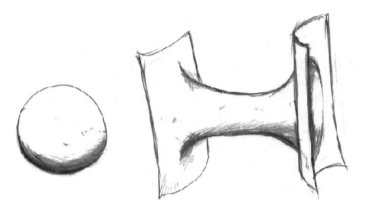

**Figure 6.**  Two typical topological elements: a droplet and a passage.

Another important example of topological transformations are topological fluctuations that should accompany every phase transition in systems with internal surfaces [76–82]. Near the phase boundaries we expect that the topology of the surface fluctuates. It follows from the liquid nature of the surfactant surfaces, which makes them susceptible to large thermal fluctuations in particular topological fluctuations. The latter occurs when two surfaces in the process of thermal undulations fuse [Fig. 7(a)], forming a passage between them and therefore changing their topology. The reverse situation is also possible when two pieces of a surface disconnect in a process of thermal fluctuations. The passages are rare far away from the phase boundaries of the ordered phases of surfactants, but are quite abundant close to these boundaries. For example, the microemulsion–lamellar phase transition [76] must be accompanied by topological fluctuations, since microemulsion is a bicontinuous disordered structure characterized by large and negative Euler characteristic, whereas the ideally ordered lamellar phase (consisting of parallel layers) has zero Euler characteristic (in the system with periodic boundary conditions). Therefore the change of topology should force the appearance of passages and/or droplets. In Fig. 7(b) we see the lamellar phase very close to the transition to the disordered microemulsion. The Euler characteristic is large and negative. The clear experimental evidence of such fluctuations has been recently found in FRAP experiments for surfactant systems [82]. The five-fold increase of the diffusion coefficient perpendicular to layers in the lamellar phase near the lamellar-sponge phase boundary indicated the appearance of the connections (passages) between the layers [82]. The topological fluctuations should give a clear signature in the off-specular scattering of neutrons or X-rays in a form of a

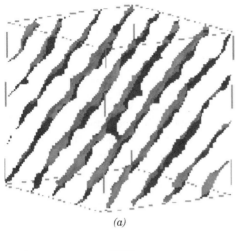

(a)

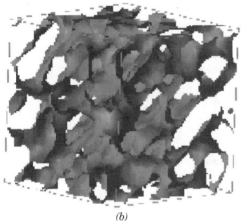

(b)

**Figure 7.** Topological fluctuations of the lamellar phase at different points of the phase diagram. (a) Single fusion between the lamellae by a passage (this configuration is close to the topological disorder line). (b) Configuration close to the transition to the disordered micro-emulsion phase; the Euler characteristic is large and negative.

scattering peak at nonzero value of the scattering wavevector [83]. It has been also found in computer simulations that due to such fluctuations the phase transition should be marked by a peak in the variance of the Euler characteristic and a jump in the average Euler characteristic [76]. It means that the Euler characteristic and its variance can be used for the study of phase transitions in the systems with internal surfaces in the same way as the internal energy and heat capacity are usually used for the same studies of phase transitions. As we can see, the Euler characteristic provides information about the average topology of surfaces and about topological fluctuations and can be used to locate phase transitions.

## G. Dynamical Scaling Hypothesis and Growth Exponents

The systems undergoing phase transitions (like spinodal decomposition) often exhibit scaling phenomena [1–4]; that is, a morphological pattern of the domains at earlier times looks statistically similar to a pattern at later times apart from the global change of scale implied by the growth of $L(t)$—the domain size. Quantitatively it means, for example, that the correlation function of the order parameter (density, concentration, magnetization, etc.)

$$g(\mathbf{r}, t) = g(\mathbf{r}/L(t)) \tag{1}$$

where

$$L(t) \sim t^n \tag{2}$$

(the characteristic length scale in the system) scales algebraically with time $t$, with the exponent $n$ different for different universality classes [1]. The Fourier transform of the correlation function gives the scattering intensity that can be represented by the following scaling form:

$$S(q, t) = L^d(t) Y[qL(t)] \tag{3}$$

where $q$ is the scattering wavevector and $Y$ is the scaling function. Assuming the scaling hypothesis, we can derive all the scaling laws for different morphological measures such as the Euler characteristic, $\chi(t)$, surface area, $S(t)$, the distribution of the mean, $P_H(H, t)$, and Gaussian, $P_K(K, t)$, curvatures. The scaling hypothesis implies the following scaling laws for any phase separating/ordering symmetric system irrespective of the universality class:

$$S(t) \sim L(t)^{-1} \tag{4}$$

$$\chi(t) \sim L(t)^{-d} \tag{5}$$

$$P_H(H, t) = P_H^*[HL(t)]/L(t) \tag{6}$$

$$P_K(K, t) = P_K^*[KL(t)^{(d-1)}]/L(t)^{(d-1)} \tag{7}$$

where $d$ is the dimensionality of the system. The first law follows from the congruence of the domains [84]. The scaling law (5) follows from the Gauss–Bonnet theorem [85]

$$\chi = \gamma_T \int K(S)\, dS \tag{8}$$

where $\int dS$ denotes the integral over the surface and $\gamma_T$ is twice the inverse of the volume of a $(d-1)$-dimensional sphere of the unit radius ($\gamma_T = 1/2\pi$ for $d = 3$). Since

$$K(t) \sim L(t)^{-d+1} \tag{9}$$

and $S(t) \sim L(t)^{-1}$, we find scaling [Eq. (5)].

The probability densities $P_H(H, t)$ and $P_K(K, t)$ are normalized to unity. Equation (6) is a simple consequence of the scaling of the mean curvature:

$$H(t) \sim L(t)^{-1} \tag{10}$$

The last relation results from the scaling [Eq. (9)] of the Gaussian curvature. For $d = 2$ the scalings (6) and (7) are equivalent. It should be noted that the four scaling laws given by Eqs. (4)–(7) are robust and apply to any symmetric system exhibiting phase ordering/separating kinetics.

As we can see, the exponent $n$ for the algebraic growth of $L(t)$ determines the behavior of all other quantities at the late stage of spinodal decomposition. Let us discuss once again the spinodal decomposition from the point of view of scaling and growth of $L(t)$. When the homogeneous binary $AB$ mixture (polymer blend) above its critical point is suddenly cooled (quenched) below its critical temperature, it ceases to be in the thermodynamical equilibrium and starts to demix. The homogeneous state can now be either a metastable or unstable state. In the case of the metastable state the process of demixing requires, in the first place, nucleation of droplets of the minority phase, say $A$-rich phase. Then the droplets starts to grow. At first they grow independently and their radius, $L(t)$ changes with time, $t$, according to the formula $L(t) \sim t^{1/2}$. This behavior has been observed in binary polymer blends [86]. At the later stage a significant fraction of A molecules disappear from the homogeneous mixture (most of them form droplets) and competitive growth starts. Small droplets decrease in size and the A molecules diffuse toward large droplets that grow. This mechanism is known as "evaporation–condensation" mechanism and is described by the Lifshitz–Slyozov–Wagner (LSW) law [1] for the growth rate of large droplets, $L(t) \sim t^{1/3}$. When the system is quenched into the thermodynamically unstable region, the demixing proceeds via the spinodal decomposition mechanism. Early stages of this process are described by the Cahn–Hilliard theory. According to this theory, the system becomes unstable with respect to small fluctuations of wavevector $q$ smaller than some value $q_0$. The key prediction of the theory is the exponential growth of the scattering intensity $S(q, t)$ in time with a well-defined maximum at $q_{max} = q_0/\sqrt{2}$. Interpenetrating $A$-rich and $B$-rich domains of the size of $L \sim 1/q_{max}$ are formed. The mixtures of simple liquids undergoing spinodal decomposition do not remain for a long time in their unstable early configuration, contrary to the high-molecular-weight polymer blends. The latter are very viscous liquids and the whole kinetics of phase separation is very slow, allowing a detailed experimental observation of the process [87]. The late-stage configuration depends on the initial volume fractions, but not on the early-stage mechanism of demixing. In the late stage we have either (a) a system of droplets of one of the phases in the sea of another or (b) an interpenetrating network of A-rich and B-rich domains that coarsen in time. The growth mechanism in the first case combines the evaporation–condensation described by the LSW law with the Binder–Stauffer mechanism of collisions and coalescence driven by the thermal fluctuations. Both mechanisms lead to the same scaling law, $L(t) \sim t^{1/3}$. In the case of bicontinuous morphology, if the coarsening proceeds via the flow induced by the surface tension of the interface, the scaling law $L(t) \sim t$ holds. However, one should be very careful with all the scaling laws, since it may happen that before reaching the late stage of spinodal decomposition we may encounter several intermediate regimes with different scaling or even with no scaling at all.

## H. Sheared Morphologies

Our everyday experience gives examples how shearing of immiscible liquids decreases the characteristic lengthscale at which they are incompatible. If two liquids have different viscosities, a continuous matrix in equilibrium is usually formed by the less viscous fluid, while a more viscous phase is dispersed in droplets of nearly spherical shapes (the volume fraction of the less viscous phase must be sufficiently large to allow this geometrically). The droplet deformation and breakup in the flow are controlled by two dimensionless quantities [88]: (1) the ratio of viscous to capillary forces, or capillary number $Ca \equiv \dot{\gamma}\eta_s a/\Gamma$ ($a$ denotes the droplet size, $\Gamma$ is the surface tension, $\dot{\gamma}$ is the shear rate, and $\eta_s$ is the viscosity), and (2) the viscosity ratio. On shearing, the average size of the droplets decreases [89]. However, if the droplet viscosity is $\gtrsim 4$ times larger than matrix viscosity, the droplet deformation remains small and no breakup is observed. On the other hand, when the matrix phase is more viscous than the droplet phase, the droplets deforms greatly before breakup. The shearing of immiscible blend could not only reduce the average droplet size. Depending on the shear rate, viscosity ratio, and volume fractions of the components, a topological transformation between droplet and bicontinuous morphology can be induced [90]. The condition for bicontinuity can be expressed by the following empirical law:

$$\frac{\eta_1(\dot{\gamma})}{\eta_2(\dot{\gamma})} \sim \frac{f_1}{f_2} \tag{11}$$

where $f_1$ and $f_2$ are the phase volume fractions, and viscosities $\eta_1$ and $\eta_2$ are shear-dependent. This condition also defines the point of the phase inversion.

The topological transformations in an incompatible blend can be described by the dynamic "phase diagram" that is usually determined experimentally at a constant shear rate. For equal viscosities, a bicontinuous morphology is observed within a broad interval of the volume fractions. When the viscosity ratio increases, the bicontinuous region of the phase diagram shrinks. At large viscosity ratios, the droplets of a more viscous component in a continuous matrix of a less viscous component are observed practically for all allowed geometrically volume fractions.

These phenomena, attributed to any immiscible mixtures, are of an extreme importance for the polymer industry, because the most common technological process in which two distinct polymers are combined into a single material is an extrusion [91]—that is, an extensive mechanical mixing at high temperatures. Due to the strong incompatibility between different polymer species, the mixing at the molecular scale rarely occurs during the extrusion process. After the extrusion the blend is rapidly cooled down below the glass transition or crystallization temperature. Because the process of the phase separation and coalescence are very slow in high polymers, the resulting material will preserve morphology that has been developed during the processing, even after a further treatment, for example, by a press molding. The properties of a composite would combine the properties of the raw materials in a proportion determined by the blend morphology and the state of the interface, which makes the polymers blending a major technology for tuning material properties. For example, if one component of the blend exhibits a rubber elasticity, while the other is brittle, the resulting blend would be (a) elastic if the rubber polymer forms a continuous matrix or (b) brittle if the matrix phase is brittle. The mixing rules in the case of the bicontinuous morphology are peculiar, and no general quantitative description has been developed so far. Sometimes, the blending may have a synergetic effect on certain properties as, for example, in the case of high-impact polystyrene blends [92].

Due to the anisotropic nature of the shear flow, a considerable anisotropy is often observed in sheared morphologies [93,94], which also reflects on anisotropic material properties. An isotropic droplet morphology exposed to a shear flow can be transformed into a bicontinuous pattern, as shown in Figs. 8(a) and 8(b). The other consequence of such transformations is appearance of a certain direction along which the domains are aligned. Such anisotropy decreases significantly the percolation threshold for the minor component, which may also have some benefits for various applications—for example, for designing conductive polymer blends [94,95]. For a disperse morphology, not only the size but also the shape and the spatial arrangement of droplets modify significantly material properties—for example, the blend viscosity. If shear accompanies crystallization, the crystallite morphology could be significantly altered, which would modify some material properties: The optical rotational power of the

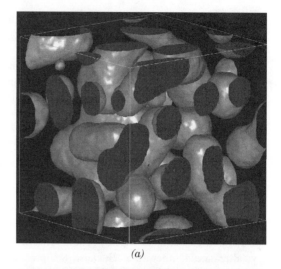

(a)

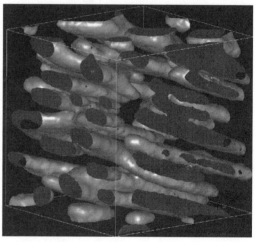

**Figure 8.** The isotropic droplet morphology (a) is transformed into the bicontinuous anisotropic morphology (b) under the shear.

(b)

birefringence, for example, could be increased a thousandfold [96]. A quantitative prediction of morphology development under shear for arbitrary volume fractions remains a challenging task, despite a considerable progress in theory [98,99,212] and simulations [99–101]. A detailed experimental investigation of the interface morphology by means of the LSCM is a necessity in order to establish the theory and simulation methods at this level.

Imposing the shear on self-assembling systems may induce ordered morphologies that are not observable in equilibrium [102]. For example, the lamellar phase in surfactant solutions can "roll" into multilayered vesicles forming the "onion" mesophase [103]. Shear also strongly affects kinetics of the ordering transitions [104,105] which can be used to produce high-ordered structures (without defects).

## I. Isosurfaces in Amorphous and Crystalline Polymers

One-component dense polymers at room temperatures are inhomogeneous even on the lengthscale less than a few repeated units. Depending on their chemical architecture, processing history, and other factors, they could be in an amorphous, glassy, or semicrystalline state. Their microscopic morphology determines most of their mechanical properties, along with the others, the gas solubility and gas diffusion. Different polymers of comparable densities and at the same temperature could have diffusion constants for the same gas up to several hundredfold different [106]. Dependence of the diffusion constant on the sort of a diffusant is peculiar and cannot be described by the simple diffusion relations. Different chemical architectures of polymers naturally lead to different microscopic structures in the amorphous state. The spatial characteristics of the volume available to the diffusant in the polymer matrix and the spatial distribution of this volume depend on the kind of polymer and diffusant, temperature, and overall polymer density. A realistic microscopic structure of amorphous polymers can be simulated by combining the Monte Carlo (MC) and molecular dynamics (MD) techniques. However, simulations of the diffusion itself is rather formidable task, since time intervals that can be probed within the MD simulation does not allow one to make a definitive statement about diffusion constant. Instead, one can investigate microstructures resulted from the MC/MD simulations and specify the morphology of the isosurfaces that confine the volume available to a diffusant [107]. Combining the knowledge about morphology of the free-volume clusters and the knowledge of their thermal motion, the diffusion constant for gas molecules can be estimated [108]. The free volume distribution in dense polymers can be confirmed experimentally by the positron-annihilation spectroscopy [109]. Computer simulations have shown that specifically for a diffusion constant problem, the shape of the clusters and their mutual spatial arrangement are very important [110]. The glass transition itself can be identified by the rapid free-volume shrinking.

The volume inside the semicrystalline polymers can be divided between the crystallized and amorphous parts of the polymer. The crystalline part usually forms a complicated network in the matrix of the amorphous polymer. A visualization of a single-polymer crystallite done [111] by the Atomic Force Microscopy (AFM) is shown in Fig. 9. The most common morphology observable in the semicrystalline polymer is that of a spherulitic microstructure [112], where the crystalline lamellae grows more or less radially from the central nucleus in all directions. The different crystal lamellae can nucleate separately

**Figure 9.**  Visualization of a single-polymer crystallite done by atomic force microscopy [111].

and independently from each other or develop from one single lamellae crystal by continuous branching and fanning out. The resulting 3D morphology is very complicated, with lots of interconnections and irregular dendrites. The lack of a quantitative description of such crystalline networks lead to a common approximation used, for example, in mechanical properties simulations [113,114], in which a single spherulite is treated as an continuous object with some homogeneous properties. However, this common approximation is rather crude, and a realistic model of the crystalline structure based on the experimental 3D reconstruction has to be developed.

## II.  MESOSCOPIC SYSTEMS

### A.  Simple Mesoscopic Model for Surfactant Systems

The Landau–Ginzburg model of a ternary mixture of oil, water, and surfactant studied here was proposed by Teubner and Strey [115] on the basis of the scattering peak in the microemulsion phase. Later it was refined by Gompper and Schick [116]. Its application to various bulk and surface phenomena is described in detail in Ref. 117.

The Landau–Ginzburg free energy functional in the form given by Gompper and Schick is as follows:

$$F[\phi] = \int d^3r(|\triangle\phi|^2 + g(\phi)|\nabla\phi|^2 + f(\phi)) \tag{12}$$

where

$$g(\phi) = g_2\,\phi^2 + g_0 \tag{13}$$

$$f(\phi) = (\phi^2 - 1)^2(\phi^2 + f_0) \tag{14}$$

Here $\phi$, the scalar order parameter, has the interpretation of a normalized difference between the oil and water concentrations; $g_0$ is the strength of surfactant and $f_0$ is the parameter describing the stability of the microemulsion and is proportional to the chemical potential of the surfactant. The constant $g_0$ is solely responsible for the creation of internal surfaces in the model. The microemulsion or the lamellar phase forms only when $g_0$ is negative. The function $f(\phi)$ is the bulk free energy and describes the coexistence of the pure water phase ($\phi = -1$), pure oil phase ($\phi = 1$), and microemulsion ($\phi = 0$), provided that $f_0 = 0$ (in the mean-field approximation). One can easily calculate the correlation function $\langle\phi(r)\phi(0)\rangle - \langle\phi(r)\rangle\langle\phi(0)\rangle$ in various bulk homogeneous phases. In the microemulsion this function oscillates, indicating local correlations between water-rich and oil-rich domains. In the pure water or oil phases it should decay monotonically to zero. This does occur, provided that $g_2 > 4\sqrt{1 + f_0} - g_0$. Because of the $\phi$, $-\phi$ (oil–water) symmetry of the model, the interface between the oil-rich and water-rich domains is given by

$$\phi(\mathbf{r}) = 0 \tag{15}$$

This equation defines the internal surfaces in the system. The model has been studied in the mean field approximation (minimization of the functional) [21–23,117] and in the computer simulations [77,117,118]. The stable phases in the model are: oil-rich phase, water-rich phase, microemulsion, and ordered lamellar phase. However, as was shown in Refs. 21–23 there is an infinite number of metastable solutions of the minimization procedure:

$$\frac{\delta F[\phi]}{\delta\phi} = 0 \tag{16}$$

which correspond to the local minima of the functional $F$. The spatial distribution of the field $\phi$ in these metastable solutions is such that the surface $\phi = 0$ is the triply periodic surface (in many cases a minimal surface). In particular, from Eq. (16) one finds solutions with P, D, G, I-WP, and F-RD surfaces. Apart from such simple solutions, one also finds very complex

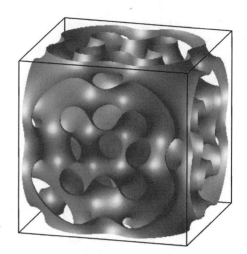

**Figure 10.** The surface SCN1 of the same symmetry as P surface generated by minimizing functional (12). The genus of this structure is 45.

structures. Three most complex are shown in Figs. 10–12. The one shown in Fig. 10 is the surface of the same symmetry as P but of more complex topology (i.e., its genus is 45 per unit cell, while for P we have $g = 3$). The second surface (Fig. 11) has genus 73 per unit cell and the same symmetry as the D surface ($g = 9$). Finally the most complex surface known to date is shown in Fig. 12. It has the same symmetry as the G gyroid surface ($g = 5$), but it has $g = 157$ per unit cell. The procedure that leads to the generation of such complex surfaces is briefly described below and in detail in Refs. 21–23.

In real space, the functional $F[\phi]$ is discretized first on the cubic lattice to find its minima. Thus the functional $F[\phi(\mathbf{r})]$ becomes a function $F(\{\phi_{i,j,k}\})$ of

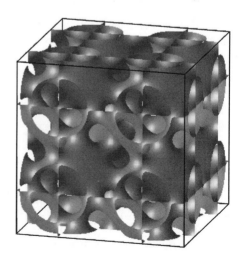

**Figure 11.** The surface CD of the same symmetry as D surface generated by minimizing functional (12). The genus of this structure is 73.

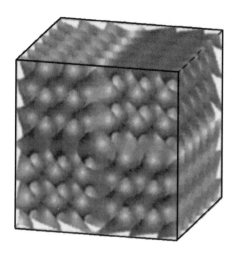

**Figure 12.** The surface GX5 of the same symmetry as G surface generated by minimizing functional (12). The genus of this structure is 157.

$N^3$ variables, where $L = (N - 1)h$ is the linear dimension of the cubic lattice, $h$ is the lattice spacing, and $\{\phi_{i,j,k}\}$ stands for the set of all variables of the function. Each variable $\phi_{i,j,k}$ represents the value of the field $\phi(\mathbf{r})$ at the point $\mathbf{r} = (i, j, k)h$, and $i, j, k = 1, \ldots, N$. With $N = 129$ one eventually deals with $F$ as a function of over 2 million variables.

The periodicity is incorporated into the functional by the boundary conditions: $\phi_{1,j,k} = \phi_{N,j,k}$, $\phi_{2,j,k} = \phi_{N+1,j,k}$, $\phi_{3,j,k} = \phi_{N+2,j,k}$, $\phi_{0,j,k} = \phi_{N-1,j,k}$, $\phi_{-1,j,k} = \phi_{N-2,j,k}$, and similarly in $y$ and $z$ directions. The points outside the unit cell, given by the periodic boundary conditions, enter the functional through the calculations of derivatives of points at the boundary and near the boundary of the lattice—that is, when at least one of the indices $i, j, k$ is equal to $1, 2, N - 1, N$.

The symmetry of the structure is imposed on the field $\phi(\mathbf{r})$ by building the field inside a unit cell of smaller polyhedron, replicating it by reflections, translations, and rotations [21–23]. This procedure reduces the number of independent variables one order of magnitude for G structure and two orders of magnitude for D structure.

The minimization in the Fourier space is much easier [119], because in the Fourier space the field $\phi(\mathbf{r})$ for periodic structures is represented by the Fourier series

$$\phi(\mathbf{r}) = A_0 + \sum_i \sum_{\mathbf{k}^i} \left(A_i \cos\left(\mathbf{k}^i \cdot \mathbf{r}\right) + B_i \sin\left(\mathbf{k}^i \cdot \mathbf{r}\right)\right) \tag{17}$$

where $\mathbf{k}^i$ are the reciprocal lattice vectors in the $i$th shell, and $A_i$ and $B_i$ are the amplitudes of the $i$th shell. The shell means here a set of reciprocal lattice vectors of the same length related to each other by the symmetry operations characteristic for a given lattice. The approximation by the Fourier series ensures that the function $\phi(\mathbf{r})$ is continuous at the unit cell boundaries. Next the functional $F[\phi]$ is minimized with respect to the amplitudes $A_i$ and $B_i$. As a result, one gets the nodal surface given by the equation $\phi(\mathbf{r}) = 0$. The larger the number of shells included, the better the quality of approximation.

From the computational point of view the Fourier space approach requires less variables to minimize for, but the speed of calculations is significantly decreased by the evaluation of trigonometric function, which is computationally expensive. However, the minimization in the Fourier space does not lead to the structures shown in Figs. 10-12. They have been obtained only in the real-space minimization. Most probably the landscape of the local minima of $F$ as a function of the Fourier amplitudes $A_i$ is completely different from the landscape of $F$ as a function of the field $\phi_{i,j,k}$ in real space. In other words, the basin of attraction of the local minima representing surfaces of complex topology is much larger in the latter case. As far as the minima corresponding to the simple surfaces are concerned (P, D, G etc.), both methods lead to the same results [21–23,119].

In this simple model characterized by a single scalar order parameter, the structures with periodic surfaces are metastable. It simply means that we need a more complex model including the surfactant degrees of freedom (its polar nature) in order to stabilize structures with P, D, and G surfaces. In the Ciach model [120–122] indeed the introduction of additional degrees of freedom stabilizes such structures.

## B. Mesoscopic Models of Copolymer Systems

The most intriguing property of block copolymer systems is their self-assembling into well-defined, ordered structures. A chemical architecture of copolymer chains together with the overall component compositions comprise the input parameters for mesoscopic models. The main aim of these models is to predict quantitatively the system morphology in equilibrium. The standard model [36] of block copolymers considers a monodisperse melt of $n$ diblock macromolecules, each of which contains a block of $N_A$ A-type monomers attached to the second block of $N_B$ B-type monomers. The architecture of the polymer molecules is specified by their composition, $f = N_A/(N_A + N_B)$. Possible conformations of the polymer chains are described by the monomer density probability function that also specifies connections between monomers. Within the random flight model [123], in which monomers are joined by the freely rotating bonds ($\mathbf{u}_i$) of fixed length, the distribution function is given as

$$W(\mathbf{r}) = \prod_{i=1}^{N} \frac{\delta(|\mathbf{u}_i| - l)}{4\pi l^2} \tag{18}$$

where $\delta(x)$ is the Dirac delta function, and $N = N_A + N_B$ is the index of polymerization. Sometimes it is convenient to use a continuous representation of the distribution function:

$$P[\mathbf{r}_\alpha; s_1, s_2] \sim \exp\left( -\frac{3}{2Nl^2} \int_{s_1}^{s_2} ds \left| \frac{d\mathbf{r}(s)}{ds} \right|^2 \right) \tag{19}$$

where $s$ measures the distances along the chain.

The interaction Hamiltonian with the specified short-range interactions between the monomers is given by the following expression [124,125]:

$$H_I = \frac{\rho_0}{k_B T} \int \frac{d\mathbf{q}}{(2\pi)^3} \left[ \frac{1}{2} w_{AA} |\hat{\phi}_A(\mathbf{q})|^2 + \frac{1}{2} w_{BB} |\hat{\phi}_B(\mathbf{q})|^2 \right.$$
$$\left. + w_{AB} \hat{\phi}_A(\mathbf{q}) \hat{\phi}_B(-\mathbf{q}) \right] \tag{20}$$

where $\rho_0 = n(N_A + N_B)/V$ is the number density of monomers in the system, and $\hat{\phi}_A(\mathbf{q})$ and $\hat{\phi}_B(\mathbf{q})$ are the Fourier transforms of the microscopic concentration operators [126]:

$$\hat{\phi}_A(\mathbf{q}) = \frac{1}{\rho_0} \sum_{\alpha=1}^{n} \sum_{i=1}^{N_A} \exp(\mathbf{q}\mathbf{r}_i^\alpha)$$
$$\hat{\phi}_B(\mathbf{q}) = \frac{1}{\rho_0} \sum_{\beta=1}^{n} \sum_{i=N_A+1}^{N_A+N_B} \exp(\mathbf{q}\mathbf{r}_i^\beta) \tag{21}$$

This part of the Hamiltonian leads to a microphase separation of block copolymers, provided that $w_{AA} + w_{BB} - 2w_{AB} = -2k_B T\chi < 0$. Here $\chi$ is the usual Flory–Huggins parameter. The origin of the interaction parameters $w_{AA}$, $w_{BB}$, and $w_{AB}$ is the van der Waals attraction between the monomers and the steric repulsive forces. The Flory–Huggins parameter for a real system is usually determined phenomenologically, and its temperature dependence has the following form: $\chi = A/T + B$, where $A$ and $B$ are empirical constants and $T$ is the absolute temperature. Within the continuous-chain model, the interaction Hamiltonian preserves the same form (21), but the sum over the monomers in the microscopic operator definitions is replaced by the integral.

The mesoscopic description is introduced by defining functions $\phi_A(\mathbf{q})$ and $\phi_B(\mathbf{q})$ that have the meaning of averaged over some mesoscopic volume values of the microscopic concentration operators. The conditional partition function, $Z(\phi_\gamma)$ $(\gamma = A, B)$, is the partition function for the system subject to the constraint that the microscopic operators $\hat{\phi}_\gamma(\mathbf{q})$ are fixed at some prescribed values [127] of $\phi_\gamma(\mathbf{q})$, that is,

$$Z(\phi_\gamma) = N_0 \prod_{\alpha=1}^{n} \int D\mathbf{r}^\alpha W(\mathbf{r}^\alpha) \prod_{\gamma=A,B} \delta[\phi_\gamma(\mathbf{q}) - \hat{\phi}_\gamma(\mathbf{q})] \exp\left(-\frac{H_I}{k_B T}\right) \quad (22)$$

Here $N_0$ is a constant, $D\mathbf{r}^\alpha$ denotes the measure $\frac{1}{V} d\mathbf{r}_0^\alpha d\mathbf{r}_1^\alpha \ldots d\mathbf{r}_N^\alpha$, the interaction Hamiltonian $H_I$ is given by Eq. (20), and $W(\mathbf{r}^\alpha)$ is the distribution function, Eq. (18).

For infinitely small $\chi$ there is no interaction in the system and the chains are mixed uniformly. The average values of $\phi_A(\mathbf{q})$ and $\phi_B(\mathbf{q})$ are equal to $f$ and $1 - f$, respectively. To describe the different phases in our system, we specify an order parameter $\Psi(\mathbf{r})$ as [36]

$$\Psi(\mathbf{r}) = \langle (1 - f)\phi_A(\mathbf{r}) - f\phi_B(\mathbf{r}) \rangle \quad (23)$$

where $\langle \cdots \rangle$ denotes a thermal average. $\Psi(\mathbf{r})$ has a vanishing value at any point in the isotropic phase, and it is nonzero in the ordered phase. Assuming that the system is incompressible, that is,

$$\phi_A(\mathbf{r}) + \phi_B(\mathbf{r}) = 1 \quad (24)$$

then

$$\Psi(\mathbf{r}) = \langle \phi_A(\mathbf{r}) - f \rangle \quad (25)$$

The partition function, $Z(\phi_\gamma)$, cannot be calculated exactly. It could be rewritten using the integral representation of the functional Dirac delta function and evaluated within the saddle place approximation. The calculations lead to the following expression [36,126,128]:

$$Z[\Psi] = \exp\Big\{ -\sum_{n=2}^{\infty} \frac{1}{n!} \int \frac{d\mathbf{q}_1}{(2\pi)^3} \cdots \int \frac{d\mathbf{q}_n}{(2\pi)^3} \, \Gamma_n(\mathbf{q}_1, \ldots, \mathbf{q}_n)$$
$$\times \, \delta(\mathbf{q}_1 + \cdots + \mathbf{q}_n) \, \Psi(-\mathbf{q}_1) \cdots \Psi(-\mathbf{q}_n) \Big\} \quad (26)$$

where the vertex functions $\Gamma_n$ for $n > 2$ are known functions arising from the ideal chain conformations and the interaction Hamiltonian, $H_I$, is included only in $\Gamma_2$. The full partition function is defined as

$$Z = \int D\Psi \, Z[\Psi] \quad (27)$$

but the LG free energy $F[\Psi]$ within the mean-field approximation (neglecting fluctuations) is simply

$$F[\Psi] = -k_B T \ln Z[\Psi] \quad (28)$$

In order to calculate the phase properties of the melt, that expansion is cut off at some lower order. The order parameter $\Psi(\mathbf{r})$ in the ordered phase which is characterized by the set $\{\mathbf{Q}_{k,m}\}$ of the wavevectors in the reciprocal space can be expanded as

$$\Psi(\mathbf{r}) = \sum_{m=1}^{\infty} \frac{1}{\sqrt{n_m}} A_m \sum_{k=1}^{n_m} \{\exp[i(\mathbf{Q}_{k,m}\mathbf{r} + \phi_{k,m})] + \text{c.c.}\} \tag{29}$$

where $A_m$ and $\phi_{k,m}$ are the amplitudes and phases of this expansion in the ordered phase, $m$ numbers the shells, and $n_m$ is the number of wavevectors in the $m$th shell. Their equilibrium values are determined by the minimization of the free energy, $F[\Psi(\mathbf{r})]$, with respect to $A_m$ and $\phi_{k,m}$ with additional constraints imposed on $\phi_{k,m}$ by the symmetry group of the ordered phase. The ordered phase with the smallest free energy for a given $\chi N$ is the stable one. Considering only the first shell for the reciprocal representation of the symmetry groups, only three stable morphologies can be found: lamellar, bcc, and cylindrical (Fig. 13). Including the second shell into consideration, the stability region of the double-gyroid phase (Fig. 5) can be calculated.

The LG free energy depends on the architecture of the copolymer chains by means of vertex functions $\Gamma_n(\mathbf{q}_1, \ldots \mathbf{q}_n)$. They are expressed in terms of density operator momenta $S_{i,\ldots,j} = \langle \phi_i(\mathbf{q}_1) \cdots \phi_j(\mathbf{q}_n) \rangle_0$ $(i, j = A, B)$ averaged within the ideal chain statistics. The generalization of the theory to more complex chain architectures is straightforward [42]. Let $\{A\}$ and $\{B\}$ denote the positions of the monomers of type $A$ and type $B$ consequently in the chain. For example, a chain $AABABAA$ would have the following position of $\{A\} = \{1,2,4,6,7\}$ and $\{B\} = \{3,5\}$. The sum over the $A$ or $B$ monomers in Eq. (21) can be replaced by the discrete summation over set $\{A\}$ or $\{B\}$ or, in the continuous limit, by the integration with a monomer distribution function $g(x)$ or $1 - g(x)$, that is,

$$\sum_{i=\{A\}} \cdots \rightarrow N \int_0^1 \cdots g(x)\, dx \tag{30}$$

$$\sum_{j=\{B\}} \cdots \rightarrow N \int_0^1 \cdots [1 - g(x)]\, dx \tag{31}$$

The monomer distribution function $g(x)$ gives the probability to find a monomer of type $A$ at the distance $x \cdot N$ from the beginning of the chain. It should satisfy the following conditions: $0 < g(x) < 1$ for every $x$. The composition, $f$, is now given by

$$f = \int_0^1 g(x)\, dx \tag{32}$$

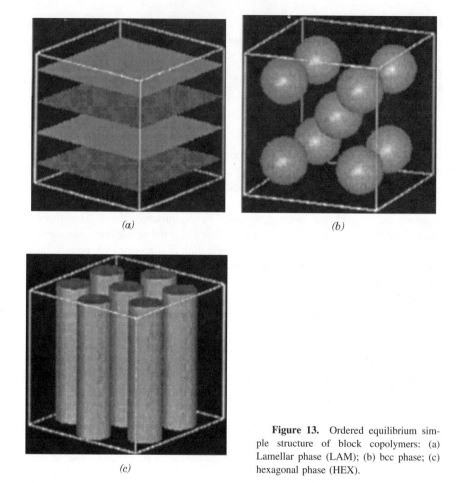

**Figure 13.** Ordered equilibrium simple structure of block copolymers: (a) Lamellar phase (LAM); (b) bcc phase; (c) hexagonal phase (HEX).

The phase diagram of the tapered copolymer melt computed within the two shell approximation [42] is shown in Fig. 14. The distribution function of $A$ monomer along the tapered copolymer chains is

$$g(x) = \frac{1}{2}(1 - \tanh(c_1 \pi (x - f_0))) \qquad (33)$$

where $c_1$ and $f_0$ determine sharpness and position of the "interface" between the blocks of $A$ and $B$ monomers. The distribution function $g(x)$ is shown in the insert of Fig. 14 for $c_1 = 3, f_0 = 0.5$ (solid line), and $f_0 = 0.3$ (dashed line). In the limit of $c_1 \to \infty$ the diblock architecture is recovered. Changing $f_0$ one can shift the position at which $g(x) = 0.5$ and model asymmetric melts. The melt composition $f$ in this case equals to $f_0$.

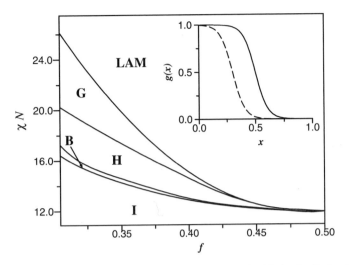

**Figure 14.** The phase diagram of the gradient copolymer melt with the distribution functions $g(x) = \frac{1}{2}(1 - \tanh(c_1\pi(x - f_0)))$ shown in the insert of this figure for $c_1 = 3, f_0 = 0.5$ (solid line), and $f_0 = 0.3$ (dashed line). $x_i$ gives the position of $i$th monomer from the end of the chain in the units of the linear chain length. $\chi$ is the Flory–Huggins interaction parameter, $N$ is a polymerization index, and $f$ is the composition ($f = \int_0^1 g(x)\,dx$). The Euler characteristic of the isotropic phase (**I**) is zero, and that of the hexagonal phase (**H**) is zero. For the bcc phase (**B**), $\chi_{\text{Euler}} = 4$ per unit cell; for the double gyroid phase (**G**), $\chi_{\text{Euler}} = -16$ per unit cell; and for the lamellar phases (**LAM**), $\chi_{\text{Euler}} = 0$.

The Euler characteristic of the ordered phases can be computed as follows. At a given point of the phase diagram the equilibrium structure is characterized by the set of wavevectors $\mathbf{Q}_{k,m}$, phases $\phi_{k,m}$, and amplitudes $A_n$, with the latter obtained from the minimization. The order parameter distribution can be generated on a lattice by using Eq. (29). It is convenient to choose the lattice size as an integer number of main harmonics of the ordered structure, $2\pi/q^*$, where $\mathbf{q}^*$ denotes the absolute maximum position in the scattering intensity. The surface is defined as points where $\Psi(x, y, z) = 0$. The Euler characteristic is then computed by using one of the algorithms described in Section III.G.

The phase diagram in Fig. 14 can be analyzed by using the Euler characteristic. The disordered phase contains no surfaces, and therefore the Euler characteristic is zero. The bcc phase [Fig. 13(b)] within the two-shell approximation is expressed as

$$\Psi_{\text{BCC}}(x, y, z) = A_1\left[\cos\frac{x}{\sqrt{2}}\cos\frac{y}{\sqrt{2}} + \cos\frac{y}{\sqrt{2}}\cos\frac{z}{\sqrt{2}} + \cos\frac{z}{\sqrt{2}}\cos\frac{z}{\sqrt{2}}\right]$$
$$+ A_2[\cos(\sqrt{2}x) + \cos(\sqrt{2}y) + \cos(\sqrt{2}z)] \tag{34}$$

where amplitudes $A_1$ and $A_2$ are determined by minimizing Eq. (28) (e.g., $A_1 = 0.0345$, $A_2 = 0.00304$). After choosing the lattice size to be a period of the main harmonic—that is, $x = 2\pi\sqrt{2}\frac{(i-0.5N)}{N} i = 1,\ldots,N$ ($N$ is the lattice size)—we found that the Euler characteristic of the bcc phase in a unit cell is $+4$. This means that the ordered phase is comprised from two disconnected surfaces per unit cell, which is exactly the definition of the spherical mesophase of the bcc symmetry. Please note that the form of the basis functions in Eq. (34) does not guarantee automatically the spherical morphology of the ordered phase. Thus when $A_1 \neq 0$ and $A_2 = 0$, $\chi_{Euler} = -12$; and when $A_1 = 0$, and $A_2 \neq 0$, $\chi_{Euler} = -32$ (eight minimal P surfaces [23] in a unit cell).

The Euler characteristic of the hexagonal phase [Fig. 13(c)] is 0, since the Euler characteristic of each cylinder is zero. The double gyroid phase (Fig. 5) within the two-shell approximation is represented as

$$\Psi_{DG}(x,y,z) = A_1 \left[ \cos\frac{x}{\sqrt{6}} \sin\frac{y}{\sqrt{6}} \sin\frac{2z}{\sqrt{6}} + \cos\frac{y}{\sqrt{6}} \sin\frac{z}{\sqrt{6}} \sin\frac{2x}{\sqrt{6}} \right.$$
$$\left. + \cos\frac{z}{\sqrt{6}} \sin\frac{x}{\sqrt{6}} \sin\frac{2y}{\sqrt{6}} \right] + A_2 \left[ \cos\frac{2x}{\sqrt{6}} \cos\frac{2y}{\sqrt{6}} \right.$$
$$\left. + \cos\frac{2x}{\sqrt{6}} \cos\frac{2z}{\sqrt{6}} + \cos\frac{2y}{\sqrt{6}} \cos\frac{2z}{\sqrt{6}} \right] \tag{35}$$

The amplitude values are determined by minimizing Eq. (28) (e.g., $A_1 = 0.152$, $A_2 = -0.038$). The Euler characteristic of the DG phase is $-16$ per unit cell ($-8$ for each G surface). Similarly to the bcc phase, the first two basis functions of the Ia3d symmetry group do not guarantee automatically the DG morphology. For example, if amplitudes $A_1$ and $A_2$ are taken of the same signs, the S1 periodic surface of the Euler characteristic $-52$ is obtained [42] (Fig. 15). The mean curvature distribution (Section III.F) computed for the interface generated in accordance with Eq. (35) clearly indicates that the DG interface is not a minimal surface, which has been also confirmed experimentally by TEMT [8] (the interface is shifted from the minimal contour of G surface). Finally, the Euler characteristic of the lamellar phase [Fig. 13(a)] is again zero, because the Euler characteristic of a flat layer is 0. The volume fraction of the minority phase gradually increases as the bcc phase ($f_m = 0.352$) transforms into the hexagonal phase ($f_m = 0.401$), the double gyroid phase ($f_m = 0.43$), and finally, the lamellar phase ($f_m = 0.5$). The minority phase volume fraction of the bcc, HEX, DG, and LAM phases depends weakly on the average composition $f$ and the Flory–Hugging parameter $\chi$. This may be explained by a very scarce number of basis functions used to represent the considered symmetry groups.

At very asymmetric compositions, the free-energy difference between the double-gyroid and hexagonally perforated lamellar (HPL) phases becomes very

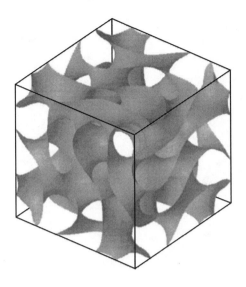

**Figure 15.** The periodic S1 surface of the Ia3d symmetry. The Euler characteristic of this surface is $-52$ per unit cell.

small. Therefore the HPL phase could be observed experimentally. In this case, the phase transformations occur as I-bcc-HEX-HPL-HML-LAM on decreasing temperature (HML abbreviates the hexagonally modulated lamellar phase). The HPL phase is represented within the two-shell approximations as

$$\Phi_{HPL}(x, y, z) = A_1 \left[ \cos(x) + \cos \frac{x - \sqrt{3}y}{2} + \cos \frac{x + \sqrt{3}y}{2} \right] + A_2 \cos(dz) \quad (36)$$

where $d$ specifies the relative period of the lamellar ordering (compared to the hexagonal one). We took the amplitudes $A_1$ and $A_2$ obtained from minimization of the Eq. (28) (the HPL phase is still metastable) in order to calculate the Euler characteristic. There are two types of the HPL structures that can be modeled by using Eq. (36). The first structure, Fig. 16(a), is bicontinuous, the Euler characteristic in a unit hexagonal cell is $-8$ (e.g., at $A_1 = 0.154$, $A_2 = 0.17$). The two hexagonal sublattices of channels interpenetrate the lamellae; thus, in a unit cell, there is one big passage filled with the majority component and surrounded by the six smaller passages filled with the minority component (only half of each small channel is in a unit cell). In the second structure, shown in Fig. 16(b), the alternated stack of lamellae is penetrated by the hexagonally arranged columns that are rich in one of the components (the phase volume fractions are nearly symmetric); thus the structure is only monocontinuous (the true catenoid phase). The Euler characteristic in the latter case is $-2$—that is, one passage is per one unit hexagonal cell (e.g., at $A_1 = 0.154$, $A_2 = 0.3$).

The Euler characteristic density of the bicontinuous HPL structure is close to the Euler characteristic density of the DG phase. Also, the free-energy costs of

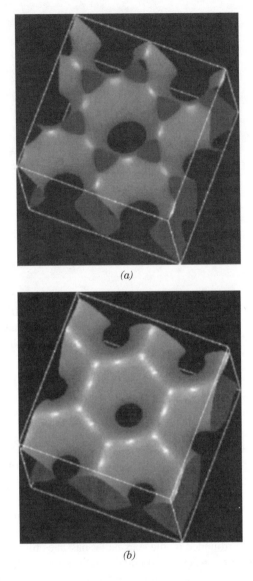

*(a)*

*(b)*

**Figure 16.** The bicontinuous (a) and monocontinuous (b) HPL structures (the metastable structures of diblock copolymers).

the bicontinuous HPL phase are similar to the costs of the DG phase. Because there are small topological and free-energy barriers, these structures could be easily transformed one into another by applying external fields. Their phase volume fractions are also similar ($f_m^{DG} = 0.43$, $f_m^{HPL_l} = 0.46$). The second (monocontinuous) type of the HPL structure is topologically and energetically close to the lamellar (or HML) phase.

The obvious disadvantage of this simple LG model is the necessity to cut off the infinite expansion (26) at some order, while no rigorous justification of doing that can be found. In addition, evaluation of the vertex function for all possible zero combinations of the reciprocal wave vectors becomes very awkward for low symmetries. Instead of evaluating the partition function in the saddle point, the minimization of the free energy can be done within the self-consistent field theory (SCFT) [38–41]. Using the integral representation of the delta functionals, the total partition function, $Z$ [Eq. (22)], can be written as

$$Z = N_0 \int D\phi_A \, DU_A \, D\phi_B \, DU_B \, D\Xi \exp\{-F/k_B T\} \qquad (37)$$

where $N_0$ is a normalization constant,

$$F/nk_B T \equiv -\ln Q + V^{-1} \int d\mathbf{r} \, [\chi N \phi_A \phi_B - U_A \phi_A - U_B \phi_B - \Xi(1 - \phi_A - \phi_B)] \qquad (38)$$

$$Q \equiv \int D\mathbf{r}_\alpha P[\mathbf{r}_\alpha; 0, 1] \exp\left\{ -\int_0^f ds U_A(\mathbf{r}_\alpha) - \int_f^1 ds U_B(\mathbf{r}_\alpha) \right\} \qquad (39)$$

The incompressibility constraint $\delta(\phi_A + \phi_B - 1)$ has been explicitly included in partition function (37), and the continuous chain model, Eq. (19), is being used. $Q$ is the partition function of an independent single chain subjected to the fields $U_A$ and $U_B$, and it can be evaluated exactly. One writes the partition function as $Q = \int d\mathbf{r} q(\mathbf{r}, 1)$, where

$$q(\mathbf{r}, s) = \int d\mathbf{r}_\alpha P[\mathbf{r}_\alpha; 0, s] \delta(\mathbf{r} - \mathbf{r}_\alpha(s)) \exp\left\{ -\int_0^s dt \, [\theta(t) U_A(\mathbf{r}_\alpha(t)) \right. $$
$$\left. + (1 - \theta(t)) U_B(\mathbf{r}_\alpha(t))] \right\} \qquad (40)$$

is the end-segment distribution function ($\theta$ denotes the step function). This distribution function satisfies the modified diffusion equation,

$$\frac{\partial q}{\partial s} \equiv 1/6Nl^2 \nabla^2 q - U_A(\mathbf{r})q, \qquad \text{if } s < f$$
$$\frac{\partial q}{\partial s} \equiv 1/6Nl^2 \nabla^2 q - U_B(\mathbf{r})q, \qquad \text{if } s > f \qquad (41)$$

and the initial condition, $q(\mathbf{r}, 0) = 1$. A second end-segment distribution function, $q^\dagger(\mathbf{r}, s)$, is defined in a similar way.

Within the mean-field SCFT, the integral in Eq. (37) is approximated by the extremum of the integrand. Thus the equilibrium free energy, $F = -k_B T \ln Z$, is a functional of the fields $\tilde{\phi}_A, \tilde{\phi}_B, u_A, u_B, \xi$ at which $F$ attains its minimum. From the definition of $F$ follows that these fields should satisfy the set of self-consistent equations

$$u_A(\mathbf{r}) = \chi N \tilde{\phi}_B(\mathbf{r}) + \xi(\mathbf{r}) \tag{42}$$

$$u_B(\mathbf{r}) = \chi N \tilde{\phi}_A(\mathbf{r}) + \xi(\mathbf{r}) \tag{43}$$

$$\tilde{\phi}_A(\mathbf{r}) + \tilde{\phi}_B(\mathbf{r}) = 1 \tag{44}$$

$$\tilde{\phi}_A(\mathbf{r}) = -\frac{V}{Q} \frac{\delta Q}{\delta u_A} \tag{45}$$

$$\tilde{\phi}_B(\mathbf{r}) = -\frac{V}{Q} \frac{\delta Q}{\delta u_B} \tag{46}$$

Finally, the $A$-monomer density, Eq. (45), can be rewritten in terms of the end-segment functions:

$$\tilde{\phi}_A(\mathbf{r}) = \frac{V}{Q} \int_0^f ds\, q(\mathbf{r}, s)\, q^\dagger(\mathbf{r}, s) \tag{47}$$

The equilibrium morphology is calculated in the following way. At a given point of the phase diagram a set of possible candidates for the equilibrium morphology is specified. For a given symmetry group, all functions of position, $g(\mathbf{r})$, are expanded as $\sum_i g_i f_i(\mathbf{r})$, where $f_i(\mathbf{r})$ are the orthonormal basis functions, each possessing the symmetry of the phase being considered. They are also chosen to be eigenfunction of the Laplacian operator. For a given set of the field amplitudes, $u_{A,i}$ and $u_{B,i}$, the density amplitudes $\tilde{\phi}_{A,i}$ and $\tilde{\phi}_{B,i}$ are determined by solving the diffusion equations (41) and using (47) (and similar equations for $q^\dagger(\mathbf{r}, s)$ and $\tilde{\phi}_B(\mathbf{r})$). The field amplitudes $u_{A,i}$ and $u_{B,i}$ are adjusted in repeated iterations to satisfy Eqs. (42)–(44). The phase diagram obtained from the SCFT calculation and from the standard LG theory are quantitatively similar. The SCFT equations can be also solved in the Fourier space [129].

In all considered above models, the equilibrium morphology is chosen from the set of possible candidates, which makes these approaches unsuitable for discovery of new unknown structures. However, the SCFT equation can be solved in the real space without any assumptions about the phase symmetry [130]. The box under the periodic boundary conditions in considered. The initial quest for $u_\gamma(\mathbf{r})$ is produced by a random number generator. Equations (42)–(44) are used to produce density distributions $\tilde{\phi}_\gamma(\mathbf{r})$ and pressure field $\xi(\mathbf{r})$. The diffusion equations are numerically integrated to obtain $q$ and $q^\dagger$ for $0 \le s \le 1$. The right-hand size of Eq. (47) is evaluated to obtain new density profiles. The volume fractions at the next iterations are obtained by a linear mixing of new and old solutions. The iterations are performed repeatedly until the free-energy change becomes lower than some threshold value. The final density configuration corresponds to either stable or metastable equilibrium. Each independent run can be considered as a quench; however, the simulated dynamic does not correspond to the dynamics of a real system. This method is very efficient in predicting the possible ordered morphologies of the multiblock copolymers [130].

The mean-field SCFT neglects the fluctuation effects [131], which are considerably strong in the block copolymer melt near the order–disorder transition [132] (ODT). The fluctuation of the order parameter field can be included in the phase-diagram calculation as the one-loop corrections to the free-energy [37,128,133], or studied within the SCFT by analyzing stability of the ordered phases to anisotropic fluctuations [129]. The real space SCFT can also applied for a confined geometry systems [134], their dynamic development allows to study the phase-ordering kinetics [135].

## C. TDGL and CHC Models for Phase Separating/Ordering Systems

In this section we present mesoscopic models, based on the coarse-grained Landau–Ginzburg-type free energy functional, which describe a time evolution of the phase separating/ordering system in terms of an order parameter field $\phi(\mathbf{r}, t)$. The order parameter represents, for example, the magnetization density of a ferromagnet or the local volume fraction of a selected component in a binary mixture. The dynamics is modeled by the time-dependent Ginzburg–Landau (TDGL) equation when the order parameter is not conserved, or by the Cahn–Hilliard–Cook (CHC) equation if the order parameter is conserved. In the theory of critical phenomena [136] the TDGL and CHC models are also referred to as the model A and B, respectively.

### 1. Nonconserved Order Parameter (TDGL Model)

The simplest physical example of the dynamics with the nonconserved order parameter is an Ising ferromagnet quenched from a temperature above its critical temperature $T_c$ to a temperature $T$ below $T_c$. After lowering the temperature, the system is brought into thermodynamically unstable two-phase region and starts to evolve toward one of the equilibrium states. Because both (spins up or down) phases are equally likely to appear, the system contains the domains of both phases. During the phase separating process the domains coarsen and the system orders over larger and larger lengthscales.

If we denote by $\phi(\mathbf{r}, t)$ the magnetization density, the dynamics of the system is governed by the following stochastic equation:

$$\frac{\partial \phi(\mathbf{r}, t)}{\partial t} = -M \frac{\delta F[\phi]}{\delta \phi} + \zeta(\mathbf{r}, t) \tag{48}$$

where $M$ is a phenomenological kinetic coefficient (the mobility), and $F[\phi]$ is the free energy functional; the stochastic term $\zeta(\mathbf{r}, t)$ represents thermal fluctuations. It is assumed to be given by uncorrelated Gaussian noise [2,4,1] with zero mean, satisfying the fluctuation–dissipation relation [1]

$$\langle \zeta(\mathbf{r},t)\zeta(\mathbf{r}',t')\rangle = k_B TM \delta(\mathbf{r}-\mathbf{r}')\delta(t-t') \qquad (49)$$

Equation (60) is called the time-dependent Ginzburg–Landau (TDGL) equation.

The functional derivative in Eq. (60) represents deterministic relaxation of the system toward a minimum value of the free-energy functional $F[\phi(\mathbf{r},t)]$, which is usually taken to have the form of the coarse-grained Landau–Ginzburg free energy

$$F[\phi(\mathbf{r},t)] = k_B T \int d\mathbf{r} \left[ \frac{C}{2} |\nabla\phi(\mathbf{r},t)|^2 + f(\phi(\mathbf{r},t)) \right] \qquad (50)$$

The gradient-squared term in the above equation represents the energy of the interface separating the $\pm$ phases; the constant $C$ can be interpreted as a measure of the interaction range. The bulk potential $f(\phi)$ has the Landau–Ginzburg (GL) "$\phi^4$" (double-well) structure

$$f(\phi) = -\frac{1}{2}r\phi^2 + \frac{1}{4}u\phi^4 \qquad (51)$$

with degenerate minima at $\phi = \pm\sqrt{r/u}$, which correspond to the two equilibrium states; here $r$ and $u$ are positive phenomenological constants. The parameter $r$ is proportional to the quench depth and is commonly set equal to the reduced temperature, $r = (T_c - T)/T_c$. The TDGL equation with the bulk potential given by (51) leads to the following kinetic equation governing the time evolution of the field $\phi(\mathbf{r},t)$:

$$\frac{\partial}{\partial t}\phi(\mathbf{r},t) = C\,\nabla^2\phi(\mathbf{r},t) + r\phi(\mathbf{r},t) - u\phi^3(\mathbf{r},t) + \zeta(\mathbf{r},t) \qquad (52)$$

which can be integrated numerically.

The kinetics of the nonconserved order parameter is determined by local curvature of the phase interface. Lifshitz [137] and Allen and Cahn [138] showed that in the late kinetics, when the order parameter saturates inside the domains, the coarsening is driven by local displacements of the domain walls, which move with the velocity $v$ proportional to the local mean curvature $H$ of the interface. According to the Lifshitz–Cahn–Allen (LCA) theory, typical time $t$ needed to close the domain of size $L(t)$ is $t \sim L(t)/v = L(t)/H(t)$, where $H(t)$ is the characteristic curvature of the system. Thus, under the assumption that $H(t) \sim 1/L(t)$, the LCA theory predicts the growth law $L(t) \sim t^{1/2}$. The late scaling with the growth exponent $n = 0.5$ has been confirmed for the nonconserved systems in many 2D simulations [139–141].

The kinetics of the system of the nonconserved order parameter has been studied for almost four decades, yet it is still far from being well understood. It is worth mentioning here that the growth law $L(t) \sim t^{1/2}$ has not been confirmed for any 3D system so far. In Section III.F.4 we demonstrate, however, that the method developed recently in Ref. 142, based on the geometry and topology of the interface separating the domains, is a useful tool in studying the process of the phase separating/ordering kinetics and gives a new insight into the LCA scaling.

### 2. Conserved Order Parameter (CHC Model)

A homogeneous A–B binary mixture after a rapid change of external conditions (quench) can be driven into the thermodynamically unstable state that will start the phase separation process. By the rapid enhancement of concentration fluctuations the domains rich in A or B component are formed shortly after the quench. These domains grow with time, changing the lengthscale of the phase separation from the microscopic molecular scale of the very beginning times to the macroscopic, comparable with the system size scale of the final stages of this process. Therefore, the time-dependent mesoscopic models that cover the most interesting, intermediate regime of the growth have become a convenient framework for modeling these phenomena. Governed by the same general principles, the time-dependent mesoscopic models have been successfully used to study the decomposition kinetics in both simple and complex mixtures [1,2,4].

Let us consider a dynamically symmetric binary mixture described by the scalar order parameter field $\phi(\mathbf{r})$ that gives the local volume fraction of component A at point $\mathbf{r}$. The order parameter $\phi(\mathbf{r})$ should satisfy the local conservation law, which can be written as a continuity equation [143]:

$$\frac{\partial \phi(\mathbf{r}, t)}{\partial t} = -\nabla \mathbf{J}_A(\mathbf{r}) + \zeta(\mathbf{r}, t) \tag{53}$$

where $\nabla \mathbf{J}_A(\mathbf{r})$ is a local flux of A component and the stochastic term $\zeta(\mathbf{r}, t)$ represents thermal noise [1,2,4]. Equation (53), governing the dynamics of the system with a conserved order parameter, is referred to as the Cahn–Hilliard–Cook (CHC) equation.

Let us assume that the molecular transport is governed only by the differences in the chemical potential (diffusion) and neglect a possible order parameter transport by the hydrodynamic flow [1,144,157]. Then, one can postulate a linear relationship between the local current and the gradient of the local chemical potential difference $\mu(\mathbf{r})$ [146,147] as

$$\mathbf{J}(\mathbf{r}) = -\int \frac{\Lambda(\mathbf{r} - \mathbf{r}')}{k_B T} \nabla' \mu(\mathbf{r}') \, d\mathbf{r} \tag{54}$$

Here $\Lambda(\mathbf{r} - \mathbf{r}')$ is the Onsager coefficient that specifies the transport properties of the considered system at a certain timescale and lengthscale, and which is nonlocal in general. The local chemical potential difference $\mu(\mathbf{r})$ can be found in a standard way as a functional derivative of the coarse-grained free energy functional $F[\phi]$:

$$\mu(\mathbf{r}) = \frac{\delta F[\phi]}{\delta \phi(\mathbf{r})} \tag{55}$$

Finally, the noise term in Eq. (53) should satisfy the appropriate fluctuation–dissipation relation [1]. In this way, all information about specific properties of the system enters into the dynamic equation (53) via the free-energy functional and Onsager coefficient.

The simplest free-energy functional that describes an inhomogeneous mixture can be written in the same form as in the case of the nonconserved order parameter system, Eq. (50). In this expression, $f(\mathbf{r})$ has the meaning of the homogeneous (bulk) free energy of mixing, and the square gradient term in Eq. (50) measures the free-energy cost of the inhomogeneities (interface). In the case of a symmetric homopolymer mixture, the equilibrium free energy can be written in the Flory–Huggins (FH) form [148]:

$$f(\phi) = \frac{1}{N}\{\phi \ln(\phi) + (1 - \phi)\ln(1 - \phi)\} + \chi\phi(1 - \phi) \tag{56}$$

where $N$ is the polymerization index and $\chi$ is the FH interaction parameter. This expression has been originally derived to describe the system of polymer chains on a lattice but can also be used in the course-grained models. In this case, the FH interaction parameter measures effectively, averaged over some mesoscopic lengthscale relative affinity between $A$ and $B$ components, and can be determined phenomenologically from experiments. The only specific polymer feature of the FH free energy is its dependence on the polymerization index $N$, which simply refers to the fact that $N$ monomers form one macromolecule. Therefore the coarse-grained free energy of mixing for a simple binary mixture must be given by the same expression, (56), but with $N = 1$. On the other hand, near the critical point, the free-energy can be written in the standard LG form with its homogeneous part

$$f(\phi) = -\frac{1}{2}r\tilde{\phi}^2 + \frac{1}{4}u\tilde{\phi}^4 \tag{57}$$

where $r$ and $u$ are positive phenomenological constants [149] and $\tilde{\phi} = \phi - \phi_c$ ($\phi_c$ is the critical value of the order parameter). Note that the potential $f(\phi)$ has the same form as the LG coarse-grained potential used in the case of the noncon-served order parameter kinetics, given by Eq. (51).

The square-gradient term in Eq. (50) can be derived from the Landau-type free-energy functional expansion by identifying that term with the lowest-order inhomogeneous correction. That formalism implies that the expansion of the inhomogeneous system free energy above that of the reference homogeneous system has been made [150]; therefore, coefficients of this expansion are constants (evaluated for the reference uniform system). A more detailed form of $C$ can be guessed from the shape of the equilibrium correlation function that in the case of the polymer blend (within the RPA) is [148]

$$\left(\frac{\delta^2 F}{\delta\phi^2}\right)_q \equiv S^{-1} = \frac{1}{N\phi_0 D(q^2 R_g^2)} + \frac{1}{N(1-\phi_0)D(q^2 R_g^2)} - 2\chi \qquad (58)$$

where $R_g$ is the polymer chain radius of gyration, $D(x)$ is the Debye function, $\phi_0$ is the average volume fraction of $A$ component, and the incompressibility constrain has been imposed. By expanding the Debye function in the limit of small $q$ as

$$D(x) = (2/x)(1 - (1 - e^{-x})/x) \approx 1 - x/3 \qquad (59)$$

where $x = (qR_g)^2$, the following form of $C$ can be found [151]:

$$C = \frac{1}{18}\frac{\sigma^2}{\phi_0(1-\phi_0)} \qquad (60)$$

Here $\sigma$ is the statistical Kuhn segment length [148] such that $R_g = \sqrt{N}\sigma/\sqrt{6}$. The truncation of the infinite Landau-type expansion at the lowest-order term is justified when the order parameter gradients are small, which requires a smooth variation of the order parameter through the interface. In the case of the polymer mixtures, this truncation can be made when the interface width is larger than the polymer chain radius of gyration ($q^2 R^2 \ll 1$). The higher-order corrections for Eq. (50) can be derived in a systematic way [150,151].

An alternative way to find out the expression for $C$ is to assume that the same form of the structure factor, Eq. (58), will be valid also locally for the inhomogeneous system, if the average volume fraction $\phi_0$ in Eq. (58) is replaced by its local values $\phi(\mathbf{r})$. Guided by this assumption, one can allow the coefficient at the gradient term to be dependent on the local volume fractions and write down $C$ as

$$C(\phi) = \frac{1}{18}\frac{\sigma^2}{\phi(\mathbf{r})[1-\phi(\mathbf{r})]} \qquad (61)$$

Is has been argued [152] that this concentration dependence of $C$ describes the lost of the conformation entropy related to some specific chain conformations at the interface. The free-energy functional in the form of Eq. (50) with the FH bulk free energy and the nonconstant square gradient coefficient (61) was postulated by de Gennes [146], and subsequently it has been widely used by others [9,143,147,152–155] to model the phase separation phenomena in polymer mixture. It must been stressed that no rigorous derivation of the above expression can be made from the standpoint of the traditional Landau-type analysis [150]. The first derivation of the Flory–Huggins–de Gennes (FHG) free energy has been made by Tang and Freed [150] within the framework of the density functional theory.

The Onsager coefficient for a simple binary mixture is usually written as

$$\Lambda(\mathbf{r} - \mathbf{r}') = M\delta(\mathbf{r} - \mathbf{r}') \tag{62}$$

where $M$ is a phenomenological mobility. For a polymer blend the form of the Onsager coefficient depends on the scale that is used to consider the phase separation. If the minimal lengthscale detectable in the simulations (mesh size) is larger than the radius of gyration of a polymer coil, then the nonlocal Onsager coefficient [143,147] can be approximated as

$$\Lambda(\mathbf{r} - \mathbf{r}') = DN\phi(1 - \phi)\delta(\mathbf{r} - \mathbf{r}') \tag{63}$$

where $D$ is the self-diffusion constant of polymer chains. In the last expression, the concentration dependence of the Onsager coefficient originates from the zero total current divergence constraint that must be satisfied for any incompressible system [146]. Therefore, the Onsager coefficient for any simple mixtures that can be considered as incompressible must also be concentration-dependent.

Equation (53) has to be solved numerically. The field $\phi(\mathbf{r})$ at time $t$ is represented by the discrete set $\phi_{i,j,k}(t)$ on a lattice, and the time evolution is obtained by performing iterations over $t$. The initial conditions are generated by assigning the average volume fraction $\phi_0$ to each lattice point, and the interface is defined as $\phi(\mathbf{r}) = \phi_0$. An example of the interface evolution for symmetric ($\phi_0 = 0.5$) quench is shown in Fig. 17(a–c). The domains grow in time, while their connectivity decreases (indicated by the change of the interface Euler characteristic). For some asymmetric compositions ($\phi_0 \neq 0.5$) the droplets are formed at the late times from the initially bicontinuous structures [Fig. 18(a–c)].

Let now consider a system where in addition to the diffusion flux due to the chemical potential differences, there is also a certain flow field $\mathbf{v}(\mathbf{r}, t)$. The equation for the temporal change of the order parameter field in this case is [1,4,157]

$$\frac{\partial\phi(\mathbf{r}, t)}{\partial t} = \Lambda\nabla^2\mu(\mathbf{r}) - \nabla \cdot [\phi(\mathbf{r}, t)\mathbf{v}(\mathbf{r}, t)] + \zeta(\mathbf{r}, t) \tag{64}$$

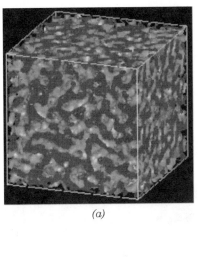

(a)

$\tau = 6.23$

Euler characteristic = $-4208$

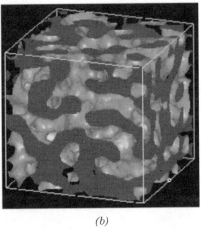

(b)

$\tau = 135.1$

Euler characteristic = $-428$

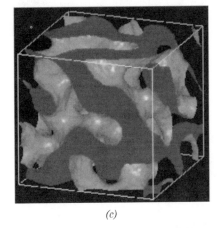

(c)

$\tau = 1125$

Euler characteristic = $-74$

**Figure 17.** Different stages of the spinodal decomposition in a symmetric mixture ($\phi_0 = 0.5$); $\tau$ is the dimensionless time. The Euler characteristic is negative, which indicates that the surfaces are bicontinuous. The Euler characteristic increases with dimensionless time. This indicates that the surface connectivity decreases.

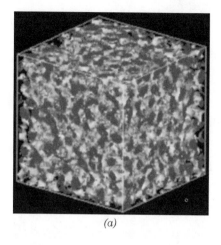

(a)

$\tau = 1.88$

Euler characteristic = −3764

$\tau = 29.69$

Euler characteristic = +132

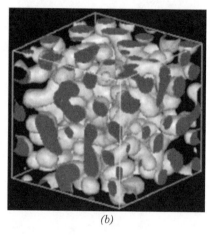

(b)

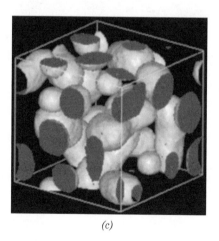

(c)

$\tau = 247.25$

Euler characteristic = +36

**Figure 18.** Different stages of the spinodal decomposition in an asymmetric mixture ($\phi_0 = 0.5$); $\tau$ is the dimensionless time. The Euler characteristic is initially negative, which indicates that morphology is bicontinuous. After a certain time the Euler characteristic becomes positive, which indicates that the transition to dispersed morphology occurred. For a dispersed morphology the Euler characteristic equals twice the droplet number.

For simplicity, the Onsager coefficient is assumed to be a constant. The velocity field should satisfy the Navier–Stokes (NS) equation, which in the general case has the following form [158,159]:

$$\rho \frac{\partial \mathbf{v}}{\partial t} + \rho(\mathbf{v} \cdot \nabla)\mathbf{v} = \mathbf{F}_\phi - \nabla p + \eta\nabla^2\mathbf{v} + \xi \tag{65}$$

where $\rho$ and $\eta$ are the fluid density and viscosity, respectively (they are assumed to be the same), and $p$ is the pressure field. $F_\phi$ is the force density associated with a local stress created by the inhomogeneous order parameter [1,157]:

$$F_\phi = -\phi(\mathbf{r})\nabla\mu(\mathbf{r}) \tag{66}$$

The noise terms in Eqs. (64) and (65) satisfy the following fluctuation–dissipation relations:

$$\langle\zeta(\mathbf{r},t)\,\zeta(\mathbf{r}',t')\rangle = -2k_B\Lambda\nabla^2\delta(\mathbf{r}-\mathbf{r}')\delta(t-t') \tag{67}$$

$$\langle\xi_i(\mathbf{r},t)\,\xi_j(\mathbf{r}',t')\rangle = -2k_B\eta\,\delta_{i,j}\nabla^2\delta(\mathbf{r}-\mathbf{r}')\delta(t-t') \tag{68}$$

The system of equations (64)–(68) can be solved numerically by eliminating pressure from the incompressibility condition

$$\nabla \cdot \mathbf{v}(\mathbf{r}) = 0 \tag{69}$$

For low-Reynolds-number fluids the second term in the right-hand side of the Navier–Stokes equation can be neglected. Additionally, assuming that the viscous relaxation occurs more rapidly than the change of the order parameter, the acceleration term in Eq. (65) can be also omitted. Such approximations are validated in the case of polymer blends, for which they become exact in the limit of infinite polymer length, $N \to \infty$. After these approximations, the NS equation can be easily solved in the Fourier space [160].

Morphology of the interface determines the growth exponents in the hydrodynamic regime of the phase separation. Siggia [161] has argued that for a bicontinuous morphology, the essential mechanism of domain growth is the hydrodynamic transport of the fluid along the interface driven by the surface tension [1]. In this case, the average domain size increases linearly with time, $L(t) \sim (\sigma/\eta)t$, where $\sigma$ is a surface tension. In contrast, when the minority phase consists of droplets, this mechanism tends to make droplet more spherical. The growth of the domains occurs by coalescence that can be described by the Binder–Staffer theory ($L(t)^{1/3} = 6f_m(k_BT/5\pi\eta)t$, where $f_m$ is the minority phase volume fraction [162]). The patterns emerging from the hydrodynamic coarsening in 2D are characterized by the double-separated domains [158,163] (Fig. 19). When the hydrodynamic motion is much faster than the concentration diffusion, the secondary phase transition can be induced within the domains that have

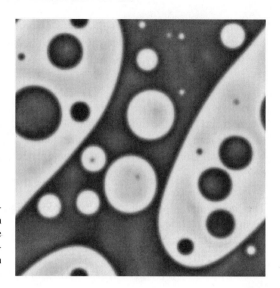

**Figure 19.** Double-phase-separated morphology reported from the 2D simulations of the phase separation in the hydrodynamic regime [158]. This figure has been kindly provided by Dr. T. Araki.

been coarsen by flow. The double-separated coarsening due to the hydrodynamic flow has not been yet observed in 3D simulations. A characteristic feature of the hydrodynamic patterns in 3D is a transient presence of the droplets that are disconnected from the main bicontinuous domain.

The above model assumes that both components are dynamically symmetric, that they have same viscosities and densities, and that the deformations of the phase matrix is much slower than the internal rheological time [164]. However, for a large class of systems, such as polymer solutions, colloidal suspension, and so on, these assumptions are not valid. To describe the phase separation in dynamically asymmetric mixtures, the model should treat the motion of each component separately ("two-fluid models" [98]). Let $\mathbf{v}_1(\mathbf{r}, t)$ and $\mathbf{v}_2(\mathbf{r}, t)$ be the velocities of components 1 and 2, respectively. Then, the basic equations for a viscoelastic model are [164–166]

$$\frac{\partial \phi}{\partial t} = -\nabla \cdot (\phi \mathbf{v}) - \nabla \cdot [\phi(1 - \phi)(\mathbf{v}_1 - \mathbf{v}_2)] \tag{70}$$

$$\mathbf{v}_1 - \mathbf{v}_2 = -\frac{1 - \phi}{\kappa} \left[ \nabla \cdot \hat{\Pi} - \nabla \cdot \hat{\sigma}^{(1)} + \frac{\phi}{1 - \phi} \nabla \cdot \hat{\sigma}^{(2)} \right] \tag{71}$$

$$\rho \frac{\partial \mathbf{v}}{\partial t} \approx -\nabla \cdot \hat{\Pi} + \nabla p + \nabla \cdot \hat{\sigma}^{(1)} + \nabla \cdot \hat{\sigma}^{(2)} \tag{72}$$

where $\kappa$ is a friction coefficient [164]. The average velocity $\mathbf{v}$ is

$$\mathbf{v} = \phi \mathbf{v}_1 + (1 - \phi) \mathbf{v}_2 \tag{73}$$

which satisfies the incompressibility condition (69). The expression for the osmotic tensor $\hat{\Pi}$ originates in the free-energy functional form (50):

$$\nabla \cdot \hat{\Pi} = \phi \nabla \left[ \frac{\partial f}{\partial \phi} - C\nabla^2 \phi \right] \tag{74}$$

The total force $\mathbf{F} = \nabla \cdot \hat{\sigma}$ acting on the mixture is comprised from the forces $\mathbf{F}_i$, which act on the $i$th component only:

$$\hat{\sigma} = \mathbf{F}_1 + \mathbf{F}_2 = \nabla \cdot \hat{\sigma}^{(1)} + \nabla \cdot \hat{\sigma}^{(2)} \tag{75}$$

Explicit forms for the stress tensors $\hat{\sigma}^{(i)}$ are deduced from the microscopic expressions for the component stress tensors and from the scheme of the total stress devision between the components [164]. Within this model almost all essential features of the viscoelastic phase separation observable experimentally can be reproduced [165] (see Fig. 20): existence of a frozen period after the quench; nucleation of the less viscous phase in a droplet pattern; the volume shrinking of the more viscous phase; transient formation of the bicontinuous network structure; phase inversion in the final stage.

The shear effects on the phase separation can be also considered within the TDGL models. The velocity profile imposed by the simply shear flow is described (in 2D) as [167,168]

$$\mathbf{v} = \gamma y \, \mathbf{e}_x \tag{76}$$

where $\gamma$ is the spatially homogeneous shear rate, and $\mathbf{e}_x$ is a unit vector in the flow direction. This flow field can be imposed in the dynamic equation (64), which is then solved numerically (this model neglects hydrodynamic fluctuations). Standard periodic boundary conditions are used in the flow direction, while in the $y$ direction the Lees–Edwards scheme [169] is adopted. The point $(x, y)$ is identified with the point $(x + \gamma L \Delta t, y + L)$, where $L$ is the size of the lattice and $\Delta t$ is the time discretization interval ($\gamma$ has to be chosen in such way that $\gamma L \Delta t$ is an integer). The patterns resulting from the simulation are characterized by a large anisotropy [Fig 8(b)]. However, the above model describes a very abstract system. In order to simulate realistic morphological transformations under shear, the hydrodynamic equations of the fluid motion should be implemented under appropriate boundary conditions, and Eq. (76) should be used only to generate starting configuration for the simulations. It has been shown recently that the Lattice–Boltzmann approach is also very effective for shear simulations [163,170–172].

Phase ordering in block copolymers can be described by the same dynamic equation as in the case of homopolymer blends [Eqs. (53)–(55)] with the LG

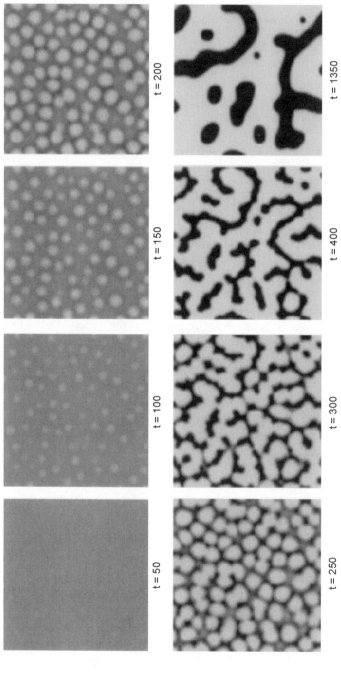

**Figure 20.** An example of the pattern evolution during the viscoelastic phase separation [165]: There is a frozen period after the quench; nucleation of the less viscous phase in a droplet pattern; the volume shrinking of the more viscous phase; transient formation of the bicontinuous network structure; phase inversion in the final stage. This figure has been kindly provided by Dr. T. Araki.

inhomogeneous free energy functional in form [173–175]

$$F[\Psi(\mathbf{r})] = \int d\mathbf{r} \left\{ -\frac{\tau}{2} \Psi(\mathbf{r})^2 + \frac{b}{2} [\nabla \Psi(\mathbf{r})]^2 + \frac{v}{3} \Psi(\mathbf{r})^3 + \frac{u}{4} \Psi(\mathbf{r})^4 \right\}$$
$$+ \frac{c}{2} \int d\mathbf{r}_1 \int d\mathbf{r}_2 \, G(\mathbf{r}_1 - \mathbf{r}_2) \Psi(\mathbf{r}_1) \Psi(\mathbf{r}_2) \qquad (77)$$

where $\Psi(\mathbf{r})$ is the order parameter [Eq. (23)], $\tau$ measures the relative distance from the ODT point, and phenomenological coefficients $b, c, u$, and $v$ are derived from the microscopic theory [36]. The last nonlocal term in Eq. (77) describes increased entropic elasticity that appears during the segregation in a block copolymer molecules and that confines the domain's dimension. It is usually required for function $G$ to satisfy Laplace's equation $\Delta G(\mathbf{r}, \mathbf{r}') = -\delta(\mathbf{r} - \mathbf{r}')$ and therefore give a local contribution into dynamic equation. It has been shown [175] that starting from one ordered mesophase and making a jump into the stability region of the other ordered phase, the order–order transition can be simulated. It should be stressed, however, that the characteristic domain size of block copolymer mesophase is of the order of the polymer chains radius of gyration, $R_g$. Therefore, to describe a realistic ordering process, the theory must operate at the level smaller than $R_g$, which essentially requires a nonlocal expression for the Onsager coefficient [143,147] and an exact expression for the free energy density [135]. In order to describe dynamics of the phase transition in the ternary mixtures, such as surfactant solutions or its polymer equivalent—mixtures of $A$–$B$ homopolymers with diblock (or random) copolymer—more than one order parameter must be introduced [176–178]. Consequently, the LG free-energy functional must be modified.

### D. Models of Reaction–Diffusion Systems

Spontaneous pattern formation can be observed in chemical reactions when the diffusion of the species plays a role. In the simplest case we put two chemicals, called an activator and an inhibitor, in a dish and allow them to react and diffuse. If we make the inhibitor diffuse faster than the activator, the system becomes unstable. The diffusion constant of the reactants can be altered by some control parameter—for example, by the temperature. The system's instability manifests as spatially inhomogeneous *steady* concentration patterns of the reactants, corresponding to the activator-rich and inhibitor-rich domains. The mechanism described above, which leads to the pattern formation in the reaction–diffusion system, was proposed in 1952 by Turing [179]. The stationary regular patterns emerging in the system are called the Turing patterns. The value of the control parameter at which the system gets unstable and the patterns occur is called the bifurcation point, and Turing's idea was to analyze the system close to this point.

The experimental evidence of the two-dimensional Turing patterns was discovered in the early 1990 in the chloride–iodide–malonic acid (CIMA) reaction in gel media [180–184]. These experiments demonstrated the existence of a range of stable regular patterns. Depending on the initial conditions, the patterns observed in the CIMA reaction can have (i) hexagonal structure, (ii) lamellar stripe structure, or (iii) irregular (turbulent) structure [183]. Examples of the 2D Turing patterns observed in the CIMA reaction are shown in Fig. 21. The gray-scale level corresponds to the concentration of iodide $I_3^-$. There are two types of the hexagonal Turing patterns observed: a triangular lattice (h0) composed of separated spots, shown in Fig. 21(a), and a honeycomb lattice (hπ). Due to defects (disclinations and dislocations) resulting from the nucleation processes the lamellar structure does not have a form of parallel stripes; instead the so-called labyrinthine patterns composed of meandering stripes are observed in real systems [Fig. 21(b)]. By changing the control

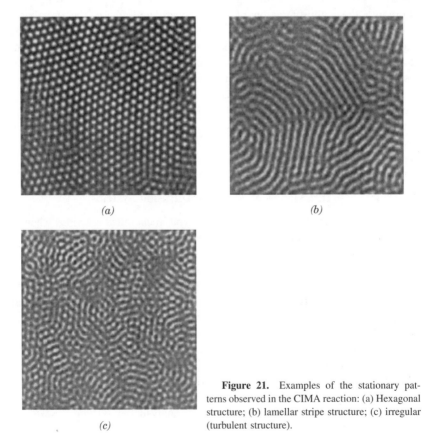

(a)                                            (b)

(c)

**Figure 21.** Examples of the stationary patterns observed in the CIMA reaction: (a) Hexagonal structure; (b) lamellar stripe structure; (c) irregular (turbulent structure).

parameters (the temperature and the concentration of the reagents), each pattern may convert reversely into another.

At the mesoscopic level, the reaction–diffusion system is described by a set of partial differential equations,

$$\frac{\partial c_i(\mathbf{r}, t)}{\partial t} = f_i(\dots, c_j, \dots) + D_i \nabla^2 c_i(\mathbf{r}, t) \tag{78}$$

where $c_i(\mathbf{r}, t)$ denote concentrations of the species, the functions $f_i$ represent the reactive processes which take place in the system, and $D_i$ are the diffusion coefficients. Usually, $f_i$ are highly nonlinear functions of the concentrations of the species. Since the seminal work of Turing, a number of specific reaction–diffusion models based on Eq. (78) have been developed. Among them, the best known are the Brusselator [185], the Lengley–Epstein model [186] of the CIMA system, the Schnackenberg model [187], or the Oregonator [188]. The oldest an perhaps the most celebrated realization of the reaction–diffusion system is the Brusselator, having originated in the research lab of Prigogine in Brussels. The Brusselator is a simple abstract model that comprises the four linked reactions: $A \rightarrow X$ $(a)$, $2X + Y \rightarrow 3X$ $(b)$, $B + X \rightarrow Y + C$ $(c)$, and $X \rightarrow D$ $(d)$. Assuming that the concentrations $c_A$ and $c_B$ of the reactants $A$ and $B$ are maintained at the constant values, the model is described by the following system of two non-linear equations for the concentrations $c_X$ and $c_Y$:

$$\begin{aligned}
\frac{\partial c_X}{\partial t} &= k_a c_A + 2 k_b c_X^2 c_Y - k_c c_X c_B - k_d c_X D_X \nabla^2 c_X \\
\frac{\partial c_Y}{\partial t} &= -k_b c_X^2 c_Y + k_c c_X c_B + D_Y \nabla^2 c_y
\end{aligned} \tag{79}$$

where $k_i$ are nonnegative rate constants of the four reactions $(a)$–$(d)$. The Turing patterns are obtained as the isoconcentration surface of one of the species, say X, from the condition $c_X(\mathbf{r}) = \text{const.}$

The set of the reaction–diffusion equations (78) can be solved by different methods, including bifurcation analysis [185,189–191], cellular automata simulations [192,193], or numerical integration [194—197]. Recently, two-dimensional Turing structures were also successfully studied by Mecke [198,199] within the framework of integral geometry. In his works he demonstrated that using morphological measures of patterns facilitates their classification and makes possible to describe the pattern transitions quantitatively.

So far, there is no experimental evidence of the Turing patterns in three-dimensional reaction–diffusion systems. However, the formation of 3D Turing structures has been investigated for the Brusselator by De Wit et al. [194,197]. In their simulations, the authors obtained four types of 3D Turing patterns: (i) the hexagonally packed cylinders (hpc), (ii) body-centered cubic lattice (bcc), (iii) lamellar structure composed of parallel walls, and (iv) the Scherk surfaces. In three dimensions, the hexagonal patterns (hpc) may have either hpc0 or hpcπ structures. They are, respectively, 3D extensions of the 2D triangular (h0) and the honeycomb (hπ) lattices. Similarly, there are two types of the bcc patterns. In the first type, called the bcc0 lattice, the maxima of concentrations occupy a body-centered grid while the minima are arranged in filaments possessing cubic symmetry. In the second type, the bccπ lattice, the maxima and minima are interchanged. The Scherk structure, shown in Fig. 22, belongs to the class of doubly periodic minimal surfaces. In numerical integration of the Brusselator the Scherk surface was obtained by applying the periodic boundary conditions along the $x$ and $y$ directions and the no-flux boundary conditions along the $z$ direction.

### E.  Model of CDW in a Two-Dimensional Electron Liquid

Another example of systems where the pattern formation is observed is a two-dimensional electron liquid in a weak magnetic field with partially filled upper Landau levels. In such systems the uniform distribution of the charge density at the upper Landau level is unstable against the formation of a charge density

**Figure 22.**  A doubly periodic Scherk surface.

wave [200–202] (CDW). The CDW state corresponds to the two-dimensional patterns, composed of domains with the filling factor equal to one or zero. The shape of these domains can be studied by using the morphological tools presented in the following sections.

Consider a clean two-dimensional electron liquid in a weak magnetic field with the lowest Landau levels (LLs) completely filled and the upper LLs partially filled. Assume that the interaction between electrons is much smaller than the inter-Landau level spacing, $\hbar\omega_c$ (here $\omega_c$ is the cyclotron frequency). Under this condition the electron–electron interactions do not destroy the Landau quantization. This means that the LLs do not mix and that the system can be described in terms of the highest LL only. Additionally, it is assumed that the number $N$ of the LLs is very big, $N \gg 1$. This implies that the repulsive interaction between a pair of electrons at the upper level is screened by the lower levels. The resulting interaction potential $u(x)$ between two electrons is a function of the distance $x$ separating centers of their cyclotron orbits. The radius $R_c$ of the cyclotron orbit is given by $R_c = \sqrt{2N + 1}\,l$, where $l$ is the magnetic length $l = \hbar/\sqrt{m_e\omega_c}$. The interaction potential can be well approximated by the "box-like" function $u(x) = u_0\Theta(2R_c - x)$, where $\Theta$ stands for the Heaviside function. In other words, there is an analogy between the partially filled LLs discussed and the gas of "soft repulsive rings" having the radii equal to the cyclotron radius $R_c$. The form of the interaction potential makes the uniform distribution of the electron density unstable and gives rise to the appearance of the CDW patterns. The morphology of these patterns is controlled by the parameter $\bar{v}_N$, which describes filling of the upper ($N$th) LL, $\bar{v}_N = v - 2N$, where $v$ is the *total* filling factor $v = k_F^2 l^2$ ($k_F$ stands for the Fermi wavevector of the 2D electron liquid in zero magnetic field). In the CDW state the charge density is distributed among domains with the filling factor equal to one or zero in such a way that the average occupancy of the upper level is equal to $\bar{v}_N$. The value of the filling parameter $\bar{v}_N$ is a fraction from the interval $[0,1]$. Note also that due the electron–hole symmetry it is sufficient to consider the parameter $\bar{v}_N$ varying in the range $0 < \bar{v}_N \leq \frac{1}{2}$.

For small values of the parameter $\bar{v}_N$ the ground state of the partially filled LL corresponds to the CDW pattern having the hexagonal structure, composed of separated domains ("bubbles"). These domains contain about $3\bar{v}_N N$ electrons, and their centers are separated by the distances ranging from two to approximately three cyclotron radii. The "bubble" structure is also called the "super" Wigner crystal. (Note that the classical Wigner crystal is formed for sufficiently small filling parameter, $\bar{v}_N \ll 1/N$, when the rings centered at neighboring lattice sites do not overlap; each site is then occupied by a single electron.) If we start to increase the filling of the upper level, the radii of the domains and the distance between them successively grow. Eventually, if the filling parameter $\bar{v}_N$ approaches some critical value $\bar{v}_N^\star$, the domains merge into parallel "stripes." The spatial period of the stripe structure is roughly equal to $3R_c$. The value of the parameter $\bar{v}_N^{\star}$ at which the transition from the "bubble" pattern to the "stripe" structure occurs depends on the number the LLs and is equal to 0.4 in the limit of large $N$.

## III.  MORPHOLOGICAL MEASURES

The most common form of data obtained from the computer simulations of mesoscopic models—numerical solution of the equations, experimental measurements such as laser scanning confocal microscopy (LSCM) [5], thermo electron micro-topography (TEMT) [8], magnetic resonance imaging, radio isotope topography, nanotomography [203], and others—is a set of scalar data arranged on a 3D lattice. Sometimes, experiments could bring directly desirable information about system properties. For example, location of the scattering intensity maxima detects some characteristic lengthscale in the system. However, in most of the cases, in order to extract a certain morphological feature, the raw data requires a further transformation. One may either determine the bulk properties of the field configuration $\phi(\mathbf{r})$ by introducing some correlation functions or investigate the interface morphology (morphology of the isosurface $\phi(\mathbf{r}) = \text{const}$). The last approach is validated when the interface is well-defined.

### A.  Digital Patterns

The most common data containing morphological information are 2D images. They can be obtained almost for all systems of interest by using different experimental techniques. Morphology of the system—that is, size, shape, and spatial arrangement of inhomogeneities—is usually made detectable for experimental devices by introducing some contrast-enhancement agent. As a result of the measurement, the spatial morphology of the system is projected onto the image plane by specifying intensity (or set of intensities) at each point of the image. A 3D image can be obtained by collecting 2D images shifted along the normal direction [5] (LSCM). The 3D image is then reconstructed by computationally stacking the images. The other possibility is to collect a set of images taken at different angles to the surface plane [8] (TEMT). The 3D representation is reconstructed in this case by using the filtered back-projection algorithm [204]. A digital image is created by mapping positions within the image onto a grid and attaching the quantized intensity to each grid point. The quantized intensity is usually determined by dividing the intensity range into a fixed number of bins and assigning to each grid point the number of the bin that most closely matches the intensity [205]. The most extreme form of quantization, by using only two bins, results in black-and-white (binary) images.

Binary digital images are obtained from the set of scalar data by thresholding. In the intensity level of a raw image, $\phi(\mathbf{r})$ varies between $\phi_{\min}$ and $\phi_{\max}$, and $\phi_{thr}$ is a threshold value. The thresholding operation results the binary image $P(\mathbf{r})$:

$$P(\mathbf{r}) = \begin{cases} 1, & \phi(\mathbf{r}) \geq \phi_{thr} \\ 0, & \phi(\mathbf{r}) < \phi_{thr} \end{cases}$$

This procedure can be applied for any scalar field, for example, for the local volume fraction configuration generated during the TDGL simulations. Mathematically, morphology of the binary images can be completely described in terms of the Minkowski functionals [199,205–208]. In two dimensions the Minkowski functionals are proportional to the area $A^{(d=2)}$ covered by the black pixels, the boundary length $L$, and the Euler characteristic $\chi^{(d=2)}$. In three dimensions the four independent Minkowski functionals are proportional to the volume $V$, surface area $A^{(d=3)}$, integral mean curvature $H$, and Euler characteristic $\chi^{(d=3)}$. Here, the Euler characteristic is defined in the same way as in algebraic topology: $\chi^{(d=2)}$ equals the number of connected components minus the number of holes, and $\chi^{(d=3)}$ is given by the number of connected components minus the number of passages plus the number of cavities [205].

These morphological measures can be easily calculated for binary images [199,205,207]. In the 2D case, the area of white (or black) pixels is calculated just by counting the number of pixels. For a pixel configuration shown in Fig. 23, the area of white pixels is $A_w^{(d=2)} = 29a^2$ and the area of the black pixels is $A_b^{(d=2)} = 13a^2$. It is convenient to specify the area fraction as $f_A = A_b/A_{total}$, where $A_{total}$ is a total area. The boundary length $L$ can be defined as a number of edges between the black and white pixels. For the configuration shown in Fig. 23, $L = 24$. The boundary length within the above definition does not converge to the continuous boundary length in the limit of vanishing lattice spacing. Therefore, in physical applications, the boundary length density $u = L/N$ is usually used ($N$ is a total pixel number). The Euler characteristic is calculated by summing up local "curvatures" along the black/while pixels boundary. For a square lattice, the local curvatures $\tau_c$ could have one of three values $\{-1, 0, +1\}$ (the Euler characteristic here is not normalized) depending on the bending of the boundary; that is, $\tau_c = \pm 1$ if the boundary is curved, and $\tau_c = 0$ if it is a straight line [199] [compare to Eqs. (120), (127), and (8)]. The pixel area, boundary length, and Euler characteristic are threshold-dependent quantities. The generalization of the above definition for various 3D lattices is straightforward [209,210] (see also Section III.F.4). In the following sections we shall compare the efficiency of the digital pattern method to others methods.

## B. Simplex Decomposition

The main idea of the simplex decomposition method is to divide the lattice, on which the scalar field is specified, into small subunits and approximate the

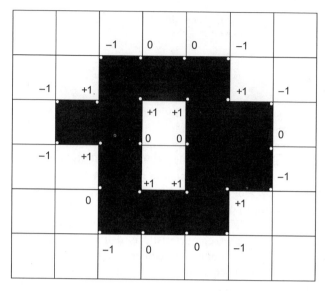

**Figure 23.** The 2D illustration of the digital pattern analysis for computing the Euler characteristic. The local curvature variables $\tau \epsilon \{-1, 0, +1\}$ are assigned to the each lattice site at the boundaries of black pixels. The Euler characteristic is a sum of local curvature variables $(\chi = 8(+1) + 8(-1) + 8(0) = 0)$.

isosurface (interface) inside each of them by simple polygons. The values of the field at the lattice points are used directly to draw the surface; therefore, no information is lost in contrast to the thresholding procedure of the digital pattern analysis.

The most intuitive way to construct the simplex decomposition of the lattice is to divide it into small cubes [21–23]. The surface inside each small cube is triangulated—represented by the interconnected triangles. One of the possible realizations of this scheme (marching cube algorithm [211]) is shown in Fig. 24. The surface $\phi(x, y, z) = \phi_0$ is detected at the edge of the simplex, If the field has different signs (compared to $\phi_0$ level) at the edge ends: If $\phi(\mathbf{x}_1 = [i, j, k]\Delta x) = \phi_{i,j,k} < \phi_0$ and $\phi(\mathbf{x}_2 = [i + 1, j, k]\Delta x) = \phi_{i+1,j,k} > \phi_0$, then the point $\mathbf{x}_0$, for which $\phi(\mathbf{x}_0) = \phi_0$, must lie between the point $\mathbf{x}_1 = [i, j, k]\Delta x$ and $\mathbf{x}_2 = [i + 1, j, k]\Delta x$ ($\Delta x$ is a mesh size) . Moreover, location of $\mathbf{x}_0$ depends on the values of the field at the point $\mathbf{x}_1$ and $\mathbf{x}_2$. It can be found by the linear interpolation:

$$\mathbf{x}_0 = \left( i + \frac{\phi_{i,j,k} - \phi_0}{\phi_{i,j,k} - \phi_{i+1,j,k}}, j, k \right) \Delta x \qquad (80)$$

Once the locations of the field intersection with the simplex edges is found, the isosurface inside the simplex is represented by triangles.

Being the most intuitive, the cube lattice decomposition scheme suffers from arbitrary choices. Thus, the 14 cases shown in Fig. 24 do not explore all the possibilities; and, when the field values are noisy, this could lead to ambiguous situations. An example is shown in Fig. 25 where for a given field configuration two different surface representations can be drawn. In the case Fig. 25(a) there are two surfaces inside the cube, while in Fig. 25(b) (the same as case 7 in Fig. 24) the number of surfaces is 3. The two cases shown in Fig. 25 specify the different number of connections between the triangles, which has immediate consequences on the Euler characteristic value. To resolve this problem, one can use the field values at the vertices of the cube to divide the cube further, into smaller cubes, then find the field values at the vertices of the new cubes by the interpolation and apply the surface approximation scheme again [76,79]. In such a way the number of ambiguous situations can be reduced significantly; however, it cannot eliminate, in principle, all of them.

An alternative way to resolve the ambiguous surface approximations is to divide the original lattice not into cubes, but into more primitive subunits. Each cube can be divided into six five-vertex pyramids as is shown in Fig. 26. Now, the surface can be drawn unambiguously by using the pyramid classification scheme shown in Fig. 27. Depending of the field values at the pyramid vertices and at the center of the four-vertex face, the surface is represented by the triangle, tetragons, pentagons and hexagons. Case $e$ in Fig. 27 can be divided into two further cases—that is, cases $e_1$ and $e_2$ shown in Fig. 28, which slightly modify the area of the isosurface but have no consequences for the Euler characteristic. Note that in order to employ such a decomposition scheme, the field values at the center of the cube (CC) and at the centers of its faces (FC) should be found by the linear interpolation. Using the interpolation for this purpose is justified, because it operates at the same level of accuracy as a surface location procedure (it is also a consequence of the linear approximation).

A single lattice unit can be also divided into the simplest possible units, the primitive four-vertex simplexes (tetrahedrons). One of the possible realizations of this procedure is shown in Fig. 29. The other decomposition schemes where a single cube is divided into five or seven four-vertex pyramids [212,213] are also possible. When the tetrahedron decomposition is employed, the surface inside the simplex can be represented by only two cases: a triangle or a tetragon, as shown in Fig. 30. Note that by combining two neighboring four-vertex simplexes a single five-vertex pyramid can be formed. However, two cases shown in Fig. 30 cannot reproduce all the possible surface localizations defined within the five-vertex decomposition scheme (case $g$ in Fig. 27 and case $e_2$ in Fig. 28). The choice of the decomposition scheme is dictated by the morphological measure which has to be calculated, required speed of calculation, and other computational resources. In the following sections we shall discuss this point for each morphological measure. Here, we note that the five-vertex

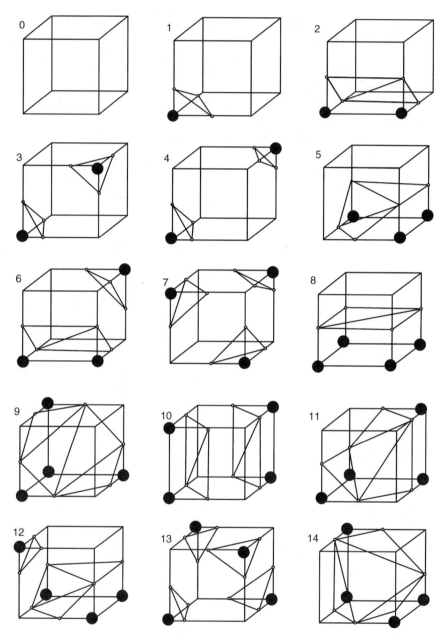

**Figure 24.** Fourteen cases of the polygonal approximation for the isosurface inside a cube used by the marching cube algorithm [211].

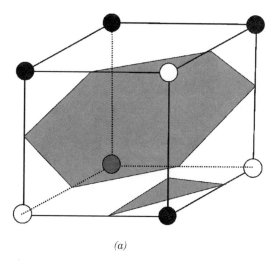

(a)

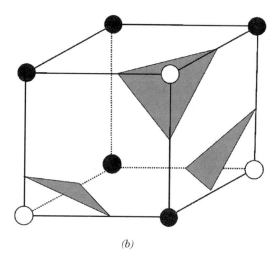

(b)

**Figure 25.** Locations of the intersections between the isosurface and the cube edges are the same in both figures, (a) and (b) (also as in case 7 Fig. 24), but the value of the Euler characteristic depends on the way the points are connected. This demonstrates the ambiguity of the cubic decomposition method.

(pyramid) and four-vertex (tetrahedron) decomposition can be applied to any of the irregular lattices and therefore is very general.

## C. Fourier Transform Analysis

The beam of the radiation passing through the studied object interact with the material that induces a certain scattering pattern of the reflected radiation detectable by the radiation sensors. This scattering is caused by the

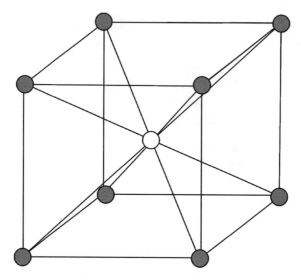

**Figure 26.** A cube is decomposed into six five-vertex pyramids. The gray circles are the field values at the lattice, the "CC" circle is a four-edge vertex, and the "FC" circles denote the centers of the four-vertex faces of the pyramids.

inhomogeneities of the scattering cross section of the material. The scattering cross section is determined by some local values of the material property the radiation is sensitive to—for example, its density or polarization [214]. In the first Born approximation the amplitude of the scattered wave is given by the Fourier transform of that property. If the inhomogeneous scattering property at position $\mathbf{r}$ and time $\tau$ is represented by the scalar field $\phi(\mathbf{r}, \tau)$, and its Fourier transform at the wavevector $\mathbf{k}$ is $\phi(\mathbf{k}, \tau)$, then the observed scattering intensity is

$$I(\mathbf{k}, \tau) = |\phi(\mathbf{k}, \tau)|^2 \tag{81}$$

where the proportionality constant has been ignored. For isotropic materials the structure factor depends only on the absolute value of the wavevector,

$$S(k, \tau) = \langle I(\mathbf{k}, \tau) \rangle \tag{82}$$

The inverse of the typical domain size is proportional to the location $k_m(\tau)$ of the peak of the spherically averaged structure factor.

The same analysis can also performed for the data obtained in computer experiments. The structure factor is calculated as [53]

$$S(\mathbf{k}, \tau) = \left\langle \frac{1}{L^3} \sum_{\mathbf{r}} \sum_{\mathbf{r}'} e^{i\mathbf{k}\mathbf{r}} [\phi(\mathbf{r} + \mathbf{r}', \tau)\phi(\mathbf{r}', \tau) - \phi_0^2] \right\rangle \tag{83}$$

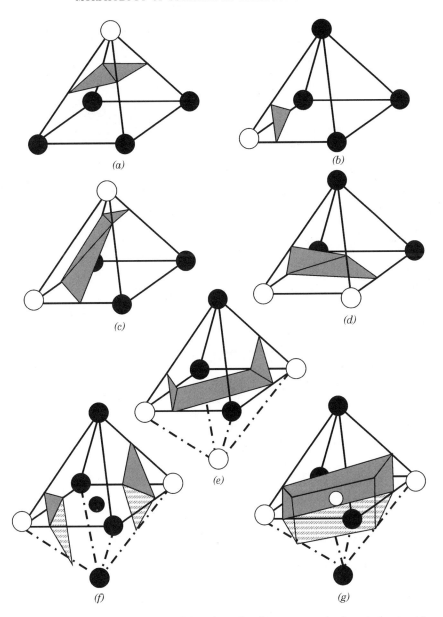

**Figure 27.** Seven possible cases of the polygonal surface representation in a single pyramid. The Euler characteristic is calculated as a sum of the number of faces and the number of vertices minus the number of edges of the polygons. The black and white circles represent points with higher and lower values relative to the threshold one. The gray area is the schematic representation of the surface inside a pyramid [225].

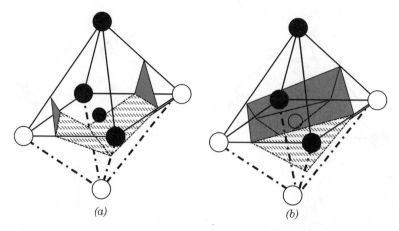

(a)                                          (b)

**Figure 28.** Case (e) from the pyramid classification procedure, Fig. 27, can be divided into two further cases (a) and (b). This improvement does not change the total Euler characteristic.

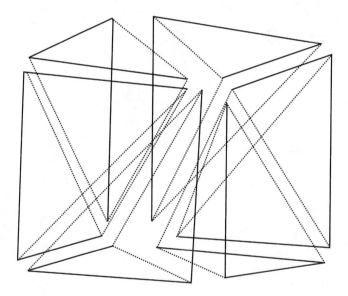

**Figure 29.** Decomposition of a cubic lattice cell into six four-vertex pyramids.

where the sums run over all lattice sites ($L$ is a lattice size), and the **k** vectors belong to the first Brillouin zone in the reciprocal space; that is, $\mathbf{k} = (2\pi/L\Delta x)\vec{v}$, $\vec{v} \equiv (v_x, v_y, v_x)$, and $0 \leq v_x, v_y, v_z \leq L - 1$. The spherically averaged structure factor is calculated as

$$S(k, \tau) = \sum_{k-(\Delta k/2)<|k|\leq k+(\Delta k/2)} S(k, \tau)/n(k, \Delta k) \qquad (84)$$

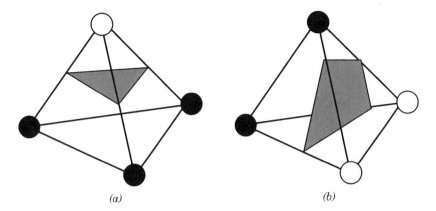

*(a)*          *(b)*

**Figure 30.** Two cases of the polygonal surface representation in a single triangular pyramid (tetrahedron). The black and white circles represent the point with higher and lower values in comparison with the threshold value. The gray area is sthe schematic representation of the surface inside the pyramid (see Fig. 29).

where

$$n(k, \Delta k) = \sum_{k-(\Delta k/2) < |k| \le k+(\Delta k/2)} 1 \tag{85}$$

denotes the number of lattice points in a spherical shell of width $\Delta k$ centered around $k$. A convenient value for the discretized Brillouin zone in lattice simulations is [152] $\Delta k = 2\pi/L\Delta x$. The maximum $k_m$ of the structure factor $S(k, \tau)$ is calculated by fitting the points near the maximum to a certain probe function (e.g., parabola) and finding the maximum for this fitted function. Due to the finite discretization of the lattice, the accuracy of the determination of $k_m(\tau)$ is rather small from simulation data. More precise information about the characteristic domain size is obtained by considering the real-space pair correlation function, defined as

$$G(\mathbf{r}, \tau) = \sum_{\mathbf{k}} e^{-i\mathbf{k}\mathbf{r}} S(\mathbf{k}, \tau) \tag{86}$$

The normalized spherically averaged correlation function is

$$g(r, \tau) = G(r, \tau)/G(0, \tau) \tag{87}$$

such that $g(0, \tau) = 1$. A quantitative measure of the domain size is the location of the first zero in the correlation function, $R_g(\tau)$. In the case of the system of the nonconserved order parameter, the domain size $L(\tau)$ can also be estimated by fitting the data to the appropriate analytical formulas [215–217] for the correlation function. The most successful formula for the correlation function

$g(r, \tau) = \langle \text{sgn}(\phi(\mathbf{r}, \tau)\text{sgn}(\mathbf{0}, \tau) \rangle$ was proposed by Ohta, Jasnow, and Kawasaki [216],

$$g(r, \tau) = \frac{2}{\pi} \arcsin \left[ \exp(-r^2/L(\tau)^2) \right] \tag{88}$$

for the nonconserved order parameter evolution [Eq. (52)].

The other useful operation available within the Fourier space analysis is a cross-sectioning of images, which allows us to obtain quantitative information about their similarity or differences [218]. The correlation function $C(\mathbf{r})$ is calculated as

$$C(\mathbf{r}) = \int_V \left\{ \left( \int_V I_1(\mathbf{r_1})\exp(-i\mathbf{kr_1}) \, d\mathbf{r_1} \right) \right.$$
$$\left. \times \left( \int_V I_2(\mathbf{r_2})\exp(-i\mathbf{kr_2}) \, d\mathbf{r_2} \right) \right\} \times \exp(-i\mathbf{kr}) \, d\mathbf{k} \tag{89}$$

This type of analysis is useful for studying the phase separation phenomena, also in detecting influences of external fields [218] (strain, etc.). In particular, by investigating speckles on scattering patterns, a certain information about interface fluctuations can be obtained [214].

The advantage of the Fourier space analysis is in providing the information directly observable in scattering experiments. However, it is almost impossible to calculate (or measure) certain morphological features by using this technique—for example, the Euler characteristic of the interface. The one possibility is to try to reconstruct the real-space morphology by using the inverse Fourier transform for a set of 2D intensity projections taken at all possible angles. Nevertheless, the spherically averaged structure factor contains *no* information about the domain connectivity. To establish this fact, the spherically averaged structure factor has been computed for a droplet morphology [Fig. 18(c)]. Then, we generate by using the random number generator a 3D set of the Fourier components that produce exactly the same spherically averaged structure factor (we employ the Gaussian distribution for both real and imaginary part). The generated set has been reversed to the real space. Obtained morphology was bicontinuous. This proves that the spherically averaged structure factor does not contain sufficient information to judge about the type of morphology.

## D. Surface Area

To calculate the surface area of a 3D object by using the digital pattern analysis, the pixels neighboring opposite phase must be classified as one of the following cases: vertex, open line segment, open square, and open cube [205,208]; and they must be counted one, by one, multiplied by the weight assigned to each vertex type. The weights (and the number of pixel types) depend on the lattice type being considered [208]. This method provides very fast but highly inaccurate estimation of the surface area density. The digitizing procedure for one pixel increases the surface area by the factor between $\sqrt{2}$ and $\sqrt{3}$. Therefore, the computed total surface area per unit volume is always larger than the real area by the factor of $\approx 1.6$. Thus, the area per unit cell of the P, D, and G minimal surfaces computed by the digital pattern analysis on $128^3$ lattice are [208] 3.67 (2.345), 6.00 (3.838), and 4.85 (3.092), respectively, where these values computed by accurate methods [23] are given in the brackets. In principle, by knowing the curvature distribution, the obtained surface area can be corrected. Nevertheless, an exact calculation of the surface area within this method is not possible.

The surface area of a 3D object sometimes can be measured by analyzing its 2D image. The perimeter length density of the domains extracted from the 2D image is proportional to the surface area density of the 3D object. However, to determine the proportionality coefficient itself, a certain model of the 3D object must be assumed [219,220] (for example, a system of spheres of a certain average size and polydispersity, which can be deduced from the 2D domain area distribution). A principal difficulty is that the model itself cannot be derived on the basis of the 2D images: The cross section of the bicontinuous and droplet morphology can be very similar. Assuming the same model for 3D morphology, the perimeter length can be used as a relative measure of the surface area.

The surface area per unit volume can be also measured in scattering experiments. For the fully developed system containing the domains of size $L$ and the interface of the intrinsic width $\xi$ ($\xi \ll L$), the Porod law [1] predicts the following asymptotic for the structure factor $S(\mathbf{k})$

$$S(\mathbf{k}) \sim \frac{1}{L k^{d+1}}, \qquad kL \gg 1 \qquad (90)$$

where $d$ is the space dimension. The prefactor $1/L$ up to the constant measures the interface area per unit volume.

The most precise estimation of the surface area can be made within one of the simplex decomposition scheme. The total area of the triangulated isosurface (isosurfaces) $\phi(\mathbf{r}) = \phi_0$ is obtained by summing up the areas of the triangles. When more than one surface is present in the studied object, it is necessary to separate different surfaces in order to calculate morphological measures for each surface. To separate points that belong to different surfaces, one has to choose an arbitrary starting point at one of the surfaces and follow the connections between the points to find the rest of the points that belong to that surface. The original set of points is then reduced by the points that have been already classified, and the procedure is repeated until all sets of the connected points will be separated. The description of the optimal algorithms can be found at http://www.ichf.edu.pl/mfialkowski/morph.html. After the separation procedure is completed, the area of each surface is calculated by summing up areas of the triangles belonging to this surface. These sets of points can be also used to calculate the Euler characteristic, curvature distributions, and enclosed volume for each surface.

**Example.** Although the surface area itself very often represents a desirable information about the system morphology, it can sometimes be used as a measure of the characteristic lengthscale in the system. Thus, if the volume fraction of the domains ($f_m$) separated by the surface is constant (or nearly constant), the following geometrical relation holds regardless of the domain size and arrangement:

$$\frac{S \cdot L}{V} \approx f_m \tag{91}$$

where $L$ is the characteristic lengthscale in the system. In such a case, the inverse of the surface area density $\Sigma^{-1} = (S/V)^{-1}$ is proportional to the average domain size. The advantage of using the surface area density as a measure of the characteristic lengthscale is that it is not so much sensitive to the finite lattice size or spacing in contrast to the maximum of the scattering intensity or the first zero of the pair correlation function. In Fig. 31 the simulated growth of the domains during the spinodal decomposition in the hydrodynamic regime is indicated by the increase of the first zero of the pair correlation function [Fig. 31(a)] and by the decrease of the surface area density [Fig. 31(b)]. In order to extract the growth exponent, a large number of independent runs should be performed in the Fig. 31(a) while a smaller number of those would lead to the same result in Fig. 31(b).

### E.  Volume Fraction

The volume fraction of the domains enclosed by the surface $\phi(\mathbf{r}) = \phi_0$ can be easily estimated by calculating the number of points for which the local volume fraction $\phi(\mathbf{r})$ is greater than $\phi_0$. The volume fraction of the $A$-type domains (with $\phi(\mathbf{r}) > \phi_0$) is given by the ratio of that number to the total number of points on the lattice. This algorithm is an exact implementation of the digital pattern method. It provides rather accurate and computationally cheap information. The other advantage of this method is that it does not require a 3D input data because it simply counts the points regardless to their spatial arrangement. Thus the volume fraction can be detected from 2D images.

However, in some special cases, the lost of information due to the thresholding procedure may cause a noticeable systematic error, because each lattice point such that $\phi(\mathbf{r}) > \phi_0$ contributes the same volume fraction $1/L^3$ regardless of the field magnitude. Consider an asymmetric binary mixture undergoing the phase separation. The local volume fraction distribution $P(\phi)$ has maxima at the equilibrium volume fractions, $\phi_A^{(1)} < \phi_A^{(2)}$, and is asymmetric relatively to

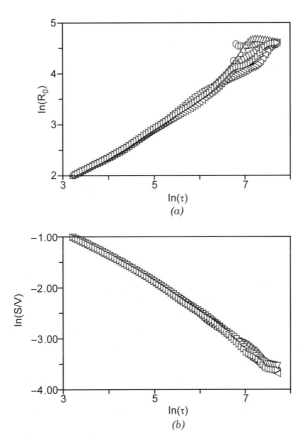

**Figure 31.** The domain growth during the phase separation process reflected by the shift of the first zero in the pair correlation function (a) and by the surface area reduction (b). Although the surface area and first zero of the pair-correlation functions are equivalent lengthscales, the time dependence of the surface area is less affected by the finite lattice size affects.

the average volume fraction $\phi_0$: $P(\phi_A^{(1)}) > P(\phi_A^{(2)})$. The thermal noise is modulated by adding random numbers of certain variance [Eqs. (67) and (68)] to each lattice point. Due to the asymmetry of the local volume fraction distribution, the number of points with $\phi(\mathbf{r}) > \phi_0$ and $\phi(\mathbf{r}) < \phi_0$ depends on the noise amplitude. Because each lattice point has equal contribution to the total volume fraction, even if $[\phi(\mathbf{r}) - \phi_0]$ is very small (noise induced), the volume fraction of $A$-type domains increases with the noise amplitude.

The volume confined by the surface $\phi(\mathbf{r}) = \phi_0$ can be calculated more precisely if the surface has been triangulated within one of the simplex decomposition schemes. Having the surface represented by the polygons inside a simplex, the volumes of the geometrical object specified by the polygons can be easily computed. The total volume is obtained by summing up over all simplexes. The tetragonal simplex decomposition is the most convenient for this purpose, because it offers the simplest geometrically volume decomposition inside the simplex.

The other way to calculate the volume inside the triangulated surface is to use the Ostrogradski–Gauss theorem. It relates the surface integral from a vector field $\mathbf{j}$ to a volume integral from its divergence:

$$\int_\Sigma \mathbf{j} \cdot d\mathbf{S} = \int_v \operatorname{div} \mathbf{j} \, dV \tag{92}$$

Substituting $\mathbf{r}$ instead of the vector field, the volume inside the surface is obtained as

$$V_s = 3 \int_\Sigma \mathbf{r} \, d\mathbf{S} = 3 \sum_{i = \{\text{triangles}\}} \int_{\Sigma_i} \mathbf{r} \, d\mathbf{S} = 3 \sum_{i = \{\text{triangles}\}} f(\{i\}) \tag{93}$$

The integral in the above expression is replaced by the sum of integrals over triangles that represent a given surface. The integral over a single triangle can be easily performed analytically. This method is very useful when a volume inside each of the disconnected surfaces have to be found. It works very good in the case of the disperse morphology, but to use it for characterization of the hyperbolic surfaces ($\chi_{Euler} \leq 0$) under the periodic boundary conditions, the virtual continuations of the surfaces on the sides of the lattice box must be introduced. The application of the volume fraction calculations in relation to the percolation threshold problem will be described in Section III.G.

### F.   Mean, Gaussian, and Principal Curvatures

The curvature at a given point on a surface is characterized by the maximum and minimum radii, $R_1$ and $R_2$, of the circles in mutually perpendicular planes perpendicular to the plane tangent to the surface at that point, best approximating the curves formed by the intersection of the surface with these planes [221], (Fig. 1). The principal curvatures, $k_1$ and $k_2$ are defined as

$$\begin{aligned} k_1 &= 1/R_1 \\ k_2 &= 1/R_2 \end{aligned} \tag{94}$$

The mean curvature is defined as

$$H = (k_1 + k_2)/2 \tag{95}$$

and the Gaussian curvature as

$$K = k_1 k_2 \tag{96}$$

When $k_1 = -k_2$ at every point, the surface is called minimal, which implies nonpositive $K$. The surface is elliptic if $K > 0$, hyperbolic if $K < 0$, and parabolic if $K = 0$.

The local properties of the surface can be described quantitatively by making statistics of the curvatures. The distribution functions $P(H)$, $P(K)$, and $P(k_1, k_2)$ are constructed to give the probability of finding a point at the surface which is characterized by the mean curvature, $H$, the Gaussian curvature, $K$, or the set of two principal curvatures $k_1$ and $k_2$. Because the curvatures are the surface defined quantities, each surface point where the curvatures are being determined must be assigned with a certain surface area. Values of the probability function simply correspond to the parts of the surface area assigned to the curvatures from some arbitrary defined intervals. If the surface has been triangulated, the curvature at the mass centers of each triangle (given by the one-third of the sum of the curvatures at the triangle vertices) is assigned by the area of the triangle. The curvature distribution functions are usually normalized to unity.

### 1. Direct Discretization of Derivatives

The mean, Gaussian, and principal curvatures are well-defined quantities from the standpoint of the differential geometry. If the surface is given by the *equation*

$$\phi(x, y, z) = \phi_0 \tag{97}$$

then $H$, $K$, $k_1$, and $k_2$ at point $\mathbf{r}_0$ can be expressed by the combination of the partial derivatives of the function $\phi(x, y, z)$ at that point. The mean curvature $(H)$ is given by the divergence of the unit vector normal to the surface, $\mathbf{n}(\mathbf{r})$,

$$H(\mathbf{r}) = -\frac{1}{2} \nabla \mathbf{n}(\mathbf{r}) = -\frac{1}{2} \nabla \frac{\nabla \phi(\mathbf{r})}{|\nabla \phi(\mathbf{r})|} \tag{98}$$

and the Gaussian curvature $(K)$ by the formula [23]

$$K(\mathbf{r}) = \frac{1}{2} \{ -(\partial_i n_j)^2 + [\nabla \mathbf{n}(\mathbf{r})]^2 \} \tag{99}$$

In numerical calculations the following formulas are used:

$$H = -\frac{1}{2\sqrt{\phi_x^2 + \phi_y^2 + \phi_z^2}} \frac{B}{A} \tag{100}$$

$$K = \frac{1}{\phi_x^2 + \phi_y^2 + \phi_z^2} \frac{C}{A} \tag{101}$$

where $A$, $B$, and $C$ are

$$A = -(\phi_x^2 + \phi_y^2 + \phi_z^2) \tag{102}$$

$$\begin{aligned}B = \phi_x^2(\phi_{yy} + \phi_{zz}) + \phi_y^2(\phi_{xx} + \phi_{zz}) + \phi_z^2(\phi_{xx} + \phi_{yy}) \\ - 2\phi_x\phi_y\phi_{xy} - 2\phi_x\phi_z\phi_{xz} - 2\phi_y\phi_z\phi_{yz}\end{aligned} \tag{103}$$

$$\begin{aligned}C = \phi_x^2(\phi_{yz}^2 - \phi_{yy}\phi_{zz}) + \phi_y^2(\phi_{xz}^2 - \phi_{xx}\phi_{zz}) \\ + \phi_z^2(\phi_{xy}^2 - \phi_{xx}\phi_{yy}) + 2\phi_x\phi_z(\phi_{xz}\phi_{yy} - \phi_{xy}\phi_{yz}) \\ + 2\phi_x\phi_y(\phi_{xy}\phi_{zz} - \phi_{xz}\phi_{yz}) + 2\phi_y\phi_z(\phi_{yz}\phi_{xx} - \phi_{xy}\phi_{xz})\end{aligned} \tag{104}$$

The principal curvatures are computed as

$$\begin{aligned}k_1 = H + \sqrt{H^2 - K} \\ k_2 = H - \sqrt{H^2 - K}\end{aligned} \tag{105}$$

This formulas are best used when the analytical expression for the function $\phi(x, y, z)$ is known. In the lattice case, all derivatives must be replaced by their finite element equivalents [23]. For example,

$$\frac{\partial\phi(\mathbf{r})}{\partial x} \rightarrow \frac{\phi_{i+1,j,k} - \phi_{i-1,j,k}}{2\Delta x} \tag{106}$$

$$\begin{aligned}\frac{\partial^2\phi(\mathbf{r})}{\partial x^2} \rightarrow \frac{1}{12\Delta x^2}(-\phi_{i+2,j,k} + 16\phi_{i+1,j,k} \\ - 30\phi_{i,j,k} + 16\phi_{i-1,j,k} - \phi_{i-2,j,k})\end{aligned} \tag{107}$$

$$\begin{aligned}\frac{\partial^2\phi(\mathbf{r})}{\partial x\partial y} \rightarrow \frac{1}{2\Delta x\Delta y}(\phi_{i+1,j,k} + \phi_{i-1,j,k} + \phi_{i,j+1,k} + \phi_{i,j-1,k} \\ - 2\phi_{i,j,k} - \phi_{i+1,j+1,k} - \phi_{i-1,j-1,k})\end{aligned} \tag{108}$$

To calculate the derivatives of the field $\phi$ at the surface points that do not lie on the lattice, the linear interpolation is used. For example, the second derivative $\phi_{xx}$ at the point $\bar{\mathbf{x}}_0 = (i + \frac{\phi_{i,j,k} - \phi_0}{\phi_{i,j,k} - \phi_{i+1,j,k}}, j, k)\Delta x$ considered above [Eq. (80)] is

$$\phi_{xx}(\bar{\mathbf{x}}_0) = \phi_{xx}(\bar{\mathbf{x}}_1) + [\phi_{xx}(\bar{\mathbf{x}}_2) - \phi_{xx}(\bar{\mathbf{x}}_1)]\frac{\phi_{i,j,k} - \phi_0}{\phi_{i,j,k} - \phi_{i+1,j,k}} \tag{109}$$

These schemes require the calculations of the second and mixed derivatives, which normally result in poor accuracy when the computations are performed on discrete data. For noisy data, computed values of $H$ and $K$ depend on the finite element scheme used to calculate the first, second, and mixed derivatives.

To end, note that the method described above can be easily adopted to calculate the curvature of the interface in two dimensions. The interface is then a curve obtained from the condition $\phi(x, y) = \phi_0$, and the curvature $k$ at the point $\mathbf{r}$ is given by $k = -\frac{1}{2}\nabla\mathbf{n}(\mathbf{r})$, with $\mathbf{n}(\mathbf{r})$ being the unit vector normal to the curve. The value of the curvature is calculated as $k = 2H$, where $H$ is given by Eq. (100) with the derivatives $\phi_z$, $\phi_{zz}$, $\phi_{xz}$, and $\phi_{yz}$ set equal to zero.

## 2. Method Based on the First and Second Fundamental Forms of the Differential Geometry

In this method the local curvatures are calculated by using the first and second fundamental forms of differential geometry [7]. The surface is parameterized near the point of interest (POI) as $\mathbf{p}(u, v)$ (see Fig. 32). The coordinates $(u, v)$ are set arbitrary on the surface in such a way that POI is located at $\mathbf{p}(u, v) = (0, 0)$. The first and the second forms of the differential geometry are expressed as

$$I = E\,du\,du + 2F\,du\,dv + G\,dv\,dv \tag{110}$$

and

$$II = L\,du\,du + 2M\,du\,dv + N\,dv\,dv \tag{111}$$

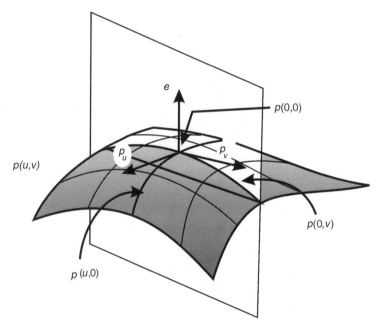

**Figure 32.** Schematic diagram of a surface expressed in a parametric form $\mathbf{p}$ $(u, v)$ and a "sectioning plane," which is comprised of $\mathbf{e}$ and $\mathbf{p}$ $(0, v)$. $\mathbf{p}$ $(0, 0)$ is a point of interest at which the local curvatures are determined [7].

where parameters $E$, $F$, $G$, $L$, $M$, and $N$ are related to the partial derivatives of the local parametric equation of the surface:

$$E = \mathbf{p}_u \cdot \mathbf{p}_u, \qquad F = \mathbf{p}_u \cdot \mathbf{p}_v, \qquad G = \mathbf{p}_v \cdot \mathbf{p}_v$$
$$L = \mathbf{p}_{uu} \cdot \mathbf{e}, \qquad M = \mathbf{p}_{uv} \cdot \mathbf{e}, \qquad N = \mathbf{p}_{vv} \cdot \mathbf{e} \tag{112}$$

Here $\mathbf{e}$ is a unit vector normal to the surface at the POI defined by $\mathbf{e} = \mathbf{p}_u \times \mathbf{p}_v / |\mathbf{p}_u \times \mathbf{p}_v|$, and subscripts of $\mathbf{p}$ denote the partial derivatives. Thus the mean, $H$, and Gaussian, $K$, curvatures are expressed as

$$H = \frac{EN + GL - 2FM}{2(EG - F^2)}$$
$$K = \frac{LN - M^2}{EG - F^2} \tag{113}$$

The surface is first "sectioned" by the plane that contains $\mathbf{p}_u$ and $\mathbf{e}$. The intersection between the plane and the surface is defined as $\mathbf{p}(u, 0)$. Parameters $E$ and $L$ are determined from $\mathbf{p}(u, 0)$. The surface then is cut by the second plane, which defines $\mathbf{p}(0, v)$. Now $F$, $G$, and $N$ can be computed. By eliminating $M$ from Eqs. (112) and (113), the following expression is obtained:

$$f(i, K, H) \equiv 0 = 4F_i^2 \left\{ L_i N_i - K(E_i G_i - F_i^2) \right\}$$
$$- \left\{ E_i N_i + G_i L_i - 2H(E_i G_i - f_i^2) \right\}^2 \tag{114}$$

Here the subscript $i$ denotes $i$th set of the curvilinear coordinates. The local mean and Gaussian curvatures are determined by using nonlinear regression fitting after a number of sections at a given point has been made [this corresponds to different sets of the local coordinates $(u, v)$].

In principle, this method is more accurate than the direct derivative discretization method described in the previous subsection, because the derivative calculations are made several times for the same surface point. Nevertheless, the method suffers essentially the same problems related to the finite derivative discretization on the lattice as in the previous case.

### 3.  Parallel Surface Method

The parallel surface method (PSM) has been invented to measure the average interface curvature (and the Euler characteristic) from the 3D data images [222]. First, a parallel surface to the interface is formed by translating the original interface along its normal by an equal distance everywhere on the surface (see Fig. 33). The change of the surface area at the infinitely small parallel shift of the surface is

$$A(t) = A(0)(1 + 2\langle H \rangle t + \langle K \rangle t^2) \tag{115}$$

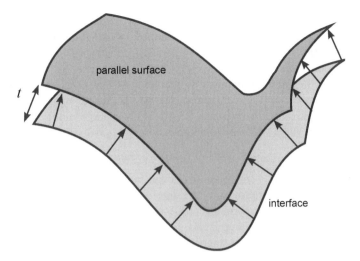

**Figure 33.** Parallel surface with the displacement $t$ from the original surface [222].

where $t$ is the displacement, and $\langle H \rangle$ and $\langle K \rangle$ are the averaged surface mean and Gaussian curvature:

$$\langle H \rangle = \frac{\int_S H(\mathbf{r}) \, dS}{\int_S dS} \tag{116}$$

$$\langle K \rangle = \frac{\int_S K(\mathbf{r}) \, dS}{\int_S dS} \tag{117}$$

A(0) and A(t) are the total surface area before and after the parallel shift. Equation (115) is exactly the definition of the averaged mean, $\langle H \rangle$, and Gaussian, $\langle K \rangle$, curvatures. From Eq. (140), $\langle H \rangle$ and $\langle K \rangle$ can be deduced from the area variation of the parallel surfaces with the parallel displacement $t$ as a variant.

In order to apply the PSM, the surface must first be triangulated by using one of the simplex decomposition scheme. The vectors normal to each vertex of the polygons are calculated as

$$\mathbf{n} = \frac{\sum_i S_i \mathbf{n}_{tri}^i}{\left| \sum_i S_i \mathbf{n}_{tri}^i \right|} \tag{118}$$

where $\mathbf{n}_{tri}$ and $S_i$ denote the normal vector, and the area of the $i$th neighboring triangle, respectively. The parallel shift it constructed by multiplying all the normal vectors by factor $t$ and shifting each vertex by the displacement vectors. Connecting the top of the displacement vectors makes the imaginary surface parallel to the original one. By performing the triangulation procedure for the parallel surface, its area $A(t)$ at $t$ is obtained by summing up areas of all triangles. In some cases, the area of parallel surface has to be assigned a "negative" value [222].

This method essentially requires triangulation of the surface and gives only approximate value of the curvatures due to the fitting character of their determination. In order to increase the accuracy a large number of parallel shifts is usually made which is sometimes computationally exhausting.

### 4. Surface-Based Formulas

The computation of the curvatures from the bulk field $\phi$ using the standard differential geometry has proven to be rather imprecise. The errors produced by the use of the approximate formulas (100)–(104) are especially big if the spatial derivatives of the field $\phi$ have sharp peaks at the phase interface. This is a common situation in the late-stage kinetics of the phase separating/ordering process, when the order parameter is saturated and the domains are separated by thin walls. Here, to calculate the curvatures, we propose a much more accurate method. It is based on the observation that the local curvatures are quantities that can be inferred solely from the shape of the interface, without appealing to the properties of the bulk field $\phi$.

Consider a polyhedron that is a discrete representation of the phase interface $\phi(\mathbf{r}) = \phi_0$ obtained in the triangulation procedure. For each vertex of the polyhedron, we can define the angle deficit by

$$T_i = 2\pi - \sum_{j=1}^{m} \alpha_i^j \tag{119}$$

where $m$ is the number of triangles that meet at $i$th vertex and $\alpha_i^j$ is the angle between the two edges of $j$th triangle at this vertex. The Gaussian curvature at $i$th vertex is given by

$$K_i \approx T_i/S_i \tag{120}$$

where $S_i$ is one-third of the area of the triangles. To prove this formula, let us first show that the integral from the Gaussian curvature over the surface region $\Sigma_i$ (over triangles sharing the same vertex) is

$$\int_{\Sigma_i} K(S)\, dS = T_i \tag{121}$$

The total angle deficit of the polyhedron, $T = \sum_{\{v_i\}} T_i$ is related to the number of its vertices $\#V$, faces $\#F$, and edges $\#E$ (Cartesian theorem) as

$$T = 2\pi(\#V + \#F - \#E) \tag{122}$$

On the other hand, the total integral from the Gaussian curvature can be expressed by using the Gauss theorem:

$$2\pi(\#V + \#F - \#E) = \int_{\Sigma} K(S)\, dS \tag{123}$$

By using the fact that

$$\int_{\Sigma} K(S)\, dS = \sum_{\{v_i\}} \int_{\Sigma_i} K(S)\, dS \tag{124}$$

and comparing Eqs. (122) to (123), Eq. (121) is deduced. Now, by assuming that the Gaussian curvature is constant within the region $\Sigma_i$, formula (120) is obtained from Eq. (121).

The integral from the mean curvature $H$ over the surface region $\Sigma_i$ can be written as

$$\int_{\Sigma_i} H(S)\, dS = \tilde{H}_i \tag{125}$$

where

$$\tilde{H}_i = \frac{1}{4} \sum_{j=1}^{m} l_i^j \theta_i^j \tag{126}$$

Here $l_i^j$ is the length of the edge of $j$th triangle, and $\theta_i^j$ is the angle between two adjacent triangles $j$ and $j+1$. Assuming again the constancy of the mean curvature within the region $\Sigma_i$, its value can be evaluated as

$$H_i \approx \tilde{H}_i / S_i \tag{127}$$

Note that in Eq. (126) the angles $\theta_i^j$ can have either sign, depending on the orientation of the surface of the polyhedron.

The curvature $k$ of the interface in two dimensions is calculated in a similar way. Consider a polygon consisting of vertices $v_i$ connected by edges $l_i$. Let us denote by $l_i$ the length of the edge between $(i-1)$th and $i$th vertex. The curvature $k_i$ at the $i$th vertex can be then approximated as follows:

$$k_i \approx 2(2\pi - \theta_i)/(l_i + l_{i+1}) \tag{128}$$

where $\theta_i$ denotes the angle between two edges which meet at the vertex $v_i$. Note also that the two-dimensional version of the Gauss–Bonnet theorem (8) implies that the sum $\sum_i \theta_i$ taken over all vertices gives $\pm 2\pi$, where the sign depends on the orientation of the polygon boundary.

In order to check the formulas, we have used the integral form for the mean curvature [Eq. (125)] and the Gauss–Bonnet theorem (8) relating the Gaussian curvature $\langle K \rangle$ and the Euler characteristic. For a representative snapshot of the order parameter configuration taken from the Langevin-type simulations, the relative error was less than $10^{-6}$; additionally, the tests for several types of minimal surfaces generated in a cube of size $32 \times 32 \times 32$ gave similar accuracy.

The formulas (120) and (127) define a set of the *local curvature* variables that can be used for the digital pattern analysis in the case of an arbitrary lattice. Calculation of the curvature distribution is, in principle, impossible within the digital pattern methods.

**Example 1.**    Distributions of the mean and Gaussian curvatures were successfully used in the analysis of the phase ordering/separating system in Ref. 142. In the paper cited, a 3D system of the nonconserved order parameter was investigated by using the TDGL equation. It was found that the system exhibits two scaling regimes: the early regime where the characteristic domain size $L(t)$ scales with $t^{1/2}$ and the intermediate regime where $L(t) \sim t^{2/5}$. (The late stage with the growth exponent $n = 0.5$ was not observed due to finite-size effects.)

The early stage is governed by the saturation of the order parameter inside the domains. Thus the phase interface follows the bulk evolution, and the exponent $n = 0.5$ results simply from the linearized TDGL equation. The scaling relations (4)–(7) are satisfied in the early stage. Scaling of the mean $H$ and Gaussian $K$ curvatures is shown in Figs. 34(a) and 34(b), respectively.

In the intermediate regime the order parameter saturates inside the domains and the kinetics of the system is driven by the local curvature of the interface. Figure 35(a) and 35(b) show the scaling of the mean and Gaussian curvatures in this regime. As seen, the scaling relations (4)–(7), based on a single lengthscale, do not hold in the intermediate regime. Instead, the curvatures $H(t)$ and $K(t)$ scale independently with two different lengthscales, $H(t) \sim L_H(t)^{-1}$ and $K(t) \sim L_K(t)^{-2}$, with $L_H(t) \sim t^{2/5}$ and $L_K(t) \sim t^{3/10}$. The quantity $L_H(t)$ scales like the characteristic size of the domains and describes the *geometrical* properties of the interface. The second lengthscale, $L_K(t)$, determines the Euler characteristic [see Eq. (8)] and describes the evolution of the *topology* of the interface.

The existence of the two lengthscales in the intermediate regime has a simple physical interpretation and can be explained in terms of the LCA theory, which links the velocity of the interface with its local curvature. Namely, one can show [223] that the quantity $p(t) = (L_H(t)/L_K(t))^2$ estimates the average number of necks piercing the surface of a sphere of radius $L_H(t)$ [$L_H(t)$ is the characteristic

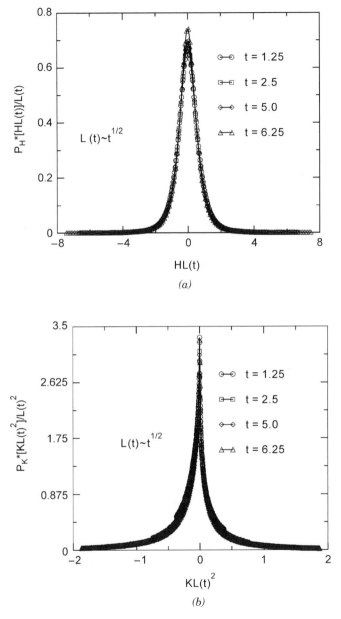

**Figure 34.** At the early stages of the phase separation, the scaling of the curvature distributions is in accordance with the dynamic scaling hypothesis [Eqs. (4)–(7)] (the order parameter is nonconserved).

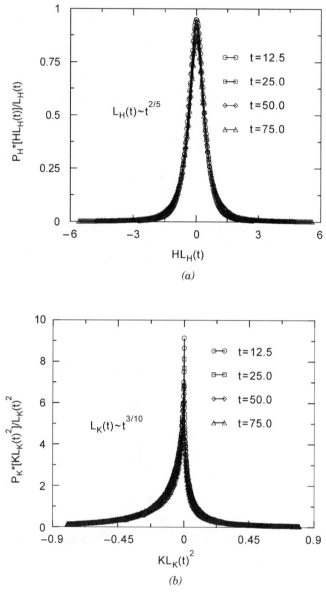

**Figure 35.** The scaling relations (4)–(7) do not hold in the intermediate regime of the phase separation. The crossover between early and intermediate regime occurs when the order parameter saturates inside the domains (the order parameter is nonconserved).

size of the domain]. In the early regime $p(t) \sim 1$, while in the intermediate regime $p(t) \sim t^{1/5}$, indicating decoupling between the domains and connections joining them. Because in the intermediate regime the average mean curvature is equal to zero and its distribution is peaked at $H = 0$ [Fig. 35(a)], the interface possesses large patches of minimal-like (saddle-like) shape. Furthermore, the apparency of the domain–neck decoupling process indicates that these areas are localized mainly at the necks connecting the domains. In other words, from the distributions of the curvatures we learn that in the intermediate regime the necks are in a "partially frozen" state and evolve slower (with the exponent $n = 0.3$) compared to the domains following the evolution with $n = 0.4$.

**Example 2.** An example of the curvature scaling for the conserved order parameter system is shown in Fig. 36. The mean and Gaussian curvature distributions collected at several time points during the phase separation process and rescaled with the interface area density do not overlap. The system considered was a homopolymer blend of a symmetric composition [9] ($\phi_0 = 0.5$). On the other hand, the curvature scaling very similar to the one shown in Fig. 34 has been observed in experiments [7]. It must be stressed that the curvature scaling is observed only if the thermal motions of the interface have been smoothed out. Thus, during the measurement by using the LSCM, the thermal undulation are averaged out during the measurement time; in addition, the characteristic size of the thermal undulation is usually smaller than the microscope resolution. The opposite is true in computer simulations. The noise term is usually generated and added to the local field configuration at each iteration step of the numerical solution. Small local undulations appear on the interface. Their size does not depend on the total surface area (if the characteristic domain size is much larger than the size of undulations). The similar effect takes place in the real systems. The geometry of the capillary waves at the interface does not scale linearly with the domain curvature and, at the late SD stages, can be considered as a constant. If before each measurement the system would have been frozen, the effect of capillary waves would show up on the curvature distributions.

The following numerical experiment illustrates the effect of thermal undulations on the shape of the curvature distributions. We select an arbitrary configuration of the field $\phi$ at one of the late times of the simulation. Then, we "switch off" the noise term [Eq. (65)] and continue the SD simulation. This results in a very rapid relaxation of the surface undulations (smoothening of the interface). The distribution functions of the mean curvatures corresponding to several time points of this relaxation process are shown in Fig. 37. After switching off the noise term, the maximum of the curvature distribution increases 2.5-fold during the short time interval. The Euler characteristic density and the average domain size do not change appreciably, while the surface area decreases by 3%. In order to obtain the curvature scaling from the computer

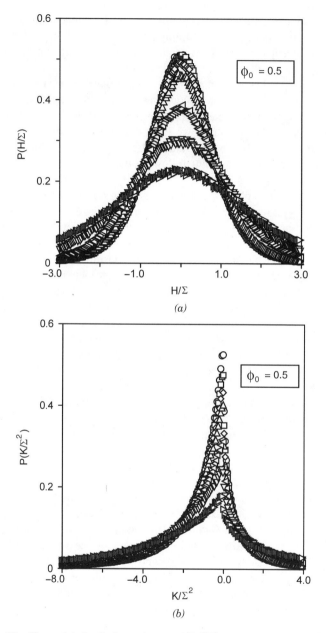

**Figure 36.** The scaled distributions of mean, $P(H/\sum)$ (a), and Gaussian, $P(K/\sum^2)$ (b), curvatures scaled with the inteface area density, $\sum$ computed at several time intervals of the spindal decomposition of a symmetric blend. There is no scaling at the late times because the amplitude of the thermal undulations does not depend on the average growth of the domains, and therefore the scaled curvature distributions functions broaden with rescaled time.

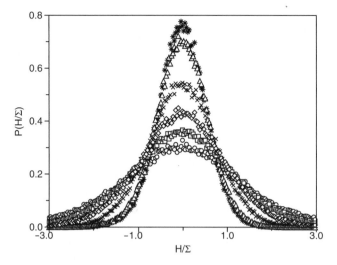

**Figure 37.** The maximum of the mean curvature distribution scaled with the interface density increases very rapidly (up 2.5 times) within a short time interval, $\tau_a$, after the noise term has been "switched off" in the simulation. The Euler characteristic and the average domain size, $R_0$, remain constant, and the surfaces area decreases by 3%. This illustrates that the curvature distributions are very sensitive to the thermal undulations of the interface. The times are $\tau_a = 0.0$, 0.032, 0.085, 0.225, 0.896, 2.05 from bottom to top at $H/\sum = 0$.

simulations, one has to either eliminate the noise generation from the numerical scheme (see Fig. 34) or perform averaging over several field configurations *before* the curvature determination.

## G.   Surface Topology: The Euler Characteristic

The analysis of three-dimensional objects of complex geometrical shapes is facilitated by computing the Euler characteristic. The Euler characteristic is a morphological measure that describes pattern connectivity. A pattern in the three-dimensional scalar field (i.e., a three-dimensional distribution of a scalar quantity) can be formed by drawing the surface at which the field has some constant value—that is, by drawing the isosurfaces. In this way the domains with higher or lower values of the field are defined. The Euler characteristic of a closed surface is related to the genus of the surface, $g$, as $\chi = 2(1 - g)$. The genus has a very simple interpretation: For a closed surface, it is the number of holes in the surface. Thus, a sphere has $g = 0$ ($\chi_{Euler} = 2$), a torus has $g = 1$ ($\chi_{Euler} = 0$), and a pretzel has $g = 2$ ($\chi_{Euler} = -2$). A large and negative Euler characteristic indicates a highly interconnected morphology. In contrast, a disconnected domain morphology is characterized by the large and positive Euler characteristic, because the Euler characteristic of a system of the disconnected surfaces is equal to the sum of the Euler characteristic for the individual surfaces. Thus, the calculation of the Euler characteristic allows us to characterize the type and complexity of the pattern which is useful in automatic pattern analysis, processing, and recognition systems. Also, the Euler characteristic is useful in materials science because it facilitates the study of the percolation problems as the crossover between the positive and negative Euler characteristic signifies percolation. If the scalar field describes a spatial distribution of concentration, the Euler characteristic determines mechanical properties, diffusion transport, and current conductivity of the system.

There are important advantages in employing the Euler characteristic in the study of complex three dimensional objects. Consider a time-dependent spatial distribution of component concentration in a system undergoing phase separation (Figs. 17 and 18), which is an example of a complex three-dimensional object often encountered in the field of the material science. A traditional way to characterize the morphology of such an object is to compute the scattering pattern. However, on the basis of the scattering pattern, it is almost impossible to distinguish between the strongly interconnected "bicontinuous" (Fig. 17) and disconnected "droplet" [Fig. 18(b,c)] morphologies. In contrast to the above, it is straightforward to automatically distinguish both types morphologies by computing the Euler characteristic, which is large and negative for the interconnected morphologies in Fig. 17 while it is positive for the droplet morphologies found in Fig 18(b,c). Thus, by computing the Euler characteristic, one may characterize pattern connectivity in three–dimensional objects in a simple and inexpensive way. Moreover, in the three-dimensional systems comprising disconnected objects such as in Fig. 18(b,c) the Euler characteristic is useful in estimating the number of separate objects without actually separating the objects. A similar reasoning can be applied to any complex three-dimensional object. Some examples of complex morphologies are porous materials, zeolites, composite materials, biological tissue, brain maps, and irregular spatial–temporal patterns occurring, for example, in chemical reaction–diffusion systems.

The Euler characteristic is usually calculated in two ways: by using the Gauss–Bonnet theorem [7,85,207,210,222] [Eq. (8)] or by combining the Cartesian and Gauss theorems [Eqs. (122) and (123)], which is also called the Euler formula [23,76,224].

### 1.   Techniques Based on the Gauss–Bonnet Theorem

The Euler characteristic, $\chi$, of a closed surface is related to the local Gaussian curvature $K(\mathbf{r})$ via the Gauss–Bonnet theorem [Eq. (8)]. A number of different schemes have been proposed to calculate the local curvatures and the integral in Eq. (8).

The simplest way to calculate the Euler characteristic is to disregard the surface integral in Eq. (8) and calculate the average Gaussian curvature $\langle K \rangle$

[Eq. (117)] by summing up the local curvatures and dividing the sum over the number of sampling points. Then the Euler characteristic is

$$\chi = \frac{1}{2\pi} \langle K \rangle S \tag{129}$$

where $S$ is the total surface area. In a more elaborate approach, a certain surface area is assigned to each sampling point, and the surface integral in Eq. (8) is performed numerically. The calculation of the Euler characteristic in this case reduces to the surface integral evaluation by the Monte Carlo method.

Producing a reasonably good accuracy for analytically defined surfaces, this scheme of calculation is very inaccurate when the field $\phi$ is specified by the discrete set of values (the lattice scalar field). The surface in this case is located between the lattice sites of different signs. The first, second, and mixed derivatives can be evaluated numerically by using some finite difference schemes, which normally results in poor accuracy for discrete lattices. In addition, the triangulation of the surface is necessary in order to compute the integral in Eq. (8) or calculate the total surface area $S$. That makes this method very inefficient on a lattice in comparison to the other methods.

Much better results are obtained by using the parallel surface method (Section III.F.3), because the integral methods are used to determine $\langle K \rangle$. Nevertheless, PSM gives only approximate estimation of the Euler characteristic and is extremely time-consuming in comparison to the methods described below.

To end this section, note that the Cartesian formula (120) is suitable to calculate the *local* Gaussian curvature while an exact value of the Euler characteristic is obtained from the Gauss–Bonnet theorem.

### 2. Digital Pattern Method

In the digital pattern method the array of real data is transformed into the pixel pattern by using some thresholding procedure. After that the lattice is covered by black and white pixels. The local curvature variables, which can be calculated for an arbitrary lattice by using formulas (120) and (127), are assigned to the corners of each black pixel [207] (see, for example, Fig. 23). The Euler characteristic is given by the sum of the curvature variables over the boundary between black and white domains. However, the loss of the information caused by the thresholding procedure leads to some specific situations when the domain boundaries can be drawn ambiguously. Figure 38 represents one of these cases. The actual morphology of the interface is determined by the field values at the lattice sites. But after the thresholding procedure, the boundaries of the domain cannot be found and the information about their connectivity is lost. For the two-dimensional systems, these ambiguous situations can be avoided by considering hexagonal pixels [210] (see Fig. 39). However, the hexagonal pixels

**Figure 38.** An example when the Euler characteristic cannot be correctly determined by using the digital pattern analysis, because the boundaries of the domains after the thresholding procedure cannot be specified.

can be specified only for a specific kind of simulations (for example, the Lattice–Boltzmann gas automata) and cannot be easily designed/anticipated in experiments. In the 3D case this problem can be solved in principle by covering the lattice with a regular polyhedron other than cubes, but this is not very practical because most of the simulations are done on a cubic lattice.

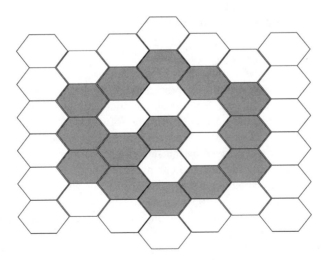

**Figure 39.** By using the 2D hexagonal lattice the specific situations of undetermined domain boundaries (see Fig. 38) are eliminated.

In conclusion, the digital pattern method is very efficient computationally but fails to calculate the Euler characteristic for the cases similar to the one shown in Fig. 38. Such cases are typical, however, when the data are noisy or when the field values are distributed around the threshold value.

### 3.   Techniques Based on the Euler Formula

The computation of the Euler characteristic based on Eq. (8) is not practical, particularly when the system is represented by a set of points on the lattice. The practical way of computing $\chi$ is related to the coverage of the surface with polygons. Then, the calculation of the Euler characteristic is straightforward when it is based on the Euler formula:

$$\chi = \#F + \#V - \#E \qquad (130)$$

where $\#F, \#V$, and $\#E$ are the number of faces $(\#F)$, vertices $(\#V)$, and edges $(\#E)$ of all polygons cut by the surface. The practical realization of this procedure can be done by decomposing the lattice into simplexes [225] (Section III.B). Formula (130) is then applied to each simplex, and the "local" Euler characteristic are specified in this way (one should also keep in mind that some vertices, faces, and edges are shared with the neighboring simplexes). The total Euler characteristic is obtained by summing up over all nonempty simplexes, because the Euler characteristic is an additive measure.

The best scheme for the Euler characteristic calculations is that which allows us to consider the maximum number of possible connections within the same interpolation approximation. By using the five-vertex simplex and seven polygonal surface representations (shown in Fig. 27), all the cases specified for the marching cube algorithm (Fig. 24) and for the tetragonal simplex decomposition (Figs. 29 and 30) can be reproduced. In addition, this decomposition scheme detects a few very specific surface configurations [Figs. 27(g) and 28(b)] that cannot be reproduced within any other schemes. Therefore, we recommend using this scheme for a precise calculation of the Euler characteristic. Nevertheless, when the surface is smooth—that is, when it is very unlikely that more than one surface is present in a single lattice cell—the four-vertex scheme might be more practical due to its simplicity (especially when the surface has to be triangulated). In fact, the triangulation of the surface is not required for the Euler characteristic calculation: Only the number of vertices, faces, and edges is needed. That makes the calculations by using the simplex decomposition very efficient in comparison to other methods.

**Example 1.**   Under the phase separation process, the time dependence of the interface area or the maximum wavevector does not exhibit any specific behavior that could be directly related to the morphological transformations that

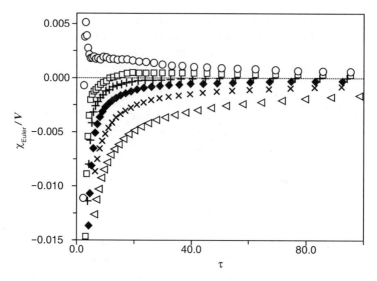

**Figure 40.** Time evolution of the Euler characteristic density for different average volume fractions, $\phi_0 = 0.5$, 0.4, 0.375, 0.36, 0.35, and 0.3, quenched from the homogeneous state binary mixture. The negative Euler characteristic corresponds to the bicontinuous morphology, while the positive Euler characteristic detects a system of disconnected droplets.

take place in the mixture. However, the time evolution of the Euler character-istic density, $\chi_{\mathrm{Euler}}/V$, is very sensitive to them. Figure 40 shows the time evolution of the Euler characteristic density for different average volume fractions, $\phi_0 = 0.5$, 0.4, 0.375, 0.36, 0.35, and 0.3, quenched from the homo-geneous state binary mixture. For the blends of $\phi_0 = 0.5$, 0.4, and 0.375, the Euler characteristic increases monotonically from the initially large negative value and remains negative. The negative Euler characteristic indicates the bicontinuous interface in these blends. The time evolution of the morphology is characterized by the growth of the average domain size accompanied by the decrease of the domain connectivity. In more asymmetric blends, the Euler characteristic approaches zero at the late times ($\phi_0 = 0.36$) or becomes positive for $\phi_0 = 0.35$ and 0.3. The positive Euler characteristic implies a disconnected droplet morphology, while close-to-zero values correspond to some intermedi-ate morphology comprised from disconnected domains of complex shapes (the system is still far from the equilibrium configuration where only one big spherical droplet is formed). For the quenches of $\phi_0 = 0.35$, one observes a characteristic maximum in the evolution of the Euler characteristic. In this case the Euler characteristic becomes positive at the relatively late times, which illustrates transformation of the bicontinuous interface formed at the early stages of the SD into the droplets. This transformation is relatively "slow" in contrast to the rapid droplet formation that occurs at the early stages of the SD in the more asymmetric blends ($\phi_0 = 0.3$). In this case, at the moment of interface formation, the blend morphology is already disperse.

For the symmetric system ($\phi_0 = 0.5$) the scaling exponent for the Euler characteristic has been found in accordance with the dynamic scaling hypothesis: $\chi \sim L(t)^{-3}$ (see Section I.G). The homogeneity index, $HI$, of the interface defined as [222]

$$HI = \left[ -\frac{S^3}{2\pi\chi_{\text{Euler}}V^2} \right]^{1/2} \tag{131}$$

in this case is time-independent, which formulates the dynamic scaling hypothesis for the interface shape at the level of the integral geometry quantities. However, in asymmetric cases, the scaling exponent for $\chi/V$ depends on the quench conditions (if the blend morphology remains bicontinuous). When the bicontinuous to droplet transformation occurs, the temporal behavior of the Euler characteristic cannot even be described by the single power law. This simply indicates that there is no self-similarity of the interface at different stages of the phase separation and that the dynamic scaling fails in this case.

It has been proposed recently [210] to use the Minkowski functionals to define the scaling length $l$ for the 2D systems as $l_\chi := (\chi_{\text{Euler}}(\tau)/L^2)^{-1/2}$ and $l_\Sigma := \Sigma(\tau)^{-1}$, where $L^2$ is a "volume" of the 2D system. A similar scaling length could be defined for the symmetric 3D systems which possess the bicontinuous interface; that is, $l_\chi := (-\chi_{\text{Euler}}(\tau)/V)^{-1/3}$ and $l_\Sigma := \Sigma(\tau)^{-1}$. However, this definition cannot be applied for the asymmetric blends where the change of the Euler characteristic is not universal.

**Example 2.** A possibility of the droplet formation during the spinodal decomposition of a binary mixture depends on two factors: the average blend composition $\phi_0$ and the quench depth. It is convenient to regard the transformation from interconnected to disperse morphology as a percolation transition [10]. One assumes that the transformation can be described by considering only geometrical features of the system, particularly studying the minority domain volume fraction $f_m$. At some special value of $f_m$ (at the percolation threshold value) the morphological transition occurs. It is important to distinguish the average $A$-component volume fraction (composition), $\phi_0$, and $f_m$. The former quantity is a constant, while $f_m$ can change during the simulations. In the very long time limit, $f_m$ approaches its equilibrium value

$$f_m^{\text{eq}} = \left(\phi_0 - \phi_A^{(1)}\right) / \left(1 - 2\phi_A^{(1)}\right) \tag{132}$$

where $\phi_A^{(1)}$ is the equilibrium volume fraction of the $A$ component in the minority phase (due to the symmetry of the phase diagram, only blends of $\phi_0 \leq 0.5$ are considered). This assumption can be tested by calculating both the

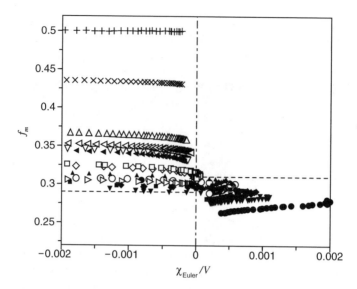

**Figure 41.** The percolation threshold determination for polymer blends undergoing the phase separation. Minority phase volume fraction, $f_m$, is plotted versus the Euler characteristic density for several simulation runs at different quench conditions, $f_m^{(eq)} = 0.225, \ldots, 0.5$. The bicontinuous morphology ($\chi_{Euler} < 0$) has not been observed for $f_m < 0.29$, nor has the droplet morphology ($\chi_{Euler} > 0$) been observed for $f_m > 0.31$. This observation suggests that the percolation occurs at $f_m = 0.3 \pm 0.01$.

Euler characteristic and the minority domain volume fraction. Because the Euler characteristic is the direct measure of the domain connectivity, we identify the domain percolation at the point where the Euler characteristic attains zero value.

In Fig. 41 we plot the minority phase volume fraction, $f_m$, versus the Euler characteristic density for a large number of simulation runs performed at different quench conditions. For the symmetric blends ($\phi_0 = 0.5$), $f_m = 0.5$ and is independent of time and $\chi_{Euler}/V$ is always negative. For the asymmetric blends, $f_m$ decreases with time and $\chi_{Euler}/V$ may change the sign. We have not observed the bicontinuous morphology for $f_m < 0.29$, nor have we observed the droplet morphology for $f_m > 0.31$. This observation suggests that the percolation occurs at $f_m = 0.3 \pm 0.01$ and that the percolation threshold is not very sensitive to the quench conditions (noise intensity).

Thus, one could expect to find a droplet morphology at those quench conditions at which the equilibrium minority phase volume fraction (determined by the lever rule from the phase diagram) is lower than the percolation threshold. However, the time interval after which a disperse coarsening occurs would depend strongly on the quench conditions (Fig. 40), because the volume fraction of the minority phase approaches the equilibrium value very slowly at the late times.

**Example 3.** There is a specific difference in the ways the Euler characteristic is calculated within the digital pattern and simplex decomposition methods. The former one counts the sign of the surface orientation by specifying whether a given point is convex or concave with respect to the pixel arrangement. If a drop of white pixels is surrounded by the "sea" of black pixels, it is characterized by the positive Euler characteristic, and in the opposite case (black drop surrounded by white pixels) the Euler characteristic would be negative (the sign has been chosen arbitrary). On the other hand, by using the simplex decomposition method, one characterizes the surface itself regardless of the surface orientation. For example, the Euler characteristic of the black/white pixel configuration shown in Fig. 23 would be zero if calculated by counting the local curvatures (digital pattern method) and would be $+4$ if calculated by the simplex decomposition method (two disconnected droplets one inside the other). Such "double separated" morphologies can be observed in real systems, for example, during the spinodal decomposition of binary mixtures in the hydrodynamic regime [158,163,210] (Fig. 19) and in ternary mixtures [48,49,130,226].

The disagreement of those two methods can be used to characterize quantitatively the "double-separated" morphology. One should simply calculate the Euler characteristic once by using the Euler formula (130), and then calculate it a second time by using the digital pattern method . The difference between obtained quantities will give the number of double-separated structures.

## H.   Measures of Disperse Morphology

The disperse morphology, comprised from a large number of disconnected domains, is characterized by the large and positive Euler characteristic. Usually, the domains have simple shapes, and if the Euler characteristic of each domain is $+2$, the morphology is droplet-like. The total Euler characteristic of the droplet morphology is exactly the double number of droplets, and the Euler characteristic density in this case is twice the droplet number density. The total interfacial area and the total droplets volume can be calculated by using one of the methods described in Sections III. D and III. E. The average droplet surface area, $\langle S_{\text{drop}} \rangle$, and the average droplet volume, $\langle V_{\text{drop}} \rangle$, can be defined as

$$S_{\text{total}} = \langle S_{\text{drop}} \rangle \frac{\chi_{\text{Euler}}}{2} \tag{133}$$

$$V_{\text{total}} = \langle V_{\text{drop}} \rangle \frac{\chi_{\text{Euler}}}{2} \tag{134}$$

The average droplet area and the average droplet volume can be related as

$$\langle V_{\text{drop}} \rangle = \frac{\langle \tilde{C} \rangle \langle S_{\text{drop}} \rangle^{3/2}}{6\sqrt{\pi}} \tag{135}$$

which defines the average droplet compactness, $\langle \tilde{C} \rangle$ [218]. Similarly, the compactness of a single droplet $\tilde{C}$ can be defined. For a sphere $\tilde{C} = 1$, and for a flat layer of vanishing thickness $\tilde{C} \to 0$.

In order to characterize quantitatively the polydisperse morphology, the shape and the size distribution functions are constructed. The size distribution function gives the probability to find a droplet of a given area (or volume), while the shape distribution function specified the probability to find a droplet of given compactness. The separation of the disconnected objects has to be performed in order to collect the data for such statistics. It is sometimes convenient to use the quantity $v^{1/3} = [V_{\text{droplet}}/V]^{1/3}$ as a dimensionless measure of the droplet size. Each droplet itself can be further analyzed by calculating the "mass center" and principal "inertia momenta" from the scalar field distribution inside the droplet [110]. These data describe the droplet anisotropy.

**Example.**    The transformation of the bicontinuous morphology into the disperse droplet pattern is observed in many mesoscopic systems. For example, in the asymmetric diblock copolymer melt, the interconnected structures such as a double gyroid phase (Fig. 5) can be transformed into the spherical morphology (bcc or cps phase) by increasing the temperature. The pathway of such transformation involves formation of the cylindrical morphology at the intermediate stage [227]. Similarly, the sponge phase in the surfactant solutions transforms into micelles via the cylindrical mesophases [228]. During the phase separation process, the morphology of the asymmetric binary mixture could be either bicontinuous or disperse, and, at some quench conditions (when the final equilibrium minority phase volume fraction is slightly below the percolation threshold), this transformation takes place dynamically (Fig. 40). In order to find the pathway of such transformation, the domain sizes and shapes have been analyzed [10] in the region around the percolation transition ($\chi_{\text{Euler}} \approx 0$).

In order to find correlations between the shapes and sizes of the domains, the droplet surface area density, $\Sigma$, has been plotted as a function of droplet compactness, $\tilde{C}$, for each droplet detected in the system [Fig. 42(a)]. The droplets which have low compactness have sizes bigger than average. When the transformation from bicontinuous to disperse morphology is slow [this case shown in Fig. 42(a)], the power law dependencies between the size and compactness of the droplets can be observed in the low compactness limit. The double-logarithmic plot of these dependencies [Fig. 42(b)] reveals that they are parabolic (the slopes in the low compactness limit are $-2. \pm 0.1$). Consider now a cylindrical (highly elongated) domain of length $l$ and cross-section radius $R$ (Fig. 43). The area of this domain is $S \approx 2\pi Rl$, and its compactness is $C \approx \frac{3}{\sqrt{2}} \sqrt{R/l}$. Now, if we keep the cross-section radius $R$ constant and increase the length of the domain $l$, the domain area would increase as its compactness decreases: $S(l) \sim C(l)^{-2}$. This matches exactly to what is shown in Fig. 42: The

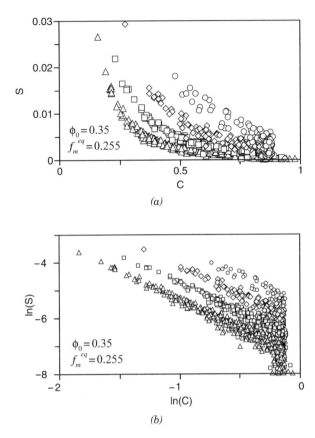

**Figure 42.** (a) The droplet surface area density, $S$, as a function of droplet compactness, $\tilde{C} = 6\pi V/S^{3/2}$, plotted for each droplet detected in the system at the four time steps of the spinodal decomposition of the asymmetric mixture. (b) The parabolic slopes of these dependences indicate the decomposition of cylindrical droplets illustrated in Fig. 43.

droplets are formed due to the decomposition of the highly elongated (cylindrical) domains (see Fig. 43). Thus, also during the phase separation, the transformation from bicontinuous to droplet pattern involves formation of the cylindrical domains, which suggests that it is a universal feature of such transformations.

## I. Measures of Anisotropic Patterns

The usual methods for characterization of anisotropic patterns involve the calculations of the scattering patterns and their further analysis [167,168] in the Fourier space. Thus, for an isotropic system, the scattering intensity $S(\mathbf{k})$ is symmetric, its maxima of the same amplitude are arranged at the circle $\mathbf{k} = |k|$

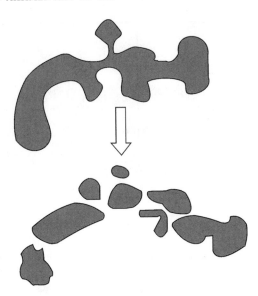

**Figure 43.** Schematic illustration of the cylinder-droplets transformation. See also Fig. 42.

(in 2D projection) under the condition that there is only one lengthscale in the system. Any anisotropy will change the symmetry of the scattering pattern, and the maxima of different amplitudes will be arranged at the ellipse, which also specifies the direction of the maximum anisotropy. But, sometimes the measures based on the Fourier transform cannot be easily used, for example, when the simulated pattern in not periodic, which is often the case for the sheared morphology simulations [170] (due to the Lee–Edwards boundary condition). On the other hand, the lengthscales derived from derivatives do not require the periodicity. Define a tensor [170]

$$d_{\alpha\beta} = \frac{\sum_{\mathbf{r}} \partial_\alpha^D \phi(\mathbf{r}) \partial_\beta^D \phi(\mathbf{r})}{\sum_{\mathbf{r}} \phi^2(\mathbf{r})} \tag{136}$$

where $\partial_\alpha^D$ is the symmetric discrete derivative in direction $\alpha$ (2D case is considered). The tensor can be diagonalized to give two eigenvalues $\lambda_1$ and $\lambda_2$ and angle $\theta^*$:

$$\lambda_1 = \frac{d_{xx} + d_{yy}}{2} + \sqrt{\frac{(d_{xx} - d_{yy})^2}{4} + d_{xy}^2} \tag{137}$$

$$\lambda_2 = \frac{d_{xx} + d_{yy}}{2} - \sqrt{\frac{(d_{xx} - d_{yy})^2}{4} + d_{xy}^2} \tag{138}$$

$$\theta = \tan^{-1}\left(\frac{d_{yy}}{d_{xy} - \lambda_2}\right) \tag{139}$$

Two eigenvalues give two orthogonal lengthscales:

$$R_1^* = \frac{1}{\lambda_1 L_w}, \qquad R_2^* = \frac{1}{\lambda_2 L_w} \qquad (140)$$

where $L_w$ is the interface width [1,2,170], which could, in principle, be anisotropic.

The total contour length of the interface is

$$L_I = \sum_i |\mathbf{l}_i| \qquad (141)$$

where $\mathbf{l}_i$ are small contour segments of the size of the lattice unit. The preferred direction is extracted by defining the vector

$$\mathbf{D} = \hat{R}^{-1} \left( \sum_i \hat{R}(\mathbf{l}_i) \right) \qquad (142)$$

where operator $\hat{R}$ is

$$\hat{R}(\mathbf{r}) = |\mathbf{r}|(\cos(2\theta), \sin(2\theta)) \qquad (143)$$

and $\theta$ is the angle between the argument of $\hat{R}$ and the $x$ axis. The vector $\mathbf{D}$ is zero for isotropic closed contours and points the average direction of the interface for anisotropic closed contours. Two lengthscales and an angle can be defined from these measures [170]:

$$R_1^\circ = \frac{L_x L_y}{L_I + |\mathbf{D}|}, \qquad R_2^\circ = \frac{L_x L_y}{L_I - |\mathbf{D}|} \qquad (144)$$

$$\theta^\circ = \cos^{-1}\left(\frac{\mathbf{r}\mathbf{D}}{|\mathbf{D}|}\right) \qquad (145)$$

where $L_x$ and $L_y$ are the lattice dimensions. The above definitions can easily be generalized for 3D case.

In order to study the connectivity of the anisotropic morphology, the 2D cuts of the 3D scalar field can be made at different angles to, for example, an $XY$ plane. The 2D Euler characteristic can be calculated for each angle of the sectioning plan, and the anisotropic connectivity can be specified in this way.

Among the methods presented above, we have chosen the best implemented in computer codes (algorithms for 2D and 3D morphology) available at http://www.ichf.edu.pl/mfialkowski/morph.html. The 3D program uses the tetragonal simplex decomposition (Section III.B) with a consequent triangulation scheme to calculate the surface area by summing up areas of all triangles, calculate the local curvatures by using the surface-based formula (Section III.F.4), calculate the Euler characteristic by both summing up local curvatures (Section III.G.1) and applying the Euler formula (Section III.G.3), and calculate the volume fraction (Section III.E) by using the digital pattern method and by the simplex decomposition. All disconnected surfaces are separated and the volume, Euler characteristic, surface area, and curvature distributions can be computed for each surface.

## IV. OUTLOOK

In two articles published recently in PNAS [229,230], the mesoscopic systems are discussed from the point of view of some new-yet-undiscovered principles that govern the emergent and protected properties of matter at the mesoscopic scale. On this scale intermediate between the microscopic and macroscopic scales, one observes a universe of phenomena and structures that in many cases are insensitive to the underlying microscopic dynamics. The belief held for centuries was that the rules governing the behavior of physical or chemical systems can be invented (inferred) from a single theory (Theory of Everything) describing the world at the most fundamental level. Indeed the great success of the Newtonian theory of gravity and motion laid a solid foundation for such belief. Yet the second half of this century revealed a new world of complex systems that seems to defy this common belief. The rules that govern the complexity are yet to be discovered. They do not follow simply from the underlying microscopic dynamics but rather emerge as a property of the system as a whole. Whether we consider the problem of protein folding or the formation of periodic surfaces in surfactant systems, we are faced with the intrinsic difficulty of stating *a priori* whether a given Hamiltonian system will produce desired results. In fact there is more in the problem. It appears that some of the properties are protected; that is, they emerge irrespective of the details of the underlying microscopic dynamics. This observation led some physicists to suspect [229,230] that there are higher organizing principles that govern the behavior of complex adaptive matter at the mesoscopic level. In order to give an example of such emergent properties and the underlying new rules related to them, let us consider a very recent article [231] concerning phase transitions. In this article it has been shown that information about a phase transition in a system described by the potential energy $H(\mathbf{r}_1 \cdots \mathbf{r}_N)$ for $N$ particles can be inferred from the topology changes of the $(3N - 1)$-dimensional hypersurface (in $3N$-dimensional configurational phase space) given by

$$H(\mathbf{r}_1, \ldots \mathbf{r}_N) = E \qquad (146)$$

The topology changes as we change $E$, and in particular it has been shown [231] that the Euler characteristic of this hypersurface, as a function of $E$, changes slope at the transition. This would be one of the rules that emerges as the property of the Hamiltonian hypersurface irrespective of the precise form of the Hamiltonian. We hope that our work focused on the surface aspect of the mesoscopic systems is also a step in this direction. As we described in this chapter, surfaces are ubiquitous at the mesoscopic scale, and their study may be helpful in the quest for new rules operating at the mesoscale. At our home page http://www.ichf.edu.pl/Dep3.html free software for the morphological analysis is available [232].

## Acknowledgments

This work is supported by the KBN grant 2P03B12516 and 5P03B09421. RH acknowledges with appreciation the hospitality of Ecole Normale Superieure and the stipendship from the French Ministry of Education. AA appreciates interesting discussions with Prof. H. Tanaka, Dr T. Araki, Prof. T. Hashimoto, and Dr. K. Moorthi.

## References

1. A. J. Bray, *Adv. Phys.* **43**, 357 (1994).

2. J. S. Langer, in *Solids far from Equilibrium*, C. Godreche, ed., Cambridge University Press, New York, 1992, p. 297.

3. K. Binder, *Rep. Prog. Phys.* **50**, 783 (1987).

4. J. D. Gunton, M. San Miguel, and P. Sahni, in *Phase Transitions and Critical Phenomena*, C. Domb and J. L. Lebowitz, eds., Academic Press, New York, Vol. 8, p. 267 (1983).

5. H. Jinnai, Y. Nishikawa, T. Koga, and T. Hashimoto, *Macromolecules* **28**, 4782 (1995).

6. H. Jinnai, T. Koga, Y. Nishikawa, T. Hashimoto, and S. T. Hyde, *Phys. Rev. Lett.* **78**, 2248 (1997).

7. H. Jinnai, Y. Nishikawa, and T. Hashimoto, *Phys. Rev. E* **59**, R2554 (1999).

8. H. Jinnai et al., *Phys. Rev. Lett.* **84**, 518 (2000).

9. A. Aksimentiev, K. Moorthi, and R. Hołyst, *J. Chem. Phys.* **112**, 6049 (2000).

10. A. Aksimentiev, R. Hołyst, and K. Moorthi, *Macromol. Theory Simul.* **9**, 661 (2000).

11. J .C .C. Nitsche, *Lectures on Minimal Surfaces*, Vol. 1. Cambridge University Press, New York, 1989.

12. D. Hoffman, *Nature* **384**, 28 (1996).

13. V Luzzati and P. A. Spegt, *Nature* **215**, 701 (1967).

14. V. Luzzati, A. Tardieu, and T. Gulik-Krzywicki, *Nature* **217**, 1028 (1968).

15. V. Luzzati, T. Gulik-Krzywicki, and A. Tardieu, *Nature* **218**, 103 (1968).

16. V. Luzzati, A. Tardieu, T. Gulik-Krzywicki, E. Rivas, and F. Reiss-Husson, *Nature* **220**, 485 (1968).

17. A. H. Schoen, *Infinite Periodic Minimal Surfaces without Self-Intersections*, NASA Technical Note No. D-5541 (1970).

18. J. Charvolin and J. F. Sadoc, *J. Physique* **48**, 1559 (1987).

19. B. Lindman, K. Shinoda, U. Olsson, D. Anderson, D. Karlström, and H. Wenneström, *Colloids Surf.* **38**, 205 (1989).

20. P. Barois, D. Eidam, D., and S. T. Hyde, *J. Phys., Colloq. France* **51**, C7-25 (1990).

21. W. Góźdź and R. Hołyst *Macromol. Theory Simul.* **5**, 321 (1996).

22. W. Góźdź and R. Hołyst *Phys. Rev. Lett.* **76**, 2726 (1996).

23. W. Góźdź and R. Hołyst *Phys. Rev. E* **54**, 5012 (1996).

24. G. Arvidson, I. Brentel, A. Khan, G. Lindblom, and K. Fontell, *Eur. J. Biochem.* **152**, 753 (1985).

25. K. Larsson, *J. Phys. Chem.* **93**, 7304 (1989).

26. M. Clerck, A. M. Levelut, and J. F. Sadoc, *J. Phys. Colloq. France* **51**, C7-97 (1990).

27. D. C. Turner, Z.-G. Wang, S. M. Gruner, D. A. Mannock, and R. N. McElhaney, *J. Phys. II France* **2**, 2039 (1992).

28. R. G. Larson, *J. Phys. II France* **6**, 1441 (1996).

29. D. M. Anderson and H. Wennerström, *J. Phys. Chem.* **94**, 8683 (1990).

30. T. Landh, *J. Phys. Chem.* **98**, 8453 (1994).

31. M. Kahlweit, R. Strey, and P. Firman, *J. Phys. Chem.* **90**, 671 (1986).

32. P. Ström and D. M. Anderson, *Langmuir* **8**, 691 (1992).

33. P. Ström, *J.Colloid Interface Sci.* **154**, 184 (1992).

34. P. Puvvada, S. B. Qadri, and B. R. Ratna, *Langmuir* **10**, 2972 (1994).

35. P. J. Maddaford and C. Toprakcioglu, *Langmuir* **9**, 2868 (1993).

36. L. Leibler, *Macromolecules* **13**, 1602 (1980).

37. G. H. Fredrickson and E. Helfand, *J. Chem. Phys.* **87**, 697 (1987).

38. M. W. Matsen and M. Schick, *Phys. Rev. Lett.* **72**, 2660 (1994).

39. M. W. Matsen and M. Schick, *Macromolecules* **27**, 4014 (1994).

40. M. W. Matsen and M. Schick, *Curr. Opin. Coll. Int. Sci.* **1**, 329 (1996).

41. M. W. Matsen and F. S. Bates, *Macromolecules* **29**, 1091 (1996).

42. A. Aksimentiev and R. Hołyst, *J. Chem. Phys.* **111**, 2329 (1999).

43. G. E. Molau, *Block Copolymers*, S. L. Aggarawal, ed., Plenum Press, New York, 1970.

44. E. L. Thomas et al., *Macromolecules* **19**, 2197 (1986); D. S. Herman et al., *Macromolecules* **20**, 2940 (1987).

45. H. Hasegawa et al., *Macromolecules* **20**, 1651 (1987).

46. E. L. Thomas et al., *Nature* **334**, 598 (1988).

47. D. A. Hajduk, P. E. Harper, S. M. Gruner, C. C. Honeker, G. Kim, E. L.Thomas, and L. J. Fetters, *Macromolecules* **28**, 2570 (1995).

48. Y. Mogi et al., *Macromolecules* **25**, 5408 (1992).

49. R. Stadler et al., *Macromolecules* **28**, 3080 (1995).

50. Y. Bohbot-Raviv and Z. G. Wang, *Phys. Rev. Lett.* **85**, 3428 (2000).

51. C. T. Kresge, M. E. Leonowicz, W. J. Rothe, J. C. Vartuli, and J. C. Beck, *Nature* **359**, 710 (1992).

52. M. Dubois, Th. Gulik-Krzywicki, and B. Cabane, *Langmuir* **9**, 673 (1993).

53. A. Monnier et al., *Science* **261**, 1299 (1993).

54. D. D. Archibald and S. Mann, *Nature* **364**, 430 (1993).

55. H. P. Lin and C. Y. Mou, *Science* **273**, 765 (1996).

56. S. Schacht, Q. Huo, I. G. Voigt-Martin, G. D. Stucky, and F. Schuth, *Science* **273**, 768 (1996).

57. G. S. Attard et al., *Science* **278**, 838 (1997).

58. X. Feng et al., *Science* **276**, 923 (1997).

59. B. de Kruijff, *Nature* **386**, 129 (1997).

60. H. Terrones and A. L. Mackay, *Chem. Phys. Lett.* **207**, 45 (1993).

61. S. T. Hyde and M. O'Keeffe, *Philos. Trans. R. Soc. Lond. A* **354**, 1999 (1996).

62. H. G. von Schnering and R. Nesper, *Angew. Chem.* **26**, 1059 (1987).

63. J. Klinowski, A. L. Mackay, and H. Terrones, *Philos. Trans. R. Soc. Lond. A* **354**, 1975 (1996).

64. A. L. Mackay and J. Klinowski, *Comp. Math. Appl.* **12B**, 803 (1986).

65. B. E. S. Günning, *Protoplasma* **60**, 111 (1965).

66. B. E. S. Günning and M. W. Steer, *Ultrastructure and the Biology of the Plant Cell,* E. Arnold, London, (1975).

67. Y. Bouligand, *J. Phys. Colloq. France* **51**, C7-35 (1990).

68. T. Landh, *FEBS Lett.* **369**, 13 (1995).

69. T. Samorajski, J. M. Ordy, and J. R. Keefe, *J. Cell. Biol.* **28**, 489 (1966).

70. H. U. Nissen, *Science* **166**, 1150 (1969).

71. I. S. Barnes, S. T. Hyde, and B. W. Ninham, *J. Phys. Colloq. France* **51**, C7-19 (1990).

72. S. Andersson, S. T. Hyde, and H. G. von Schnering, *Z .Kristallogr.* **168**, 1 (1984).

73. A. L. Mackay and H. Terrones, *Nature* **352**, 762 (1991).

74. W. Helfrich, (1973), *Z. Naturforshung.* **28c**, 693 (1973).

75. H. Chung and M. Caffrey, *Nature* **368**, 224 (1994).

76. R. Hołyst and W. Goźdź, *J. Chem. Phys.* **106**, 4773 (1997).

77. R. Hołyst and P. Oswald, *J. Chem. Phys.* **109** 11051 (1998).

78. R. Hołyst and R. P. Oswald, *Phys.Rev.Lett* **79**, 1499 (1997).

79. R. Hołyst, *Curr. Opin. Coll. Int. Sci.* **3** , 422 (1998).

80. L. Golubović, *Phys. Rev. E* **50**, R2419 (1994).

81. T. Charitat and B. Fourcade, *J. Phys. II France* **7**, 15 (1997).

82. D. Constantin and P. Oswald, *Phys. Rev. Lett.* **85**, 4297 (2000).

83. R. Hołyst and B. Przybylski *Phys. Rev. Lett.* **85** 130 (2000).

84. P. Debye, H. R. Anderson, and H. Brumberger, *J. Appl. Phys.* **28**, 679 (1957).

85. D. Hilbert, *Geometry and the Imagination,* Chelsea Publishers, New York, 1952.

86. A. Cumming, P. Wiltzius, and F. S. Bates, *Phys. Rev. Lett.* **65**, 863 (1990).

87. F. S. Bates and P. Wiltzius, *J. Chem. Phys.* **91**, 3258 (1989).

88. R. G. Larson, *The Structure and Rheology of Complex Fluids*, Oxford University Press, Oxford, 1999.

89. P. Harold and P. Grace, *Chem. Eng. Commun.* **14**, 225 (1982).

90. S. Abdou-Sabet, R. C. Puydak, and C. P. Rader, *Rubber Chem. Techno.* **69**, 476 (1996).

91. D. Y. Moon and O. O. Park, *Adv. in Polym. Technol.* **17**, 203 (1998).

92. F. S. Bates, *Science* **251**, 898 (1991).

93. H. Verhoogt, J. van Dam, A. Posthuma de Boer, A. Draaijer, and P. M. Houpt, *Polymer* **34**, 1325 (1993).

94. C. H. Arus, M. A. Knachstedt, A. P. Roberts, and V. W. Pinczewski, *Macromolecules* **32**, 5964 (1999).

95. M. A. Knackstedt and A. P. Roberts, *Macromolecules* **29**, 1369 (1996).

96. Y. Tajitsu, R. Hosoya, T. Maruyama, M. Aoki, Y. Shikinami, and M. D. E. Fukada, *J. Mater. Sci. Lett.* **18**, 1785 (1999).

97. M. Doi and T. Ohta, *J. Chem. Phys.* **92**, 1242 (1991).

98. A. Onuki, *Europhys. Lett.* **28**, 175 (1994).

99. T. Okuzono, H. Shibuya, and M. Doi, *Phys. Rev. E.* **61**, 4100 (2000).

100. M. Loewenberg and E. J. Hinch, *J. Fluid Mech.* **321**, 395 (1996).

101. H. Xi and C. Duncan, *Phys. Rev. E* **59**, 3022 (1999).

102. P. D. Olmsted and C-Y. D. Lu, *Faraday Discuss*, **112**, 183 (1999).

103. M. E. Cates and S. T. Milner, *Phys. Rev. Lett.* **62**, 1856 (1989).

104. G. H. Fredrickson, *J. Rheol.* **38**, 1045 (1994).

105. A. N. Morozov, A. V. Zvelindovsky, and J. G. E. M. Fraaije, *Phys. Rev. E* **61**, 4125 (2000).

106. J. Brandrup and E. H. Immerhut, eds., *Polymer Handbook*, John Wiley & Sons, 1989, pp. VI/437–449.

107. A. A. Gusev, S. Arizzi, U. W. Suter, and D. J. Moll, *J. Chem. Phys.* **99**, 2221 (1993).

108. A. A. Gusev and U. W. Suter, *J. Chem. Phys.* **99**, 2228 (1993).

109. Y. C. Jean, in *Positron Annihilation in Solids* A. Dupasquier and A. P. Mills, Jr., eds., IOS, Amsterdam 1995, p.503.

110. S. Arizzi, P. H. Mott, and U. W. Suter, *J. Polym. Sci.* **30**, 415 (1992).

111. D. I. Ivanov, B. Nysten, and A. M. Jonas, *Polymer* **40**, 5899 (1999).

112. Y. Long, R. A. Shanks, and Z. H. Stachurski, *Prog. Polym. Sci.* **20** 651 (1995).

113. R. Ma. and M. Negahban, *Mech. Mater.* **21**, 25 (1995).

114. X. Jiang, R. H. Nochetto, and C. Verdi, *Comput. Methods Appl. Mech. Eng.* **125**, 303 (1995).

115. M. Teubner and R. Strey, *J. Chem. Phys.* **87**, 3195 (1987).

116. G. Gompper and M. Schick, *Phys. Rev. Lett.* **62**, 1647 (1989); *Phys. Rev. Lett.* **65**, 1116 (1990).

117. G. Gompper and M. Schick, in *Phase Transitions and Critical Phenomena*, Vol. 16, C. Domb and J. L. Lebowitz, ed., Academic Press, New York, 1994.

118. G. Gompper and M. Kraus, *Phys. Rev. E* **47**, 4301 (1993).

119. U. S. Schwarz and G. Gompper, *Phys.Rev. E* **59**, 5528 (1999).

120. A. Ciach, *J. Chem. Phys.* **104**, 2376 (1996).

121. A. Ciach, *Phys. Rev. E.* **56**, 1954 (1997).

122. A. Ciach and R .Hołyst *J. Chem. Phys.* **110**, 3207 (1999).

123. K. F. Freed, *Renormalization Group Theory of Macromolecules,"* John Wiley & Sons, New York, 1987, p. 22.

124. T. Ohta and K. Kawasaki, *Macromolecules* **19**, 2621 (1986).

125. S. F. Edwards, *Proc. Phys. Soc. (London)* **88**, 265 (1966).

126. R. Hołyst and T. A. Vilgis, *Macromol. Theory Simul.* **5**, 573–643 (1996).

127. L. D. Landau and E. M. Lifshitz, *Statistical Physics, Part 1*, 3rd edition, Pergamon Press, Oxford, 1980, p. 479.

128. A. Aksimentiev and R. Hołyst, *Macromol. Theory Simul.* **8**, 328 (1999).

129. M. Laradji, A.-C. Shi, J. Noolandi, and R. Desai, *Macromolecules* **30**, 3242 (1997).

130. F. Drolet and G. H. Fredrickson, *Phys. Rev. Lett.* **83**, 4317 (1999).

131. S. A. Brazovskii, *Sov. Phys. JETP* **41**, 81 (1975).

132. K. Almdal, J. H. Rosendale, F. S. Bates, G. D. Wignall, and G. H. Fredrickson, *Phys. Rev. Lett.* **65**, 1112 (1990).

133. G. H. Fredrickson and K. Binder, *J. Chem. Phys.* **91**, 7265 (1989).

134. G. T. Pickett and A. C. Balazs, *Macromol. Theory Simul.* **7**, 249 (1998).

135. J. G. E. M. Fraaije et al., *J. Chem. Phys.* **106**, 4260 (1997).

136. P. C. Hohenberg and B. I. Halperin, *Rev. Mod. Phys.* **49**, 435 (1977).

137. I. M. Lifshiz, *Zh. Exp. Teor. Fiz.* **42**, 1354 (1962).

138. S. M. Allen and J. W. Cahn, *Acta Metal.* **27**, 1085 (1979).

139. S. Kumar, J. Vinals, and J. D. Gunton, *Phys. Rev. B* **34** 1908 (1986).

140. G. Brown, P. A. Rikvold, M. Sutton, and M. Grant, *Phys. Rev. E* **56**, 6601 (1997).

141. G. Brown, P. A. Rikvold, and M. Grant, *Phys. Rev. E* **58**, 5501 (1998).

142. M. Fiałkowski, A. Aksimentiev, and R. Hołyst, *Phys. Rev. Lett.* **86**, 240 (2001).

143. P. Pincus, *J. Chem. Phys.* **75**, 1996 (1981).

144. E. Siggia, *Phys. Rev. A* **20**, 595 (1979).

145. T. Koga and K. Kawasaki, *Physica A* **196**, 389 (1993).

146. P. G. de Gennes, *J. Chem. Phys.* **72**, 4756 (1980).

147. K. Binder, *J. Chem. Phys.* **79**, 6387 (1983).

148. P. de Gennes, *Scaling Concepts in Polymer Physics,* Cornell University Press, Ithaca, NY, 1979.

149. T. M. Rogers, K. R. Elder, and R. C. Desai, *Phys. Rev. B* **37**, 9638 (1988).

150. H. Tang and K. F. Freed, *J. Chem. Phys.* **94**, 1572 (1991).

151. A. Z. Akcasu and I. C. Sanchez, *J. Chem. Phys.* **88**, 7847 (1988).

152. M. A. Kotnis and M. Muthukumar, *Macromolecules* **25**, 1716 (1992).

153. A. Chakrabarti, R. Toral, J. D. Gunton, and M. Muthukumar,*J. Chem. Phys.* **92**, 6899 (1990).

154. G. Brown and A. Chakrabarti, *J. Chem. Phys.* **98**, 2451 (1993).

155. H. Zhang, J. Zhang, and Y. Yang, *Macromol. Theory Simul.* **4**, 1001 (1995).

156. Y. Oono and S. Puri, *Phys. Rev. Lett.* **58**, 836 (1987).

157. T. Koga and K. Kawasaki, *Physica A,* **196**, 389 (1993).

158. H. Tanaka and T. Araki, *Phys. Rev. Lett.* **81**, 389 (1998).

159. L. D. Landau and E. M. Lifshitz, *Fluid Mechanics,* 2nd edition, Butterworth-Heinemann, Oxford, 1998, pp. 44–51.

160. W. H. Press, W. T. Vettering, S. A. Teukolsky, and B. P. Flannery *Numerical Recipes in Fortran,* 2nd edition, Cambridge University Press, 1986, pp. 848–852.

161. E. D. Siggia, *Phys. Rev. A* **20**, 595 (1979).

162. H. Tanaka, *J. Chem. Phys.* **107**, 3734 (1997).

163. A. J. Wagner and J. M. Yeomans, *Phys. Rev. Lett.* **80**, 1429 (1998).

164. H. Tanaka, *Phys. Rev. E* **56**, 4451 (1997).

165. H. Tanaka and T. Araki, *Phys. Rev. Lett.* **78**, 4966 (1997).

166. H. Tanaka, *Phys. Rev. E.* **59**, 6842 (1999).

167. F. Corberi, G. Gonnella, and A. Lamura, *Phys. Rev. Lett.* **81**, 3852 (1998).

168. F. Corberi, G. Gonnella, and A. Lamura, *Phys. Rev. Lett.* **83**, 4057 (1999).

169. A. W. Lees and S. F. Edwards, *J. Phys. C* **5**, 1921 (1972).

170. A. J. Wagner and J. M. Yeomans, *Phys. Rev. E* **59**, 4366 (1999).

171. V. M. Kendon, J-C. Desplat, P. Bladon, and M. E. Cates, *Phys. Rev. Lett.* **83**, 576 (1999).

172. M. E. Cates, V. M. Kendon, J-C. Desplat, and P. Bladon, *Faraday Discuss.* **112**, 1 (1999).

173. E. L. Aero, S. A. Vakulenko, and A. D. Vilesov, *J. Phys. France* **51**, 2205 (1990).

174. M. Bahiana and Y. Oono, *Phys. Rev. A* **41**, 6763 (1990).

175. S. Qi and Z-G. Wang, *Phys. Rev. Lett.* **76**, 1679 (1996).

176. G. Pätzold and K. Dawson, *Phys. Rev. E* **52**, 6908 (1995).

177. T. Kawakatsu, K. Kawasaki, M. Furusaka, H. Okabayashi, and T. Kanaya, *J. Chem. Phys.* **99**, 8200 (1993).

178. L. Kielhorn, and M. Muthukumar, *J. Chem. Phys.* **110**, 4079 (1999).

179. A. M. Turing, *Philos. Trans. R. Soc. London B* **273**, 37 (1952).

180. V. Castets, E. Dulos, J. Boissonade, and P. De Kepper, *Phys. Rev. Lett.* **64**, 2953 (1990).

181. P. De Kepper, V. Castets, and E. Dulos, J. Boissonade, *Physica D* **49**, 161 (1991).

182. Q. Ouyang and H. L. Swinney, *Nature* **352**, 610 (1991).

183. Q. Ouyang and H. L. Swinney, *Chaos* **1**, 411 (1991).

184. Q. Ouyang, Z. Noszticzius, and H.L. Swinney, *J. Phys. Chem.* **96**, 6773 (1992).

185. G. Nicolis and I. Prigogine, *Self-Organization in Non-Equilibrium Systems*, Wiley-Interscience, New York, 1977.

186. I. Lengley and I. R. Epstein, *Science* **251**, 650 (1991).

187. J. Schnackenberg, *J. Theor. Biol.* **81**, 389 (1979).

188. R. J. Field, *J. Chem. Phys.* **63**, 7929 (1975).

189. H. Othmer, *Annals NY Acad. Sci.* **316**, 64 (1979).

190. A. Rovinsky and M. Metzinger, *Phys. Rev. A* **46**, 6315 (1992).

191. T. K. Callahan and E. Knobloch, *Nonlinearity* **10**, 1179 (1997).

192. J. R. Weimar and J. P. Boon, in *Lattice Gas Automata and Pattern Formation*, L. Lawniczak and R. Kapral, eds., Field Institute, Canada, 1994.

193. J. R. Weimar, *Simulation with Cellular Automata*, Logos, Berlin, 1997.

194. A. De Wit, G. Dewel, P. Brockmans, and D. Walgreaf, *Physica D* **61**, 289 (1992).

195. V. Dufiet and J. Boissonade, *J. Chem. Phys.* **96**, 664 (1992).

196. O. Jensen, E. Mosekilde, P. Brockmans, and G. Dewel, *Physica Scripta* **53**, 243 (1996). *J. Phys. Chem.* **96**, 732 (1984).

197. A. De Wit, P. Brockmans, and G. Dewel, *Proc. Nat.–Acad. Sci. USA* **94**, 12768 (1997).

198. K. R. Mecke, *Phys. Rev. E* **53**, 4794 (1996).

199. K. R. Mecke, *Acta Phys. Pol. B* **28**, 1747 (1997).

200. A. A. Koulakov, M. M. Folger, and B. I. Shklovskii, *Phys. Rev. Lett.* **76**, 499 (1995).

201. M. M. Folger, A. A. Kulakov, and B.I. Shklovskii, *Phys. Rev. B* **54**, 1853 (1995).

202. I. L. Aleiner and L. I. Glazman, *Phys. Rev. B* **52**, 11296 (1995).

203. R. Megerle, *Phys. Rev. Lett.* **85**, 2749 (2000)

204. J. Frank, *Electron Tomography*, Plenum, New York, 1992.

205. K. Michielsen and H. De Raedt, *Comput. Phys. Commun.* **132**, 94 (2000).

206. K. R. Mecke and H. Wagner, *J. Stat. Phys.* **64**, 843 (1991).

207. K. R. Mecke, *Int. J. Mod. Phys. B* **12**, 861 (1998).

208. K. Michielsen and H. De Raedt, *Phys. Rep.* (in press).

209. C. N. Likos, K. R. Mecke, and H. Wagner, *J. Chem. Phys.* **102**, 9350 (1995).

210. V. Sofonea and K. R. Mecke, *Eur. Phys. J. B* **8**, 99 (1999).

211. W. E. Lorensen and H. E. Cline, *Comput. Graphics* **21**, 163 (1987).

212. A. Doi and A. Koide, *IEICE Trans. Commun. Electr. Inf. Syst.* **E74**(1), 214 (1991).

213. A. Gueziec and R. Hummel, *IEEE Trans. Visual. Comput. Graphics* **1**(4), 328 (1995).

214. G. Brown, P. A. Rikvold, M. Sutton, and M. Grant, *Phys. Rev. E* **60**, 5151 (1999).

215. K. Kawasaki, M. C. Yalabik, and J. D. Gunton, *Phys. Rev. A* **17**, 455 (1978).

216. T. Ohta, D. Jasnow, and K. Kawasaki, *Phys. Rev. Lett.* **49**, 1223 (1982).

217. G. F. Mazenko, *Phys. Rev. Lett.* **63**, 1605 (1989); G. F. Mazenko, *Phys. Rev. B* **42**, 4487 (1990).

218. H. Tanaka, T. Hayashi, and T. Nishi, *J. Appl. Phys.* **59**, 3627 (1986).

219. L. M. Cruz-Orive, *J. Microsc.* **107**, 235 (1976).

220. J. C. Russ, *Practical Stereology*, Plenum Press, New York, 1986, Chapter 4.

221. P. M. Chaikin and T. C. Lubensky, *Principles of Condensed Matter Physics,* Cambridge University Press, New York, 1995.

222. Y. Nishikawa, H. Jinnai, T. Koga, T. Hashimoto, and S. T. Hyde, *Langmuir* **14**, 1242 (1998).

223. M. Fiałkowski and R. Hołyst, *Acta Phys. Pol. B* **32**, 1579 (2001).

224. D. A. Hoffman, *J. Phys. (Paris) Colloq.* **51**, c7-197 (1990).

225. A. Aksimentiev, T. Biben, R. Hołyst, and K. Moorthi, Japanese patent application N. 2000-221562.

226. N. M. Maurits, G. J. A. Sevink, and J. G. E. M. Fraaije, *Macromolecules* **32**, 7674 (1999).

227. K. Kimishima, T. Koga, and T. Hashimoto, *Macromolecules* **33**, 968 (2000).

228. T. Tlusty, S. A. Safran, and R. Strey, *Phys. Rev. Lett.* **84**, 1244 (2000).

229. R. B. Laughlin and D. Pines, *Proc. Natl. Acad. Sci.* **97**, 28 (2000).

230. R. B. Laughlin, D. Pines, J. Schmalian, B. P. Stojkovic, and P. Wolynes, *Proc. Natl. Acad. Sci.* **97**, 32 (2000).

231. R. Franzosi, M. Pettini, and L. Spinelli, *Phys. Rev. Lett.* **84**, 2774 (2000).

232. Free software for morphological analysis in 2D and 3D systems is available at http://www.ichf.edu.pl/Dep3.html or http://www.ichf.edu.pl/mfialkowski/morph.html.

# INFRARED LINESHAPES OF WEAK HYDROGEN BONDS: RECENT QUANTUM DEVELOPMENTS

OLIVIER HENRI-ROUSSEAU, PAUL BLAISE, and DIDIER CHAMMA

*Centre d'Etudes Fondamentales, Université de Perpignan, Perpignan, France*

## CONTENTS

*Advances in Chemical Physics, Volume 121*, Edited by I. Prigogine and Stuart A. Rice.
ISBN 0-471-20504-4 © 2002 John Wiley & Sons, Inc.

# I.  INTRODUCTION

We focus our review on the dynamical properties of hydrogen bonds X–H$\cdots$Y
which have been widely studied by means of infrared spectroscopy. Indeed, the
infrared (IR) spectra of hydrogen bonds (H bonds) appeared to be a very useful
tool because the broad stretching band $\nu_S$ (X–$\overrightarrow{\text{H}}\cdots$Y) is very informative,
containing complete information on the electronic and consequently nuclear

TABLE I
Main Characteristics of the $\nu_S$ (X–H$\cdots$Y) Infrared Band

---

**Lower average frequency.** Compared to the frequency of the free X–H stretching, the average frequency of the $\nu_S$ (X–$\overrightarrow{H}\cdots$Y) band is strongly lowered. As an example, the $\nu_S$ (OH) wavenumber of the gaseous acetic acid monomer at 430 K is decreased from 3583 cm$^{-1}$ to 2875 cm$^{-1}$ when in solid state (90 K [1]. Stronger hydrogen bonds show higher frequency shifts.

**Broadened band.** The $\nu_S$ (OH) band goes from 30 cm$^{-1}$ width in the case of gaseous acetic acid monomer, and it increases up to 400 cm$^{-1}$ for hydrogen-bonded dimers.

**Complex and asymmetric structures, which are sometimes almost identical in gas and condensed**

**Greatly enhanced intensity of the band**

**Deuteration.** The average frequency, the width, and the intensity of the band are lowered upon isotopic substitution of the proton by a deuterium.

**Frequency dependence on the temperature.** The average frequency shifts with a change in the temperature: The average wavenumber of the adipic acid crystals increases by 30 cm$^{-1}$ from 0 to 300 K [2]. Greater shifts have been observed such as 70 cm$^{-1}$ for CH$_3$OH, or 52 cm$^{-1}$ for CH$_3$OD, between 170 K and 280 K [3,4].

---

dynamics. The same is true of Raman spectra, but owing to the weakness of its bands and owing to experimental hurdles in obtaining good Raman spectra, far less experimental material is available and then most theories are focused on infrared spectra. Nevertheless, the essential ideas of the latter can be applied to Raman spectra with little adaptation.

## A. Infrared Spectral Properties of Hydrogen Bonds

The infrared bandshape of the $\nu_S$ (X–$\overrightarrow{H}\cdots$Y) exhibits a rich and somewhat complex stucture whose main well-known properties are gathered in Table I. As shown in this table, the $\nu_S$ (X–$\overrightarrow{H}\cdots$Y) band exhibits important features by comparison with the free $\nu_S$ (X–$\overrightarrow{H}$) band. Depending on the properties of the hydrogen bond, the average wavenumber red shifts may be as large as some thousands cm$^{-1}$, and the intensity may increase up to a thousandfold. The changes of the various properties are interrelated, and it is the interaction energy—often termed *strength*—that is usually taken as the reference parameter for correlations of the most conspicuous spectroscopic properties involving the bandwidth and, more generally, the bandshape. This band width is known to increase from around 15 wavenumbers (cm$^{-1}$) to several thousands.

The large increase of the high-frequency bandwidth is but one challenge for the nuclear dynamics theories of hydrogen bonding, which are the subject of this chapter. Other challenges are the band asymmetry, the puzzling appearance of subsidiary absorption maxima and minima, such as windows, and the peculiar isotope effects on frequency shifts.

The wide range of changes in the hydrogen bond properties allows for a conventional classification into weak, medium strong, and very strong hydrogen bonds. However, there are no obvious discontinuities in the evolution of properties upon an increase of the interaction energy, and the limits between the classes are set deliberately. For the purpose of theoretical treatments of band-shapes, a different discriminator is more useful; it is based on the one-dimensional proton potential function. Thus hydrogen bonds can be classified into those with a single, asymmetric potential minimum (roughly corresponding to weak and medium strong hydrogen bonds) and those with low (or without) barrier in the potential function—in other words, hydrogen bonds with appreciable tunneling effects and those in which tunneling may be neglected.

There are several mechanisms that influence the evolution of the high-frequency bandshape. They may be separated into intrinsic mechanisms and medium-related or extrinsic mechanisms. Bandshapes in the low-pressure gas phase are obviously owing exclusively to mechanisms determined by the nuclear dynamics of the isolated hydrogen-bonded complex; this is also true, with a good approximation, of dilute noble gas matrices. Intermolecular collisions are the only complementary broadening effect under these conditions. In dilute solution, the medium (inert solvent, for instance) takes the important role of the thermal bath. In more concentrated solution and pure liquid state, several other mechanisms arise such as differences in the metrics of association. These are absent in crystalline solids, but the correlation field splitting and phonon coupling effects have to be considered instead. A complete theory of bandshapes of hydrogen-bonded systems has to cover the above phenomeno-logy. Obviously, the main task is to find the explanation for the inherent band broadening, but the theory also has to provide for inclusion of (a) the bandshaping mechanism operating in the condensed phases and (b) the effects of tunneling that appear with low barrier proton potentials. Clearly, the theory has to be quantitative and should permit the reconstruction of spectra with the use of physically accessible parameters.

## B.  Intrinsic Anharmonicity of the Fast Mode

At the basis of the interpretation of the lineshapes of the $\nu_S$ ($X\overrightarrow{-H}$) stretching mode, there is the assumption of a strong anharmonicity [5,6]. The first fundamental one is an anharmonic coupling between the high $X\overrightarrow{-H}\cdots Y$ frequency mode and the slow $\overleftarrow{X-H}\cdots\overrightarrow{Y}$ mode—that is the H-bond bridge. That is at the basis of the strong anharmonic coupling theory which is assuming a dependence of the angular frequency of the fast mode on the slow mode coordinate [7,8]. Another fundamental anharmonicity that is supposed in the area of an H bond [9], is related to the possibility of Fermi resonances between the first excited state of the high-frequency mode and the first harmonic of some bending modes or of a combination of the first excited states of two bending modes. There is also the intrinsic anharmonicity of the H-bond bridge [10–13]. That is because the potential of this slow mode is characterized by a very low frequency so that, at room temperature, several levels must be thermally occupied and not only the ground state. As a consequence, because the potential is of Morse nature, its anharmonicity—which is more and more sensitive with excitation of the energy levels—cannot be ignored. Besides the potential governing the motion of the proton is more generally an asymmetric double well. Of course, such a potential is susceptible to induce tunneling effects in special situations [14–19].

At last, because of the whole strong anharmonicity involved in the H-bond system, the possibility of electrical anharmonicity cannot be *a priori* ignored [20].

There is also the possibility of Davydov coupling, which is likely to appear when there are double or multiple H-bond systems [7,21–23]. It is responsible for cooperative effects between neighboring hydrogen bonds in cyclic hydrogen bonded dimers, or more generally in hydrogen-bonded chains in solids [10,24–34].

At last, we shall quote several studies dealing with a *dynamical* anharmonicity which arise when one goes beyond the Born–Oppenheimer approximation [35–39].

We give in Fig. 1 a schematic illustration of the main anharmonicities that are assumed to take place in hydrogen bonds.

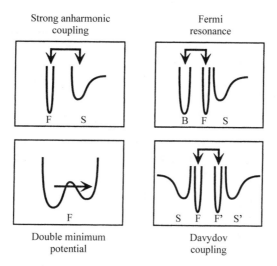

**Figure 1.** Main anharmonicities of the fast mode. F, fast mode; S, slow mode (intrinsic anharmonic Morse potential); B, bending mode.

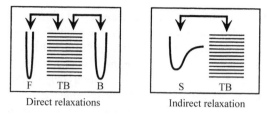

Direct relaxations                Indirect relaxation

**Figure 2.** The two relaxational mechanisms in hydrogen bonding. F, fast mode; S, slow mode; B, bending mode; TB, thermal bath.

## C.  Relaxational Mechanisms

On the other hand, one has to take into account the influence of the surrounding which must induce an irreversible evolution of the H-bond system when its fast mode is excited: the fast mode may be directly damped by the medium that is the direct relaxation mechanism. It may be also damped through the slow mode to which it is anharmonically coupled, that is the indirect relaxation mechanism. A schematical illustration of these two damping mechanism is given in Fig. 2. Of course, the role played by damping must be more important for H bonds in condensed phase.

The possibility of predissociation of the H-bond bridge has been also studied by Coulson and Robertson [40,41] and, separately, by Ewing [42,43], both works leading to negative conclusions. More recently, Mühle et al. [44,45] obtained opposite conclusions, supported by experimental studies [46].

## D.  The Theoretical Challenge

As a matter of fact, the ideal theory should treat quantum mechanically, on equal grounds, the intrinsic anharmonicities of the fast mode (asymmetric double well potential) and of the slow mode (Morse potential), as well as the strong anharmonic coupling between these slow and fast modes, together with the direct and indirect damping mechanisms, and should allow the introduction of multiple Fermi resonances and Davydov coupling when required. This is a project that may appear somewhat formidable. There have been many attempts to take into account quantum mechanically some aspects of this question. Each attempt unfortunately involves a variable number of approximations, among which the adiabatic approximation (allowing us to separate the fast motion of the high frequency mode from the slow one of the H-bond bridge) will be extensively studied in this chapter.

## II.  THEORETICAL FRAMEWORK

The present review relies on our precedent survey in this area [8] and will suppose the knowledge of other, older reviews [5,47–49]. We shall consider here quantum theories that all (i) assume, following Maréchal and Witkowski [7], a dependence of the angular frequency of the fast mode on the coordinate of the H-bond bridge, (ii) are deeply interconnected, and (iii) reduce to the Maréchal and Witkowski model.

## A. The Linear Response Theory

The theoretical studies usually obtained the infrared transition probabilities from the diagonalization of a total Hamiltonian which did not account for relaxational mechanisms. The theoretical spectra are then composed of Dirac delta peaks that are not fully suitable for comparison with experimental spectra.

The linear response theory [50,51] provides us with an adequate framework in order to study the dynamics of the hydrogen bond because it allows us to account for relaxational mechanisms. If one assumes that the time-dependent electrical field is weak, such that its interaction with the stretching vibration $X\text{–}\overrightarrow{H}\cdots Y$ may be treated perturbatively to first order, linearly with respect to the electrical field, then the IR spectral density may be obtained by the Fourier transform of the autocorrelation function $G(t)$ of the dipole moment operator of the X–H bond:

$$I(\omega) = \int_{-\infty}^{+\infty} G(t)\exp(-i\omega t)\,dt = 2\,\mathrm{Re}\left[\int_{0}^{+\infty} G(t)\exp(-i\omega t)\,dt\right] \quad (1)$$

The autocorrelation function $G(t)$ may be written, in a general form:

$$G(t) = \mathrm{tr}(\hat{\rho}(0)\hat{\mu}^{\dagger}(0)\hat{\mu}(t)) \quad (2)$$

where tr is the trace operator. Note that the knowledge of the total Hamiltonian $\hat{H}$ of the considered system is needed in order to express the Boltzmann operator $\hat{\rho}(0)$, and it is also needed in order to derive the dipolar momentum operator $\hat{\mu}(t)$, at any time $t$, from $\hat{\mu}(0)$ (taken at initial time) in an Heinsenberg picture—that is, by using the time evolution operator $\exp(-i\hat{H}t/\hbar)$ of the system:

$$\hat{\mu}(t) = \exp(i\hat{H}t/\hbar)\,\hat{\mu}(0)\exp(-i\hat{H}t/\hbar) \quad (3)$$

The total Hamiltonian $\hat{H}$ should include all the vibrational modes involved in the hydrogen bond dynamics, as well as the various couplings taking place between these modes. At last, $\hat{H}$ would include the relaxation mechanisms.

## B. The Main Mechanism: Strong Coupling Theory of Anharmonicity

The simplest hydrogen bond $X\text{–}H\cdots Y$ model may be viewed as composed of two oscillators. The first one corresponds to the stretching $X\text{–}\overrightarrow{H}\cdots Y$ of the valence bond X-H. We will refer to this mode as the "fast mode" of the

TABLE II
Parameters Associated with the Two Elementary Quantum Oscillators
Forming an Hydrogen Bond

| | Fast Mode $\overleftrightarrow{X-H}\cdots Y$ | Slow Mode $\overleftarrow{X-H}\cdots\overrightarrow{Y}$ |
|---|---|---|
| Angular frequency | $\omega_0^{eff}$ | $\omega_{oo}$ |
| Reduced mass | $m$ | $M$ |
| Dimensions position coordinate | $\tilde{q} = \frac{1}{\sqrt{2}}(b^\dagger + b)$ | $\tilde{Q} = \frac{1}{\sqrt{2}}(a^\dagger + a)$ |
| Conjugate moment | $\tilde{p} = \frac{1}{\sqrt{2}}(b^\dagger - b)$ | $\tilde{P} = \frac{1}{\sqrt{2}}(a^\dagger - a)$ |
| Commutation rule | $[b, b^\dagger] = 1$ | $[a, a^\dagger] = 1$ |

$a$ and $b$ ($a^\dagger$, $b^\dagger$) are the usual dimensionless lowering and raising operators.

hydrogen bond. The second one corresponds to the stretching $\overleftarrow{X-H}\cdots\overrightarrow{Y}$ of the hydrogen bridge, sometimes called the intermonomer mode, to which we will refer as the "slow mode" of the hydrogen bond. The slow mode frequency is 10 to 20 times slower than that of the fast mode. The parameters associated with these two vibrational modes are shown in Table II.

The creation and annihilation operators defined in Table II obey the following eigenvalues equations:

$$b^\dagger b|\{k\}\rangle = k\,|\{k\}\rangle, \qquad a^\dagger a\,|(m)\rangle = m\,|(m)\rangle \qquad (4)$$

with the orthonormality relations

$$\langle\{k\}|\{l\}\rangle = \delta_{kl}, \qquad \langle(m)|(n)\rangle = \delta_{mn} \qquad (5)$$

The cornerstone of the strong anharmonic coupling theory relies on the assumption of a modulation of the fast mode frequency by the intermonomer distance. This behavior is correlated by many experimental observations, and it is undoubtly one of the main mechanisms that take place in a hydrogen bond. Because the intermonomer distance is, in the quantum model, represented by the dimensionless position coordinate $\tilde{Q}$ of the slow mode, the effective angular frequency of the fast mode may be written [52,53]

$$\omega_0^{eff}(\tilde{Q}) = \omega_o + \alpha_0\,\omega_{oo}\,\tilde{Q} + \beta_0\,\omega_{oo}\,\tilde{Q}^2 + \cdots \qquad (6)$$

where appear the dimensionless coupling parameters $\alpha_o$ and $\beta_o$. It is of importance to note that $\omega_o$ is the angular frequency of the free X–H stretching motion lowered by a term that depends on both the H-bond strength and the equilibrium position of the slow mode ( because the establishment of the H bond weakens the force constant of the X–H bond).

It is then reasonable to introduce in the same way the equilibrium position $\tilde{q}_e$ of the fast mode [52,53]:

$$\tilde{q}_e(\tilde{Q}) = f_o \tilde{Q} + g_o \tilde{Q} \tag{7}$$

where the modulation parameters $f_o$ and $g_o$ are dimensionless.

The potential energy $\hat{U}$ of the fast mode should be assumed to be harmonic, in a first approximation:

$$\hat{U}(\tilde{q}, \tilde{Q}) = \frac{1}{2}\frac{\hbar}{\omega_o}\left[\omega_o^{\text{eff}}(\tilde{Q})\right]^2\left[\tilde{q} - \tilde{q}_e(\tilde{Q})\right]^2 \tag{8}$$

Although this potential is written within the harmonic approximation, it is extrinsically anharmonic when one assumes the modulation hypothesis (6) and (7). Then, $\hat{U}$ expands into

$$
\hat{U}(\tilde{q}, \tilde{Q}) = \frac{\hbar}{\omega_o}
\begin{bmatrix}
\frac{1}{2}\beta_o^2\omega_{oo}^2\tilde{Q}^4 \\
+\alpha_o\omega_{oo}^2\beta_o\tilde{Q}^3 \\
+\left(\frac{1}{2}\alpha_o^2\omega_{oo}^2 + \omega_o\beta_o\omega_{oo}\right)\tilde{Q}^2 \\
+\omega_o\alpha_o\omega_{oo}\tilde{Q} \\
+\frac{1}{2}\omega_o^2
\end{bmatrix}\tilde{q}^2
$$

$$
+ \frac{\hbar}{\omega_o}
\begin{bmatrix}
-\beta_o^2\omega_{oo}^2\,g_o\tilde{Q}^6 \\
-\left(2\alpha_o\omega_{oo}^2\beta_o\,g_o + \beta_o^2\omega_{oo}^2f_o\right)\tilde{Q}^5 \\
-\left(2\alpha_o\omega_{oo}^2\beta_of_o + 2\omega_o\beta_o\omega_{oo}\,g_o + \alpha_o^2\omega_{oo}^2\,g_o\right)\tilde{Q}^4 \\
-\left(2\omega_o\beta_o\omega_{oo}f_o + 2\omega_o\alpha_o\omega_{oo}\,g_o + \alpha_o^2\omega_{oo}^2f_o\right)\tilde{Q}^3 \\
-\left(\omega_o^2\,g_o + 2\omega_o\alpha_o\omega_{oo}f_o\right)\tilde{Q}^2 \\
-\omega_o^2f_o\tilde{Q}
\end{bmatrix}\tilde{q}
$$

$$
+ \frac{\hbar}{\omega_o}
\begin{bmatrix}
+\frac{1}{2}\beta_o^2\omega_{oo}^2\,g_o^2\tilde{Q}^8 \\
+\left(\alpha_o\omega_{oo}^2\beta_o\,g_o^2 + \beta_o^2\omega_{oo}^2f_o\,g_o\right)\tilde{Q}^7 \\
+\left(2\alpha_o\omega_{oo}^2\beta_of_o\,g_o + \frac{1}{2}\beta_o^2\omega_{oo}^2f_o^2 + \omega_o\beta_o\omega_{oo}\,g_o^2 + \frac{1}{2}\alpha_o^2\omega_{oo}^2\,g_o^2\right)\tilde{Q}^6 \\
+\left(2\omega_o\beta_o\omega_{oo}f_o\,g_o + \alpha_o^2\omega_{oo}^2f_o\,g_o + a_o\omega_{oo}^2\beta_of_o^2 + \omega_o\alpha_o\omega_{oo}\,g_o^2\right)\tilde{Q}^5 \\
+\left(\frac{1}{2}\omega_o^2\,g_o^2 + 2\omega_o\alpha_o\omega_{oo}f_o\,g_o + \frac{1}{2}\alpha_o^2\omega_{oo}^2f_o^2 + \omega_o\beta_o\omega_{oo}f_o^2\right)\tilde{Q}^4 \\
+\left(\omega_o\alpha_o\omega_{oo}f_o^2 + \omega_o^2f_o\,g_o\right)\tilde{Q}^3 \\
+\frac{1}{2}\omega_o^2f_o^2\tilde{Q}^2
\end{bmatrix}
\tag{9}
$$

Note that founder theoretical treatment of the strong anharmonic coupling has been done by Maréchal and Witkowski [7] in the simplest case, obtained by neglecting the terms in $\beta_o$, $f_o$, and $g_o$.

The general potential $\hat{U}$ (8) has not been used before 1999 [52] because its numerical matrix representation requires huge basis sets, incompatible with the common computers. In order to avoid this situation, an approximation has been undertaken in previous studies: the adiabatic approximation [54,55]. Following an idea of Stepanov [56], Maréchal and Witkowski assumed that the fast mode follows adiabatically the slow intermonomer motions, just as the electrons are assumed to follow adiabatically the motions of the nuclei in a molecule. It has been shown [57] that the adiabatic approximation is only suitable for very weak hydrogen bonds, as discussed in the next section.

## III.   VALIDITY OF THE ADIABATIC APPROXIMATION

### A.   The Common Background

#### 1.   The Spectral Density Within the Linear Response Theory

We shall modify in the present section the general definition of the spectral density (1), according to

$$I(\omega) = \int_{-\infty}^{+\infty} G(t) \exp(-\gamma t) \exp(-i\omega t)\, dt \tag{10}$$

The supplementary term $\exp(-\gamma t)$ is added in order to account for the direct damping mechanism of the fast mode, in the spirit of the Rösch and Ratner results [58]. Indeed, the Hamiltonians used in this section do not account for relaxational mechanisms, which will be discussed later in Section VI.

#### 2.   The Anharmonic Hamiltonian of the Hydrogen Bond

We restrict the study of the adiabatic approximation to the case where the modulation parameters $\beta_o$, $f_o$, and $g_o$, defined in Eqs. (6) and (7), are neglected. The effective angular frequency of the fast mode reduces to:

$$\omega_o^{\text{eff}}(\tilde{Q}) = \omega_o + \alpha_o\, \omega_{oo}\, \tilde{Q} \tag{11}$$

Then, the strong anharmonic coupling Hamiltonian is

$$\hat{H}(\tilde{q}, \tilde{Q}) = \frac{1}{2}\, \hbar\omega_{oo}\, \tilde{P}^2 + \frac{1}{2}\hbar\omega_o \tilde{p}^2 + \frac{1}{2}\hbar\omega_{oo}\, \tilde{Q}^2 + \frac{1}{2}\hbar\frac{[\omega_o^{\text{eff}}(\tilde{Q})]^2}{\omega_o}\tilde{q}^2 \tag{12}$$

and it expands into

$$\hat{H}(\tilde{q}, \tilde{Q}) = \hbar\omega_{oo}\left[\frac{\tilde{P}^2 + \tilde{Q}^2}{2}\right] + \hbar\omega_o\left[\frac{\tilde{p}^2 + \tilde{q}^2}{2}\right] + \hbar\,\alpha_o\, \omega_{oo}\, \tilde{Q}\tilde{q}^2 + \frac{1}{2}\hbar\alpha_o^2\frac{\omega_{oo}^2}{\omega_o}\tilde{Q}^2\,\tilde{q}^2 \tag{13}$$

Note that anharmonic Hamiltonians similar to (13) have been extensively studied [59,60].

Let us define the product base

$$\{|\{k\}, (m)\rangle\} \equiv \{|\{k\}\rangle \, |(m)\rangle\} \tag{14}$$

Then, the eigenvalue equation of the representation of the Hamiltonian (13) is

$$\hat{H}|\Psi_l\rangle = \hbar\omega_l|\Psi_l\rangle \tag{15}$$

with

$$|\Psi_l\rangle = \sum_k \sum_m \left\{ C_{l,m}^{\{k\}} \right\} |\{k\}, (m)\rangle \tag{16}$$

### 3. The Autocorrelation Function in the Standard Approach

In the present situation the general autocorrelation function (2) becomes

$$
G_{\text{std}}(t) \propto \sum_s \sum_r \sum_m \sum_n \exp\left\{ -\frac{\hbar\omega_s}{k_B T} \right\} \left\{ C_{s,m}^{\{0\}} \right\}^{-1} \left\{ C_{m,r}^{\{1\}} \right\} \left\{ C_{r,n}^{\{1\}} \right\}^{-1} \left\{ C_{n,s}^{\{0\}} \right\}
$$
$$
\times \exp\{ -i(\omega_s - \omega_r) t \} \tag{17}
$$

where the expansion coefficients are defined in Eq. (16) and where the angular frequencies are given by the eigenvalues of Eq. (15).

### B. The Adiabatic Approximation

In order to consider the Maréchal and Witkowski approach [7], it is suitable to write the basic fundamental Hamiltonian (12) according to the equations

$$\hat{H}(\tilde{q}, \tilde{Q}) = h(\tilde{q}, \tilde{Q}) + \hbar\omega_{oo} \left[ \frac{\tilde{P}^2 + \tilde{Q}^2}{2} \right] \tag{18}$$

$$h(\tilde{q}, \tilde{Q}) = \hbar(\omega_o + \alpha_o \omega_{oo} \tilde{Q}) \left[ \frac{\tilde{p}^2 + \tilde{q}^2}{2} \right] \tag{19}$$

The eigenvalue equation of the last Hamiltonian is

$$h(\tilde{q}, \tilde{Q})|\phi_k(\tilde{q}, \tilde{Q})\rangle = \left( k + \frac{1}{2} \right) \hbar \left( \omega_o + \alpha_o \omega_{oo} \tilde{Q} \right) |\phi_k(\tilde{q}, \tilde{Q})\rangle \tag{20}$$

where the $|\phi_k(\tilde{q}, \tilde{Q})\rangle$ are the eigenstates of the fast mode which depend parametrically on the slow mode coordinate $Q$. By combining the above equations, one gets the equations

$$\left[\hbar\omega_{oo}\frac{\tilde{P}^2+\tilde{Q}^2}{2}+\left(k+\frac{1}{2}\right)\hbar\left(\omega_o+\alpha_o\omega_{oo}\tilde{Q}t\right)\right]|\chi_n^{\{k\}}(\tilde{Q})\rangle=\hbar\omega_n^{\{k\}}|\chi_n^{\{k\}}(\tilde{Q})\rangle$$

$$(21)$$

where the $|\chi_n^{\{k\}}(\tilde{Q})\rangle$ are the eigenstates of the slow mode.

The Hamiltonian (18) may be partioned as usually into an adiabatic part and a diabatic correction:

$$\{\hat{H}(\tilde{q},\tilde{Q})\}=\{\hat{H}_{\text{adiab}}(\tilde{q},\tilde{Q})\}+\{\hat{H}_{\text{diab}}(\tilde{q},\tilde{Q})\}\qquad(22)$$

with respectively:

$$\{\hat{H}_{\text{adiab}}(\tilde{q},\tilde{Q})\}=\sum_k\sum_n\left\{\langle\phi_k(\tilde{q},\tilde{Q})|\left[\hbar\omega_{oo}\frac{\tilde{P}^2+\tilde{Q}^2}{2}+\left(k+\frac{1}{2}\right)\right.\right.$$
$$\times\hbar\left(\omega_o+\alpha_o\omega_{oo}\tilde{Q}\right)\Big]|\phi_k(\tilde{q},\tilde{Q})\rangle$$
$$+\langle\chi_n^{\{k\}}(\tilde{Q})|\left[\hbar\omega_{oo}\frac{\tilde{P}^2+\tilde{Q}^2}{2}\right]|\chi_n^{\{k\}}(\tilde{Q})\rangle\bigg\}$$
$$\times|\phi_k(\tilde{q},\tilde{Q})\rangle|\chi_n^{\{k\}}(\tilde{Q})\rangle\langle\chi_n^{\{k\}}(\tilde{Q})|\langle\phi_k(\tilde{q},\tilde{Q})|\qquad(23)$$

and

$$\{\hat{H}_{\text{diab}}(\tilde{q},\tilde{Q})\}=\sum_k\sum_l\sum_m\sum_n\left\{\hbar\omega_{oo}\langle\chi_n^{\{k\}}(\tilde{Q})|\tilde{P}|\chi_m^{\{l\}}(\tilde{Q})\rangle\langle\phi_k(\tilde{q},\tilde{Q})|\tilde{P}|\phi_l(\tilde{q},\tilde{Q})\rangle\right.$$
$$+\langle\Psi_n^{\{k\}}(\tilde{Q})|\langle\phi_k(\tilde{q},\tilde{Q})|\left[\frac{1}{2}\hbar\omega_{oo}\tilde{P}^2\right]|\phi_l(\tilde{q},\tilde{Q})\rangle|\chi_m^{\{l\}}(\tilde{Q})\rangle\bigg\}$$
$$\times|\phi_l(\tilde{q},\tilde{Q})\rangle|\chi_n^{\{k\}}(\tilde{Q})\rangle\langle\chi_m^{\{l\}}(\tilde{Q})|\langle\phi_k(\tilde{q},\tilde{Q})|\qquad(24)$$

Now, recall that for weak hydrogen bonds the high-frequency mode is much faster than the slow mode because $\omega_o\approx20\,\omega_{oo}$. As a consequence, the quantum adiabatic approximation may be assumed to be verified when the anharmonic coupling parameter $\alpha_o$ is not too strong. Thus, neglecting the diabatic part of the Hamiltonian (22) and using Eqs. (18) to (20), one obtains

$$\{\hat{H}_{\text{adiab}}(\tilde{q},\tilde{Q})\}=\sum_k\left[\hbar\omega_{oo}\frac{\tilde{P}^2+\tilde{Q}^2}{2}+\left(k+\frac{1}{2}\right)\hbar\left(\omega_o+\alpha_o\omega_{oo}\tilde{Q}\right)\right]$$
$$\times|\phi_k(\tilde{q},\tilde{Q})\rangle\langle\phi_k(\tilde{q},\tilde{Q})|\qquad(25)$$

## C. The Effective Hamiltonians of the Slow Mode in Different Representations

### 1. The Adiabatic Hamiltonian as a Sum of Effective Hamiltonians

Then the adiabatic Hamiltonian (25) may be written as a sum of effective Hamiltonians defined by the equations

$$\{\hat{H}_{\text{adiab}}(\tilde{q}, \tilde{Q})\} = \sum_k |\{k\}\rangle \{\hat{H}^{\{k\}}\} \langle\{k\}| \tag{26}$$

$$|\{k\}\rangle \equiv |\phi_k(\tilde{q}, \tilde{Q})\rangle \quad \text{and} \quad |\chi_n^{\{k\}}\rangle \equiv |\chi_n^{\{k\}}(\tilde{Q})\rangle \tag{27}$$

$$\hat{H}^{\{k\}} = \left(k + \frac{1}{2}\right)\hbar\omega_o + \left[\hbar\omega_{oo}\frac{\tilde{P}^2 + \tilde{Q}^2}{2} + \left(k + \frac{1}{2}\right)\hbar\alpha_o\omega_{oo}\tilde{Q}\right] \tag{28}$$

### 2. Representation I of the Effective Hamiltonians

Passing to the Boson operators by aid of Table II, and after neglecting the zero-point-energy of the fast mode, we obtain a quantum representation we shall name I, in which the effective Hamiltonians of the slow mode corresponding respectively to the ground and first excited states of the fast mode are

$$\hat{H}_I^{\{0\}} = \hbar\omega_{oo}\left\{\left[a^\dagger a + \frac{1}{2}\right] + \frac{1}{2}\alpha_o\left[a^\dagger + a\right]\right\} \tag{29}$$

$$\hat{H}_I^{\{1\}} = \hbar\omega_{oo}\left\{\left[a^\dagger a + \frac{1}{2}\right] + \frac{3}{2}\alpha_o\left[a^\dagger + a\right] + \frac{\omega_o}{\omega_{oo}}\right\} \tag{30}$$

The eigenvalue equation of the representation of the effective Hamiltonian operators (28) in the base of the number occupation operator of the slow mode is characterized by the equation

$$H_I^{\{k\}}|\chi_{j\,I}^{\{k\}}\rangle = \hbar\{\omega_j^{\{k\}}\}_I|\chi_{j\,I}^{\{k\}}\rangle \tag{31}$$

$$|\chi_{j\,I}^{\{k\}}\rangle = \sum_n \{D_{j,n}^{\{k\}}\}_I|(n)\rangle \tag{32}$$

### 3. Representation II of the Effective Hamiltonians

If we want to remove the driven term in the potential of the slow mode, when $k = 0$ (ground state of the fast mode), it is suitable to perform the following phase transformation:

$$\hat{H}_{II}^{\{k\}} = e^{\frac{1}{2}\alpha_o\left[a^\dagger + a\right]}\,\hat{H}_I^{\{k\}}\,e^{-\frac{1}{2}\alpha_o\left[a^\dagger + a\right]} \tag{33}$$

In this new representation, which we shall name II, the transformed effective Hamiltonians corresponding respectively to the ground and first excited states of the fast mode become

$$\hat{H}_{II}^{\{0\}} = \hbar\omega_{oo}\left[a^\dagger a + \frac{1}{2}\right] \tag{34}$$

$$\hat{H}_{II}^{\{1\}} = \hbar\omega_{oo}\left[\left[a^\dagger a + \frac{1}{2}\right] + \alpha_o\left[a^\dagger + a\right] - \alpha_o^2 + \frac{\omega_o}{\omega_{oo}}\right] \tag{35}$$

The ground-state effective Hamiltonian is diagonal with eigenvalues $\hbar\omega_{oo}\left[n + \frac{1}{2}\right]$, whereas the excited state one is that of a driven quantum harmonic oscillator that must lead to coherent states.

### 4. Representation III of the Effective Hamiltonians of Interest

On the other hand, one may perform on the effective Hamiltonian in representation II the canonical transformations:

$$\hat{H}_{III}^{\{k\}} = \exp\left\{k\alpha_o\left[a^\dagger + a\right]\right\}\hat{H}_{II}^{\{k\}}\exp\left\{-k\alpha_o\left[a^\dagger + a\right]\right\} \tag{36}$$

In the new representation, which we shall name III, one has

$$\hat{H}_{III}^{\{0\}} = \hat{H}_{II}^{\{0\}} \tag{37}$$

$$\hat{H}_{III}^{\{1\}} = \hbar\omega_{oo}\left[\left(a^\dagger a + \frac{1}{2}\right) - 2\alpha_o^2 + \frac{\omega_o}{\omega_{oo}}\right] \tag{38}$$

The corresponding eigenvalue equations are therefore

$$\hat{H}_{III}^{\{0\}}|\chi_{n\ III}^{\{0\}}\rangle = \hbar\omega_{oo}\left[n + \frac{1}{2}\right]|\chi_{n\ III}^{\{0\}}\rangle \tag{39}$$

$$\hat{H}_{III}^{\{1\}}|\chi_{n\ III}^{\{1\}}\rangle = \hbar\omega_{oo}\left[\left(n + \frac{1}{2}\right) - 2\alpha_o^2 + \frac{\omega_o}{\omega_{oo}}\right]|\chi_{n\ III}^{\{1\}}\rangle \tag{40}$$

Owing to Eq. (36), the eigenvectors of the two effective Hamiltonians are connected by the transformation

$$|\chi_{n\ III}^{\{1\}}\rangle = \exp\left\{\alpha_o\left[a^\dagger + a\right]\right\}|\chi_{n\ III}^{\{0\}}\rangle \tag{41}$$

### D. The Three Autocorrelation Functions and the Spectral Densities

Now, we must look at the autocorrelation function. Of course, these functions, which may be expressed indifferently in any one of the three quantum representations I, II, or III, are irrelevant to the choice of the representation so that the spectral density must be the same. This is because the physical properties must be invariant with respect to the canonical transformations allowing to pass from one representation to another. Thus, we may write for the adiabatic autocorrelation function

$$G_{\text{adiab}}(t) = G_I(t) = G_{II}(t) = G_{III}(t) \tag{42}$$

Although equivalent, their expressions may be more of less tractable. We shall treat them separately. In the first place, we shall consider the representation I that will have later the merit to allow us to obtain an expression for the autocorrelation function having a form similar to that of the above standard approach. Aside from this representation I, we shall also recall the other quantum representations II and III still used in the literature [8,61].

### 1. The Autocorrelation Function in Representation I

Now, let us look at the autocorrelation function of the dipole moment operator within the adiabatic approach. In the representation I it is

$$G_I(t) \propto \text{tr}\left[ \sum_m \sum_n \exp\left\{ -\frac{\hat{H}_I^{\{0\}}}{k_B T} \right\} |\{0\}, (m)\rangle \langle (m), \{1\}| \right.$$
$$\left. \times \exp\left\{ i\frac{\hat{H}_I^{\{1\}}}{\hbar} t \right\} |\{1\}, (n)\rangle \langle (n), \{0\}| \times \exp\left\{ -i\frac{\hat{H}_I^{\{0\}}}{\hbar} t \right\} \right] \tag{43}$$

where the effective Hamiltonians are given, respectively, by Eqs. (37) and (38). Such an autocorrelation function has not been used in the study of the spectral densities of hydrogen bond, because it can be only numerically computed. After some calculations, one obtains

$$\{G_I(t)\} \propto \sum_l \sum_s \sum_m \sum_r \exp\left\{ -\frac{\hbar\left\{ \omega_l^{\{0\}} \right\}_I}{k_B T} \right\} \left\{ D_{l,s}^{\{0\}} \right\}_I^{-1} \left\{ D_{s,m}^{\{1\}} \right\}_I \left\{ D_{m,r}^{\{1\}} \right\}_I^{-1} \left\{ D_{r,l}^{\{0\}} \right\}_I$$
$$\times \left\{ \exp\left\{ -i\left( \left\{ \omega_l^{\{1\}} \right\}_I - \left\{ \omega_m^{\{0\}} \right\}_I \right) t \right\} \right\} \tag{44}$$

which contains the expansion coefficients given by Eq. (32), along with the eigenvalues given by Eq. (31). The numerical spectral densities computed with this autocorrelation function will be later compared to that obtained in the standard approach and also to the others obtained with representation II and III that we shall now consider.

## 2. The Autocorrelation Function in Representation II

It may be shown [8] that the adiabatic autocorrelation function in representation II may be written

$$G_{//}(t) \propto \text{tr}\left[\exp\left\{-\frac{\hat{H}_{//}^{\{0\}}}{k_B T}\right\}\{U(t)\}^+ \exp\left\{-i\frac{\hat{H}_{//}^{\{0\}}}{\hbar}t\right\}\right] \qquad (45)$$

Here, $\{U(t)\}$ is the reduced time evolution operator of the driven damped quantum harmonic oscillator. Recall that representation II was used in preceding treatments, taking into account the indirect damping of the hydrogen bond. After rearrangements, the autocorrelation function (45) takes the form [8]

$$G_{//}(t) \propto \exp\left\{i\alpha_0^2 \sin(\omega_{oo}\,t)\right\} \exp\left\{i(\omega_o - 2\alpha_0^2\omega_{oo})\,t\right\}$$
$$\times \exp\left\{\alpha_0^2(1 + 2\langle n\rangle)[\cos(\omega_{oo}\,t) - 1]\right\} \qquad (46)$$

where $\langle n\rangle$ is the thermal average of the mean number occupation,

$$\langle n\rangle = \left[\exp\left\{\frac{\hbar\omega_{oo}}{k_B T}\right\} - 1\right]^{-1} \qquad (47)$$

It has been shown that the autocorrelation function (46) is the limit situation in the absence of damping of the autocorrelation function of the hydrogen bond within the indirect damping [8]. It must be emphasized that this autocorrelation function does not require for computation any particular caution because of its analytic character. Thus, it may be considered as a numerical reference for the computation involving representation I or III.

## 3. The Autocorrelation Function in Representation III

Now, we may recall the representation III of the autocorrelation function because its Fourier transform leads to the well-known Franck–Condon progression of delta Dirac peaks appearing in the pioneering work of Maréchal and Witkowski [7]. In this representation III, the general autocorrelation function (2) takes the form

$$G_{III}(t) \propto \text{tr}\left[\exp\left\{-\frac{\hbar\omega_{oo}}{k_B T}\left(a^\dagger a + \frac{1}{2}\right)\right\} \exp\left\{-\alpha_0[a^\dagger + a]\right\}\right.$$
$$\times \exp\left\{i\left[\left[a^\dagger a + \frac{1}{2}\right] - 2\alpha_0^2 + \frac{\omega_o}{\omega_{oo}}\right]\omega_{oo}\,t\right\}$$
$$\left.\times \exp\left\{\alpha_0[a^\dagger + a]\right\} \exp\left\{-i\left[a^\dagger a + \frac{1}{2}\right]\omega_{oo}t\right\}\right] \qquad (48)$$

It may be shown that the autocorrelation function III may be written after some manipulations, in the final form [8]

$$
G_{III}(t) \propto \sum_m \sum_n \exp\left\{-\frac{m\hbar\omega_{oo}}{k_B T}\right\} \left|\Gamma_{m,n}(\alpha_o)\right|^2 \exp\left\{i(m-n)\omega_{oo}t\right\}
$$
$$
\times \exp\left\{-2i\alpha_o^2\omega_{oo}t\right\} \exp\left\{i\omega_o t\right\} \tag{49}
$$

Equation (49) contains the Franck–Condon factors that are the matrix elements of the translation operator involved in the canonical transformation (36) with $k = 1$ that are given for $m \geq n$ by

$$
\Gamma_{m,n}(\alpha) =
$$
$$
\exp\left\{-\frac{\alpha^2}{2}\right\}(-1)^{n-m}\,\alpha^{n-m}\sqrt{m!}\sqrt{n!}\,\sum_{k=m-n}^{m}\left\{\frac{(-1)^k\alpha^{2k}}{(m-k)!\,k!\,(n-m+k)!}\right\}\quad \text{for } m \geq n
$$
$$
\tag{50}
$$

or, for $n \geq m$ by an expression which is the same except the $m$ and $n$ indices must be permuted and $(-1)^{m-n}$ removed.

### 4. Numerical Equivalence Between the Standard and the Three Adiabatic Spectral Densities

Figure 3 illustrates the equivalence between the three representation I, II, and III, which was stated by Eq. (42).

### 5. The Situation in the Absence of Damping

By Fourier transform of the representation III of the undamped adiabatic autocorrelation function (49), one obtains the Franck–Condon progression

$$
I_{\text{adiab III}}^\circ(\omega) \propto \sum_m \sum_n \exp\left\{-\frac{m\hbar\omega_{oo}}{k_B T}\right\} \left|\Gamma_{m,n}(\alpha_o)\right|^2 \delta(\omega - [m-n]\omega_{oo} - 2\alpha_o^2\omega_{oo} - \omega_o)
$$
$$
\tag{51}
$$

We may recall and emphasize that the autocorrelation function obtained in the three representations I, II, and III must be equivalent, from the general properties of canonical transformation which must leave invariant the physical results. Thus, because of this equivalence, the spectral density obtained by Fourier transform of (43) and (45) will lead to the same Franck–Condon progression (51).

On the other hand, the undamped autocorrelation function (17) we have obtained within the standard approach avoiding the adiabatic approximation must lead after Fourier transform to spectral densities involving very puzzling Dirac delta peaks given by

$$
I_{\text{std}}^\circ(\omega) \propto \sum_m \sum_n \sum_l \sum_p \exp\left\{-\frac{\hbar\omega_l}{k_B T}\right\} \left\{C_{l,m}^{\{0\}}\right\}^{-1}\left\{C_{m,p}^{\{1\}}\right\}\left\{C_{p,n}^{\{1\}}\right\}^{-1}
$$
$$
\times \left\{C_{n,l}^{\{0\}}\right\} \delta(\omega - [\omega_l - \omega_p]) \tag{52}
$$

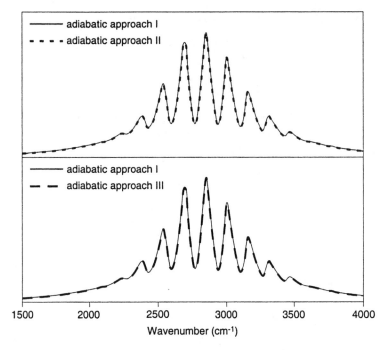

**Figure 3.** Numerical equivalence between the three representations, I, II, and III. Within the adiabatic approximation, this figure shows the numerical equivalence between the Fourier transforms of $G_I$ given by Eq. (44), $G_{II}$ given by Eq. (46) and $G_{III}$ given by Eq. (49).

The question remains to know if, when infinite bases are used, the approximated adiabatic spectral density (51) given by the Franck–Condon progression is narrowing the standard one (52) involving very puzzling Dirac delta peaks. However, lineshapes are actually not Dirac delta peaked but are involving some broadening because of the influence of the surrounding.

### 6.  Numerical Equivalence Between the Standard and Adiabatic Spectral Densities

In Fig. 4 we compare the adiabatic (dotted line) and the stabilized standard spectral densities (continuous line) for three values of the anharmonic coupling parameter and for the same damping parameter. Comparison shows that for $\alpha_o = 1$, the adiabatic lineshapes are almost the same as those obtained by the exact approach. For $\alpha_o = 1.5$, this lineshape escapes from the exact one. That shows that for $\alpha_o > 1$, the adiabatic corrections becomes sensitive. However, it may be observed by inspection of the bottom spectra of Fig. 4, that if one takes for the adiabatic approach $\omega_{oo} = 165 \, \text{cm}^{-1}$ and $\alpha_o = 1.4$, the adiabatic lineshape simulates sensitively the standard one obtained with $\omega_{oo} = 150 \, \text{cm}^{-1}$ and $\alpha_o = 1.5$.

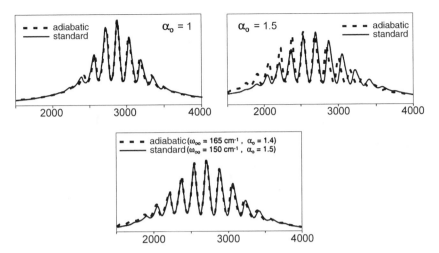

**Figure 4.** Accuracy of the lineshapes with respect to the adiabatic approximation. When unspecified, the parameters are $\omega_o = 3000\,\text{cm}^{-1}$, $\omega_{oo} = 150\,\text{cm}^{-1}$, $\gamma = 30\,\text{cm}^{-1}$, and $T = 300$ K.

These results may be generalized — that we have verified numerically — as follows: at 300 K, for any given damping parameters, the adiabatic lineshape remains very close to the exact standard one, when the dimensionless anharmonic coupling parameter $\alpha_o$ is increased from 0 to 1. Beyond, i.e. for $\alpha_o > 1$, the adiabatic lineshape progressively escapes from the exact one. This discrepancy may be removed for a great part by introducing an effective anharmonic coupling parameter $\alpha_o^{\text{eff}} < \alpha_o$ and an effective angular frequency of the slow mode $\omega_o^{\text{eff}} > \omega_o^o$, leaving only some frequency shifts on the high and low frequency tails which are involving the smallest intensities. The diabatic corrections become sensitive for larger values of $\alpha_o$ when the temperature is lowered and for smaller values of this parameter when the temperature is raised. At last, the magnitude of the damping parameter does not affect the results for a range lying between 0.05, that is a lowest limit, to 1, the increase of the damping smoothing the adiabatic corrections when they are occurring.

### E.   Theoretical Connection Between the Standard and the Adiabatic Approaches

Now, it may be of interest to look at the connection between the autocorrelation functions appearing in the standard and the adiabatic approaches. Clearly, it is the representation I of the adiabatic approach which is the most narrowing to that of the standard one [see Eqs. (43) and (17)] because both are involving the diagonalization of the matricial representation of Hamiltonians, within the product base built up from the bases of the quantum harmonic oscillators corresponding to the separate slow and fast modes. However, among the differences, there is the fact that the adiabatic approach involves effective Hamiltonians whereas the standard one ignores them. Thus, in order to find a deeper connection, it will be suitable to write the standard approach within an effective Hamiltonian formalism. For this purpose, it is possible to use either the quasi-degenerate perturbation theory or the wave operator procedure, because the frequency of the fast mode is very large with respect to that of the slow mode. For simplicity, we shall use here the wave operator procedure because all the ingredients necessary for its calculation are known. The wave operator procedure supposes here a large gap in the diagonal matrix elements with respect to the off ones, and it requires small truncated bases in order to avoid intruder states but not too small in order to get stabilized spectral densities. As a consequence, this approach will hold only for small anharmonic coupling parameters.

### 1.   From Adiabatic to Effective Hamiltonian Matrices Through the Wave Operator Procedure

Let us look at the standard Hamiltonian (13). Its representation restricted to the ground state and the first excited state of the fast mode may be written according to the wave operator procedure [62] by aid of the four equations

$$\hat{H}_{\text{eff}} = P^\circ \hat{H} \, \Omega \tag{53}$$

$$\Omega = P[P^\circ P P^\circ]^{-1} \tag{54}$$

$$P^\circ = \sum_{k=0}^{1} \sum_{m=0}^{N^{\circ\circ}-1} \{|\{k\}, (m)\rangle \, \langle (m), \{k\}|\} \tag{55}$$

$$P = \sum_{r=0}^{2N^{\circ\circ}-1} |\Psi_r\rangle \, \langle \Psi_r| \tag{56}$$

Here, $\Omega$ and $P^\circ$ are, respectively, the wave operator and the projector on the model space given, whereas $P$ is the projector on the target space. Recall that the $|\Psi_r\rangle$ are the eigenvectors of the Hamiltonian given by Eq. (13) whereas the states $|\{k\}, (m)\rangle$ are defined by Eq. (14) and that $N^{\circ\circ}$ is the dimension of the slow mode base. Note that, owing to the properties of the effective Hamiltonian, the $2N^{\circ\circ}$ eigenvalues of the matrix must be narrowing the lowest $2N^{\circ\circ}$ eigenvalues of the full Hamiltonian matrix of dimension $[N^\circ N^{\circ\circ}]^2$. At last, recall also that the effective Hamiltonian matrix is not Hermitian.

The numerical computations by aid of Eqs. (13) and (53)–(56) show that for all $m$ and $n$, when the dimensionless anharmonic coupling parameter is small (i.e., $\alpha_o \leq 0.5$), we obtain

$$\langle \{0\}, (n)|\hat{H}_{\text{eff}}|(m), \{1\}\rangle = \langle (n)|\hat{H}_{\text{eff}}^{\{0,1\}}|(m)\rangle \simeq 0$$

As a consequence, the effective Hamiltonian matrix restricted to the ground state and the first excited state of the fast mode has a block structure.

The eigenvalue equations of the two diagonal blocks of the effective Hamiltonian matrix is characterized by the equations

$$\left\{\hat{H}_{\text{eff}}^{\{k,k\}}\right\}|\Psi_r^{\{k\}}\rangle = \hbar\omega_r^{\{k\}}|\Psi_r^{\{k\}}\rangle \tag{57}$$

$$|\Psi_r^{\{k\}}\rangle = \sum_m \left\{B_{r,(m)}^{\{k\}}\right\}|(m)\rangle \tag{58}$$

Note that because the effective Hamiltonian matrix is not Hermitian, the eigenvectors are not orthogonal. However, when $\alpha_o$ is small, the orthogonality properties are satisfactorily verified.

### 3. The Standard Spectral Density Within the Effective Hamiltonian Procedure

The autocorrelation function of the hydrogen bond within the effective Hamiltonian expression of the standard approach takes the form

$$G_{\text{std}}^{\text{eff}}(t) \propto \text{tr}\left[\exp\left\{-\frac{\{\hat{H}_{\text{eff}}^{\{0,0\}}\}}{k_B T}\right\}\sum_m\sum_n|\{0\},(m)\rangle\langle(m),\{1\}|\exp\left\{i\frac{\{\hat{H}^{\{1,1\}}\}}{\hbar}t\right\}\right.$$

$$\left. \times |\{1\},(n)\rangle\langle(n),\{0\}|\exp\left\{-i\frac{\{\hat{H}^{\{0,0\}}\}}{\hbar}t\right\}\right] \tag{59}$$

Then, one obtains

$$G_{\text{std}}^{\text{eff}}(t) \propto \sum_l\sum_s\sum_m\sum_r\exp\left\{-\frac{\hbar\omega_l^{(0)}}{k_B T}\right\}\left\{B_{l,s}^{\{0\}}\right\}\left\{B_{s,m}^{\{1\}}\right\}\left\{B_{m,r}^{\{1\}}\right\}\left\{B_{r,l}^{\{0\}}\right\}$$

$$\times \exp\left\{-i\left(\left\{\omega_l^{\{1\}}\right\} - \left\{\omega_m^{\{0\}}\right\}\right)t\right\} \tag{60}$$

where the expansion coefficients are defined by Eq. (58). Thus, it is observed that this expression of the autocorrelation function of the hydrogen bond in the standard approach, which holds only for small anharmonic coupling parameter $\alpha_o$, has the same structure as that [Eq. (44)] obtained in the adiabatic approach within representation III. For the case of small $\alpha_o$ values, the standard spectral density in the presence of direct damping is given by the Fourier transform of the approximate Eq. (60),

$$I_{\text{std}}^{\text{eff}}(\omega) = \int G_{\text{std}}^{\text{eff}}(t)\, e^{-\gamma t}\exp\{-i\omega t\}\, dt \tag{61}$$

which has the same formal structure as the adiabatic spectral density within representation I. Because Eq. (61) holds only for small $\alpha_o$, the corresponding adiabatic spectral density is exact. Thus, in this situation an agreement must be

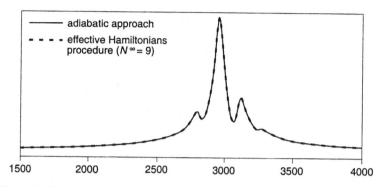

**Figure 5.** Illustration of the equivalence between the spectral densities obtained within the adiabatic approximation and those resulting from the effective Hamiltonian procedure, using the wave operator. Common parameters: $\alpha_0 = 0.4$, $\omega_0 = 3000\,\mathrm{cm}^{-1}$, $\omega_{00} = 150\,\mathrm{cm}^{-1}$, $\gamma = 30\,\mathrm{cm}^{-1}$, and $T = 300$ K.

expected between the stabilized lineshape computed by aid of the approximate standard Eq. (61) and the corresponding adiabatic one computed by means of Eq. (44), which is then exact. Figure 5 illustrates this for $\alpha_0 = 0.4$, $\gamma = 0.2\omega_{00}$, and $T = 300$ K. It must be observed that the stabilization conditions of the lineshape computed with Eq. (61) is sharp with respect to the dimension of the truncated bases because bases that are too small do not allow stabilizations, whereas ones that are too large introduce intruder states in the wave operator (54) and in the effective Hamiltonian which lead to numerical drifts.

## IV.  FERMI RESONANCES: "EQUAL DAMPING" TREATMENT

Bratos and Hadži have developed another origin of the anharmonicity of the fast mode $\overrightarrow{\mathrm{X-H}}\cdots\mathrm{Y}$, the Fermi resonance, which is supported by several experimental studies [1,3,63–70]. Widely admitted for strong hydrogen bonds [67], the important perturbation brought to the infrared lineshape by Fermi resonances has also been pointed out in the case of weaker hydrogen bonds [53,71–73].

In this section we shall give the connections between the nonadiabatic and damped treatments of Fermi resonances [53,73] within the strong anharmonic coupling framework and the former theory of Witkowski and Wójcik [74] which is adiabatic and undamped, involving implicitly the exchange approximation (approximation later defined in Section IV.C).

Note that Maréchal [48,75] proposed an original *peeling-off* procedure, but in a very different perspective, whose purpose is to eliminate from a given experimental spectrum the features brought by multiple Fermi resonances, thus revealing the intrinsic informations on the hydrogen bond dynamics.

TABLE III
Parameters Associated with a Bending Mode

|                                    | Bending Mode |
| ---------------------------------- | ------------ |
| Angular frequency                  | $\omega_\delta$ |
| Dimensionless position coordinate  | $\tilde{q}_\delta = \frac{1}{\sqrt{2}}(d^\dagger + d)$ |
| Conjugate moment                   | $\tilde{p}_\delta = \frac{1}{\sqrt{2}}(d^\dagger - d)$ |
| Commutation rule                   | $[d, d^\dagger] = 1$ |

$d$ and $d^\dagger$ are the usual dimensionless lowering and raising operators.

In order to develop a model for the Fermi resonance mechanism, we need to define at least a third vibrational mode — that is, a bending mode located in the X part of the hydrogen bond X–H$\cdots$Y — to which the fast mode of the hydrogen bond is coupled. We gathered in Table III the parameters associated with this bending mode. A basis set for this bending mode may be constructed on the eigenvectors $|[u]\rangle$, defined by the eigenvalues equation:

$$d^\dagger d \, |[u]\rangle = u \, |[u]\rangle, \quad \text{with} \quad \langle[u]|[v]\rangle = \delta_{uv} \tag{62}$$

where the square brackets, used inside a ket, will always refer to a state of the bending mode. Recall that the state $|\{k\}\rangle$ refers to the $k$th state of the fast mode and that similarly $|(m)\rangle$ refers to the $m$th state of the slow mode [see Eqs. (4) and (5)].

### A.   The "Equal Damping" Treatment

It is of importance to note that we shall consider, in the present section, that the fast and bending modes are subject to the same quantitative damping. Indeed, the damping parameter of the fast mode $\gamma_0$ and that of the bending mode $\gamma_\delta$ will be supposed to be equal, so that we shall use in the following a single parameter, namely $\gamma \, (= \gamma_0 = \gamma_\delta)$. This drastic restriction cannot be avoided when going beyond the adiabatic approximation.

We shall study in Section V the role played independently by $\gamma_0$ and $\gamma_\delta$, within the adiabatic approximation.

### B.   General Nonadiabatic Treatment

#### 1.   Potential of the Fermi Coupling

We may split the Hamiltonian $\hat{H}_{gf}$ (general nonadiabatic treatment of Fermi resonances), which describes the three undamped modes, as follows:

$$\hat{H}_{gf} = \hat{T}_{gf} + \hat{U}_{gf} \tag{63}$$

where the kinetic part $\hat{T}_{gf}$ describes the three quantum harmonic modes:

$$\hat{T}_{gf} = \frac{1}{2}\hbar\omega_o\tilde{p}^2 + \frac{1}{2}\hbar\omega_{oo}\tilde{p}^2 + \frac{1}{2}\hbar\omega_\delta\tilde{p}_\delta^2 \tag{64}$$

The potential part $\hat{U}_{gf}$ is

$$\hat{U}_{gf} = \hat{U}_1(\tilde{q},\tilde{Q}) + \hat{U}_2(\tilde{Q}) + \hat{U}_3(\tilde{q}_\delta) + \hat{H}_{gf}^\delta(\tilde{q},\tilde{q}_\delta) \tag{65}$$

where $\hat{U}_1$, $\hat{U}_2$, and $\hat{U}_3$ are the potentials for the fast, slow, and bending modes, respectively. The last term $\hat{H}_{gf}^\delta$, which refers to the Fermi resonance mechanism, accounts for the coupling between the fast and bending modes.

In the strong anharmonic coupling framework, the fast mode potential $\hat{U}_1$ is, according to Eq. (8),

$$\hat{U}_1(\tilde{q},\tilde{Q}) = \frac{1}{2}\frac{\hbar}{\omega_o}\left[\omega_o^{eff}(\tilde{Q})\right]^2\left[\tilde{q}-\tilde{q}_e(\tilde{Q})\right]^2 \tag{66}$$

Remember that it is extrinsically anharmonic when one assumes the modulation hypothesis (6) and (7).

Using the harmonic approximation, the potentials $\hat{U}_2$ and $\hat{U}_3$ may be written

$$\hat{U}_2(\tilde{Q}) = \frac{1}{2}\hbar\omega_{oo}\tilde{Q}^2 \tag{67}$$

and

$$\hat{U}_3(\tilde{q}_\delta) = \frac{1}{2}\hbar\omega_\delta\tilde{q}_\delta^2 \tag{68}$$

The potential $\hat{U}_3$ of the bending mode has been generally approximated by a harmonic potential [22,23,53,71–73]. Extension of this model should go beyond the harmonic approximation used in the description of the three vibrational modes, by substituting to the previous potentials (66), (67), and (68) a Morse-type potential [13].

At last, the Fermi coupling potential $\hat{H}_{gf}^\delta$ may be expressed beyond the exchange approximation by introducing the dimensionless Fermi coupling parameter $\Lambda$:

$$\hat{H}_{gf}^\delta(\tilde{q},\tilde{q}_\delta) = \hbar\Lambda\left[\tilde{q}-\tilde{q}_e(\tilde{Q})\right]\tilde{q}_\delta^2 \tag{69}$$

which expands according to

$$\hat{H}_{gf}^\delta(\tilde{q},\tilde{q}_\delta) = \hbar\Lambda\tilde{q}\tilde{q}_\delta^2 - \hbar\Lambda f_o\tilde{Q}\tilde{q}_\delta^2 - \hbar\Lambda g_o\tilde{Q}^2\tilde{q}_\delta^2 \tag{70}$$

As a matter of fact, looking at the two negative terms in the previous equation, the combination of the Fermi resonance mechanism with the dependence of $\tilde{q}_e$ on $\hat{Q}$ [as given by Eq. (7)] results in an effective coupling between the slow and bending modes. This coupling may lead to noticeable spectral features as the Fermi resonance becomes stronger and as the angular frequencies become closer to $\omega_\delta$ and $\omega_{oo}$.

### 2.  The Autocorrelation Function of the Dipole Moment

Focusing on the $|\{0\}\rangle \rightarrow |\{1\}\rangle$ infrared transition of the fast mode, the dipole moment operator of this mode is given at initial time by

$$\hat{\mu}(0) = \sum_{m=0}^{\infty} |\{0\}\rangle |[0]\rangle |(m)\rangle \langle (m)| \langle [0]| \langle \{1\}| \equiv \sum_{m=0}^{\infty} |\{0\}, [0], (m)\rangle \langle (m), [0], \{1\}|$$

$$(71)$$

where we introduced a bra and ket shortcut notation. At time $t$, within the Heisenberg picture, this operator becomes

$$\hat{\mu}(t) = \exp(i\hat{H}_{gf}t/\hbar)\, \hat{\mu}(0)\, \exp(-i\hat{H}_{gf}t/\hbar) \qquad (72)$$

The density operator at initial time is given by

$$\hat{\rho}(0) = \epsilon \exp\{-\hat{H}_{gf}/(k_B T)\} \quad \text{with} \quad \epsilon = 1 - \exp\{\hbar\omega_{oo}/(k_B T)\} \qquad (73)$$

where $k_B$ is the Boltzmann constant and $T$ is the absolute temperature. According to (2), the autocorrelation function $G_{gf}(t)$ is obtained by performing the trace operation over the eigenvectors of $\hat{H}_{gf}$ which are defined by the following eigenvalues equation:

$$\hat{H}_{gf} |\Phi_v^{gf}\rangle = \hbar \omega_v^{gf} |\Phi_v^{gf}\rangle \qquad (74)$$

Thus, we find

$$G_{gf}(t) \propto \sum_{v_1}\sum_{v_2}\sum_{m}\sum_{n} \exp\{-\hbar\, \omega_{v_1}^{gf}/(k_B T)\} \exp\{i\,(\omega_{v_2}^{gf} - \omega_{v_1}^{gf})\, t\}$$
$$\times \langle \Phi_{v_1}^{gf}|\{1\}, [0], (m)\rangle \langle (m), [0], \{0\}|\Phi_{v_2}^{gf}\rangle \langle \Phi_{v_2}^{gf}|\{0\}, [0], (n)\rangle$$
$$\times \langle (n), [0], \{1\}|\Phi_{v_1}^{gf}\rangle \qquad (75)$$

The eigenvalues $\omega_v^{gf}$ and the four sets of scalar products may be computed by full diagonalization of the Hamiltonian $\hat{H}_{gf}$. Then, the spectral density of a medium-strength H bond involving a Fermi resonance is

$$I_{gf}(\omega) = \int_{-\infty}^{+\infty} G_{gf}(t) \exp(-\gamma t) \exp(-i\omega t)\, dt \qquad (76)$$

where we incorporated the direct relaxation mechanism of the excited states of the fast and bending modes through a time-decaying exponential (of time constant $1/\gamma$), according to the results obtained by Rösch and Ratner within the adiabatic approximation [58] (see Section VI).

We must stress that the use of a single damping parameter $\gamma$ supposes that the relaxations of the fast and bending modes have the same magnitude. A more general treatment of damping has been proposed [22,23,71,72]; however, this treatment (discussed in Section IV.D) requires the use of the adiabatic approximation, so that its application is limited to very weak hydrogen bonds.

Some spectra have been calculated [53] using the spectral density $I_{gf}(\omega)$. Additional spectra may be found at the following URL: http://hbond.univ-perp.fr/.

### 3. Non-resonant Fermi Resonances

The spectral density (76) involves 10 physical parameters: the angular frequencies of the three modes, $\omega_o$, $\omega_{oo}$ and $\omega_\delta$, the four modulation parameters $\alpha_o$, $\beta_o$, $f_o$ and $g_o$, the Fermi coupling parameter $\Lambda$, the damping parameter $\gamma$, and finally the absolute temperature $T$. The great polymorphism of the theoretical spectral bandshapes has been already quoted [53,73], and we shall emphasize here the features coming under varying the bending mode frequency $\omega_\delta$. More precisely, noticeable perturbations of the theoretical lineshape are obtained for a gap $(\omega_o - 2\omega_\delta)$ between the fast mode frequency and the second overtone of the bending which varies over a range that is as large as 10 times the Fermi coupling parameter $\Lambda$.

Figure 6 illustrates this behavior by showing eight spectra (full lines) for which $\omega_\delta$ is varied from $810 \, \text{cm}^{-1}$ to $1740 \, \text{cm}^{-1}$, with $\Lambda = 120 \, \text{cm}^{-1}$. The unperturbed spectrum (computed for $\Lambda = 0$) is superimposed by using dashed lines. Because we used for all calculations the value $\omega_o = 3000 \, \text{cm}^{-1}$, the gap $(\omega_o - 2\omega_\delta)$ ranges from $1380 \, \text{cm}^{-1}$ to $(-480) \, \text{cm}^{-1}$. One may notice that, even when far from resonance, the spectrum involving the Fermi coupling differs significantly from the one that ignores it. The efficient role played by the Fermi resonances in cases that are far from resonance is a consequence of the large width of the $\nu_S$ ($X-\overset{\rightarrow}{H}\cdots Y$) band, which in turn comes mainly from the strong anharmonic coupling hypothesis.

Then, in complex molecules where many bending modes may participate to a Fermi resonance, we can expect the $\nu_S$ ($X-\overset{\rightarrow}{H}\cdots Y$) bandshape to be perturbed in a very complex way. It is also of importance to note that the perturbation due to a sole Fermi resonance is delocalized on the whole $\nu_S$ ($X-\overset{\rightarrow}{H}\cdots Y$) band.

### 4. Neglecting the Parameters $\beta_o$, $f_o$, and $g_o$

We shall name *standard* the nonadiabatic treatment of Fermi resonances for which the modulations of the fast mode frequency and equilibrium position by

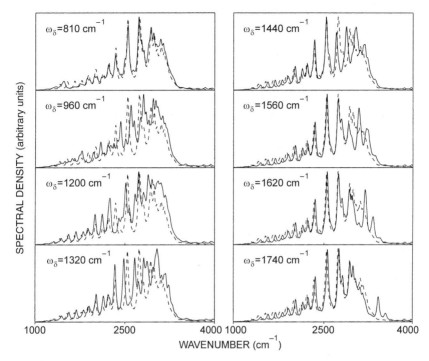

**Figure 6.** The Fermi resonances are efficient over a wide frequency range. This series shows eight spectra involving a Fermi resonance (full lines) for different values of $\omega_\delta$. The unperturbed spectrum (dashed lines) for which no Fermi resonance occurs (i.e., $\Lambda = 0$) is also given as a comparison. All spectra were computed from Eq. (76). Common parameters: $\alpha_o = 1.41$, $\beta_o = -0.4$, $f_o = 0.26$, $g_o = 0.026$, $\Lambda = 120\,\mathrm{cm}^{-1}$ (except dashed lines: $\Lambda = 0$), $\omega_o = 3000\,\mathrm{cm}^{-1}$, $\omega_{oo} = 150\,\mathrm{cm}^{-1}$, $\gamma = 22.5\,\mathrm{cm}^{-1}$, and $T = 300$ K.

the slow mode motion are neglected, except the linear dependence of its frequency. In this situation, the hydrogen bond involving a Fermi resonance is characterized by the *standard* Hamiltonian $\hat{H}_{sf}$, given by

$$\hat{H}_{sf} \equiv \hat{H}_{gf(\beta_o = f_o = g_o = 0)} = \hat{H}_{sf}^\circ + \hat{H}_{sf}^\delta \tag{77}$$

$\hat{H}_{sf}^\circ$ is the Hamiltonian of the three vibrational modes:

$$\hat{H}_{sf}^\circ = \frac{1}{2}\hbar\omega_o\tilde{p}^2 + \frac{1}{2}\hbar\omega_{oo}\tilde{P}^2 + \frac{1}{2}\hbar\omega_\delta\tilde{p}_\delta^2 + \frac{1}{2}\hbar\omega_o\tilde{q}^2 + \hbar\omega_{oo}\alpha_o\tilde{Q}\tilde{q}^2$$
$$+ \frac{1}{2}\hbar\frac{\omega_{oo}^2}{\omega_o}\alpha_o^2\tilde{Q}^2\tilde{q}^2 + \frac{1}{2}\hbar\omega_{oo}\tilde{Q}^2 + \frac{1}{2}\hbar\omega_\delta\tilde{q}_\delta^2 \tag{78}$$

whereas $\hat{H}_{sf}^\delta$ characterizes the Fermi coupling between the fast and bending modes, respectively linear and quadratic with respect to the fast and bending modes coordinates:

$$\hat{H}_{sf}^{\delta}(\tilde{q},\tilde{q}_{\delta}) \equiv \hat{H}_{gf}^{\delta}(\tilde{q},\tilde{q}_{\delta},f_{o}=g_{o}=0) = \hbar\Lambda\tilde{q}\tilde{q}_{\delta}^{2} \tag{79}$$

Defining the eigenvalues equation for the Hamiltonian (77)

$$\hat{H}_{sf}|\Phi_{v}^{sf}\rangle = \hbar\,\omega_{v}^{sf}\,|\Phi_{v}^{sf}\rangle \tag{80}$$

the *standard* spectral density $I_{sf}(\omega)$ may be written as the Fourier transform of the autocorrelation function of the fast mode dipolar momentum:

$$I_{sf}(\omega) \equiv I_{gf}(\omega,\beta_{o}=f_{o}=g_{o}=0) = \int_{-\infty}^{+\infty} G_{sf}(t)\exp(-\gamma t)\exp(-i\omega t)\,dt \tag{81}$$

with

$$
\begin{aligned}
G_{sf}(t) &\propto \sum_{v_{1}}\sum_{v_{2}}\sum_{m}\sum_{n} \exp\left\{-\hbar\,\omega_{v_{1}}^{sf}/(k_{B}T)\right\}\exp\left\{i\left(\omega_{v_{2}}^{gf}-\omega_{v_{1}}^{gf}\right)t\right\} \\
&\quad \times \langle\Phi_{v_{1}}^{sf}|\{1\},[0],(m)\rangle\langle(m),[0],\{0\}|\Phi_{v_{2}}^{sf}\rangle\langle\Phi_{v_{2}}^{sf}|\{0\},[0],(n)\rangle \\
&\quad \times \langle(n)[0],\{1\}|\Phi_{v_{1}}^{sf}\rangle
\end{aligned}
\tag{82}
$$

## C. The Exchange Approximation

Making use of the raising and lowering operators defined in Tables II and III, Hamiltonians (78) and (79) become finally

$$
\begin{aligned}
\hat{H}_{sf}^{\circ} &= \hbar\omega_{o}\left[b^{\dagger}b+\frac{1}{2}\right] + \hbar\omega_{oo}\left[a^{\dagger}a+\frac{1}{2}\right] + \hbar\omega_{\delta}\left[d^{\dagger}d+\frac{1}{2}\right] \\
&\quad + \hbar\omega_{oo}\left(\frac{\alpha_{o}^{2}}{4}\frac{\omega_{oo}}{\omega_{o}}\right)\left[b^{\dagger}+b\right]^{2}\left[a^{\dagger}+a\right]^{2} + \hbar\omega_{oo}\left(\frac{\alpha_{o}}{2}\right)\left[b^{\dagger}+b\right]^{2}\left[a^{\dagger}+a\right]
\end{aligned}
\tag{83}
$$

and

$$\hat{H}_{sf}^{\delta} = \frac{\hbar\Lambda}{2\sqrt{2}}\left[b^{\dagger}+b\right]\left[d^{\dagger}+d\right]^{2} \tag{84}$$

If we expand the expression (84), we obtain

$$\hat{H}_{sf}^{\delta} = \frac{\hbar\Lambda}{2\sqrt{2}}\left[b^{\dagger}d^{2}+b\,d^{\dagger^{2}}\right] + \frac{\hbar\Lambda}{2\sqrt{2}}\left[b^{\dagger}d^{\dagger^{2}}+b\,d^{2}+2(b^{\dagger}+b)\left(d^{\dagger}d+\frac{1}{2}\right)\right] \tag{85}$$

The *exchange approximation* consists in ignoring the second part of this expansion [73]. The two first ignored terms are composed of simultaneous triple excitations $(b^\dagger d^{\dagger 2})$ or triple desexcitations $(b d^2)$, which may appear in the spectra outside the $\nu_S$ (X–H$\cdots$Y) band, at $\omega_o + 2\omega_\delta$ (near 6000 cm$^{-1}$), and may have a low intensity at room temperature. The other ignored term drives the fast mode and depends on the number operator $d^\dagger d$ of the bending mode: the higher the excited state of the bending mode, the more driven the fast mode. This last term may not be neglected, as will be discussed in the following. Even in the case of pure Fermi resonances, discussed in Section V.B, without H-bond, the driven part of the Hamiltonian of the fast mode gives a noticeable effect on the infrared lineshape [73].

More explicitly, the exchange approximation needs to keep in the total Hamiltonian the simple coupling term

$$\{\hat{H}^\delta_{sf}(\tilde{q}, \tilde{q}_\delta)\}^{ex} = \frac{\hbar \Lambda}{2\sqrt{2}} [b^\dagger d^2 + b d^{\dagger 2}] \tag{86}$$

so that the standard Hamiltonian $\{\hat{H}_{sf}\}^{ex}$ within the exchange approximation is

$$\{\hat{H}_{sf}\}^{ex} = \hat{H}^o_{sf} + \{\hat{H}^\delta_{sf}(\tilde{q}, \tilde{q}^\delta)\}^{ex} \tag{87}$$

Following the steps shown in section IV.B which remain valid within the exchange approximation, one may easily obtain the expression of the resulting spectral density $\{I_{sf}\}^{ex}$:

$$\{I_{sf}(\omega)\}^{ex} = \int_{-\infty}^{+\infty} \text{tr}\left[ \exp\left( \frac{-\{\hat{H}_{sf}\}^{ex}}{k_B T} \right) \hat{\mu}^\dagger_{sf}(0) \exp\left( \frac{i\{\hat{H}_{sf}\}^{ex} t}{\hbar} \right) \hat{\mu}_{sf}(0) \right.$$
$$\left. \times \exp\left( \frac{-i\{\hat{H}_{sf}\}^{ex} t}{\hbar} \right) \right] \exp(-\gamma t) \exp(-i\omega t) \, dt \tag{88}$$

Now we may look, as is usually done when applying the adiabatic approximation, at the representation of the Hamiltonian $\{\hat{H}_{sf}\}^{ex}$ within the reduced base (89) spanned by [cf. Eq. (71)]

$$|\psi^f_o(m)\rangle = |\{0\}, (m), [0]\rangle \tag{89a}$$

$$|\psi^f_1(m)\rangle = |\{1\}, (m), [0]\rangle \qquad m = 0, 1, 2, \ldots \tag{89b}$$

$$|\psi^f_2(m)\rangle = |\{0\}, (m), [2]\rangle \tag{89c}$$

Within this reduced base, the coupling Hamiltonian $\{\hat{H}^\delta_{sf}\}^{ex}$ describes the energy exchange between the states $|\{0\}, [2]\rangle$ and $|\{1\}, [0]\rangle$. This kind of exchange coupling is, as a matter of fact, the one used in the adiabatic frame. Indeed, the role played by $\{\hat{H}^{\delta}_{\mathrm{sf}}\}^{\mathrm{ex}}$ is similar to the action of the operator $\hat{l}$ in the adiabatic Hamiltonian $\hat{H}_{\mathrm{f}}$ [given later by Eq. (98)], which implies the equivalence

$$\hat{l} = \frac{\hbar\Lambda}{2}\left[|\{1\},[0]\rangle\langle[2],\{0\}| + |\{0\},[2]\rangle\langle[0],\{1\}|\right] \tag{90}$$

We shall compare later the spectral density $\{I_{\mathrm{sf}}\}^{\mathrm{ex}}$ and the adiabatic one $I_{\mathrm{f}}$ [given later by Eq. (103)] that supposes implicitly the exchange approximation. We may expect no difference between their lineshape in cases for which the adiabatic approximation is valid.

### D. The Adiabatic Treatment Within the Exchange Approximation

Applying the adiabatic approximation, we restrict the representation of the Hamiltonian to the reduced base (89). Within this base, the Hamiltonian that describes an undamped H bond involving a Fermi resonance may be split into effective Hamiltonians whose structure is related to the state of the fast and bending modes:

$$\hat{H}^{\circ}_{\mathrm{f}} = \begin{bmatrix} \hat{H}^{\{0\}} & 0 & 0 \\ 0 & \hat{H}^{\{1\}}_{\circ} & \hat{l} \\ 0 & \hat{l} & \hat{H}^{[2]}_{\circ} \end{bmatrix} \tag{91}$$

A physical picture of the Fermi coupling within the exchange approximation is given in Fig. 7.

The effective Hamiltonian $\hat{H}^{\{0\}}$, related to the ground states $|\{0\}\rangle$ and $|[0]\rangle$ of the fast and bending modes, is the Hamiltonian of a quantum harmonic oscillator characterizing the slow mode:

$$\hat{H}^{\{0\}} = \hbar\omega_{\mathrm{oo}}\left(a^{\dagger}a + \frac{1}{2}\right) \tag{92}$$

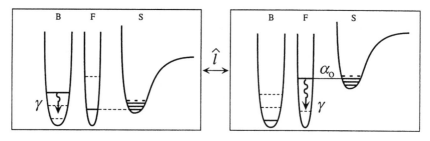

**Figure 7.** The Fermi resonance mechanism within the adiabatic and exchange approximations. F, fast mode; S, slow mode; B, bending mode.

The effective Hamiltonian $\hat{H}_\circ^{\{1\}}$ holds for a single excitation of the fast mode and involves, according to some unitary transformations, a driven term that describes the intermonomer motion. Within the sub-base (89b), it may be given by

$$\hat{H}_\circ^{\{1\}} = \hbar\omega_\circ + \hbar\omega_{oo}\left[a^\dagger a + \alpha_\circ(a^\dagger + a) - \alpha_\circ^2 + \frac{1}{2}\right] \tag{93}$$

Similarly, the effective Hamiltonian $\hat{H}_\circ^{[2]}$ holds for a double excitation of the bending mode, but involves the nondriven Hamiltonian of the slow mode (92). Within the sub-base (89c), it may be written

$$\hat{H}_\circ^{[2]} = \hat{H}^{\{0\}} + 2\hbar\omega_\delta \tag{94}$$

The Fermi coupling operator $\hat{l}$ is given by (90). Owing to the eigenvalue equation

$$\begin{bmatrix} \hat{H}_\circ^{\{1\}} & \hat{l} \\ \hat{l} & \hat{H}_\circ^{[2]} \end{bmatrix} |\zeta_v^f\rangle = \hbar\,\theta_v^f\,|\zeta_v^f\rangle \tag{95}$$

we have shown [71] that the spectral density takes the form

$$I_f(\omega, \gamma = \gamma_\circ = \gamma_\delta) \propto \int_{-\infty}^{+\infty} \sum_m \sum_v \exp\left\{-\left(m+\frac{1}{2}\right)\hbar\,\omega_{oo}/(k_BT)\right\}$$
$$\times \exp\left\{i\left[\theta_v^f - \left(m+\frac{1}{2}\right)\omega_{oo}\right]t\right\}$$
$$\times \langle\zeta_v^f|\psi_1^f(m)\rangle\langle\psi_1^f(m)|\zeta_v^f\rangle \exp\{-\gamma t\}\exp\{-i\omega t\}dt \tag{96}$$

We shall see later in Section V.A the generalization $I_f(\omega)$ of this spectral density [see Eq. (103)], in a situation where the damping of the fast and bending modes are treated separately ($\gamma_\circ \neq \gamma_\delta$).

### E. Discussion on the Adiabatic and Exchange Approximations

Our purpose is to compare the lineshapes of the spectral densities $I_{sf}$, $\{I_{sf}\}^{ex}$, and $I_f$, computed respectively from (81), (88) and (96).

First let us restrict the comparison to very weak H bonds because in such a case we can expect that the adiabatic approximation is fulfilled. We made some sample calculations shown in Figs. 8(c) and 8(d), with a small dimensionless anharmonic coupling parameter, $\alpha_\circ = 0.6$. The figures displays $\{I_{sf}\}^{ex}$ [dashed line (d)] and $I_{sf}$ [dashed line (c)] at 300 K. Each one is compared with the adiabatic spectra $I_f$ (superimposed full lines). Note that the adiabatic spectra (c)

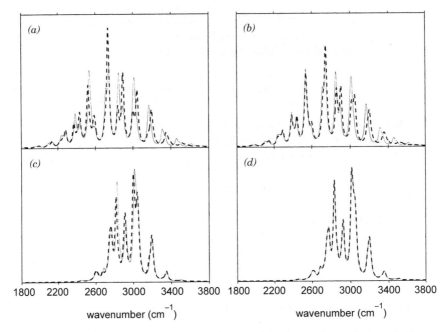

**Figure 8.** Comparison between the adiabatic spectral density and the standard one (with or without the exchange approximation). (a) and (c) display the spectral density $I_{sf}$ from Eq. (81), using dashed lines. (b) and (d) display the spectral density $\{I_{sf}\}^{ex}$ from Eq. (88), within the exchange approximation, using dashed lines. Comparison is made with the adiabatic spectral density $I_f$ (thin plain lines) obtained from (96). Spectra (a) and (b) are computed with $\alpha_o = 1.2$, whereas spectra (c) and (d) are computed with $\alpha_o = 0.6$. Common parameters: $\Lambda = 120\,\text{cm}^{-1}$, $\omega_o = 3000\,\text{cm}^{-1}$, $\omega_\delta = 1440\,\text{cm}^{-1}$, $\omega_{oo} = 150\,\text{cm}^{-1}$, $\gamma_o = \gamma_\delta = 60\,\text{cm}^{-1}$, and $T = 300\,\text{K}$.

and (d) are identical. As may be seen in Fig. 8(c), the shape of the spectral densities $I_{sf}$ and $I_f$ are not fully similar. The details of these spectra display some differences that however, disappear in Fig. 8(d) when applying the exchange approximation. It clearly reveals the implicit use of the exchange approximation when deriving $I_f$ in the adiabatic studies of Fermi resonances [22,23,71,72,74]. Nevertheless, recall that the adiabatic model of damped Fermi resonances [71,72] must not be simply reduced to a limiting case of the present one (i.e., assuming very weak H bonds and applying the exchange approximation), because it allows a more general treatment of damping.

We kept the same structure in Figs. 8(a) and 8(b), but the spectra were computed for a greater value $\alpha_o = 1.2$. As may be seen, some differences between the adiabatic and nonadiabatic spectral densities appear in all cases, whether applying the exchange approximation or not. Within the exchange approximation, Fig. 8(b), these discrepancies may be safely attributed to the nonadiabatic corrections that are involved by $\{I_{sf}\}^{ex}$. It is worth noting that the changes induced by these corrections appear to be here more unforeseeable than those mentioned in Section III. Indeed, we may observe here the decrease and the increase of the splitting of some sub-bands, which were not observed in the absence of Fermi resonances, as well as a shift of the band toward higher frequencies, and an increase of the gap between the main sub-bands. Note that these main sub-bands are reminiscent of the Franck–Condon intensity distribution that appears within the adiabatic approximation when ignoring the Fermi resonances. We have shown previously [22,23,72] that the lineshape details are strongly sensitive to very small changes of the gap $(\omega_o - 2\omega_\delta)$ between the fast mode frequency and the overtone of the bending mode. As a matter of fact, the shift of the average frequency of the band which is induced by the adiabatic corrections is responsible for a change of this gap because the overtone $2\omega_\delta$ is unaltered. It implies that the splitting perturbations of the sub-bands may be viewed as the consequence of cooperative effects between the nonadiabatic terms and the Fermi coupling terms.

Now, comparing $I_{sf}$ and $\{I_{sf}\}^{ex}$, we observe also some splitting modifications. We may attribute them to some additional changes in the frequency gap. Indeed, as shown in Section V.B, the full treatment of the Fermi coupling mechanism leads to a displacement of the potential of the fast mode (both in energy and position) that does not appear within the exchange approximation. The fast mode then involves an effective frequency that differs from $\omega_o$, which leads to an effective gap which differs from $(\omega_o - 2\omega_\delta)$.

We may gather our results as follows:

$$\begin{cases} I_{sf}(\omega) \neq I_f(\omega), & \text{if} \quad \Lambda \neq 0 \\ \{I_{sf}(\omega)\}^{ex} \neq I_f(\omega, \gamma_o = \gamma_\delta), & \text{if} \quad \alpha_o > 1 \\ \{I_{sf}(\omega)\}^{ex} = I_f(\omega, \gamma_o = \gamma_\delta), & \text{if} \quad \alpha_o \leq 1 \end{cases} \qquad (97)$$

where the first inequality results from the implicit use of the exchange approximation in $I_f$ and stands for all values of $\alpha_o$, including zero. The two next expressions traduce the reliability of the adiabatic approximation. Note that the pivot value $\alpha_o = 1$ is crude because it may be increased insofar as the relaxations effects are strong (see Section III).

## V.  FERMI RESONANCES: "UNEQUAL DAMPING" TREATMENT

It was stated in Section IV that undertaking a nonadiabatic treatment prevents us from using different damping parameters for the fast and bending modes (i.e., $\gamma_o \neq \gamma_\delta$). We shall see in the present section the role played by these damping parameters in two situations:

- When applying the adiabatic approximation (weak hydrogen bonds)
- When neglecting the strong anharmonic coupling—that is, in the situation of a pure Fermi coupling (no hydrogen bond)

## A.  The Adiabatic Treatment Within the Exchange Approximation

We shall give here a brief summary of our previous work [71,72] that was concerned with the introduction of the relaxation phenomenon within the adiabatic treatment of the Hamiltonian (77), as was done in the undamped case by Witkowski and Wójcik [74]. Following these authors, we applied the adiabatic approximation and then we restricted the representation of the Hamiltonian to the reduced base (89). Within this base, the Hamiltonian that describes a damped H bond involving a Fermi resonance may be split into effective Hamiltonians whose structure is related to the state of the fast and bending modes:

$$
\hat{H}_f = \begin{bmatrix} \hat{H}^{\{0\}} & 0 & 0 \\ 0 & \hat{H}^{\{1\}} & \hat{I} \\ 0 & \hat{I} & \hat{H}^{[2]} \end{bmatrix}
\tag{98}
$$

The effective Hamiltonian $\hat{H}^{\{0\}}$ has already been defined by Eq. (92):

$$
\hat{H}^{\{0\}} = \hbar\omega_{oo}\left(a^{\dagger}a + \frac{1}{2}\right)
\tag{99}
$$

The damped effective Hamiltonian $\hat{H}^{\{1\}}$ holds for a single excitation of the fast mode and involves, according to some unitary transformations, a driven term that describes the intermonomer motion. Within the sub-base (89b), it may be given by

$$
\hat{H}^{\{1\}} = \hbar\omega_o + \hbar\omega_{oo}\left[a^{\dagger}a + \alpha_o(a^{\dagger} + a) - \alpha_o^2 + \frac{1}{2}\right] + i\hbar\gamma_o
\tag{100}
$$

Similarly, the damped effective Hamiltonian $\hat{H}^{[2]}$ holds for a double excitation of the bending mode, but involves the nondriven Hamiltonian of the slow mode (99). Within the sub-base (89c), it may be written

$$
\hat{H}^{[2]} = \hat{H}^{\{0\}} + 2\hbar\omega_{\delta} + i\hbar\gamma_{\delta}
\tag{101}
$$

The account for the relaxations in (100) and (101) was made through the damping parameters $\gamma_o$ of the fast mode and that $\gamma_{\delta}$ of the bending mode. The

Fermi coupling operator $\hat{l}$ is given by (90). Owing to the eigenvalues equation

$$
\begin{bmatrix} \hat{H}^{\{1\}} & \hat{l} \\ \hat{l} & \hat{H}^{[2]} \end{bmatrix} |\Phi_v^f\rangle = \hbar\,\omega_v^f\,|\Phi_v^f\rangle \tag{102}
$$

we have shown [71] that the spectral density takes the form

$$
I_f(\omega) \propto \int_{-\infty}^{+\infty} \sum_m \sum_v \exp\left\{ -\left(m+\frac{1}{2}\right)\hbar\omega_{oo}\big/(k_B T) \right\} \exp\left\{ i\left[\omega_v^f - \left(m+\frac{1}{2}\right)\omega_{oo}\right] t \right\}
$$
$$
\times \langle \Phi_v^f | \psi_1^f(m)\rangle \langle \psi_1^f(m) | \phi_v^f\rangle \, \exp\{-i\omega t\}\, dt \tag{103}
$$

One must be reminded that the eigenvalues $\omega_v^f$ are complex.

### B. The Fermi Resonance in the Absence of Hydrogen Bond

Several studies of Fermi resonances in the absence of H bond have been made [76–80]. We shall account for this situation by simply ignoring the anharmonic coupling between the fast and slow modes ($\alpha_o = 0$). The theory then describes the coupling between the fast mode and a bending mode through the potential $\hat{H}_{sf}^\delta$, with both of these modes being damped in the same way. Because $\alpha_o = 0$, the slow mode does not play any role, so that the total Hamiltonian does not refer to it:

$$
\hat{H}_{sf}(\alpha_o = 0) = \hbar\omega_o\left[b^\dagger b + \frac{1}{2}\right] + \hbar\omega_\delta\left[d^\dagger d + \frac{1}{2}\right] + \hat{H}_{sf}^\delta \tag{104}
$$

Recall that the developed expression (85) of $\hat{H}_{sf}^\delta$ is

$$
\hat{H}_{sf}^\delta = \{\hat{H}_{sf}^\delta\}^{ex} + \frac{\hbar\Lambda}{2\sqrt{2}}\left[b^\dagger d^{\dagger 2} + b\,d^2 + 2(b^\dagger + b)\left(d^\dagger d + \frac{1}{2}\right)\right]
$$

with the exchange coupling Hamiltonian $\{\hat{H}_{sf}^\delta\}^{ex}$:

$$
\{\hat{H}_{sf}^\delta\}^{ex} = \frac{\hbar\Lambda}{2\sqrt{2}}\,[b^\dagger d^2 + b\,d^{\dagger 2}]
$$

#### 1.  The Spectral Density Within the Exchange Approximation

Applying the exchange approximation and neglecting the zero-point energy terms, we may safely limit the representation of the Hamiltonian $\{\hat{H}_{sf}\}^{ex}$ within the following reduced base which accounts for the ground states of each mode and for the first (second) overtone of the fast (bending) mode:

$$
\{|\{1\}, [0]\rangle; |\{0\}, [2]\rangle\} \tag{105}
$$

The Hamiltonian $\{\hat{H}_{\text{sf}}(\alpha_o = 0)\}^{\text{ex}}$ describes then the coupling between two undamped two-levels systems, whose excited states may be nonresonant. Introducing the relaxation $\gamma_o$ and $\gamma_\delta$ of the excited states in the representation of $\{\hat{H}_{\text{sf}}(\alpha_o = 0)\}^{\text{ex}}$ within the base (105), we obtain

$$\{\hat{H}_{\text{sf}}(\alpha_o = 0, \gamma_o, \gamma_\delta)\}^{\text{ex}} \equiv \hat{H}_{\text{f}}(\alpha_o = 0)$$
$$= \hbar \begin{bmatrix} \omega_o + i\gamma^o & \Lambda/2 \\ \Lambda/2 & \omega_o + \Delta + i\gamma_\delta \end{bmatrix} \quad \text{with} \quad \Delta = 2\omega_\delta - \omega_o$$

$$(106)$$

It has been shown by Giry et al [81] that, in this situation, the spectral density is the following Fourier transform:

$$\{I_{\text{sf}}(\omega, \alpha_o = 0, \gamma_o, \gamma_\delta)\}^{\text{ex}} \equiv I_{\text{f}}(\omega, \alpha_o = 0) \propto \int_{-\infty}^{+\infty} F(t) \exp\{-i\omega t\}\, dt \quad (107)$$

where the time-dependent autocorrelation function $F(t)$ may be written

$$F(t) = \exp\left\{i(\omega_o + \Delta)\right\} \exp\left\{-\frac{\xi t}{2}\right\} \frac{1}{2(y + ix)}$$
$$\times \left\{ [(\Delta + y) - i(\zeta - x)] \exp\left\{-\frac{xt}{2}\right\} \exp\left\{\frac{it(\Delta + y)}{2}\right\} \right.$$
$$\left. - [(\Delta - y) - i(\zeta + x)] \exp\left\{\frac{xt}{2}\right\} \exp\left\{\frac{it(\Delta - y)}{2}\right\} \right\} \quad (108)$$

with the definitions

$$x^2 = \frac{\delta - S}{2}, \qquad y^2 = \frac{\delta + S}{2}, \qquad S = \Delta^2 + \Lambda^2 - \zeta^2$$
$$\delta = \sqrt{S^2 + \Lambda^2 \zeta^2}, \qquad \xi = \gamma^\delta + \gamma_o, \qquad \zeta = \gamma_\delta - \gamma_o \quad (109)$$

Dealing with the restrictive situation of equal dampings, where $\gamma_o = \gamma_\delta \equiv \gamma$, the spectral density (107) may be written as the limit of $\{I_{\text{sf}}(\omega)\}^{\text{ex}}$ [Eq. (88)] when neglecting the anharmonic coupling—that is, when $\alpha_o = 0$:

$$\{I_{\text{sf}}(\omega, \alpha_o = 0)\}^{\text{ex}} \equiv I_{\text{f}}(\omega, \alpha_o = 0, \gamma_o = \gamma_\delta) \propto \int_{-\infty}^{+\infty}$$
$$\times \left[ \sum_{v=1}^{2} \langle \Phi_v^{\text{ex}} | \{1\}, [0] \rangle \langle \{1\}, [0] | \Phi_v^{\text{ex}} \rangle \exp\{i(\Omega_v^{\text{ex}} - \omega)t\} \right.$$
$$\times \exp\{-\gamma t\} \Bigg]\, dt \quad (110)$$

where $|\Phi_v^{ex}\rangle$ and $\Omega_v^{ex}$ are the eigenkets and eigenvalues of the undamped Hamiltonian:

$$\begin{bmatrix} \omega_o & \Lambda/2 \\ \Lambda/2 & (\omega_o + \Delta) \end{bmatrix} |\Phi_v^{ex}\rangle = \hbar\Omega_v^{ex} |\Phi_v^{ex}\rangle \qquad (111)$$

Looking at (110), it appears that the spectral density $\{I_{sf}(\omega, \alpha_o = 0)\}^{ex}$ is composed of two sub-bands, as is also shown by the four sample spectra of Fig. 10 which were computed for various parameters $\Delta$ and $\Lambda$. The frequency and intensity of these two sub-bands do not depend on the temperature, and are similar to those which may be obtained within the simpler undamped treatment. Solving the eigenvalue problem (111), we obtain

$$\Omega_1^{ex} = \frac{1}{2}(\omega_o + 2\omega_\delta) + \frac{1}{2}\sqrt{\Delta^2 + \Lambda^2} \qquad (112)$$

$$\Omega_2^{ex} = \frac{1}{2}(\omega_o + 2\omega_\delta) - \frac{1}{2}\sqrt{\Delta^2 + \Lambda^2} \qquad (113)$$

$$\langle\Phi_1^{ex}|\{1\},[0]\rangle\langle\{1\},[0]|\Phi_1^{ex}\rangle = \left[\frac{1}{2} - \frac{\Delta}{2\sqrt{\Delta^2 + \Lambda^2}}\right] \qquad (114)$$

$$\langle\Phi_2^{ex}|\{1\},[0]\rangle\langle\{1\},[0]|\Phi_2^{ex}\rangle = \left[\frac{1}{2} + \frac{\Delta}{2\sqrt{\Delta^2 + \Lambda^2}}\right] \qquad (115)$$

Then, the undamped spectral density may be written

$$\{I_{sf}(\omega, \alpha_o = 0, \gamma = 0)\}^{ex} = \left[\frac{1}{2} - \frac{\Delta}{2\sqrt{\Delta^2 + \Lambda^2}}\right] \delta(\omega, -\Omega_1^{ex})$$
$$+ \left[\frac{1}{2} + \frac{\Delta}{2\sqrt{\Delta^2 + \Lambda^2}}\right] \delta(\omega, -\Omega_2^{ex}) \qquad (116)$$

Note that this expression may be considered as a consequence of the well-known Rabi equations. One may show easily that the two first moments of the undamped spectra are

$$\bar{\omega} = \omega_o \qquad (117)$$

$$\overline{\omega^2} = \omega_o^2 + \frac{\Lambda^2}{4} \qquad (118)$$

so that the second centered moment $\sigma$ of the undamped spectra is

$$\sigma = \sqrt{\overline{\omega^2} - \bar{\omega}^2} = \frac{\Lambda}{2} \qquad (119)$$

Note that within the damped case the expressions (117) and (118) of $\overline{\omega^2}$ and $\sigma$ are not valid.

We may observe that the two sub-bands of the damped spectral density (110), as well as the two peaks involved by the undamped case, must be of the same intensity in the resonant case ($\Delta = 0$), which may be verified by looking at the

Fig. 10(a). On the contrary, their intensities differ in the nonresonant case ($\Delta \neq 0$), but they must become closer as the Fermi coupling parameter $\Lambda$ increases. Note that the frequency $\Omega_E^{ex}$ of the Evans' hole which separates the two sub-bands is independent of $\Lambda$, because it is given by

$$\Omega_E^{ex} = \frac{\Omega_1^{ex} + \Omega_2^{ex}}{2} = \frac{\omega_o + 2\omega_\delta}{2} \tag{120}$$

These behaviors may be observed in Fig. 10(b), which displays three spectra computed with an increasing parameter $\Lambda$ and $\Delta = -120$ cm$^{-1}$. We shall see in the following that beyond the exchange approximation the spectral density behaves in an opposite way.

### 2. The Spectral Density of Pure Fermi Resonance, Beyond the Exchange Approximation

We have seen that the exchange approximation consists mainly in ignoring the driven part of the fast mode motion, given by

$$\frac{\hbar\Lambda}{\sqrt{2}} \left[ (b^\dagger + b) \left( d^\dagger d + \frac{1}{2} \right) \right] \tag{121}$$

In order to account for this term, the base (105) must be enlarged by adding several overtones of the fast and bending modes:

$$\{|\{0\}\rangle, |\{1\}\rangle, \cdots, |\{k_{max}\}\rangle\} \otimes \{|[0]\rangle, |[1]\rangle, \cdots, |[u_{max}]\rangle\} \tag{122}$$

Note that the stability of the spectra with respect to $k_{max}$ and $u_{max}$ must be carefully checked. Within this enlarged base, the eigenvalues equation of the total Hamiltonian (104) may be written

$$\hat{H}_{sf}(\alpha_o = 0) |\Phi_v\rangle = \hbar\Omega_v |\Phi_v\rangle \tag{123}$$

Then, neglecting the anharmonic coupling $\alpha_o$, the spectral density $I_{sf}$ [Eq. (81)] reduces to

$$I_{sf}(\omega, \alpha_o = 0) \propto \int_{-\infty}^{+\infty} \sum_{v_1} \sum_{v_2} \exp\{-\hbar\Omega_{v_1}/(k_B T)\} \exp\{i(\Omega_{v_2} - \Omega_{v_1})t\}$$
$$\times \langle\Phi_{v_1}|\{1\}, [0]\rangle\langle[0], \{0\}|\Phi_{v_2}\rangle \times \langle\Phi_{v_2}|\{0\}, [0]\rangle\langle[0], \{1\}|\Phi_{v_1}\rangle$$
$$\times \exp\{-i\omega t\} \exp\{-\gamma t\} dt \tag{124}$$

We may observe that the spectral density (124) is temperature-dependent. However, due to the magnitude of the involved frequencies (first overtone $\approx 4500$ K), this dependence is irrelevant within the experimental temperature range.

## C.  Theories Linking

Some of the connections between the theories given in this section are shown in Fig. 9.

All the theories illustrated in this figure involve the following:

- The adiabatic approximation when dealing with an hydrogen bond ($\alpha_o \neq 0$)

- The exchange approximation when dealing with Fermi resonance ($\Lambda \neq 0$)

Note that the spectral densities (a) to (e) were computed from the same expression (103) of $I_f$, by zeroing some of the physical parameters. The undamped spectra (f) to (j) were computed from the following expression:

$$I_f(\omega, \gamma = 0) \propto \sum_m \sum_\nu \exp\left\{ -\left(m + \frac{1}{2}\right)\hbar\,\omega_{oo} \Big/ (k_B T) \right\} \langle \zeta_\nu^f | \psi_1^f(m) \rangle \langle \psi_1^f(m) | \zeta_\nu^f \rangle$$

$$\times \delta\left(\omega, \left[\theta_\nu^f - \left(m + \frac{1}{2}\right)\omega_{oo}\right]\right) \tag{125}$$

where $|\zeta_\nu\rangle$ and $\theta_\nu^f$ are obtained by solving the eigenvalue equation (95).

## D.  Discussion on the Exchange Approximation in the Absence of a Hydrogen Bond

Some sample calculations are displayed in Fig. 10. As may be seen, the spectral density (124) involves two sub-bands in the $\nu_S$ (X–H) frequency region, like it is observed within the exchange approximation. Note that other submaxima appear at overtone frequencies (near $2\omega_o$, $3\omega_o$, ...) with a much lower intensity (less than 0.1% of the doublet intensity) and will not be studied here.

Let us stress that the driven term (121), which depends on the number operator of the bending mode, shifts the whole potential energy curve of the fast mode (both in energy and position). As a consequence, the energy exchange described by the Hamiltonian $\{\hat{H}_{sf}^\delta\}^{ex}$ may be viewed as to occur between the excited states of the bending mode, which are unchanged, and some effective levels that correspond to the displaced excited states of the fast mode. Hence it follows that in the resonant case, for which $\Delta = 2\omega_\delta - \omega_o = 0$, the intensities of the two sub-bands must not be the same. This feature is shown by the three asymmetric sample spectra of Fig. 10, computed for $\Delta = 0$.

Moreover, because the driven term depends on $\Lambda$, a strengthening of the Fermi coupling implies an increase of the effective frequency gap between the coupled levels, and therefore an increase of the intensity gap, that is at the opposite of the behavior observed within the exchange approximation. This

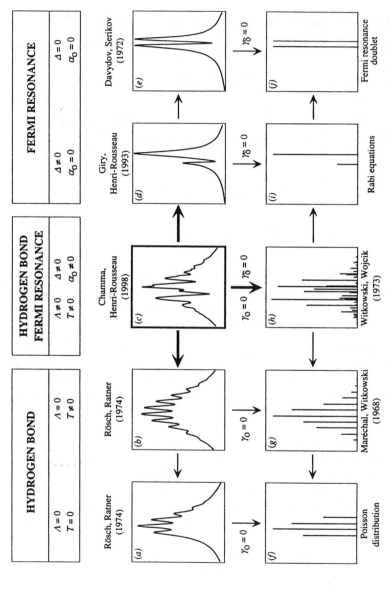

**Figure 9.** Links between several theories of Fermi resonances and hydrogen bonding. (a) Ref. 58; (b) Ref. 58: (c) Ref. 71 and 72, see Eq. (103); (d) Ref. 81, see Eq. (107); (e) Ref. 82; (f) Fourier transform of $G_{III}(t)$, Eq. (49) with $T = 0$; (g) Ref. 7, same as lineshape (f) with $T \neq 0$; (h) Ref. 74; (i) Eq. (116); (j) Eq. (116) with $\Delta = 0$.

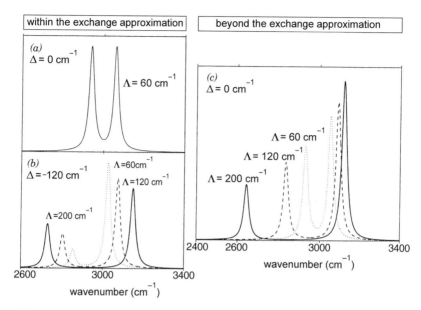

**Figure 10.** Pure Fermi coupling within or beyond the exchange approximation. Left column spectra were obtained from expression (110) of the spectral density $\{I_{sf}(\omega, \alpha_o = 0)\}^{ex}$: (a) Resonant case: $\Delta = 0, \Lambda = 60\,cm^{-1}$; (b) nonresonant case: $\Delta = -120\,cm^{-1}$ with $\Lambda = 60\,cm^{-1}$ (dotted line), $\Lambda = 120\,cm^{-1}$ (dashed line) and $\Lambda = 200\,cm^{-1}$ (full line). Right column spectra were obtained from expression (124) of the spectral density $I_{sf}(\omega, \alpha_o = 0)$: (c) Resonant case with $\Lambda = 60\,cm^{-1}$ (dotted line), $\Lambda = 120\,cm^{-1}$ (dashed line), and $\Lambda = 200\,cm^{-1}$ (full line). Common parameters: $\alpha_o = 0, \omega_o = 300\,cm^{-1}, \omega_\delta = \frac{1}{2}(\Delta + \omega_o), \omega_{oo} = 150\,cm^{-1}, \gamma_o = \gamma_\delta = 60\,cm^{-1}$.

feature may be observed in Fig. 10 where we varied the magnitude of $\Lambda$ for the resonant case $\Delta = 0$.

We may finally conclude that, with the purpose of comparison with experiments, one has to be careful and must remember the following (i) Two sub-bands of the same intensity may not be the consequence of a resonant situation $\omega_o = 2\omega_\delta$, (ii) The frequency of each submaxima is governed by the three parameters $\omega_o$, $\omega_\delta$, and $\Lambda$, and (iii) The frequency of the Evans' hole, which appears between the two sub-bands, is given within the exchange approximation by the average frequency $\frac{1}{2}(\omega_o + 2\omega_\delta)$, but it is dependent on $\Lambda$ within the full treatment of Fermi resonances.

### E. Discussion on the Two Damping Treatments

As shown in Section IV.D, it is possible, within the adiabatic approximation, to account for the general situation where the relaxation parameters of the fast mode $\gamma_o$ and of the bending mode $\gamma_\delta$ are not supposed to be equal.

Let us look at the role played by each damping parameter $\gamma_o$ and $\gamma_\delta$ in the simplest case of pure Fermi coupling ($\alpha_o = 0$). Fig. 11 displays spectral densities computed from Eq. (110) for different magnitudes of $\gamma_o$ and $\gamma_\delta$, as well as Dirac delta peaks, ignoring both relaxations, computed from Eq. (116). Because we varied only the damping parameters in this figure, the Dirac peaks are obviously identical. However, it may be observed that the lineshapes involve strong changes upon varying the two damping parameters independently. There are three limit cases:

- When both $\gamma_o$ and $\gamma_\delta$ are weak, the lineshapes (a) and (d) display the usual splitting (i.e., the doublet transitions), because this situation is close to the undamped one. However, note that these two lineshapes (a)

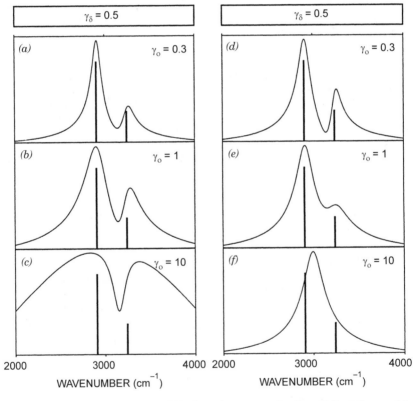

**Figure 11.** Pure Fermi coupling within the exchange approximation: relative influence of the damping parameters. Common parameters: $\omega_o = 3000\,\mathrm{cm}^{-1}$, $\Lambda = 150\,\mathrm{cm}^{-1}$, $2\omega_\delta = 3150\,\mathrm{cm}^{-1}$.

and (d) are not fully identical (different sub-band shapes and relative intensities).

• When the direct damping of the fast mode is strong, with that of the bending mode being weak, the lineshape (c) shows a broad shape involving a window that is due to the band splitting.

• In the opposite situation ($\gamma_\delta \gg \gamma_o$), lineshape (f), the band splitting vanishes such as the lineshape become close to the spectrum of an isolated fast mode (unique band, centered on the frequency $\omega_o$).

Of course, varying the damping parameters allows the bandshapes to evolve between these three limiting cases.

Owing to the above remarks, it is clear that the spectral densities resulting from the dynamical study of the two-level system, involving relaxational mechanisms, should not be viewed as the simple broadening of the infrared transitions obtained in the undamped frame.

Let us see what happens when accounting for the two damping mechanisms in the case of a hydrogen bond. Fig. 12 illustrates the drastic modifications of the bandshape which may be observed while varying the damping parameters at low temperature, $T = 30$ K. It displays three infrared lineshapes of an hydrogen bond involving a Fermi resonance, computed from Eq. (103) within the adiabatic and exchange approximations. The Dirac delta peaks, which are shown here to recall that the same undamped Hamiltonian is involved in the three situations, were computed from the corresponding undamped spectral density Eq. (125). The main characteristic of these sample spectra is the intensity balancing between the two main sub-bands composing the $\nu_S$ $(X\overset{\rightarrow}{-H}\cdots Y)$ band when passing from the situation (a) where $\gamma_\delta > \gamma_o$ to the opposite one (c) where $\gamma_\delta < \gamma_o$.

Another illustration is Fig. 13, where one can observe the complex modifications of the band structure, owing to the numerous infrared transitions enhanced by a higher temperature $T = 300$ K.

In the simplest case of no H bond, the Fermi coupling splits the free X–H stretching band into two sub-bands. In the presence of an H bond ($\alpha_o \neq 0$), a remaining spectral window, often called an Evans window, is usually expected. However, the perturbation of the Fermi resonance may be hidden inside the fine structure that is due to the modulation of the fast mode by the hydrogen-bonded bridge. The conditions for an Evans window to be well-defined is a strong coupling parameter $\Lambda$, which may happen in medium strength or strong H-bonds, as well as a moderate or strong damping parameter $\gamma$ [72]. In the general case, the spectrum of a hydrogen bond may involve several submaxima and subminima that are the consequence of a sole Fermi resonance.

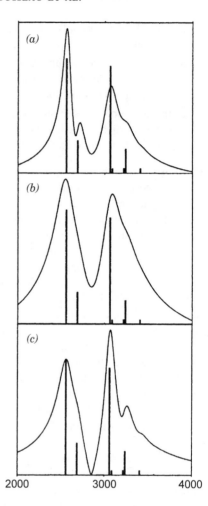

**Figure 12.** Hydrogen bond involving a Fermi resonance: damping parameters switching the intensities. The lineshapes were computed within the adiabatic and exchange approximations. Intensities balancing between two sub-bands are observed when modifying the damping parameters: (a) with $\gamma_o = 0.1$, and $\gamma_\delta = 0.8$; (b) with $\gamma_o = \gamma_\delta = 0.8$; (c) with $\gamma_o = 0.8$ and $\gamma_\delta = 0.1$. Common parameters: $\alpha_o = 1$, $\Lambda = 150\,\mathrm{cm}^{-1}$, $2\omega_\delta = 2850\,\mathrm{cm}^{-1}$, and $T = 30\,\mathrm{K}$.

Besides, we have shown elsewhere [22,23,71,72] that the term "Fermi resonances" is not fully adequate because noticeable perturbations of the $\nu_S$ ($\overrightarrow{X–H}\cdots Y$) bandshape may be obtained in nonresonant cases—that is when the Fermi coupling takes place between the fast mode (of frequency $\omega_o$) and a bending mode whose overtone frequency $2\omega_\delta$ may be very far from $\omega_o$ (see Section IV.B.3).

As a consequence of the wide frequency range over which Fermi resonances may be effective, let us conclude that, in practical situations, several Fermi couplings may be involved in a spectrum, which must lead to puzzling spectral lineshapes.

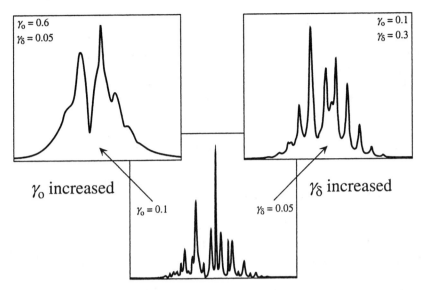

**Figure 13.** Hydrogen bond involving a Fermi resonance: relative influence of the damping parameters. Spectral densities $I_{sf}(\omega)$ computed from Eq. (81). Common parameters: $\alpha_0 = 1$, $\Lambda = 160\,\mathrm{cm}^{-1}$, $\omega_0 = 3000\,\mathrm{cm}^{-1}$, $\omega_{00} = 150\,\mathrm{cm}^{-1}$, $2\omega_\delta = 2790\,\mathrm{cm}^{-1}$, and $T = 300\,\mathrm{K}$.

## VI.  DYNAMICS OF THE HYDROGEN BOND AND THE TWO RELAXATION MECHANISMS

There are two kinds of damping that are considered within the strong anharmonic coupling theory; the direct and the indirect. In the direct mechanism the excited state of the high-frequency mode relaxes directly toward the medium, whereas in the indirect mechanism it relaxes toward the slow mode to which it is anharmonically coupled, which relaxes in turn toward the medium.

Bratos [83] has studied H-bonded complexes from a semiclassical point of view by considering within the linear response theory the indirect relaxation of the slow mode, the physical idea being that in liquids, because the surrounding molecules are acting on the hydrogen bonds, the motion associated with the low-frequency mode of the complex becomes stochastic. The slow modulation limit approach of Bratos predicts broad symmetric Gaussian shapes. Robertson and Yarwood (RY) [84] have suggested that the slow modulation limit assumed by Bratos is not always realized and that the low-frequency motion of the H-bonded complex in solution is not fully stochastic. Their major assumption was that the low-frequency stretching mode can be represented by a Brownian oscillator obeying a Langevin equation. The spectra they obtained remain symmetric, with lineshapes that are evolving from Gaussian to Lorentzian when passing from slow to fast modulation limit. Sakun [85] has improved the above RY work by considering the low-frequency mode as obeying a generalized Langevin equation, whereas Abramczyk [86] has incorporated the rotation diffusion in the RY model.

The first full quantum mechanical treatment of weak H bonds is that of Maréchal and Witkowski (MW) [7], which ignores the influence of the surroundings and which leads, within the adiabatic approximation, to Franck–Condon progression of Dirac delta peaks. Furthermore, Rösch and Ratner (RR) [58] have considered within the adiabatic approximation of MW the influence of the direct relaxation of the fast mode toward the medium. They supposed that the relaxation is due to the coupling between the fluctuating local electric field and the dipole moment of the complex. However, they did not give the corresponding numerical lineshapes. The spectral density they obtained within the linear response theory reduces to the Franck–Condon progression when the damping is missing. It has been shown recently that the Rösch and Ratner spectral density remains the same for weak H bonds when the adiabatic approximation is removed [57] (see Sections IV and V). Note also that the direct damping has been recently introduced in situations involving Fermi resonances [71,72] (see Section III). Later, Boulil et al. [87] have studied in a full quantum mechanical way the influence of the slow mode damping. The physical idea was that, within the MW model, more precisely within a quantum representation, denoted as III, according to notations of Section III.C.4, the slow mode becomes a coherent state just after excitation of the fast mode, so that, in the presence of indirect relaxation, the coherent state becomes damped. Then, using the theoretical work of Louisell and Walker [88] on the behavior of the damped coherent state, they have obtained (with the aid of a small approximation) an autocorrelation function that was involving infinite sums. The spectral density they obtained, after Fourier transform according to the linear response theory, reduces (when the damping is missing) to the MW Franck–Condon progression. Abramczyk [32,33] has proposed a theory for the case when the fluctuations of the energy levels and the Born–Oppenheimer energy potential play an important role through tunneling effect. This approach, when the tunneling effect is missing, leads to an autocorrelation function which is that of Boulil et al. times a somewhat complex autocorrelation function taking into account the fluctuations of the energy levels. For large time, this last autocorrelation function reduces to that of Boulil et al. times a decreasing exponential factor. Later, Blaise et al. [89] and Boulil et al. [90,91] reconsidered their approach in two ways. In Ref. 90, they obtained a closed form for their autocorrelation function. Besides, in Ref. 91, working in another quantum representation (representation II according to the notation of Section III.C.3), which allowed them to avoid their initial small approximation, they considered the slow mode just after excitation of the fast mode, as a driven damped quantum harmonic oscillator. Then they obtained, with the aid of the same Louisell and Walker formalism, as above, a closed autocorrelation function leading to a new accurate spectral density that reduces, as in their first approach, to the MW Franck–Condon progression, when the indirect damping is missing, and to the closed autocorrelation function they obtained in the first quantum representation, when the damping becomes very weak. They have not published the numerical Fourier transform of their two closed autocorrelation functions except some results given without comment in a review [8]. Note that the full autocorrelation function of Abramczyk [32,33], when expressed in terms of the closed autocorrelation function of Boulil et al. [91], reduces (for large time and in the absence of the slow mode damping), to that of Rösch and Ratner; and for situations where the direct damping may be neglected the function reduces to that of Boulil et al. [90].

## A.  The General Background

The Hamiltonian of the weak H bond in the absence of damping is

$$\hat{H}(\tilde{q}, \tilde{Q}) = \frac{1}{2}\hbar\omega_{oo}\,\tilde{P}^2 + \frac{1}{2}\hbar\omega_o\tilde{p}^2 + \frac{1}{2}\hbar\omega_{oo}\tilde{Q}^2 + \frac{1}{2}\hbar\frac{[\omega_o^{eff}(\tilde{Q})]^2}{\omega_o}\tilde{q}^2 \qquad (126)$$

It may be shown [8] that both semiclassical [83,84], and full quantum mechanical approaches [7,32,33,58,87] of anharmonic coupling have in common the assumption that the angular frequency of the fast mode depends linearly on the slow mode coordinate and thus may be written

$$\omega_o^{eff}(\tilde{Q}) = \omega_o + \alpha_o\omega_{oo}\tilde{Q} \qquad (127)$$

In the full quantum mechanical approach [8], one uses Eq. (22) and considers both the slow and fast mode obeying quantum mechanics. Then, one obtains within the adiabatic approximation the starting equations involving effective Hamiltonians characterizing the slow mode that are at the basis of all principal quantum approaches of the spectral density of weak H bonds [7,24,25,32,33,58, 61,87,91]. It has been shown recently [57] that, for weak H bonds and within direct damping, the theoretical lineshape avoiding the adiabatic approximation, obtained directly from Hamiltonian (22), is the same as that obtained from the RR spectral density (involving adiabatic approximation).

On the other hand, in the semiclassical approaches [83–85,92–95], one uses Eqs. (126) and (127) and assumes that the fast mode obeys quantum mechanics whereas the slow one obeys classical mechanics.

The theories of the spectral density of direct and indirect damped H bond have been reviewed recently in an extensive way [8] so that we give here a

straight presentation, begining with the quantum damped spectral densities that
are, of course, more fundamental than the semiclassical ones.

## B.  The Quantum Indirect Damped Autocorrelation Function

First, let us consider the Boulil et al.[90] spectral density, in which the slow
mode damping is treated quantum mechanically. Following MW [7] and using the
adiabatic approximation, allowing us to separate the slow and fast movements
of the slow- and high-frequency modes, we obtain effective Hamiltonians
governing the slow mode, the nature of which depends on the excitation degree
of the fast mode. In the quantum representation named III in Ref. 90, when the
fast mode is in its ground state, the slow mode is simply described by a
Boltzmann density operator involving a quantum harmonic Hamiltonian. Just
after the fast mode has jumped into its first excited state, the slow mode
becomes a coherent state density operator, which, because it does not commute
with the Hamiltonian of the slow mode, evolves with time and thus is
susceptible to be undergoing a damping caused by the medium. In another
representation (named II in Ref. 8), the slow mode—which was, before
excitation of the fast mode, simply described by a quantum harmonic oscillator,
as in representation III—becomes a quantum driven harmonic oscillator just
after excitation. Because it is driven, the slow mode becomes susceptible to be
damped by the surroundings—this case being taken into account by the
quantum model of indirect damping in representation II. (Representation III is a
special situation of it for very weak damping.)

### 1.  The Autocorrelation Function $\{G_{II}(t)\}$ in Representation II

Boulil et al. [8] have obtained the following expression for the quantum indirect
damped autocorrelation function in representation II:

$$
\begin{aligned}
\{G_{II}(t)\} \propto \ & \exp\left\{ -\frac{1}{2}|\beta|^2(1+2\langle n\rangle) \right\} \exp\left\{ i|\beta|^2 \exp(-\gamma_{oo}t/2)\sin(\omega_{oo}t) \right\} \\
& \times \exp\left\{ i(\omega_o - \alpha_o^2\omega_{oo})t \right\} \exp\left\{ -i|\beta|^2\omega_{oo}t \right\} \\
& \times \exp\left\{ \frac{1}{2}|\beta|^2(1+2\langle n\rangle)[2\exp(-\gamma_{oo}t/2)\cos(\omega_{oo}t) - \exp(-\gamma_{oo}t)] \right\}
\end{aligned}
$$

$$(128)$$

where $\langle n\rangle$ is the thermal average of the mean number occupation of the slow
mode,

$$
\langle n\rangle = \left[ \exp\left\{ \frac{\hbar\omega_{oo}}{k_B T} \right\} - 1 \right]^{-1}
$$

$$(129)$$

$\beta$ is a dimensionless parameters given by

$$\beta = \alpha_o \frac{[2\omega_{oo}^2 + i\gamma_{oo}\omega_{oo}]}{2(\omega_{oo}^2 + \frac{\gamma_{oo}^2}{4})} \tag{130}$$

### 2. The Closed-Form $\{G_{III}(t)\}$ of the Autocorrelation Function in Representation III

We may observe that for very weak damping, $\beta \approx \alpha_o$, and the above autocorrelation function reduces to

$$\{G_{II}(t)\} = \{G_{III}(t)\} \tag{131}$$

with

$$\{G_{III}(t)\} \propto \exp\left\{ -\frac{1}{2}\alpha_o^2(1 + 2\langle n\rangle)\right\} \exp\left\{i\alpha_o^2 \exp(-\gamma_{oo}t/2)\sin(\omega_{oo}t)\right\}$$

$$\times \exp\{i(\omega_o - 2\alpha_o^2\omega_{oo})t\}\exp\left\{\frac{1}{2}\alpha_o^2(1 + 2\langle n\rangle)\right.$$

$$\times [2\exp(-\gamma_{oo}t/2)\cos(\omega_{oo}t) - \exp(-\gamma_{oo}t)]\bigg\} \tag{132}$$

Note that this last expression is nothing but the closed form [90] of the autocorrelation function obtained (as an infinite sum) in quantum representation III by Boulil et al.[87] in their initial quantum approach of indirect damping. Although the small approximation involved in the quantum representation III and avoided in the quantum representation II, both autocorrelation functions are of the same form and lead to the same spectral densities (as discussed later).

### C. The Rösch and Ratner Directly Damped Quantum Autocorrelation Function

Now, consider the quantum model of RR [58] dealing with direct damping mechanism where the relaxation is due to the coupling between the fluctuating local electric field and the dipole moment of the complex. It may contribute when polar solvents are involved, so that the local electrical field fluctuations are correspondingly large. Due to vibrational anharmonicity, the expectation value of the interaction will differ, according to the fact that the fast mode is in its ground state or in its first excited state. Therefore, the energy difference between these states will depend on the local field. Fluctuations in this field will result in a broadening of the spectrum. RR [58] assumed that the thermal mean value of the electrical field and its autocorrelation function obey expressions similar to that considered for the stochastic force appearing in the generalized

Langevin equation. RR [58] have obtained for the dipole moment operator the following autocorrelation function which may be written after a correction (corresponding to a zero-point-energy they neglected) [8]:

$$\{G_{\text{Rat}}(t)\} \propto \exp\{i\alpha_0^2 \sin(\omega_{oo}t)\} \exp\{i(\omega_o - 2\alpha_0^2\omega_{oo})t\}$$
$$\times \exp\{\alpha_0^2(1 + 2\langle n \rangle)[\cos(\omega_{oo}t) - 1]\} \exp\{-\gamma_o t\} \qquad (133)$$

where $\gamma_0$ is the direct damping parameter (for which RR found an explicit expression as a function of the basic parameters, but which we shall consider as phenomenological) whereas the other symbols have the same meaning as for the Boulil et al. autocorrelation function (128). It may be observed that when the damping of the slow mode is missing in the Boulil et al. equation [Eq. (128)], or when the direct damping is missing in the Rösch and Ratner Eq. (133), the two corresponding autocorrelation functions reduce to the same expression, that is,

$$\{G_{\text{Rat}}(t, \gamma_o = 0)\} = \{G_{\text{II}}(t, \gamma_{oo} = 0)\} \qquad (134)$$

Owing to the above remarks, we shall study in the following the features of the lineshapes of weak H bonds in which there is a dephasing of the fast mode and simultaneously a damping of the slow mode to which the fast one is anharmonically coupled. The quantum spectral density is then [96]

$$\{I(\omega)\} = \int_{-\infty}^{+\infty} \{G_{\text{II}}(t)\} \exp(-\gamma_o t) \exp(-i\omega t)\, dt \qquad (135)$$

As required, Eq. (135) reduces in the absence of slow-mode damping to the Rösch and Ratner spectral density and, in the absence of dephasing of the fast mode, to that of Boulil et al. At last, when the damping parameters $\gamma_o$ and $\gamma_{oo}$ are missing, this spectral density (135) reduces to the Franck–Condon progression given by Eq. (51). On the other hand, it may be observed that for no dephasing of the fast mode and very weak damping of the slow mode, one may perform, according to Eq. (130), the approximation $\alpha_o \approx \beta$. As a consequence, for this special situation, the spectral density (135) reduces to a new one which, for $\gamma_o = 0$, is nothing else but the closed form of the spectral density in representation III [90], which is formally equivalent to the one involving infinite sums, obtained by Boulil et al. in 1988 [87] in their initial work.

### D.  The Semiclassical Robertson–Yarwood Spectral Density Involving Indirect Relaxation

The theory of Robertson and Yarwood [84] is very similar in its spirit to the initial one of Bratos [83], but it considers, following Kubo [97], the angular frequency of the fast mode as a variable $\omega(t)$ that is stochastically time-dependent through the stochastic time dependence of the slow mode to which it is anharmonically coupled. The major assumption is that the low-frequency stretching mode is classical and can be represented by the Ornstein–Uhlenbeck stochastic process [98] according to which the slow mode is a Brownian oscillator obeying the Langevin theory. That leads to a loss of the phase coherence of the high-frequency stretching mode because of the anharmonic coupling between the slow and fast modes. It must be emphasized that the demonstration of the RY model supposes that the slow mode coordinate does not commute with its conjugate momentum.

The basic Robertson and Yarwood autocorrelation function, for the underdamped situation, may be written within our present notations [8]:

$$
\begin{aligned}
G_{\text{Rob}}(t) \propto \exp\{i\omega_0 t\}\exp\Bigg\{ &-2\alpha_0^2\omega_{00}^2\langle\tilde{Q}^2\rangle \times \int_0^t \exp\Big\{ -\frac{\gamma_{00}}{2}t'\Big\}[\cos(\omega_{\text{und}}t') \\
&+ \frac{\gamma_{00}}{2\omega_{\text{und}}}\sin(\omega_{\text{und}}\,t')](t - t')\,dt'\Bigg\}
\end{aligned}
\tag{136}
$$

Here $\langle\tilde{Q}^2\rangle$ is the fluctuation of the slow mode coordinate, whereas $\omega_{\text{und}}$ is the effective angular frequency of the slow mode in the underdamped situation which is given by

$$
\omega_{\text{und}}^2 = \omega_{00}^2 - \frac{\gamma_{00}^2}{4}
\tag{137}
$$

In their article, Robertson and Yarwood consider the fluctuation appearing in Eq. (136) as given by quantum statistical mechanics, that is,

$$
\langle\tilde{Q}^2\rangle = \coth\{\hbar\omega_{00}/(2k_B T)\}
\tag{138}
$$

However, in all the rest of their approach, Robertson and Yarwood consider the slow mode $\tilde{Q}$ as a scalar obeying simply classical mechanics, because they neglect the noncommutativity of $\tilde{Q}$ with its conjugate momentum $\tilde{P}$. As a consequence, the logic of their approach is to consider the fluctuation of the slow mode as obeying classical statistical mechanics and not quantum statistical mechanics. Thus we write, in place of Eq. (138), the corresponding classical formula:

$$
\langle\tilde{Q}^2\rangle = k_B T/\hbar\omega_{00}
\tag{139}
$$

After integration of Eq. (136) and using Eq. (139), we obtain the following, after some manipulations

$$
\begin{aligned}
G_{\text{Rob}}(t) \propto \exp\{i\omega_0 t\} \exp\Bigg\{ &-2\alpha_0^2 \left[\frac{k_B T}{\hbar\omega_{oo}}\right] \omega_{oo}^2 \left[2\exp\left\{-\frac{1}{2}\gamma_{oo}t\right\}\right. \\
&\times \frac{\left(6\omega_{\text{und}}\gamma_{oo}^2 - 8\omega_{\text{und}}^3\right)\cos(\omega_{\text{und}}t) - \left(12\omega_{\text{und}}^2 - \gamma_{oo}^2\right)\gamma_{oo}\sin(\omega_{\text{und}}t)}{\left(\gamma_{oo}^2 + 4\omega_{\text{und}}^2\right)^2} \\
&+ 4\frac{\gamma_{oo}^2(\gamma_{oo}t - 3) + 4\omega_{\text{und}}^2(1 + \gamma_{oo}t)}{\left[\gamma_{oo}^2 + 4\omega_{\text{und}}^2\right]^2}\Bigg]\Bigg\}
\end{aligned}
\tag{140}
$$

The spectral density is the Fourier transform of this autocorrelation function, that is,

$$
\{I_{\text{Rob}}(\omega)\} = \int_{-\infty}^{+\infty} \{G_{\text{Rob}}(t)\} \exp(-i\omega t)\, dt \tag{141}
$$

It may be of interest to observe that at zero temperature this last expression of the Robertson and Yarwood spectral density (in which we consider the fluctuation of the slow mode as classical in contrast with the RY paper) reduces (unsatisfactorily!) to a single Dirac delta peak:

$$
I_{\text{Rob}}(\omega) = \delta(\omega - \omega_0) \tag{142}
$$

This remark, which shows that the deep nature of the semiclassical RY model is to lead to a collapse of the broadened structure at zero temperature [that was not obvious in the RY paper, because of the use of the quantum statistical expression (138) in place of the classical one (139)], will later appear to be of some interest.

### E. Narrowing of the Lineshapes by Pure Damping of the Slow Mode

#### 1. Comparison Between Lineshapes Computed by Aid of the Three Basic Models

Before beginning this study, it may be observed that in Ref. 8, the lineshapes given for situations involving indirect damping were computed by adding a direct damping with $\gamma_0 \approx 15\,\text{cm}^{-1}$. This addition was not explicit.

In this section, we shall compare the evolution of the spectral densities with increasing the damping parameter, computed from the RR, the Boulil et al., and the RY models, respectively. In Fig. 14, the spectral densities corresponding respectively to the classical indirect relaxation (left columns), the quantum damping of the slow mode (middle column), and the quantum dephasing of the fast mode (right column) are given. For all these spectral densities, the same anharmonic coupling parameter ($\alpha_0 = 1$) and the same room temperature ($T = 300$ K) have been used. The top spectral densities of Fig. 14 correspond to

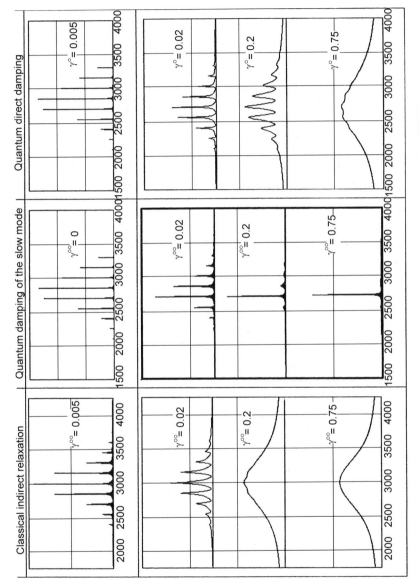

**Figure 14.** Narrowing of the lineshapes by the slow mode damping ($T = 300$ K).

situations of very weak damping. More precisely, the damping parameters have been taken as $\gamma_0 = 0.005\omega_{oo}$ for the RR (very weak direct damping), $\gamma_{oo} = 0.005$ $\omega_{oo}$ for the RY models (very weak damping of the slow mode), and $\gamma_0 = 0.005$ $\omega_{oo}$ and $\gamma_{oo} = 0$ (very small dephasing of the fast mode and zero damping of the slow mode) for the Boulil et al. model. Note that $\gamma_0 = \gamma_{oo} = 0.005\omega_{oo}$ indeed corresponds to a situation of very weak damping because the corresponding dimensioned parameters are respectively $\gamma_0 = \gamma_{oo} = 0.75\,\mathrm{cm}^{-1}$. Besides, the *ad hoc* relaxation parameters are increased from top to bottom for each model. In the left and middle columns (corresponding to the possibility of indirect relaxation), the parameter $\gamma_{oo}$ is progressively enhanced, whereas in the right column (corresponding to quantum direct relaxation), it is $\gamma_0$:

Examination of Fig. 14 shows that for the top situation of very weak damping, the spectral densities are formed of very narrowed lineshapes, somewhat evoking Dirac delta peaks: The spectral densities corresponding to quantum situations (right and middle top spectral densities) are the Franck–Condon progression of the MW model appearing in Eq. (49) and for which the peaks are weakly broadened and have intensities near those given by the Frank–Condon factors (50); these spectral densities are very near that of the semiclassical model (left top spectral density). Besides, it appears that the classical indirect relaxation spectral density and the quantum dephasing spectral densities (left and right columns) coalesce progressively in a similar way, when the corresponding damping parameter is enhanced from top to bottom, leading at the bottom to very smoothed broadened profiles. In contrast, an increase in the damping in the quantum model involving pure relaxation of the slow mode (middle column) produces a drastic collapse of the Franck–Condon progression, without broadening the fine structure, leading to a single sharp lineshape that is weakly asymmetric.

## 2. *Physical Discussion*

It must be emphasized that *the collapse of the fine structure (middle column) is in contradiction with the Robertson and Yarwood semiclassical model (left column) which predicts a broadening of the lineshapes.* This contradiction may appear surprising because it was generally assumed that the semiclassical and the quantum model of "indirect relaxation" must lead to similar situations. However, the narrowing of the high-frequency lineshape induced by the quantum damping of the slow mode to which the high-frequency mode is anharmonically coupled is essential in the Boulil et al. model (at the basis of which there is the Louisell and Walker [88] work on driven damped coherent state), and cannot be denied.

It must be observed that the deep nature of the "indirect relaxation" is not the same in the model of Robertson and Yarwood and in that of Boulil et al. [99]. Before we consider this, it is necessary to emphasize [99] that the assumption performed by Robertson and Yarwood (see above)—according to which they consider the fluctuation of the slow mode as obeying quantum statistical mechanics, although they neglect in their model the noncommutativity of the slow mode coordinate with respect to its conjugate momentum, is inconsistent—so that the logic of their approach requires us to use [see Eqs. (138) and (139)], in place of the quantum statistical fluctuation of the slow mode coordinate (which become purely quantum mechanical at zero temperature), its corresponding classical statistical fluctuation (which becomes zero at zero temperature). After this important remark, we may observe [99] that in the model of Robertson and Yarwood, there is a dephasing of the fast mode through the stochastic behavior of the slow mode to which the fast mode is anharmonically coupled, with the slow mode starting at initial time from a fluctuation that is damped and which vanishes at zero temperature. On the other hand, in the Boulil et al. model, there is a fine structure in the spectral density of the fast mode caused by the anharmonic coupling between the slow and fast mode, which induces for the slow mode a driven harmonic oscillator which, in turn, is damped by its surroundings. More precisely [99], in the Boulil et al. model, the slow mode becomes driven just after excitation of the fast mode, owing to the change in nature of the effective Hamiltonians governing the slow modes, whereas in the RY model it does not change (because the change in the effective Hamiltonians of the slow mode, after excitation of the fast mode, is ignored). Therefore, in the quantum model the surroundings act as a damping on the driven oscillator, whereas in the semiclassical model, these surroundings act only as a stochastic source for fluctuations of the slow mode coordinate which are damped, according to the fluctuation–dissipation theorem and which reduces to zero at zero temperature. As a consequence, at zero temperature, the nature of the semiclassical model is to lead to a single Dirac delta peak whereas that of the quantum model is to give (as the Maréchal and Witkowski model) a large broadening.

Now, let us look more deeply at the cause of the collapse of the fine structure of the lineshapes caused by damping of the driven harmonic oscillator: Just after the fast mode has been excited, the slow mode becomes a driven harmonic oscillator that is damped by the surroundings. In the absence of damping, the driven term of the slow mode oscillator induces a quantum coherent state that (at $T = 0$ K) may be expanded in terms of the eigenstates of the number occupation operator by means of a Poisson distribution, the Franck–Condon progression of which is the spectral manifestation. Now, when the damping of the slow mode is introduced and then progressively enhanced, the time required by the driven term of the oscillator to induce the coherent state (and thus the Poisson distribution) becomes less and less efficient, so that the Franck–Condon progression collapses. It appears that the damping of the slow mode does not affect the width of the components and the angular frequency of the components of the Franck–Condon distribution. It modifies only through the effectiveness of the anharmonic coupling parameter the relative intensities of these components. When the effective anharmonic coupling is lowered by the damping, all the intensities of the Franck–Condon progression are progressively modified in such a way that when the effective anharmonic coupling is reaching zero, only the single component corresponding to the case where there is no H bond remains. It may be observed here that for a crude model—in which one considers (in place of the driven damped harmonic oscillator appearing in the Boulil et al. model), a driven harmonic oscillator, the driven term of which is damped—the narrowing is again obtained.

As a matter of fact, this behavior of the Boulil et al. model may be viewed as *homogeneous narrowing* (homogeneous having the same meaning as in the usual *homogeneous broadening* process where each absorbing molecule is identical to the rest and is undergoing a finite lifetime because of damping) in contrast with the semi classical relaxation which is involving a broadening of the high-frequency mode through a *dephasing* process in which the fast mode is anharmonically coupled to the slow mode acting as a stochastic oscillator. At last, we may emphasize that at the basis of the *homogeneous narrowing* by damping of the slow mode, there is the reduced time evolution operator of a driven damped harmonic oscillator as calculated by Louisell and Walker [88] in the framework of a reduced density operator formalism so that both the evolution of the populations and of the coherences involved in the corresponding matrix representation of this operator are taken into account.

Now, it may be observed that if the narrowing of the lineshape by damping of the slow mode is true, there is a difficulty. In H bonds, the low-frequency mode is certainly damped because this mode, which is spectroscopically observed, is very broad. Thus, because of the anharmonic coupling, this damping must induce in the framework of the quantum model a narrowing of the lineshape characterizing the fast mode, in contradiction with the observed fact that this lineshape is broadened. But this contradiction is only apparent: As we shall see later, the presence of the dephasing of the fast mode, together with the damping of the slow mode to which the fast mode is anharmonically coupled, is canceling the narrowing induced by the damping of the slow mode.

### F.   Rösch–Ratner and Robertson–Yarwood Lineshapes

As we have found in the above section, the damping of the slow mode when it is nearly alone produces (within the Boulil et al. model) a collapse of the fine structure of the lineshapes. That is in contradiction with the RY semiclassical model which predicts a *broadening* of the lineshape. Of course, because the quantum model is more fundamental, the semiclassical model must be questioned. However, it is well known that the RY semiclassical model of indirect relaxation has the merit to predict lineshapes that may transform progressively, with all possible intermediates, from the Gaussian profiles (slow modulation limit considered initially by Bratos [83]) to Lorentzian ones (fast modulation limit). Now, the question arises as to whether if the RR model cannot lead to the same kinds of lineshapes as that of RY.

### 1. Similarity Between the Rösch and Ratner and Robertson and Yarwood Lineshapes

Now, return to Fig. 14. The right and left bottom damped lineshapes (dealing respectively with quantum direct damping and semiclassical indirect relaxation) are looking similar. That shows that for some reasonable anharmonic coupling parameters and at room temperature, an increase in the damping produces approximately the same broadened features in the RY semiclassical model of indirect relaxation and in the RR quantum model of direct relaxation. Thus, one may ask if the RR quantum model of direct relaxation could lead to the same kind of prediction as the RY semiclassical model of indirect relaxation.

Figure 15 gives the superposition of RR (full line) and RY (dotted plot) spectral densities at 300 K. For the RR spectral density, the anharmonic coupling parameter and the direct damping parameter were taken as unity ($\alpha_o = 1$, $\gamma_o = \omega_{oo}$), in order to get a broadened lineshape involving reasonable half-width ($\alpha_o = 1$ was used systematically, for instance, in Ref. 72). For the RY spectral density, the corresponding parameters were chosen $\alpha_o = 1.29$, $\gamma_{oo} = 0.85\omega_{oo}$ to fit approximately the above RR lineshape, after a suitable angular frequency shift (the RY model fails to obtain the low-frequency shift predicted by the RR model) and a suitable adjustment in the intensities that are irrelevant in the RR and RY models.

Figure 15 shows that the RR quantum direct relaxation model fits successfully the RY semiclassical relaxation one. Other computations that are not given here lead to the conclusion that such a result is generally met when:

1. The lineshapes are computed at temperatures that are not too low (the RY model cannot be applied at low temperature because it is semiclassical).

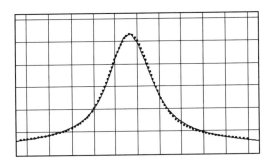

**Figure 15.** Rösch–Ratner and Robertson lineshapes at 300 K: Rösch and Ratner (direct relaxation) $\alpha_o = 1$, $\gamma_o = 1$; Robertson (indirect relaxation) $\alpha_o = 1.29$, $\gamma_{oo} = 0.85$.

2. The damping parameter is large enough to allow the smoothing of the fine structure.

3. The anharmonic coupling parameter is able to reproduce a relatively large half-width of a weak H bond.

As a consequence, one may infer that the experimental features of the lineshapes of the $\nu_S(X-H)$ that were explained by means of the RY semiclassical model of indirect relaxation would be as well-explained in terms of the RR quantum model of direct relaxation.

### 2. Slow and Fast Modulation Limits

If the previous conclusion is true, the RR model has to be able to reproduce successfully the ability of the RY model to predict lineshapes passing progressively from the Gaussian profile to the Lorentzian one, when going from the slow modulation to the fast modulation limit. Now, recall that, according to the RY model [84], the slow and fast modulation limit leading respectively to Gaussian and Lorentzian shapes must verify respectively inequalities that, when expressed within the notations of [8], give

$$\text{Slow modulation limit:} \quad \alpha_0 \sqrt{\frac{2 k_B T}{\hbar \omega_{00}}} > \frac{\omega_{00}}{\gamma} \qquad (143)$$

$$\text{Fast modulation limit:} \quad \alpha_0 \sqrt{\frac{2 k_B T}{\hbar \omega_{00}}} < \frac{\omega_{00}}{\gamma} \qquad (144)$$

where we have written $\gamma$ instead of $\gamma_{00}$, for reasons which will appear later. These inequalities predict that for damping $\gamma$ of the slow mode of the same magnitude as the corresponding angular frequency $\omega_{00}$ (a situation allowing to induce the smoothing of the fine structure), high temperature and strong anharmonic coupling $\alpha_0$ favor the Gaussian lineshape, whereas low temperature and weak anharmonic coupling lead to Lorentzian profiles. Now if, according to Fig. 15, the RR quantum model of direct damping is able to predict the same Gaussian-Lorentzian behavior as does the RY semiclassical one of indirect relaxation, then the inequalities (143) and (144) ought to hold for quantum direct damping. In order to verify such an ability for the RR model to reproduce the Gaussian–Lorentzian profiles, two lineshapes computed by means of the RR model for the situation of slow and fast modulation limits are given in Fig. 16.

The RR lineshape given at the left of the figure corresponds to room temperature ($T = 300$ K), standard anharmonic coupling parameter ($\alpha_0 = 1$, see above), and $\gamma_0 = 1.25\omega_{00}$, whereas the right RR lineshape is characterized by a low temperature ($T = 50$ K), a weak anharmonic coupling parameter ($\alpha_0 = 0.25$), and a stronger damping ($\gamma_0 = 1.5\omega_{00}$). As a consequence, the

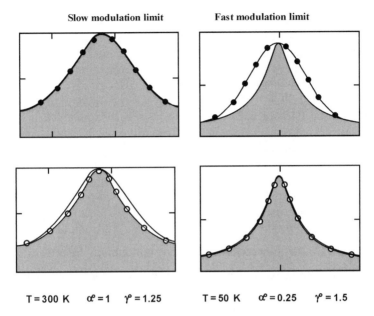

**Figure 16.** Rösch and Ratner spectral density (direct damping): Rösch and Ratner lineshapes (lines); Lorentzian fit (circles); Gaussian fit (black dots).

numerical values of the right and left terms involved in the inequalities (143) and (144) are those given in Table IV.

Clearly, after taking $\gamma_0$ in place of $\gamma$, and according to the inequalities (143) and (144), the left situation satisfies the slow modulation condition whereas the right one satisfies the fast modulation condition. As a consequence, the right and left spectra are expected to fit the Lorentzian and Gaussian profiles, respectively. In order to verify this, the right and left spectra (full lines) are both compared to pure Gaussian lineshape (filled circles) at the top of Fig. 16 and are compared to pure Lorentzian profiles (unfilled circles) at the bottom. Inspection of Fig. 16 confirms that the right spectrum satisfies the fast modulation condition and

TABLE VI
The Slow and Fast Modulation Conditions (see Fig. 16)

|  | $\alpha_0 \sqrt{(2 k_B T / \hbar \omega_{00})}$ |  | $\omega_{00}/\gamma_0$ |
|---|---|---|---|
| Left spectrum | 1.67 | > | 0.80 |
| Right spectrum | 0.17 | < | 0.67 |

therefore fits completely the Lorentzian profile whereas the left one satisfies the slow modulation condition and therefore fits successfully a Gaussian profile.

This result manifests the ability of the RR quantum model of direct relaxation to reproduce—as the RY semiclassical model of indirect relaxation does—the evolution from Gaussian to Lorentzian profiles when going from the slow to the fast modulation limit (when the damping is strong enough to produce the smoothing of the fine structure). Besides, we must keep in mind the fact that the quantum model of Boulil et al. taking into account the damping of the fast mode, which is more consistent than the RY semiclassical model of indirect relaxation, predicts a strong narrowing of the lineshape by collapse of the fine structure—in contrast to the RY model, which predicts broadening by smoothing of the fine structure. As a consequence, one may considers that many explanations in the literature based on the RY semiclassical model of indirect damping ought to be reformulated by means of the RR quantum autocorrelation function.

## G. Asymmetry, Temperature, and Isotope Effects

It is well known from experiment that the IR spectrum of weak H bonds exhibits broadening of the lineshapes, a low-frequency shift of its first moment, a half-width increase by temperature narrowing, and a low–frequency shift by D substitution. Now let us investigate how, from a theoretical viewpoint, both quantum dephasing of the fast mode and damping of the slow mode are able to reproduce these experimental trends. For this purpose, three series of two superposed lineshapes computed with the aid of Eq. (135) are reported in Fig. 17 within three columns. In each series, one of the superposed curves (the full line) refers to the situation without damping of the slow mode, whereas the other (the dotted plots) corresponds to situation where both $\gamma_0$ and $\gamma_{00}$ are simultaneously acting. For the two kinds of lineshapes, $\gamma_0$ is decreased in the same way, from top to bottom, within the three columns: It passes from $\gamma_0 = 0.8\omega_{00}$ (top situation) to $\gamma_0 = 0.6\omega_{00}$ (middle situation) and to $\gamma_0 = 0.4\omega_{00}$ (bottom situation). For the lineshapes also involving damping of the slow mode (dotted plots), $\gamma_{00}$ is decreased in the same way—that is, according to $\gamma_{00} = 0.8\omega_{00}$ (top situation), $\gamma_{00} = 0.6\omega_{00}$ (middle situation), and $\gamma_{00} = 0.4\omega_{00}$ (bottom situation) respectively. Besides, in order to simulate temperature and isotope effects, the lineshapes of the H-bonded species at 300 K characterized by an anharmonic coupling $\alpha_0 = 1.5$ and the usual angular frequency of the fast mode $\omega_0 = 3000 \, \text{cm}^{-1}$ are given in the middle column of the figure. Finally, the left column gives the corresponding lineshape with the same anharmonic coupling $\alpha_0$ and the same angular frequency $\omega_0$ but at the lower temperature of 100 K, and the right column gives the lineshapes corresponding to a D-bonded species at 300 K, for which, owing to the D substitution, the anharmonic coupling reduces to $\alpha_0 = 1.5/\sqrt{2} \approx 1$ and the angular frequency of the fast mode to $\omega_0 = 3000/\sqrt{2} \approx 2100 \, \text{cm}^{-1}$.

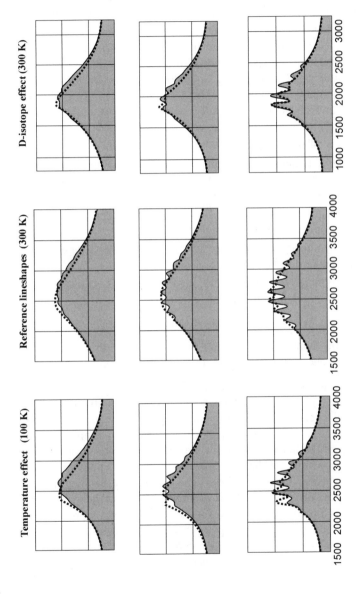

**Figure 17.** Lineshapes involving both direct and slow mode damping (dots) as compared to those involving only direct damping (lines). In each column, the *ad hoc* dampings ($\gamma_o$ for the full lines, $\gamma_o$ and $\gamma_{oo}$ for the dotted lines) are decreased (from top to bottom) from 0.8 to 0.6 and 0.4. The middle column corresponds to the H species at 300 K, the right one corresponds to the D ones at the same temperature, and the left one corresponds to the H ones but at 100 K. $\alpha_o = 1.5$, $\omega_o = 3000 \, \text{cm}^{-1}$, $\omega_{oo} = 150 \, \text{cm}^{-1}$.

Inspection of Fig. 17 shows the following trends:

1. Width of the lineshapes. The lineshapes are broadened approximately in the same way for the two situations where either the fast mode of, the slow one is damped.

2. Low-frequency shift of the lineshapes. The lineshapes of hydrogenated species (undeuterated species) are all low-frequency-shifted with respect to the basic angular frequency $\omega_0 = 3000 \, \text{cm}^{-1}$ by an amount around $450 \, \text{cm}^{-1}$ slightly smaller than $2\alpha_0^2\omega_{oo}$ [i.e., $2 \times (1.5)^2 \times 150 \, \text{cm}^{-1} = 675 \, \text{cm}^{-1}$, which would be the shift at zero temperature and in the absence of damping in the MW model].

3. Asymmetry of the lineshapes.
   a. The introduction of the damping of the slow mode in situations where there is a dephasing of the fast mode induces an asymmetry (compare the dotted to the continuous lineshapes)
   b. This asymmetry is enhanced by increase in the damping of the slow mode (compare the bottom, middle, and top lines).
   c. This asymmetry involves, for all situations, a low-frequency wing, falling more deeply than the high-frequency one.
   d. This asymmetry is involving stronger corrections at low temperature (compare the left to the middle column).

## H. Discussion

From systematic numerical investigations combining direct and indirect damping effects, one may perform the following conclusions:

1. The full quantum mechanical approach of Boulil et al., taking into account the damping of the slow mode, is in contradiction to the Robertson and Yarwood semiclassical model taking into account the stochastic behavior of the slow mode: When the fast mode damping is very weak or vanishing, an increase of the damping of the fluctuations of the slow mode produces, in the semiclassical model, a smoothing of the fine structure and thus a whole broadening of the lineshape, whereas in the Boulil et al. model the corresponding increase of the damping of the driven oscillator is destroying the fine structure before any possibility of its broadening. Of course, the Boulil et al. approach is more consistent from a theoretical viewpoint than the semiclassical one. As a consequence, when only the slow mode damping is acting, it is not possible to reproduce the smoothing of the broadened lineshape observed by experiment and thus, the "indirect mechanism" cannot be considered as the only cause of the smoothed and broadened lineshapes. Thus, the fast mode damping or other dephasing factor not necessarily of the same kind [99], must be acting everywhere.

2. When both the quantum dampings of the slow and fast modes now occur, an increase of the slow mode damping at constant (nonnegligible) fast mode damping produces a weak narrowing of the lineshapes, whereas an increase of the fast mode damping, at constant slow mode damping, leads to a broadening that is more sensitive than the narrowing by the slow mode damping in the first situation.

3. The full quantum mechanical approach of direct damping by Rösch and Ratner leads to lineshapes that are close to those obtained within the indirect relaxation mechanism, by the Robertson and Yarwood semiclassical model, although the quantum fast mode damping mechanism leads to weak asymmetric lineshapes whereas the semiclassical model cannot reproduce this asymmetry. For cases where the asymmetry of the lineshape may be ignored, both the RR quantum approach of the fast mode damping and the RY semiclassical approach of indirect relaxation lead to Lorentzian lineshape profiles in the fast modulation limit and to Gaussian ones in the slow modulation limit. Thus, owing to the above conclusion according to which the quantum theory dealing with the slow mode damping (when it is acting alone) cannot lead to broadened lineshapes, the experimental fact that many broadened nearly symmetric lineshapes are intermediate between the Lorentzian and Gaussian profiles should be explained more physically within the quantum RR mechanism.

4. When the damping of the slow mode and the dephasing of the fast mode are simultaneously present, the slow mode damping produces distortions of the features of the lineshapes (yet broadened by the direct damping): When the slow mode damping is slowly enhanced, the asymmetry of the lineshape increases with a low-frequency wing falling more deeply than the high-frequency one; besides, a stronger increase of the slow mode damping then produces an enhancement of the relative intensity of the sub-bands corresponding to the $|\{0\}, (m)\rangle \rightarrow |\{1\}, (m)\rangle$ transition of the slow mode; the greater the slow mode damping, the higher and narrower the $|\{0\}, (m)\rangle \rightarrow |\{1\}, (m)\rangle$ sub-bands.

5. It is emphasized that the two kinds of quantum relaxation mechanisms lead, in agreement with experiment, to a shift of the first moment of the spectrum toward low frequencies, whereas the RY semiclassical indirect relaxation mechanism does not produce this effect.

6. Observe that all the mechanisms—that is, the classical indirect mechanism and the two quantum ones—predict a satisfactory isotope effect when the proton of the H bond is substituted by deuterium: All the damping mechanisms induce approximately a $1/\sqrt{2}$ low-frequency shift of the first moment and a $1/\sqrt{2}$ narrowing of the breadth, which is roughly in agreement with experiment. As a consequence, the isotope effect does not allow us to distinguish between the two damping mechanisms.

7. The two quantum mechanisms lead to a broadening of the lineshapes obeying approximately a $\coth^{1/2}(cst/T)$ law, so that the temperature effect is

not suitable to distinguish between them. Note that the deep nature of the semi-classical RY model, according to which the slow mode is treated classically, leads us to conclude that the autocorrelation function must reduce to a Dirac delta peak at zero temperature in contrast with the quantum models of RR and Boulil et al. which lead satisfactorily to broadened lineshapes at the same limit temperature.

All these remarks, which may be performed keeping in mind the connections between the theoretical works dealing with the same subject, lead to the conclusion that the slow mode damping cannot reproduce the observed broadened lineshapes when the dephasing of the fast mode is ignored, but that, combined with the dephasing mechanism, it may explain satisfactorily the general observed trends in the features of the experimental lineshapes, that is (i) low-frequency shift, (ii) broadening, (iii) asymmetry, (iv) temperature broadening law, (v) isotope effect leading to narrowing and low-frequency shift, (vi) profiles intermediate between Lorentzian and Gaussian shapes for near symmetric lineshapes, and (vii) distortions in the sub-band intensities and frequencies, with respect to the Franck-Condon progression law.

According to the above conclusions, some remarks may be made regarding the nature of the damping mechanism that may be inferred to be efficient in any actual lineshape of weak H bond, according to its experimental features:

1. If the experimental lineshapes involve sub-bands that are either (a) broadened and nearly obeying the Franck–Condon progression law or (b) without sub-bands, but involving nearly symmetric profiles intermediate between Gaussian and Lorentzian, then the dephasing mechanism ought to be viewed as playing an important role (although the slow mode damping cannot be ignored because the slow mode lineshape, which is spectroscopically observed, is broadened).

2. If the experimental lineshapes do not exhibit sub-bands and are asymmetric, or if they involve sub-bands, but with intensity anomalies with respect to the Franck–Condon progression law, then, together with the dephasing mechanism, damping of the slow mode ought to be also considered as occurring in a sensitive competitive way.

These conclusions must be considered keeping in mind that the general theoretical spectral density used for the computations, in the absence of the fast mode damping, reduces [8] to the Boulil et al. spectral density and, in the absence of the slow mode damping, reduces to that obtained by Rösch and Ratner; one must also rember that these two last spectral densities, in the absence of both dampings [8], reduce to the Franck–Condon progression involving Dirac delta peaks that are the result of the fundamental work of Maréchal and Witkowski. Besides, the adiabatic approximation at the basis of the Maréchal and Witkowski approach and of the two kinds of quantum damping models is very satisfying (at least for the fast mode damping) because, as has been shown recently [57], the spectral density remains the same when the adiabatic approximation is avoided.

## VII. CONCLUSION

The present review has dealt with some recent developments on quantum theories of the IR $\nu_S$ (X–$\vec{H}\cdots$Y) lineshapes of weak H bonds. The basic tool is the strong anharmonic coupling theory of Maréchal and Witkowski [7] according to which there is an anharmonic coupling between the high-frequency mode and the hydrogen-bonded bridge. Recall that the Maréchal and Witkowski model relies on the adiabatic approximation that allows to separate the fast motion of the high-frequency mode from the slow one (H-bond bridge). We considered quantum theories that, starting from this basic model, have incorporated the direct and indirect relaxation mechanisms, and within the direct mechanism we have considered the role to be played by Fermi resonances. In all these theories, the intrinsic anharmonicities of the slow and fast modes were ignored; that is, the potentials describing these modes were assumed to be harmonic. It must be emphasized that all these quantum theories are deeply and satisfactorily interconnected and that all reduce to the basic Maréchal and Witkowski model when the incorporated physical terms are vanishing.

When Fermi resonances are ignored, the main conclusions are as follows:

1. The pure quantum approach of the strong anharmonic coupling theory performed by Maréchal and Witkowski [7] gives the most satisfactorily zeroth-order physical description of weak H-bond IR lineshapes.

2. The quantum approaches of direct and indirect dampings, respectively by Rösch and Ratner [58] and by Boulil et al. [90], may be considered as the well-behaved extension of the zeroth-order approach of Maréchal and Witkowski [7].

3. The Rösch and Ratner [58] lineshapes, which are obtained within the adiabatic approximation, are not modified when working beyond the adiabatic approximation, if the anharmonic coupling between the slow and fast modes has a low magnitude.

4. It is not possible for the indirect damping mechanism, as considered semiclassically by Robertson and Yarwood [84] and later quantum mechanically by Boulil et al. [90], to be the unique damping mechanism occurring in a hydrogen bond, because the quantum mechanism leads, at the opposite of the less rigorous semiclassical treatment, to a drastic collapse of the lineshapes.

5. The quantum model of direct damping by Rösch and Ratner [58] leads to the same kinds of profiles as does the semiclassical model of indirect damping by Robertson and Yarwood [84], with the two limit situations of slow or fast modulations leading to Gaussian or Lorentzian profiles.

6. The combined Boulil et al. (indirect damping) [90] and Rösch and Ratner models (direct damping) [58], and therefore that of Maréchal and Witkowski (no damping) [7], lead to satisfactory predictions concerning isotopic effects and bandwidth behavior upon temperature changes.

7. It is probable that both the direct and indirect damping mechanisms occur simultaneously.

On the other hand, when the indirect damping is ignored but Fermi resonances are taken into account, the most important conclusions may be the following:

1. The Fermi resonances ought to be very universal in the area of H bonds because the anharmonic coupling between the slow and fast modes and the Fermi anharmonic coupling between the first excited state of the fast mode and the first harmonic of some bending modes are acting in a cooperative way. Indeed, the influence of the Fermi resonances on the details of the lineshapes remains sensitive when the energy gap between the interacting levels is one order of magnitude greater than the Fermi resonance coupling parameter—that is really far from resonance.

2. The details of the lineshapes involving Fermi resonances are very sensitive to the relative magnitudes of the damping parameters characterizing respectively the first excited state of the fast mode (direct damping, $\gamma_o$) and the first harmonic of the bending mode ($\gamma_\delta$): The relative intensities of the sub-bands and their relative half-width may be turned upside down by changes in these damping parameters. Unfortunately, the theories that escape from the adiabatic and exchange approximations require the equality of the damping parameters.

3. The features of the lineshapes involving Fermi resonances are sensitive to the exchange approximation according to which are ignored all energy exchanges except those in which the first excited state of the fast mode exchanges energy with the ground state of the bending mode in such a way as the bending mode jumps on its first harmonic (and vice versa). The complete Fermi coupling, which accounts for higher harmonics of these modes, introduces noticeable changes in the infrared band structure, so that the exchange approximation should be overcome.

4. As a consequence of the two above points, there is presently no satisfactory theory able to incorporate the specific relaxation of the fast mode and of the bending modes in a model working beyond both the adiabatic and exchange approximations.

Note that the electrical anharmonicity is ignored in all the reviewed quantum approaches. Besides, although there is a recent article [12] which treats that of the H-bond bridge by aid of a Morse potential, the intrinsic anharmonicity (considered quantum mechanically) has not been treated in this review for concision reasons. Note that a very recent approach [100] is treating on equal grounds the anharmonic coupling between the slow and fast modes and the intrinsic anharmonicity (through an asymmetric double well potential) of the fast mode, capable of inducing a tunneling effect.

Let us finally quote a web site, http://hbond.univ-perp.fr/, which is dedicated to the theories of hydrogen bonding presented in this review.

## References

1. M. Haurie and A. Novak, *Spectrochim. Acta* **21**, 1217 (1965).

2. G. Auvert and Y. Maréchal, *J. Phys. (Paris)* **40**, 735 (1979).

3. J. Bournay and G. N. Robertson, *Chem. Phys. Lett.* **60**, 286 (1979).

4. J. Bournay and G. N. Robertson, *Mol. Phys.* **39**, 163 (1980).

5. G. Hofacker, Y. Maréchal, and M. Ratner, in *The Hydrogen Bond*, P. Schuster, G. Zundel, and C. Sandorfy, eds., Elsevier, Amsterdam, 1976, p. 297.

6. C. Sandorfy, in *The Hydrogen Bond*, P. Schuster, G. Zundel, and C. Sandorfy, eds., Elsevier, Amsterdam, 1976.

7. Y. Maréchal and A. Witkowski, *J. Chem. Phys.* **48**, 3697 (1968); see also Y. Maréchal, Ph.D. thesis, University of Grenoble, France (1968).

8. O. Henri-Rousseau and P. Blaise, in *Advances in Chemical Physics*, Vol. 103, S. Rice and I. Prigogine, eds., Wiley, New York, 1998, pp. 1–186.

9. S. Bratos and D. Hadži, *J. Chem. Phys.* **27**, 991 (1957).

10. J. L. Leviel and Y. Maréchal, *J. Chem. Phys.* **54**, 1104 (1971).

11. M. Wójcik, *Chem. Phys. Lett.* **46**, 597 (1977).

12. P. Blaise and O. Henri-Rousseau, *Chem. Phys.* **256**, 85 (1999).

13. O. Henri-Rousseau, D. Chamma and K. Belhayara, *J. Mol. Struct.*, to be published.

14. D. Hadži, *Pure Appl. Chem.* **11**, 435 (1965).

15. D. Hadži, *Chimia* **26**, 7 (1972).

16. N. Rösch, *Chem. Phys.* **1**, 220 (1973).

17. G. Robertson and M. Lawrence, *Chem. Phys.* **62**, 131 (1981).

18. G. Robertson and M. Lawrence, *Mol. Phys.* **43**, 193 (1981).

19. G. Robertson and M. Lawrence, *J. Chem. Phys.* **89**, 5352 (1988).

20. J. Bournay and Y. Maréchal, *Chem. Phys. Lett.* **27**, 180 (1974).

21. A. Witkowski, *J. Chem. Phys.* **47**, 3654 (1968).

22. D. Chamma and O. Henri-Rousseau, *Chem. Phys.* **248**, 53–70 (1999).

23. D. Chamma and O. Henri-Rousseau, *Chem. Phys.* **248**, 71–89 (1999).

24. A. Witkowski and M. Wójcik, *Chem. Phys. Lett.* **20**, 615–617 (1973).

25. A. Witkowski and M. Wójcik, *Chem. Phys. Lett.* **26**, 327–328 (1974).

26. M. Wójcik, *Int. J. Quant. Chem.* **10**, 747–760 (1976).

27. A. Witkowski, M. Wójcik, *Chem. Phys.* **21**, 385–391 (1977).
28. M. Wójcik, *Mol. Phys.* **36**, 1757–1767 (1978).
29. H. Flakus, *Acta Physica Pol.* **A53**, 287–296 (1978).
30. H. Flakus, *Chem. Phys.* **50**, 79–89 (1980).
31. A. Lami and G. Villani, *Chem. Phys.* **115**, 391 (1987).
32. H. Abramczyk, *Chem. Phys.* **144**, 305 (1990).
33. H. Abramczyk, *Chem. Phys.* **144**, 319 (1990).
34. N. Sokolov and M. Vener, *J. Chem. Phys.* **168**, 29 (1992).
35. A. Witkowski, *J. Chem. Phys.* **73**, 852 (1983).
36. Y. Maréchal, *J. Chem. Phys.* **83**, 247 (1985).
37. A. Witkowski, *Phys. Rev. A* **41**, 3511 (1990).
38. A. Witkowski and O. Henri-Rousseau, P. Blaise, *Acta Phys. Pol. A* **91**, 495 (1997).
39. O. Tapia, *J. Mol. Struct. Teochem.* **433**, 95 (1998).
40. C. Coulson and G. Robertson, *Proc. R. Soc. Lond. A* **337**, 167 (1974).
41. C. Coulson and G. Robertson, *Proc. R. Soc. Lond. A* **342**, 380 (1975).
42. G. Ewing, *Chem. Phys.* **29**, 253 (1978).
43. G. Ewing, *J. Chem. Phys.* **72**, 2096 (1980).
44. S. Mühle, K. Süsse, and D. Welch, *Phys. Lett. A* **66**, 25 (1978).
45. S. Mühle, K. Süsse, and D. Welch, *Ann. Phys.* **7**, 213 (1980).
46. M. M. Audibert, E. Palange, *Chem. Phys. Lett.* **101**, 407 (1983).
47. Y. Maréchal, in H. Ratajczak and W. J. Orville-Thomas, eds., Wiley, New York, 1980.
48. Y. Maréchal, in *Vibrational Spectra and Structure*, Vol. 16, J. R. Durig, ed., Elsevier, Amsterdam, 1987, pp. 311–356.
49. D. Hadži and S. Bratos, in *The Hydrogen Bond*, P. Schuster, G. Zundel, C. Sandorfy, eds., Elsevier, Amsterdam, 1976, p. 567.
50. R. Kubo, *J. Phys. Soc. Japan* **12**, 570 (1957).
51. R. Kubo, in *Lectures in Theoretical Physics* **I**, W. E. Brittin and L. G. Dunham, eds., Interscience, Boulder, CO, 1958.
52. O. Henri-Rousseau, P. Blaise, *Chem. Phys.* **250**, 249–265 (1999).
53. D. Chamma, *J. Mol. Struct.* **552**, 187 (2000).
54. R. M. Badger and S. H. Bauer, *J. Chem. Phys.* **5**, 839 (1937).
55. S. Barton and W. Thorson, *J. Chem. Phys.* **71**, 4263 (1979).
56. B. Stepanov, *Zh. Fiz. Khim.* **19**, 507 (1945).
57. P. Blaise, O. Henri-Rousseau, *Chem. Phys.* **243**, 229 (1999).
58. N. Rösch and M. Ratner, *J. Chem. Phys.* **61**, 3344 (1974).
59. L. Salasnich, *Meccanica* **33-4**, 397 (1998).
60. V. P. Karassiov, preprint Quantum-Physics number 9710012 (1997).
61. O. Henri-Rousseau and P. Blaise, in *Theoretical Treatments of Hydrogen Bonding*, D. Hadži, ed., Wiley, New York, 1997, pp. 165–186.
62. Ph. Durand and J. P. Malrieu, in *Ab Initio Methods in Quantum Chemistry*, K. P. Lawley, ed., Wiley, New York (1987) 321.
63. M. Haurie and A. Novak, *J. Chim. Phys.* **62**, 137 (1965).
64. M. Haurie and A. Novak, *J. Chim. Phys.* **62**, 146 (1965).

65. A. Hall and J. L. Wood, *Spectrochim. Acta* **23A**, 1257 (1967).

66. A. Hall and J. L. Wood, *Spectrochim. Acta* **23A**, 2657 (1967).

67. M. F. Claydon and N. Sheppard, *Chem. Commun.*, 1431 (1969).

68. A. Novak, *J. Chim. Phys.* **72**, 981 (1975).

69. R. L. Dean, F. N. Masri and J. L. Wood, *Spectrochim. Acta* **31A**, 79 (1975).

70. H. Wolff, H. Müller, and E. Wolff, *J. Chem. Phys.* **64**, 2192 (1976).

71. O. Henri-Rousseau and D. Chamma, *Chem. Phys.* **229**, 37–50 (1998).

72. D. Chamma and O. Henri-Rousseau, *Chem. Phys.* **229**, 51–73 (1998).

73. D. Chamma and O. Henri-Rousseau, *Chem. Phys.* **248**, 91–104 (1999).

74. A. Witkowski and M. Wójcik, *Chem. Phys.* **1**, 9–16 (1973).

75. Y. Maréchal, *Chem. Phys.* **79**, 69 (1983); *Chem. Phys.* **79**, 85 (1983).

76. M. Schwartz and C. H. Wang, *J. Chem. Phys.* **59**, 5258–5267 (1973).

77. J. L. McHale and C. H. Wang, *J. Chem. Phys.* **73**, 3601 (1980).

78. E. Weidemann and A. Hayd, *J. Chem. Phys.* **67**, 3713–3721 (1977).

79. K. Fujita and M. Kimura, *Mol. Phys.* **41**, 1203–1210 (1980); see also P. Piaggio, G. Dellepiane, R. Tubino and L. Piseri, *Chem. Phys.* **77**, 185–190 (1983).

80. A. Bródka, B. Stryczek, *Mol. Phys.* **54**, 677–688 (1985).

81. M. Giry, B. Boulil, and O. Henri-Rousseau, *C. R. Acad. Sci. Paris II* **316**, 455 (1993); see also B. Boulil, M. Giry, and O. Henri-Rousseau, *Phys. Stat. Sol. b* **158**, 629 (1990).

82. A. Davydov and A. A. Serikov, *Phys. Stat. Sol. (b)* **51**, 57 (1972).

83. S. Bratos, *J. Chem. Phys.* **63**, 3499 (1975).

84. G. Robertson and J. Yarwood, *Chem. Phys.* **32**, 267 (1978).

85. V. Sakun, *Chem. Phys.* **99**, 457 (1985).

86. H. Abramczyk, *Chem. Phys.* **94**, 91 (1985).

87. B. Boulil, O. Henri-Rousseau, and P. Blaise, *Chem. Phys.* **126**, 263 (1988).

88. W. Louisell and L. Walker, *Phys. Rev.* **137**, 204 (1965).

89. P. Blaise, M. Giry, and O. Henri-Rousseau, *Chem. Phys.* **159**, 169 (1992).

90. B. Boulil, J.-L. Déjardin, N. El-Ghandour, and O. Henri-Rousseau, *J. Mol. Struct. Teochem.* **314**, 83 (1994).

91. B. Boulil, P. Blaise, and O. Henri-Rousseau, *J. Mol. Struct. Teochem.* **314**, 101 (1994).

92. J. Bournay and G. Robertson, *Nature (London)* **275**, 47 (1978).

93. J. Yarwood and G. Robertson, *Nature (London)* **257**, 41 (1975).

94. J. Yarwood, R. Ackroyd and G. Robertson, *Chem. Phys.* **32**, 283 (1978).

95. S. Bratos and H. Ratajczak, *J. Chem. Phys.* **76**, 77 (1982).

96. P. Blaise, O. Henri-Rousseau, and A. Grandjean, *Chem. Phys.* **244**, 405–437 (1999).

97. R. Kubo, in *Fluctuations, Relaxation and Resonance in Magnetic Systems*, D. Ter Haar, ed., Oliver and Boyd, Edinburgh, 1961, p. 23.

98. G. Uhlenbeck and L. Ornstein, *Phys. Rev.* **36**, 823 (1930).

99. P. M. Déjardin, D. Chamma, and O. Henri-Rousseau, to be published.

100. N. Rekik, A. Velcescu, P. Blaise, and O. Henri-Rousseau, *Chem. Phys.* **273**, 11–37 (2001).

# TWO-CENTER EFFECTS IN IONIZATION BY ION-IMPACT IN HEAVY-PARTICLE COLLISIONS

S. F. C. O'ROURKE, D. M. McSHERRY, and D. S. F. CROTHERS

*Theoretical and Computational Physics Research Division, Department of Applied Mathematics and Theoretical Physics, Queen's University Belfast, Belfast, Northern Ireland*

## CONTENTS

## I. INTRODUCTION

In the past few years there has been a very intense effort to understand the process of single ionization in ion–atom collisions. Experiments in this area have followed largely along three main paths: recoil-ion momentum spectroscopy

*Advances in Chemical Physics, Volume 121*, Edited by I. Prigogine and Stuart A. Rice.
ISBN 0-471-20504-4   © 2002 John Wiley & Sons, Inc.

311

[1–5], projectile scattering detection in coincidence with the recoil-ion charge state [6–8], and ejected electron spectroscopy [9–15]. Due to the rapid advances in these techniques at this stage, it is essential to understand the complementary nature of recoil-ion momentum spectroscopy, projectile scattering measurements, and ejected electron spectroscopy. Each technique is an extremely powerful method of ionization research and each produces detailed information on the driving mechanisms of the ionization process, which can serve as a most stringent test for theory.

However, although there has been significant progress in this field of atomic collision physics, the theory of ionization is still rather incomplete. From the theoretical point of view the main problem is the representation of the final electronic state, where the ionized electron travels in the presence of two Coulomb potentials (target and projectile). Due to the long-range ionic tail of the Coulomb potential, the "free particle" cannot be represented by a plane wave. The difficulty arises from the fact that the Schrödinger equation for the three-body problem cannot be solved exactly. However, an exact asymptotic form can be obtained [16–20], and the main objective of this chapter is to review the continuum-distorted-wave (CDW) model [20,21], and the continuum-distorted-wave eikonal-initial-state (CDW-EIS) model [16–20,22] which at least satisfy the exact asymptotic conditions in the initial and final states. Both the approximations are based on distorted wave perturbation theory. The difference between CDW and CDW-EIS theory lies in the distortion of the initial state, which is a Coulomb wave in the former and an eikonal phase in the latter. Both of these theoretical approximations take into account the long-range nature of the Coulomb potential which has been incorporated into computationally efficient codes [23–25]. The main purpose of this chapter is to review the related effects on the electron yield due to the long-range nature of the Coulomb potential. These effects are usually referred to as two-center electron emission and are a direct manisfestation of the three-body Coulomb interactions. The continuum distorted-wave models account theoretically for the two-center effects, thus permitting a detailed study of the experimental results obtained from recoil-ion momentum spectroscopy, projectile scattering measurements, and ejected electron spectroscopy.

It should be noted, however, that gaining a deeper insight into the problem of ionization phenomena is not the only reason for steady interest in the problem. Data on charged particle impact ionization is used both for industrial applications and for fundamental scientific research. For applications it is the collisions rates and total cross sections which are usually the most relevant. But in studies focused on the understanding of collision mechanisms of ionization processes, most of the information is lost in the total cross sections due to the integration over the momenta of the ejected electrons in the exit channel. Therefore it is the singly and doubly differential cross sections which are of primary interest for fundamental scientific research into ion–atom collision theory. The more differential the cross section the greater the information that can be derived about the collision mechanisms of the ionization process.

We begin Section II with a theoretical description of the CDW and CDW-EIS theory to describe the single ionization of helium, neon, argon, and molecular hydrogen targets.

In Section III we discuss the applicability of these models for kinematically complete experiments on target single ionization in ion–atom collisions, which have been performed using the technique of recoil-ion momentum spectroscopy. The examples illustrated will include the pioneering experiments [2,4,5] of 3.6-MeV/amu $Ni^{24+}$, $Se^{28+}$, and $Au^{53+}$ ions on helium target atoms. These studies which focus on the calculations of cross sections singly differential in the longitudinal momentum highlight many of the important characteristics of the single ionization of atoms by fast heavy ions such as the post-collision interaction, which is an important two-center effect. Other studies discussed in this section include the recent work of Khayyat et al. [26] for 1 MeV/u proton–helium and 945 keV/u antiproton–helium collisions.

More details of the emission of ultralow- and low-energy electrons from fast heavy ion–atom collisions may be seen in the doubly differential cross sections as functions of the longitudinal electron velocity for increasing transverse electron velocity. Examples considered in this chapter include singly ionizing 3.6-MeV/amu $Au^{53+}$ ions on helium, neon, and argon targets. These studies are also discussed in Section III. They show that the prediction of ultralow- and low-energy electrons in fast ion-induced ionization using the CDW-EIS model depend on the details of the target potential.

Secondly we are interested in projectile scattering measurements because these provide useful information on the the binary encounter mechanism. In particular, we consider the projectile transfer momentum distributions for the single ionization of He by protons for 2 MeV, 3 MeV, and 6 MeV. The purpose of this specific study is to discuss the origin of the distinctive phenomenon, a shoulder in the cross sections around the proton scattering angle of 0.55 mrad. This shoulder is attributed to the dominant binary projectile–electron scattering contribution for $\theta_P \leq 0.55$ mrad [6]. This result has been interpreted using the CDW-EIS model with an inclusion of the internuclear interaction [27]. This work is reviewed in Section IV and compared with our version of the CDW-EIS model, which simply takes into account the projectile–electron interaction when calculating the projectile deflection.

Thirdly we are also interested in the electron spectroscopy method, which allows investigations on the two-center effects that influence electron emission. In particular, the richness of the ionization process lies in the possibility of measuring the doubly differential cross sections as a function of the electron emission angle and energy. This technique of electron emission spectroscopy is a very powerful tool in ionization research [13]. It has been used extensively since the pioneering work of Rudd and Jorgensen [28] and Rudd et al. [29], where they considered electrons ejected from He and $H_2$ targets by proton impact. It is from these types of measurements that the most detailed information about the driving mechanisms of the ionization process can be obtained.

The electron emission energy spectra of the double differential cross section exhibits several distinct characteristic regions arising due to the two-center effects. These regions can be identified by various mechanisms. The first is the well-known soft collision peak that occurs in the near-threshold region. This corresponds to the low-energy region ($E_k < 1$ eV) for emitted electrons [14]. This particular region gives the major contribution to the total ionization cross section and hence is important in applications where the total cross section is of main interest. The second is the electron capture to the continuum mechanism. Here the doubly differential cross section shows a cusp for small emitted angles at the electron energy corresponding to an electron velocity that matches the projectile velocity [9]. This mechanism occurs due to the strong interaction between the ejected electron and scattered projectile in the final continuum state. This two-center effect is the most impressive demonstration of the significance of the long-range three-body Coulomb interactions in the ionizing collision. Thirdly, another important characteristic feature of the doubly differential cross section is the binary encounter mechanism. This signature can be identified as a peak in the emission spectra centered at an electron velocity that is twice the projectile velocity. The binary encounter peak in this spectrum is due to the elastic scattering cross section of a free electron in the field of the projectile approximately folded with a velocity distribution determined by the Compton profile of the target atom and describes a quasi-elastic scattering of the target electrons in the projectile field. It has been suggested that there may be a fourth feature associated with the emission of the electrons through a "saddle-point mechanism." This process arises from the possibility that the ejected electron is stranded on the saddle point of the two-center potential between the residual target and receding ion. This saddle is formed by the combined Coulomb potentials of the collision partners having a minimum perpendicular to the internuclear axis and a maximum along it [30,31]. The saddle-point emission mechanism corresponds to an electron distribution centered at an electron velocity close to half the projectile velocity. This signature, however, has been the subject of some controversy [32–39]. These studies have made a major contribution to both our understanding of atomic collision systems and the evolution of quantum mechanics. In Section V we review the applicability of the CDW and CDW-EIS to predict these ionization features in the double differential cross section. The examples illustrated include the measurements from Lee et al. [12] of 1.5-MeV/amu $H^+$ and $F^{9+}$ on

He and measurements from Rudd et al. [40] for 1-MeV H$^+$ on Ar. In Section V we examine the characteristics of the saddle-point mechanism from the point of view of the CDW-EIS model. We consider the recent measurements of Nesbitt et al. [38] for 40-keV H$^+$ on He and H$_2$ targets, McGrath et al. [39] for 100 keV H$^+$ on He and H$_2$ targets, and McSherry et al. [41] for 80-keV H$^+$ on Ne. Finally we summarize our results in Section VI.

Atomic units will be used except where otherwise stated.

## II.  THEORETICAL DESCRIPTION OF THE CDW AND CDW-EIS MODELS

In the present section we present a theoretical description for the continuum-distorted-wave (CDW) model and the continuum-distorted-wave eikonal-initial-state (CDW-EIS) model. We restrict our discussion to the process of single ionization by charged particle impact for neutral target atoms ranging from hydrogen to argon. Our analysis is within the semiclassical rectilinear impact parameter ($\boldsymbol{\rho}$), time-dependent ($t$) formalism. We consider the problem of three charged particles where an ion of nuclear charge $Z_P$ and mass $M_P$ impinges with a collision velocity $\mathbf{v}$ on a neutral target atom with nuclear charge $Z_T$ and mass $M_T$. As $M_{T,P} \gg 1$, the motion of the nuclei can be uncoupled from that of the electron [42]. The trajectory of the projectile is then characterized by two parameters $\boldsymbol{\rho}$ and the impact velocity $\mathbf{v}$ such that $\boldsymbol{\rho} \cdot \mathbf{v} = \mathbf{0}$. The internuclear coordinate is defined by

$$\mathbf{R} = \mathbf{r}_T - \mathbf{r}_P = \boldsymbol{\rho} + \mathbf{v}t \tag{1}$$

and

$$\mathbf{r} = \frac{1}{2}(\mathbf{r}_T + \mathbf{r}_P) \tag{2}$$

where $\mathbf{r}_T$, $\mathbf{r}_P$, and $\mathbf{r}$ are the position vectors of the electron relative to the target nucleus, projectile nucleus, and their midpoint, respectively. The transition amplitude in post form may be written as

$$a_{if}(\boldsymbol{\rho}) = -i \int_{-\infty}^{\infty} dt \langle \chi_i^+ | \left( H_e - i\frac{d}{dt_{\mathbf{r}}} \right) | \chi_f^- \rangle \tag{3}$$

where

$$H_e = -\frac{1}{2}\nabla_{\mathbf{r}}^2 - \frac{Z_T}{r_T} - \frac{Z_P}{r_P} \tag{4}$$

the internuclear potential having been removed by a phase transformation [43]. We adopt an independent electron model to approximate the neutral target atom; therefore, as in any independent electron model, no explicit electron correlation in the initial state is considered. As such, the electronic Hamiltonian for the projectile/neutral target atom collision system is modified from the original monoelectronic CDW/CDW-EIS approximation, in which the electron residual target interaction $-Z_T/r_T$ is replaced by a Coulombic potential with an effective charge $-\tilde{Z}_T$. This we do as in Belkič et al. [44] by making the assumption that the emitted electron, ionized from an orbital of Roothan–Hartree–Fock energy $\varepsilon_i$, moves in a residual target potential of the form

$$V_T(\mathbf{r}_T) = -\frac{\tilde{Z}_T}{r_T} \tag{5}$$

The effective target charge is given by

$$\tilde{Z}_T = \sqrt{-2n^2\varepsilon_i} \tag{6}$$

where $\varepsilon_i$ is the binding energy of the neutral target atom, and $n$ is the principal quantum number. The initial and final CDW wavefunctions may be expressed explicitly as

$$|\chi_i^+\rangle = \varphi_i(\mathbf{r}_T)\exp\left(-\frac{1}{2}i\mathbf{v}\cdot\mathbf{r} - \frac{1}{8}iv^2t - i\varepsilon_it\right)N(v)\,_1F_1(iv; 1; ivr_P + i\mathbf{v}\cdot\mathbf{r}_P) \tag{7}$$

$$|\chi_f^-\rangle = (2\pi)^{-3/2}N^*(\varepsilon)N^*(\zeta)\exp\left(i\mathbf{k}\cdot\mathbf{r}_T - \frac{1}{2}i\mathbf{v}\cdot\mathbf{r} - \frac{1}{8}iv^2t - iE_kt\right)$$
$$\times \,_1F_1(-i\varepsilon; 1; -ikr_T - i\mathbf{k}\cdot\mathbf{r}_T)\,_1F_1(-i\zeta; 1; -ipr_P - i\mathbf{p}\cdot\mathbf{r}_P) \tag{8}$$

and the initial and final CDW-EIS wave functions are defined as

$$|\chi_i^+\rangle = \varphi_i(\mathbf{r}_T)\exp\left(-\frac{1}{2}i\mathbf{v}\cdot\mathbf{r} - \frac{1}{8}iv^2t - i\varepsilon_it\right)\exp(-iv\ln(vr_P + \mathbf{v}\cdot\mathbf{r}_P)) \tag{9}$$

$$|\chi_f^-\rangle = (2\pi)^{-3/2}N^*(\varepsilon)N^*(\zeta)\exp\left(i\mathbf{k}\cdot\mathbf{r}_T - \frac{1}{2}i\mathbf{v}\cdot\mathbf{r} - \frac{1}{8}iv^2t - iE_kt\right)$$
$$\times \,_1F_1(-i\varepsilon; 1; -ikr_T - i\mathbf{k}\cdot\mathbf{r}_T)\,_1F_1(-i\zeta; 1; -ipr_P - i\mathbf{p}\cdot\mathbf{r}_P) \tag{10}$$

Here

$$E_k = \frac{1}{2}k^2 \tag{11}$$

is the electron energy in the final continuum state. The momentum $\mathbf{p}$ of the ejected electron relative to the projectile is given by

$$\mathbf{p} = \mathbf{k} - \mathbf{v} \tag{12}$$

where $\mathbf{k}$ is the momentum of the ejected electron with respect to a reference frame fixed on the target nucleus. We note that the polar axis for reference is taken along the incident beam direction so that

$$d\mathbf{k} = k^2 dk \, \sin\theta \, d\theta \, d\phi \tag{13}$$

The spherical coordinates of the ejected electron momentum $\mathbf{k}$ are $k, \theta$, and $\phi$, where $\theta = \cos^{-1}(\hat{\mathbf{k}} \cdot \hat{\mathbf{v}})$. Because the impact velocity lies along the $Z$ axis, then $\mathbf{v} = v\,\hat{\mathbf{Z}}$. The three Sommerfeld parameters are defined by

$$\varepsilon = \frac{\tilde{Z}_T}{k} \tag{14}$$

$$v = \frac{Z_P}{v} \tag{15}$$

$$\zeta = \frac{Z_P}{p} \tag{16}$$

and

$$N(a) = \exp\left(\frac{\pi a}{2}\right)\Gamma(1 - ia) \tag{17}$$

represents the Coulomb density of states factor. Equations (7) and (9) represent the first Born initial state distorted by a CDW/eikonal phase respectively, while Eqs. (8) and (10) model the electron moving within the continuum of both the projectile and target. In the CDW/CDW-EIS approximation the residual post-interaction is the nonorthogonal kinetic energy term,

$$\left(H_e - i\frac{d}{dt_{\mathbf{r}}}\right)|\chi_f^-\rangle = [-\nabla_{\mathbf{r}_T} \ln {}_1F_1(-i\varepsilon; 1; -ikr_T - i\mathbf{k}\cdot\mathbf{r}_T) \cdot \nabla_{\mathbf{r}_P} \ln {}_1F_1$$
$$\times (-i\zeta; 1; -ipr_P - i\mathbf{p}\cdot\mathbf{r}_P)]|\chi_f^-\rangle \tag{18}$$

The active electron bound state $\varphi_i(\mathbf{r}_T)$ satisfies the Schrödinger equation

$$\left(\frac{1}{2}\nabla_{\mathbf{r}_T}^2 + \frac{\tilde{Z}_T}{r_T} + \varepsilon_i\right)|\varphi_i(\mathbf{r}_T)\rangle = 0 \tag{19}$$

The ground state $\varphi_i(\mathbf{r}_T)$ of the neutral target atom is represented by a Roothan–Hartree–Fock wavefunction that may be expanded in terms of Slater orbitals [45]. Denoting the triply differential cross section by

$$\frac{d^3\sigma(k)}{d\mathbf{k}} = \alpha(\mathbf{k}) = \int d\boldsymbol{\rho}|a_{if}(\boldsymbol{\rho})|^2 \tag{20}$$

with

$$a_{if}(\boldsymbol{\rho}) = i(\rho\upsilon)^{2i\tilde{Z}_T Z_P/\upsilon}\tilde{a}_{if}(\boldsymbol{\rho}) \tag{21}$$

we obtain for the CDW model

$$\tilde{a}_{if}(\boldsymbol{\rho}) = -i\frac{N(\varepsilon)N(\upsilon)N(\zeta)}{(2\pi)^{3/2}} \int_{-\infty}^{\infty} dt \int d\mathbf{r}\varphi_i(\mathbf{r}_T)\exp(it\Delta\epsilon - i\mathbf{k}\cdot\mathbf{r}_T)$$
$$\times {}_1F_1(i\upsilon; 1; i\upsilon r_P + i\mathbf{v}\cdot\mathbf{r}_P)[\nabla_{\mathbf{r}_T}{}_1F_1(i\varepsilon; 1; ikr_T + i\mathbf{k}\cdot\mathbf{r}_T)\cdot$$
$$\nabla_{\mathbf{r}_P}{}_1F_1(i\zeta; 1; ipr_P + i\mathbf{p}\cdot\mathbf{r}_P)] \tag{22}$$

and for the CDW-EIS model

$$\tilde{a}_{if}(\boldsymbol{\rho}) = -i\frac{N(\varepsilon)N(\zeta)}{(2\pi)^{3/2}} \int_{-\infty}^{\infty} dt \int d\mathbf{r}\,\varphi_i(\mathbf{r}_T)\exp(it\Delta\epsilon - i\mathbf{k}\cdot\mathbf{r}_T)$$
$$\times \exp(-i\upsilon\ln(\upsilon r_P + \mathbf{v}\cdot\mathbf{r}_P))[\nabla_{\mathbf{r}_{Tt}}F_1(i\varepsilon; 1; ikr_T + i\mathbf{k}\cdot\mathbf{r}_T)\cdot$$
$$\nabla_{\mathbf{r}_P}{}_1F_1(i\zeta; 1; ipr_P + i\mathbf{p}\cdot\mathbf{r}_P)] \tag{23}$$

where $\Delta\epsilon = E_k - \varepsilon_i$. Instead of $\tilde{a}_{if}(\boldsymbol{\rho})$, it is easier to calculate its two-dimensional Fourier transform $R_{if}(\boldsymbol{\eta})$ as a function of the transverse heavy-particle relative momentum transfer $\boldsymbol{\eta}$; thus the scattering amplitude is

$$R_{if}(\boldsymbol{\eta}) = \frac{1}{2\pi} \int d\boldsymbol{\rho}\exp(i\boldsymbol{\eta}\cdot\boldsymbol{\rho})\tilde{a}_{if}(\boldsymbol{\rho}) \tag{24}$$

where $\boldsymbol{\eta}\cdot\mathbf{v} = 0$. Then application of Parseval's theorem [46] gives the triple differential cross section, which can be written in the form

$$\alpha(\mathbf{k}) = 2\pi a_0^2 \int d\boldsymbol{\eta}|R_{if}(\boldsymbol{\eta})|^2 \tag{25}$$

Closed analytical expressions for the squares of the moduli of the CDW/CDW-EIS scattering amplitudes $|R_{if}(\boldsymbol{\eta})|^2$ are given in the next section for neutral target atoms ranging from hydrogen up to and including argon. From the scattering amplitude $R_{if}(\mathbf{\eta})$ as a function of the transverse momentum $\mathbf{\eta}$ we may obtain the probability that for a certain fixed value of $\mathbf{\eta}$, the electron initially in a bound state of the target will be emitted to a continuum state with momentum $\mathbf{k}$. The integration over $\phi$ gives the double differential cross section as a function of the ejected electron energy $E_k$ and angle $\theta$:

$$\frac{d^2\sigma}{dE_K d(\cos\theta)} = k \int_0^{2\pi} \alpha(\mathbf{k})\, d\phi \tag{26}$$

By further integrations over the energy or angle of the emitted electron we obtain the single differential cross section as a function of the angle and energy of the emitted electron, respectively:

$$\frac{d\sigma}{d(\cos\theta)} = \int_0^\infty k^2\, dk \int_0^{2\pi} \alpha(\mathbf{k})\, d\phi \tag{27}$$

$$\frac{d\sigma}{dE_k} = \int_0^\pi k\sin\theta\, d\theta \int_0^{2\pi} \alpha(\mathbf{k})\, d\phi \tag{28}$$

Finally the total cross section may be calculated by performing the last integration, in Eq. (27) over $d(\cos\theta)$ or the last integration in Eq. (28) over $dE_k$; that is, we obtain

$$\sigma = \int_0^\infty k\, dk \int_0^\pi k\,\sin\theta\, d\theta \int_0^{2\pi} \alpha(\mathbf{k})\, d\phi \tag{29}$$

### A.   CDW and CDW-EIS Scattering Amplitudes

In this section we consider a generalization of the original CDW-EIS model, which was originally designed to calculate the single ionization from a $1s$ orbital for the monoelectronic case of hydrogen to consider single ionization of any multielectronic atom ranging from helium up to and including argon by charged particle impact. The scattering amplitudes considered here are obtained in closed analytical form. The advantage of our analytical method is that the extension to multielectronic targets in the frozen core approximation is straightforward, and computation of the various cross sections is very fast. Numerical models of the CDW-EIS theory, requiring greater computational effort than the analytical model, have been considered by Gulyás et al. [18] using Hartree–Fock–Slater target potentials and more recently by Gulyás et al. [22] using target potentials obtained from the optimized potential method of Engel and Vosko [47].

The validity of our analytical model for the multielectronic targets lies in the description of the bound and continuum states of the target atom. The analytical form of the scattering amplitudes are derived by representing the initial-target bound state by a linear combination of Slater-type orbitals. The coefficients of these expansions are obtained using the tables of Clementi and Roetti [45]. The continuum states of the target atom are represented by a hydrogenic wavefunction with effective charge chosen from the energy of the initial bound state. This of course means that the the initial and final states correspond to different potentials and hence are not orthogonal. Let us first consider the single ionization of the argon atom by ion impact. The electronic configuration of argon is $1s^2 \, 2s^2 \, 2p^6 \, 3s^2 \, 3p^6$. Thus to provide a description of the target wavefunctions for all the atoms ranging from the monoelectronic case of hydrogen to the multielectronic atoms from helium to argon, we first require the post form of the CDW/CDW-EIS scattering amplitudes for the $1s$, $2s$, and $3s$ orbitals in the K, L, and M shells (corresponding to the values $n = 1, 2$, and 3 of the principal quantum number, respectively). Secondly, we require post forms of the CDW/CDW-EIS scattering amplitudes for the $2p$ and $3p$ orbitals in the L and M shells (corresponding to the principal quantum numbers $n = 2$ and 3, respectively). The post form of the square of the CDW/CDW-EIS scattering amplitudes for the $1s$, $2s$, and $3s$ orbitals is given by

$$|R_{if}(\boldsymbol{\eta})|^2 = \frac{|N(\varepsilon)N(\zeta)N(v)|^2}{2\pi^2\alpha^2\gamma^2v^2} Z_P^2\tilde{Z}_T^2 A|^s A_I \, {}_2F_1(iv; i\zeta; 1; \tau)$$
$$- iv^s A_{II} \, {}_2F_1(1v + 1; i\zeta + 1; 2; \tau)|^2 \tag{30}$$

where ${}^s A_I$ and ${}^s A_{II}$ have different values depending on the atom selected. For the case of atoms from sodium through argon

$$^s A_I = \sum_{\lambda=1} b_\lambda Z_\lambda^{3/2} G_\lambda B_\lambda + \sum_{\lambda=2}^{8} \frac{2}{\sqrt{90}} b_\lambda Z_\lambda^{7/2} G_\lambda (A_\lambda B_\lambda + iL_\lambda \hat{\mathbf{k}} \cdot \mathbf{q}) \tag{31}$$

and

$$^s A_{II} = \sum_{\lambda=1} b_\lambda Z_\lambda^{3/2} G_\lambda B_\lambda \Omega_\lambda + \sum_{\lambda=2}^{8} \frac{2}{\sqrt{90}} b_\lambda Z_\lambda^{7/2} G_\lambda (A_\lambda B_\lambda \Omega_\lambda + i\Omega_\lambda L_\lambda \mathbf{q} \cdot \hat{\mathbf{k}}) \tag{32}$$

where

$$A_\lambda = \frac{2(i\varepsilon - 1)(-1 - i\varepsilon)Z_\lambda(Z_\lambda - ik)}{\alpha_\lambda(\alpha_\lambda + \beta_\lambda)} + \frac{(i\varepsilon - 2)(i\varepsilon - 1)Z_\lambda^2}{\alpha_\lambda^2} + \frac{(i\varepsilon - 1)}{\alpha_\lambda}$$
$$+ \frac{(-1 - i\varepsilon)}{(\alpha_\lambda + \beta_\lambda)} + \frac{(-1 - i\varepsilon)(-2 - i\varepsilon)(Z_\lambda - ik)^2}{(\alpha_\lambda + \beta_\lambda)^2} \tag{33}$$

and

$$L_\lambda = \frac{2Z_\lambda(i\varepsilon - 1)}{\alpha_\lambda} + \frac{2(-1 - i\varepsilon)(Z_\lambda - ik)}{\alpha_\lambda + \beta_\lambda} \tag{34}$$

Quantities in the expression common to both the CDW and the CDW-EIS versions arising from the fact that the CDW and CDW-EIS model have the same final continuum target state are

$$\mathbf{q} = -\boldsymbol{\eta} - \frac{\Delta\epsilon}{\upsilon}\hat{\mathbf{v}} \tag{35}$$

$$G_\lambda = \frac{1}{\alpha_\lambda^2}\left(\frac{\alpha_\lambda}{\alpha_\lambda + \beta_\lambda}\right)^{1+i\varepsilon} \tag{36}$$

$$B_\lambda = \mathbf{q} \cdot (i\hat{\mathbf{k}}Z_\lambda + \mathbf{k} + \mathbf{q}) \tag{37}$$

$$C_\lambda = (\hat{\mathbf{p}}\upsilon - \mathbf{v}) \cdot (i\hat{\mathbf{k}}Z_\lambda + \mathbf{k} + \mathbf{q}) \tag{38}$$

$$\alpha_\lambda = \frac{1}{2}[Z_\lambda^2 + (\mathbf{q} + \mathbf{k})^2] \tag{39}$$

$$\beta_\lambda = -[\mathbf{q} \cdot \mathbf{k} + k^2(1 + i\varepsilon_\lambda)] \tag{40}$$

$$\varepsilon_\lambda = \frac{Z_\lambda}{k} \tag{41}$$

$$\alpha = \frac{1}{2}q^2 \tag{42}$$

$$\beta = \mathbf{q} \cdot \mathbf{v} \tag{43}$$

$$\gamma = \mathbf{p} \cdot \mathbf{q} + \alpha \tag{44}$$

$$\delta = \mathbf{p} \cdot \mathbf{v} - p\upsilon + \beta \tag{45}$$

The coefficients $Z_\lambda$ and $b_\lambda$ in the above expressions come from the description of the Roothan–Hartree–Fock wavefunctions of the particular atom which may range from hydrogen to argon inclusively [45]. The quantities $A$, $\Omega_\lambda$, and $\tau$ are common to both the $p$ shell and the $s$ shell in all atoms, yet they have specific values for the two theories. For the CDW model we have

$$A = \begin{cases} 1 & \text{if} \quad q^2 - 2\Delta\epsilon > 0 \\ \exp(-2\pi v) & \text{if} \quad q^2 - 2\Delta\epsilon < 0 \end{cases} \tag{46}$$

$$\Omega_\lambda = \frac{\alpha}{\gamma B_\lambda}\frac{(\gamma C_\lambda + \delta B_\lambda)}{(\alpha + \beta)} \tag{47}$$

$$\tau = \frac{\beta\gamma - \alpha\delta}{\gamma(\alpha + \beta)} \tag{48}$$

and for the CDW-EIS model

$$A = \exp(-2\pi v) \tag{49}$$

$$\Omega_\lambda = \frac{\alpha}{\beta\gamma B_\lambda}(\gamma C_\lambda + \delta B_\lambda) \tag{50}$$

$$\tau = 1 - \frac{\alpha\delta}{\beta\gamma} \tag{51}$$

Expressions for $^sA_I$ and $^sA_{II}$ for ionization from the $1s$ and $2s$ states of target atoms ranging from lithium to neon are given by

$$^sA_I = \sum_{\lambda=1}^{2} b_\lambda Z_\lambda^{3/2} G_\lambda B_\lambda + \sum_{\lambda=3}^{6} \frac{1}{\sqrt{3}} b_\lambda Z_\lambda^{5/2} G_\lambda (A_\lambda B_\lambda - i\hat{\mathbf{k}} \cdot \mathbf{q}) \tag{52}$$

$$^sA_{II} = \sum_{\lambda=1}^{2} b_\lambda Z_\lambda^{3/2} G_\lambda B_\lambda \Omega_\lambda + \sum_{\lambda=3}^{6} \frac{1}{\sqrt{3}} b_\lambda Z_\lambda^{5/2} G_\lambda (A_\lambda B_\lambda \Omega_\lambda - i\Omega_\lambda \mathbf{q} \cdot \hat{\mathbf{k}}) \tag{53}$$

and

$$A_\lambda = \frac{(1 - i\varepsilon)Z_\lambda}{\alpha_\lambda} + \frac{(1 + i\varepsilon)(Z_\lambda - ik)}{\alpha_\lambda + \beta_\lambda} \tag{54}$$

For ionization from the $1s$ states of helium the value of $^sA_I$ and $^sA_{II}$ are given by

$$^sA_I = \sum_{\lambda=1}^{5} b_\lambda Z_\lambda^{3/2} G_\lambda B_\lambda \tag{55}$$

and

$$^sA_{II} = \sum_{\lambda=1}^{5} b_\lambda Z_\lambda^{3/2} G_\lambda B_\lambda \Omega_\lambda \tag{56}$$

The values of $^sA_I$ and $^sA_{II}$ for hydrogen can be found from this by taking the value of $Z_\lambda = \tilde{Z}_T = 1 = Z_T$ and $b_\lambda = 1$. Because our theoretical treatment is based on the independent-electron model, we can easily obtain an analytical expression for the molecular hydrogen wavefunction. This may be derived by simplifying the $H_2$ target to an effective one-electron hydrogenic target with charge $\tilde{Z}_T = 1.064$, where $\frac{1}{2}\tilde{Z}_T^2$ is the single ionization of $H_2$. This approximation is valid for intermediate-to-high energy collisions where the cross sections for the $H_2$ target are essentially similar to twice the atomic-hydrogen cross sections.

Next turning to the $2p$ and $3p$ Roothan–Hartree–Fock orbitals, the post form of the square of the CDW/CDW-EIS scattering amplitudes may be given as

$$|R_{if}(\boldsymbol{\eta})|^2 = \frac{|N(\varepsilon)N(\zeta)N(\nu)|^2}{2\pi^2 v^2 \alpha^2 \gamma^2} Z_P^2 \tilde{Z}_T^2 A \times |{}^P A_I\, {}_2F_1(i\nu;i\zeta;1;\tau)$$
$$- i\nu\,{}^P A_{II}\, {}_2F_1(i\nu + 1; i\zeta + 1; 2; \tau)|^2 \tag{57}$$

Again as in the $s$ shells, ${}^P A_I$ and ${}^P A_{II}$ have different values depending on the atom selected. Thus to calculate the various ionization cross sections, the contributions from the $2p$ and $3p$ spectra depend on the values of ${}^P A_I$ and ${}^P A_{II}$, which are given by

$$
{}^P A_I = \sum_{\lambda=1} b_\lambda Z_\lambda^{3/2} G_\lambda [A_\lambda^{\hat{\mathbf{n}}} B_\lambda - Z_\lambda \hat{\mathbf{n}} \cdot \mathbf{q}] + \sum_{\lambda=2}^{8} \frac{b_\lambda Z_\lambda^{3/2}}{\sqrt{105}} \left[ \left( -\frac{d^3}{d\mu^3}(G_\lambda) \bigg|_{\mu=0} \right) B_\lambda \right.
$$
$$
\left. + 3\left( \frac{d^2}{d\mu^2}(G_\lambda) \bigg|_{\mu=0} \right) Z_\lambda \hat{\mathbf{n}} \cdot \mathbf{q} \right] \tag{58}
$$

$$
{}^P A_{II} = \sum_{\lambda=1} b_\lambda Z_\lambda^{3/2} G_\lambda [A_\lambda^{\hat{\mathbf{n}}} B_\lambda \Omega_\lambda - Z_\lambda \hat{\mathbf{n}} \cdot \mathbf{q}\Omega_\lambda] + \sum_{\lambda=2}^{8} \frac{b_\lambda Z_\lambda^{3/2}}{\sqrt{105}} \left[ \left( -\frac{d^3}{d\mu^3}(G_\lambda) \bigg|_{\mu=0} \right) B_\lambda \Omega_\lambda \right.
$$
$$
\left. + 3\left( \frac{d^2}{d\mu^2}(G_\lambda) \bigg|_{\mu=0} \right) Z_\lambda \hat{\mathbf{n}} \cdot \mathbf{q}\Omega_\lambda \right] \tag{59}
$$

where

$$
A_\lambda^{\hat{\mathbf{n}}} = \frac{(1 - i\varepsilon)}{\alpha_\lambda} Z_\lambda \hat{\mathbf{n}} \cdot (\mathbf{q} + \mathbf{k}) + \frac{(1 + i\varepsilon)}{(\alpha_\lambda + \beta_\lambda)} Z_\lambda \hat{\mathbf{n}} \cdot \mathbf{q} \tag{60}
$$

with

$$
\hat{\mathbf{n}} = \begin{cases} \mathbf{i} & \text{for } 2p_x \text{ state} \\ \mathbf{j} & \text{for } 2p_y \text{ state} \\ \mathbf{k} & \text{for } 2p_z \text{ state} \end{cases} \tag{61}
$$

Here

$$
\frac{d^2}{d\mu^2}(G_\lambda) \bigg|_{\mu=0} = \alpha_\lambda^{i\varepsilon-1}(\alpha_\lambda + \beta_\lambda)^{-1-i\varepsilon} \left[ \frac{(1 - i\varepsilon)(2 - i\varepsilon)}{\alpha_\lambda^2} (\hat{\mathbf{n}} \cdot \mathbf{q} + \hat{\mathbf{n}} \cdot \mathbf{k})^2 Z_\lambda^2 \right.
$$
$$
+ \frac{(i\varepsilon - 1)Z_\lambda^2}{\alpha_\lambda} + \frac{2(1 - i\varepsilon)(1 + i\varepsilon)Z_\lambda^2(\hat{\mathbf{n}} \cdot \mathbf{q} + \hat{\mathbf{n}} \cdot \mathbf{k})\hat{\mathbf{n}} \cdot \mathbf{q}}{\alpha_\lambda(\alpha_\lambda + \beta_\lambda)}
$$
$$
\left. + \frac{(1 + i\varepsilon)(-Z_\lambda^2)}{(\alpha_\lambda + \beta_\lambda)} + \frac{(1 + i\varepsilon)(2 + i\varepsilon)Z_\lambda^2(\hat{\mathbf{n}} \cdot \mathbf{q})^2}{(\alpha_\lambda + \beta_\lambda)^2} \right] \tag{62}
$$

and

$$
\begin{aligned}
\frac{d^3}{d\mu^3}(G_\lambda)\Big|_{\mu=0} = {}& \alpha_\lambda^{i\varepsilon-1}(\alpha_\lambda+\beta_\lambda)^{-1-i\varepsilon} Z_\lambda^3 \Bigg[ \frac{1}{\alpha_\lambda^3}(\hat{\mathbf{n}}\cdot\mathbf{q}+\hat{\mathbf{n}}\cdot\mathbf{k})^3(1-i\varepsilon)(2-i\varepsilon) \\
& \times (3-i\varepsilon) + \frac{1}{(\alpha_\lambda+\beta_\lambda)^3}(\hat{\mathbf{n}}\cdot\mathbf{q})^3(1+i\varepsilon)(2+i\varepsilon)(3+i\varepsilon) \\
& + \frac{1}{\alpha_\lambda^2}(\hat{\mathbf{n}}\cdot\mathbf{q}+\hat{\mathbf{n}}\cdot\mathbf{k})(-3(1-i\varepsilon)(2-i\varepsilon)) \\
& + \frac{1}{(\alpha_\lambda+\beta_\lambda)^2}(\hat{\mathbf{n}}\cdot\mathbf{q})(-3(1+i\varepsilon)(2+i\varepsilon)) \\
& + \frac{1}{\alpha_\lambda^2(\alpha_\lambda+\beta_\lambda)}(\hat{\mathbf{n}}\cdot\mathbf{q}+\hat{\mathbf{n}}\cdot\mathbf{k})^2(\hat{\mathbf{n}}\cdot\mathbf{q})(3(1-i\varepsilon)(2-i\varepsilon)(1+i\varepsilon)) \\
& + \frac{1}{\alpha_\lambda(\alpha_\lambda+\beta_\lambda)}(\hat{\mathbf{n}}\cdot\mathbf{q})(-3(1-i\varepsilon)(1+i\varepsilon)) \\
& + \frac{1}{\alpha_\lambda(\alpha_\lambda+\beta_\lambda)}(\hat{\mathbf{n}}\cdot\mathbf{q}+\hat{\mathbf{n}}\cdot\mathbf{k})(-3(1-i\varepsilon)(1+i\varepsilon)) \\
& + \frac{1}{\alpha_\lambda(\alpha_\lambda+\beta_\lambda)^2}(\hat{\mathbf{n}}\cdot\mathbf{q}+\hat{\mathbf{n}}\cdot\mathbf{k})(\hat{\mathbf{n}}\cdot\mathbf{q})^2(3(1-i\varepsilon)(1+i\varepsilon)(2+i\varepsilon)) \Bigg]
\end{aligned}
\tag{63}
$$

Partial contributions to the cross sections for atoms from boron to neon from the outer $L$ shell depend on the values of $^pA_I$ and $^pA_{II}$, which are given by

$$
^pA_I = \sum_{\lambda=1}^{4} b_\lambda Z_\lambda^{3/2} G_\lambda [A_\lambda^{\hat{\mathbf{n}}} B_\lambda - Z_\lambda \hat{\mathbf{n}}\cdot\mathbf{q}]
\tag{64}
$$

and

$$
^pA_{II} = \sum_{\lambda=1}^{4} b_\lambda Z_\lambda^{3/2} G_\lambda [A_\lambda^{\hat{\mathbf{n}}} B_\lambda \Omega_\lambda - Z_\lambda \hat{\mathbf{n}}\cdot\mathbf{q}\Omega_\lambda]
\tag{65}
$$

where $A_\lambda^{\hat{\mathbf{n}}}$ and $\hat{\mathbf{n}}$ are defined as above in Eqs. (60) and (61), respectively.

## III.   RECOIL-ION MOMENTUM SPECTROSCOPY

### A.   Kinematics: Longitudinal Momentum Balance

In this section we consider the final-state longitudinal momentum distributions of the ejected electron, recoil-ion, and projectile momentum transfer in the single ionization of helium by swift ion impact. We closely follow the notation of Rodriguez et al. [48]. By using the energy and momentum conservation laws for single target ionization of atoms by ion-impact, we can derive a relation between the conventional measurement of energy and angular distributions and the final-state momentum distributions of the electron, recoil-ion, and projectile. We denote $p_{e\parallel}$ and $p_{R\parallel}$, respectively, as the longitudinal momenta of the emitted electron and recoiling ion. The longitudinal projectile momentum transfer is denoted by $p_{P\parallel}$. Then in the laboratory frame the longitudinal conservation equation satisfies

$$p_{R\parallel} = p_{P\parallel} - p_{e\parallel} = \frac{E_k - \varepsilon_i}{v} - k\cos\theta \qquad (66)$$

where

$$p_{e\parallel}^2 + p_{e\perp}^2 = k^2 \qquad (67)$$

Equation (66) is valid provided that:

1. The mass $M_P$ of the projectile ion is much heavier than the electrons ($M_P \gg 1$).
2. The initial collision energy is much larger than $E_k - \varepsilon_i (M_P v^2/2 \gg E_k - \varepsilon_i)$.
3. The projectile scattering angle $\Theta$ is small ($\Theta \ll 1$).

Normally these conditions are satisfied in fast highly charged ion–atom collisions. From Eq. (66) we can derive the equations for the singly differential cross sections with respect to the components of the longitudinal momentum distributions for the electron, recoil-ion, and projectile. The longitudinal electron momentum distribution $d\sigma/dp_{e\parallel}$ for a particular value of $p_{e\parallel}$ may be derived by integrating over the doubly differential cross section with respect to the electron energy $E_k$:

$$\frac{d\sigma}{dp_{e\parallel}} = \int_{p_{e\parallel}/2}^{\infty} \frac{1}{k}\frac{d^2\sigma}{dE_k d(\cos\theta)}\, dE_k \qquad (68)$$

This integral has a singularity at $p_{e\parallel} \to v$ in the region of electron capture to the continuum (ECC). Similarly for a particular value of $p_{R\parallel}$ the longitudinal recoil momentum distribution is given by

$$\frac{d\sigma}{dp_{R\parallel}} = \int_{E_k^-}^{E_k^+} \frac{1}{k}\frac{d^2\sigma}{dE_k d(\cos\theta)}\, dE_k \qquad (69)$$

where the integration limits $E_k^{\pm} = \frac{1}{2}(k^{\pm})^2$ with

$$k^{\pm} = v\cos\theta \pm \sqrt{v^2\cos^2\theta + 2(p_{R\parallel}v - |\epsilon_i|)} \qquad (70)$$

are determined from Eq. (66). Here $k^+$ corresponds to $\theta = 0$ and $|k^-|$ corresponds to $\theta = \pi$.

The longitudinal momentum projectile transfer $d\sigma/dp_{P\parallel}$ obtained by consideration of Eq. (66) is expressed as a function of the singly differential cross section of the emitted electron from the projectile ion by

$$\frac{d\sigma}{dp_{P\parallel}} = v\frac{d\sigma}{dE_k} \qquad (71)$$

Consideration of Eq. (66) indicates that the initial value of the momentum spectrum in Eq. (71) occurs at $p_{P\parallel} = |\epsilon_i|/v$. The quantity $d\sigma/dp_{P\parallel}$ reflects the inelastic energy gain or loss of the collision process.

### 1.   Comparison Between Theory and Experiment

The longitudinal momentum balances for helium single ionization by 3.6-MeV/ amu $Ni^{24+}$, $Se^{28+}$, and $Au^{53+}$ are shown in Figs. 1 to 7. Several of the main features characterizing the dynamics of target single ionization can be deduced from this representation. First the forward–backward asymmetry of the electron and recoil-ion which according to theory (CDW, CDW-EIS and the classical trajectory Monte Carlo (CTMC) model) is a result of the long-range interaction with the receding projectile—that is, the post-collision interaction (PCI) effect. The PCI effect is sometimes referred to as two-center emission. That the PCI effect in fast heavy-ion collisions depends on the sign of the projectile charge may be seen from CTMC calculations by Moshammer et al. [2] carried out for imaginary anti-nickel helium collisions and by CDW-EIS calculations carried out by O'Rourke et al. [49] for imaginary anti-selenium helium collisions where the data are mirror images of the $Ni^{24+}$ and $Se^{28+}$ helium collisions, respectively. In particular, these calculations show that the emission characteristic of the electron and recoil-ion reflects the sign and strength of the projectile ion. Secondly, it may be seen from Figs. 1, 2, 4, and 5 that the electron and recoil-ion are preferably emitted back to back, balancing their momenta on a level that corresponds to the small momentum transferred by the heavy projectile as illustrated in Figs. 3 and 6. In Figs. 3 and 6, to make a realistic comparison with experiment for the longitudinal projectile momentum transfer, we convoluted the theoretical CDW/CDW-EIS models with a Gaussian distribution that represents the experimental resolution. Here we have only considered one width of the distribution. This is 0.22 a.u., which corresponds to the upper estimate limit of the full half-width maximum (FHWM) of the experimental resolution of 72 eV.

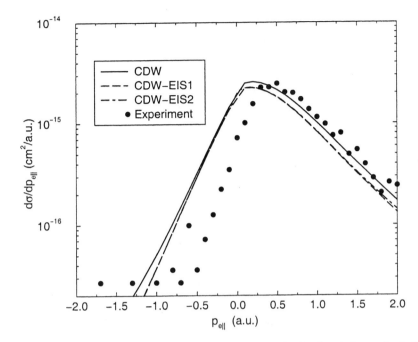

**Figure 1.** Calculated longitudinal momentum distributions for the ejected electron in single ionization of He by 3.6-MeV/amu $Ni^{24+}$ ions. Experimental data are from Moshammer et al. [2]. Theoretical results: CDW results [20], CDW-EIS1 results [20], CDW-EIS2 results [48].

Here the CDW model is in better agreement with experiment than is the CDW-EIS model. The peak of the experiment and the CDW model coincide for $Ni^{24+}$ and $Se^{28+}$ at widths of 0.19 a.u. [20] and 0.1875 a.u. [20], respectively. For the CDW-EIS model, agreement with the experimental peak occurs at widths of 0.16 a.u. for both $Ni^{24+}$ [20] and $Se^{28+}$ [49], respectively. The tail on the left-hand side of the distribution in Figs. 3 and 6 can be attributed to the electron energy distribution. Recent calculations for the longitudinal momentum balance for the three collision products in the single ionization of 3.6-MeV/amu $Au^{53+}$ on helium is shown in Fig. 7.

All the distributions obtained for the longitudinal momentum balances of 3.6-MeV $Ni^{24+}$, $Se^{28+}$, and $Au^{53+}$ result from the fact that the target atom dissociates in the strong electric field of the highly charged ion, which delivers energy but only very little momentum. Thus the action of the fast heavy projectile ions reveal similarities to photoabsorption where the electron perfectly balances the recoil-ion momentum to the extent of the negligibly small momentum transferred by the absorbed photon. In fact the electric field of a fast, highly charged projectile

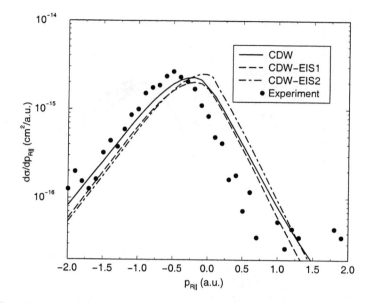

**Figure 2.**   Same as Fig. 1 but for the longitudinal momentum distribution of the recoil ion.

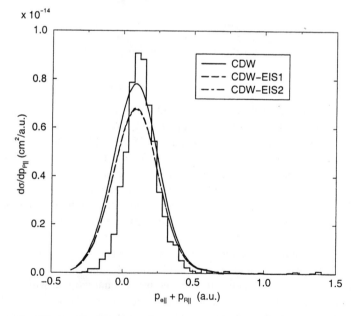

**Figure 3.**   Calculated longitudinal projectile momentum transfer distributions convoluted to a Gaussian distribution for a Gaussian width of 0.22 a.u. Experimental data (histogram) are from Moshammer et al. [2]. Theoretical results: CDW results [20], CDW-EIS1 results [20], CDW-EIS2 [48].

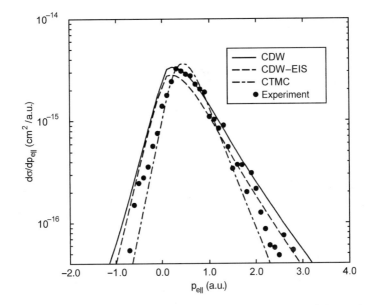

**Figure 4.** Calculated longitudinal momentum distributions for the ejected electron in single ionization of He by 3.6-MeV/amu $Se^{28+}$ ions. Experimental data are from Moshammer et al. [4]. Theoretical results: CDW results [49], CDW-EIS results [49], CTMC results [4].

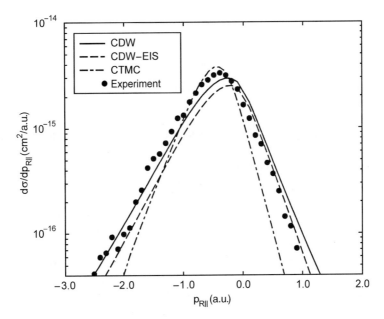

**Figure 5.** Same as Fig. 4, but for the longitudinal momentum distribution of the recoil ion.

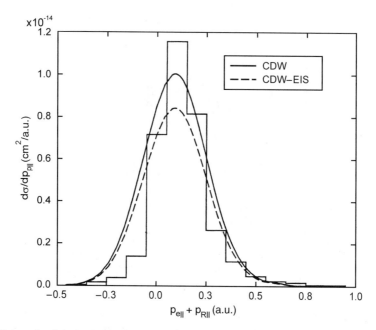

**Figure 6.** Calculated longitudinal projectile momentum transfer distributions in single ionization of He by 3.6-MeV/amu $Se^{28+}$ ions convoluted to a Gaussian distribution for a Gaussian width of 0.22 a.u. Experimental data (histogram) are from Moshammer et al. [4]. Theoretical results: CDW results [49], CDW-EIS results [49].

ion can be described as an intense electromagnetic field that contains a broadband of photon frequencies. This phenomenon is due to the fact that because very little momentum is carried by the incident photons, only those whose energy corresponds to the momentum of the electron in the bound state at the moment of absorption can interact. Thus the momenta that are obtained by the recoil-ions and electrons are directly related to the internal momenta of the electrons in the helium atom at the instant of collision with the highly charged ions.

Results from Khayyat et al. [26] are shown in Figs. 8(a), 8(b), 9(a), and 9(b). Figures 8(a) and 8(b) show cross sections for antiproton–helium collisions at an energy of 945 keV differential in longitudinal electron and recoil-ion momenta. These results are compared with longitudinal momentum distributions for 1-MeV proton–helium collisions shown in Figs. 9(a) and 9(b). In contrast to the earlier cases (3.6-MeV/amu $Ni^{24+}$, $Se^{28+}$, and $Au^{53+}$) considered for highly charged ions, Khayyat et al. [26] predict forward emission for the electrons for both protons and antiprotons. These findings are supported qualitatively by the CDW and CTMC calculations. Thus the data here indicate that the

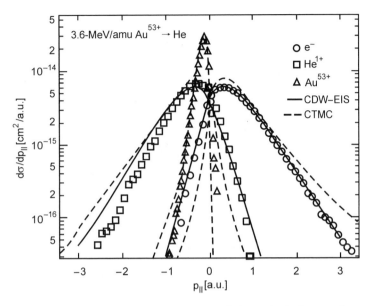

**Figure 7.** Longitudinal momentum distributions of the emitted electron, the recoiling target ion, and the projectile after the single ionization of helium by 3.6-MeV/amu Au$^{53+}$ ions. Experimental data are from Schmitt et al. [50]. Also shown are CDW-EIS calculations [50] and CTMC calculations [50].

post-collision interaction effects due to the two-center electron emission which depend on the sign of the projectile charge are very small in collisions with fast low-Z projectiles. Figures 8(a) and 8(b) also show that the projectile charge asymmetries are less than 10% for a collision energy of 1 MeV. Figures 8(b) and 9(b) illustrate that the recoil-ions have almost a symmetric distribution about the zero momentum which indicates that the recoil-ion behaves primarily as a spectator in the collision. That is, the recoil-ion momentum distribution is mainly given by its momentum in the initial state which is set free in the collision because the electron is knocked out by the projectile. The CDW model, however, predicts a small backward shift of the peak, although the CTMC calculations are in excellent agreement with the experimental data. Thus Figs. 8(a), 8(b), 9(a), and 9(b) show that in contrast to the findings for the highly charged ions considered above (3.6-MeV/amu Ni$^{24+}$, Se$^{28+}$, and Au$^{53+}$), the sign of the low-Z projectile does not play a major role in the collision dynamics above 1 MeV. However, as Khayyat et al. [26] point out, experimentally it remains a challenge to observe projectile charge asymmetries at low antiproton energies where the two-center effects are expected to play a major role in the collision process.

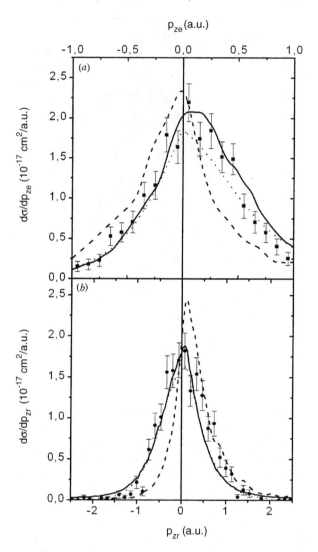

**Figure 8.** Longitudinal momentum distribution for single ionization of helium by 945-keV antiproton (data points) in comparison with proton collision (full curve). (a) Electron momentum data [26]; (b) recoil-ion data [26]. The theoretical calculations represent antiproton collisions: dotted curve, CDW results [26]; broken curve, CTMC result [26]. Here $p_{ze}$ and $p_{zr}$ are equivalent to the notation of $p_{e\|}$ and $p_{R\|}$ of Figs. 1 and 2, respectively.

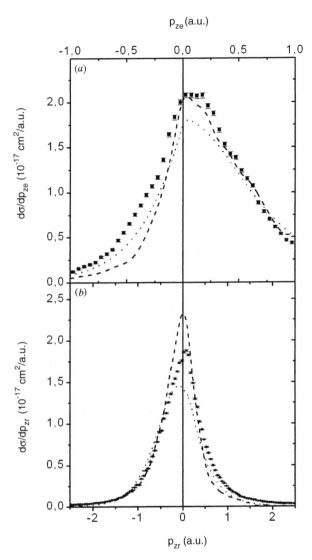

**Figure 9.** Longitudinal momentum distribution for single ionization of helium by 1-MeV proton (data points [26]) in comparison with theory: CTMC (broken curve [26]) and CDW (dotted curve [26]). Here $p_{ze}$ and $p_{zr}$ are equivalent to the notation of $p_{e\parallel}$ and $p_{R\parallel}$ of Figs. 1 and 2, respectively.

## B.  Low-Energy Electron Emission in Fast Heavy-Ion–Atom Collisions

After several decades of systematic electron spectroscopy in ion–atom collisions by many groups (for recent reviews see Refs. 13 and 51), there are only two data sets of doubly differential experimental cross sections $d^2\sigma/dE_kd\Omega$ for the emission of electrons with $E_k < 1$ eV. It has been only recently that, with entirely new and extremely efficient electron spectrometers combined with recoil-ion momentum spectroscopy [52], doubly differential cross sections for ultralow- and low-energy electrons (1.5 meV $\le E_k \le 100$ eV) have been obtained by Schmitt et al. [5]. Other recent results include doubly differential cross sections for electron emission in the low-energy continuum of neon and argon [53]. Thus in essence until these recent experiments, ultralow- and low-eneregy continuum electrons in the screened Coulomb potential of the projectile and electrons from fast heavy-ion helium collisions have remained unexplored from a theoretical perspective. Low-energy electrons are produced in collisions where the energy transferred to the target is just above the ionization threshold. It was generally assumed according to Wannier theory [54] that ultralow- and low-energy continuum did not depend on the sensitivity of the initial-state wavefunctions employed. This loss of memory of initial conditions is in agreement with low-energy electron impact ionization experiments where both the scattered and ionized electrons escape with very low energies [55]. However, recent CDW-EIS calculations [5,53] have shown that observed target specific structures in the electron continuum are attributable to the nodal structure of the initial bound state momentum distribution. This observation has been countered by Gulyás et al. [22], who have shown that these structures which occur in the low-energy continuum may be traced back to a known deficiency in the Hartree–Fock–Slater target potential. Gulyás et al. [22] have shown these structures do not occur if the target atom is described in terms of the optimized potential method where the electronic exchange interaction is treated exactly. Our own calculations [73] demonstrate that an alternative to first-principles schemes like the optimized potential method is to model the target atom using a hydrogenic wavefunction rather than using the Hartree–Fock–Slater model potentials for the target atom.

### 1.  Longitudinal Electron Velocity Distributions

Details of the emission of ultralow- and low-energy electrons can be seen in the cross sections doubly differential in the longitudinal and transverse momenta. The formula for the double differential cross section as a function of the longitudinal electron velocity $v_{e\parallel}$ for various transverse electron velocity $v_{e\perp}$ cuts can be derived from Eq. (66), noting that

$$E_k = \frac{1}{2}(p_{e\parallel}^2 + p_{e\perp}^2) \tag{72}$$

where $p_{e\perp}$ is the transverse momentum of the ejected electron. Hence

$$\frac{1}{2\pi v_{e\perp}} \frac{d^2\sigma}{dv_{e\parallel}dv_{e\perp}} = \frac{1}{2\pi} \int_0^{2\pi} \alpha(\mathbf{k})\, d\phi \tag{73}$$

The doubly differential cross sections are defined in this manner in order to correct for the increasing volume element with increasing $v_{e\perp}$. We have adopted cylindrical coordinates in velocity space with the axis along the beam propagation direction. This choice of coordinate system is well-adapted to the azithmual symmetry of the electron emission if no scattering plane is defined. Thus the cross sections will have the shape and dimension of the triply differential cross sections assuming azithmual symmetry. In Fig. 10 the doubly differential cross section is shown as a function of the longitudinal electron velocity for various transverse velocity cuts in singly ionizing 3.6-MeV/amu $Au^{53+}$ on helium collisions.

There is excellent agreement between the CDW-EIS theory and measurement. This is observed in both shape and in absolute magnitude. A strong

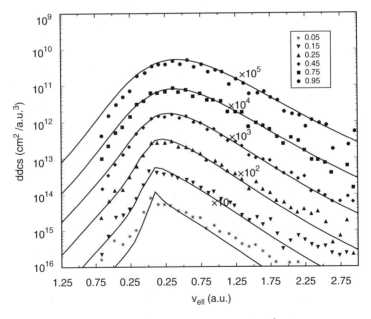

**Figure 10.** Double differential cross sections (ddcs $= \frac{1}{2\pi v_{e\perp}} \frac{d^2\sigma}{dv_{e\parallel}dv_{e\perp}}$) as a function of the longitudinal electron velocity for various transverse velocity cuts in singly ionizing 3.6-MeV/amu $Au^{53+}$ ions on He. CDW-EIS results (solid lines [5]) are shown along with the experimental data from Schmitt et al. [5]. Cross sections at different $v_{e\perp}$ are multiplied by factors of 10, respectively.

forward–backward asymmetry is found. All the longitudinal velocity distributions for different transverse velocities including their maxima are shifted toward positive velocities. These asymmetrical distributions are attributed to the two-center effects—that is, the post-collision interaction effect. The CDW-EIS model predicts these features because it is adapted to take account of the long-range Coulomb interaction of the projectile and target field. The emerging long-range potential of the $Au^{53+}$ projectile ion provides a large perturbation strength of $Z_p/v_p$ (where $Z_p$ and $v_p$ are the charge and velocity of the projectile). This attracts the emitted electron pulling it into the forward direction. With increasing $v_{e\perp}$ the electron distributions are shifted to higher values of $v_{e\parallel}$ while the electrons with very small $v_{e\perp} = 0.05$ a.u. show a maximum close to $v_{e\parallel} = 0$ a.u. It is easier to see this behavior in Fig. 11. The target cusp of ultra-low-energy

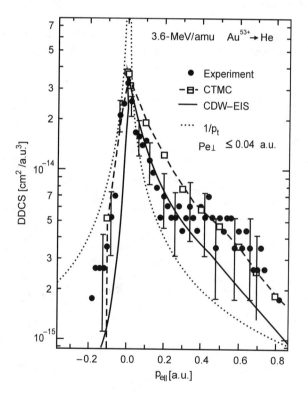

**Figure 11.** Doubly differential cross sections (DDCS $= \frac{1}{2\pi v_{e\perp}} \frac{d^2\sigma}{dv_{e\parallel} dv_{e\perp}}$) for the electrons emitted after the single ionization of helium by 3.6-MeV/amu $Au^{53+}$ ions, plotted for the electron's longitudinal momentum distributions for increasing transverse momenta. Here only one very small cut has been made in the electron's transverse momenta ($p_{e\perp} \leq 0.04$ a.u.). Experimental data and theoretical results are from Schmitt et al. [50].

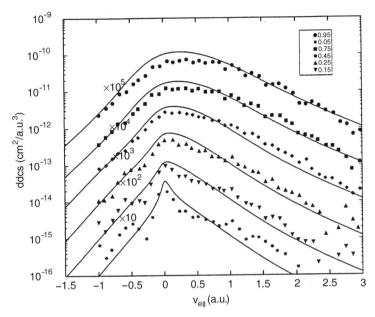

**Figure 12.** Double differential cross sections (ddcs $= \frac{1}{2\pi v_{e\perp}} \frac{d^2\sigma}{dv_{e\parallel} dv_{e\perp}}$) as a function of the longitudinal electron velocity for various transverse velocity cuts in single ionizing 3.6-MeV/amu Au$^{53+}$ ions on Ne. CDW-EIS results (solid lines, [56]) are shown along with the experimental data from Moshammer et al. [53]. Cross sections at different $v_{e\perp}$ are multiplied by factors of 10, respectively.

electrons is strongly asymmetric with a steep drop in the backward direction and a much smoother decrease in the forward direction. The position of the maximum is in good agreement with the CDW-EIS model and the CTMC model. Another characteristic feature that emerges from this study is that in the limit $v_e \rightarrow 0$, the cross section differential in electron velocity diverges as

$$\frac{d\sigma}{dv_e} = \frac{1}{v_e} \tag{74}$$

for the long-range Coulomb interaction. In Figs. 12, 13, and 14 the experimental longitudinal electron velocity distributions for different transverse electron velocities are shown for singly ionizing 3.6-MeV/amu Au$^{53+}$ on neon and argon, respectively. As in the case of helium the theoretical CDW-EIS experimental distributions show the characteristic features of single target ionization by fast heavy ions, and most of these features have been discussed above. In particular, both neon and argon show the sharp maximum at $v_{e\parallel} = 0$ as was shown in

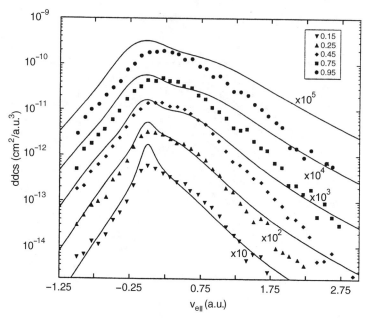

**Figure 13.** Double differential cross sections (ddcs $= \frac{1}{2\pi v_{e\perp}} \frac{d^2\sigma}{dv_{e\parallel} dv_{e\perp}}$) as a function of the longitudinal electron velocity for various transverse velocity cuts in singly ionizing 3.6 MeV/u $Au^{53+}$ ions on Ar. CDW-EIS results (solid lines [73]) are shown along with the experimental data from Moshammer et al. [53]. Cross sections at different $v_{e\perp}$ are multiplied by factors of 10, respectively.

Figs. 10 and 11 for helium. This feature is closely analogous to the one observed for $v_e = v_p$ for electron capture to the continuum mechanism or to electron loss to the continuum which is often referred to as the cusp. The electrons with $v_e = v_T = 0$ in the low-lying continuum states of the target ion also form a sharp peak known as the "target cusp."

Beyond these well-established characteristic features, however, the total width of the electron distributions are much narrower for the argon target than for neon or helium. These patterns are due to the signatures of the initial-state wavefunction. In Fig. 13, systematic discrepancy between experiment and theory appears at higher electron energies. This discrepancy occurs due to one of the basic postulates of the CDW-EIS model, namely that it is based on the independent electron model which considers there to be only one active electron. Within this model it is assumed that the entire impact parameter range contributes only to single ionization. However, particularly at low-impact parameters, single ionization competes with double and multiple ionization. By plotting the differential cross section irrespective of the degree of ionization,

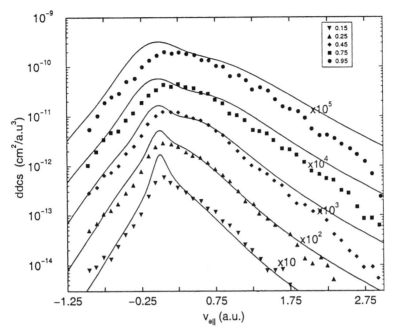

**Figure 14.** Double differential cross sections (ddcs $= \frac{1}{2\pi v_{e\perp}} \frac{d^2\sigma}{dv_{e\parallel} dv_{e\perp}}$) for electron emission due to single, double, or triple ionization of Ar by 3.6-MeV/amu Au$^{53+}$ ions. The DDCS for the specified recoil-ion charge states are added according to their relative contribution to the total cross section. CDW-EIS results (solid lines [73]) are shown along with the experimental data from Moshammer et al. [53]. The experimental data are divided by 1.4. Cross sections at different $v_{e\perp}$ are multiplied by factors of 10, respectively.

this extra contribution can be included. Figure 14 shows that this considerably improves the accord between the CDW-EIS theory and experimental data for the complete range of energies considered.

Similiar theoretical calculations using the CDW-EIS approximation for neon and argon have been carried out by Moshammer et al. [53]. Their calculations are based on using a generalized version of the CDW-EIS model which differs from our model by solving the stationary Schrödinger equation with Hartree–Fock–Slater model potentials for the target atoms. In the case of argon, their theoretical CDW-EIS results showed a shoulder effect at lower transverse velocity cuts. This effect does not appear in our model as shown in Figs. 13 and 14. This shoulder effect has also been analyzed recently by Gulyás et al. [22] using the CDW-EIS model approximation with a target potential obtained from the optimized potential method. They concluded that the shoulder effect was an artifact caused by a deficiency of the Hartree–Fock–Slater model. Both our work and the recent work of Gulyás et al. [22] suggest that it is better to avoid using the Hartree–Fock–Slater model [53] in calculations for ion–atom collisions because it can produce artificial structures in the low-energy continuum wavefunctions, especially for the heavier targets such as argon. However, all three studies [22,53,73] confirm that the target wavefunction is sensitive to the initial state.

## IV.  BINARY PROJECTILE-ELECTRON SCATTERING

The projectile scattering angle $\theta_P$ has always proved to be extremely difficult to measure experimentally due to the small deflection of the projectile. However, the singly differential cross section as a function of $\Omega_P$ contains a wealth of information on binary collisions between the projectile and the target electrons. This singly differential cross section is given by

$$\frac{d\sigma}{d\Omega_P} = \mu\upsilon \int k_f |R_{if}(\boldsymbol{\eta})|^2 \, d\mathbf{k} \tag{75}$$

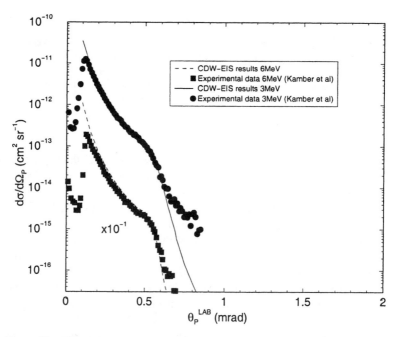

**Figure 15.**  Differential cross sections $d\sigma/d\Omega_P$ for the ionization of He by protons as a function of the scattering angle of the proton. Experimental data are from Kamber et al. [6]. Theoretical results are present CDW-EIS calculations.

where $k_f$ is the final relative momentum of the proton as given by

$$k_f = \sqrt{\mu^2 v - 2\mu\Delta\epsilon - \mu k^2} \qquad (76)$$

and $\mu$ is the reduced mass. Because the calculations are carried out in the laboratory system, a transformation from the center-of-mass frame has to be made. Following the treatment of Salin [56], it is sufficient to replace $\eta$ with

$$\eta = p_{P\perp} = v M_P \sin\theta_P^{LAB} \qquad (77)$$

where $\eta$ is the transverse momentum transfer.

In Fig. 15 we show the theoretical calculation of the singly differential cross section for the single ionization of He by proton impact. There are two impact energies considered here, 3 MeV and 6 MeV, and both are compared to the experimental results of Kamber et al. [6]. For both impact energies there appears a distinctive shoulder effect that takes place at 0.55 mrad in both the experimental data and the theoretical results. This has been attributed to the

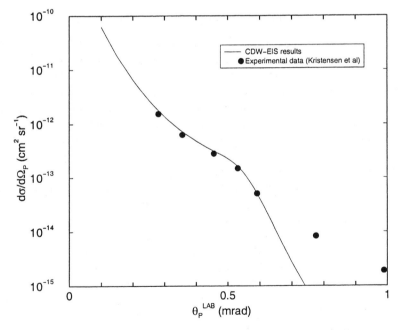

**Figure 16.**    Differential cross sections $d\sigma/d\Omega_P$ for the ionization of He by 2-MeV protons as a function of the scattering angle of the proton. Experimental data are from Kristensen and Horsdal-Pedersen [7]. Theoretical results are present CDW-EIS calculations.

binary collisions between the protons and the target electrons, in which the protons are scattered to near the maximum scattering angle for a proton, off a free electron. This maximum scattering angle is given by the ratio of the electron mass to that of the projectile. The recoiling electrons from such an event are then seen in this binary encounter ridge.

A lower impact energy of 2 MeV is considered in Fig. 16. Here the experimental results are from Kristensen and Horsdal-Pedersen [7], and again they compare well with the theoretical predictions. At this lower energy it is easier to see the effect of the internuclear interaction because for higher energies the ridge becomes more washed out by the momentum distribution of the target electrons. For larger angles the theoretical results fall off faster than the experimental results, fitting in with the idea of a maximum scattering angle for the proton. Another mechanism must therefore explain the cross sections for angles greater than 0.55 mrad. Salin [56] suggested the internuclear interaction as a possible explanation for the proton deviation, and calculations by Rodriguez [27] support this.

## V.  EJECTED ELECTRON SPECTROSCOPY

### A.  Two-Center Effects in Electron Emission

We now look at recent experimental and theoretical (CDW, CDW-EIS) results in single target ionization in ion–helium collisions from the perspective of ejected electron spectroscopy. The richness of this particular method lies in the possibility of measuring the doubly differential cross sections (DDCS). This allows one to study the momentum space of the ionized electrons and thus to explore the dynamics of a "free" electron in the presence of the projectile and target residual potentials. The main features are determined from the fact that both potentials behave asymptotically as Coulomb potentials, thus leading to long-range distortions. This explains several different effects in the observed double differential cross-section (DDCS) spectra such as the soft collision peak, the existence and asymmetry of the electron capture to the continuum (ECC) cusp, and the binary encounter peak and saddle-point emission. We briefly examine these characteristics in the examples below. In Figs. 17 and 18 the theoretical DDCS are plotted in comparison with experimental data from Lee et al. [12] for 1.5-MeV/amu $F^{9+}$ and $H^+$ ions on He as a function of the energy of electrons emitted in the forward direction. The above characteristic structures are easily seen in this spectrum. The low-energy region is attributed to the electrons produced in soft collisions. The region between the soft collision peak and the electron capture to the continuum peak is attributed to two-center electron emission. For $F^{9+}$ projectile the present CDW calculations are in better agreement than the CDW-EIS calculations over the whole electron energy range,

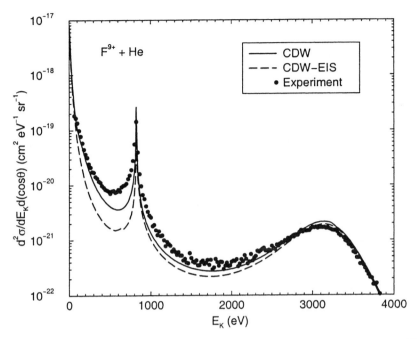

**Figure 17.** Doubly differential cross sections for the ionization of He by 1.5-MeV/amu $F^{9+}$ impact at an observation angle of $\theta = 0°$ as a function of electron energy. Experimental data are from Lee et al. [12]. Theoretical results: CDW results [57], CDW-EIS results [58].

especially on the low-energy side of the cusp where agreement between theory and experiment is much improved. At the high-energy end of the spectrum the pronounced peak is due to the binary encounter collisions. For the case of $H^+$ on He the difference between the CDW and CDW-EIS theory is marginal over the whole electron energy range. It would, however, appear from this study of these two systems that for more highly charged ions, the CDW model reproduces the asymmetry of the ECC cusp to a better degree than the CDW-EIS model. Indeed it may be that inclusion of distortion of distorted waves may be important in explaining the discrepancy between theory and experiment on the low-energy side of the cusp in the case where both models have adopted hydrogenic wavefunctions with effective charges in the final state. For CDW-EIS calculations involving the use of Hartree–Fock wavefunctions to represent the target atom bound and continuum states on the above systems, see Gulyás et al. [18].

We also apply the CDW-EIS model to the energy distributions of electrons ejected from argon by 1-MeV protons [41]. For the argon target we have calculated the contributions from various shells which are then added to obtain

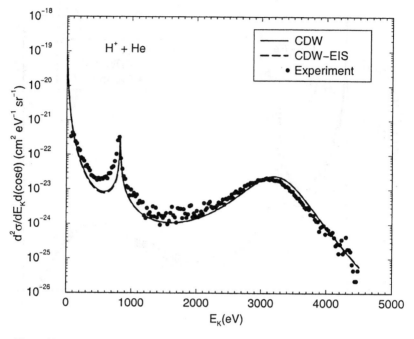

**Figure 18.**  Same as in Fig. 17 but for ionization of He by 1.5-MeV/amu $H^+$ impact.

the single ionization cross sections for various ejected electron emission angles. Comparing these values to the available experimental data raised the question of the role of the second-order processes such as double ionization, transfer ionization, and so on, because they are not distinguished in the experiments from the purely single ionization. However, at high impact energies only which we are considering here, the single ionization is dominant and multiple processes are negligible. It is only at intermediate energies that the multiple processes have an increasing role with decreasing impact energy. In Fig. 19 we present CDW-EIS calculations of doubly differential cross sections as a function of ejected electron energy for fixed electron emission angles for 1-MeV proton impact in comparison with the data of Rudd et al. [40]. Overall our theoretical results are in good agreement with the experimental data except at higher energies for 90 and 125 degrees, where the theoretical predictions fall off quickly compared to experiment. It has been suggested in the work of Madison [59] and Manson et al. [60] that this underestimation of the doubly differential cross sections at large emission angles could be due to the nonorthogonality of the initial and final states. Indeed in our analytical approximation the initial wavefunction is represented by Roothan–Hartree–Fock wavefunctions written as a linear combination of Slater-type orbitals. The continuum states, on the

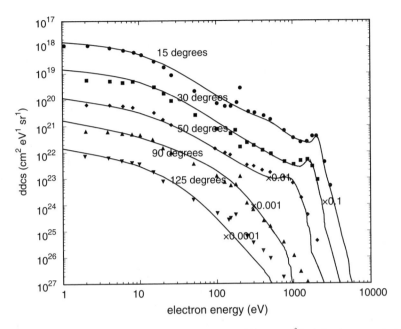

**Figure 19.** Measured double differential cross sections (ddcs $= \frac{d^2\sigma}{dE_k d(\cos\theta)}$) for electron emission at various degrees in collisions of 1 MeV protons with Ar compared to CDW-EIS predictions (solid lines [41]). Experimental data (circles) is taken from Rudd et al. [40].

other hand, are described by a hydrogenic wavefunction with an effective charge chosen from the energy of the initial bound state. Hence there is a nonorthogonality that exists due to the states corresponding to different potentials. The double differential cross section for electron emission at various degrees in collision of 1-MeV proton with argon was also considered by Gulyás et al. [18]. In this the CDW-EIS approximation was also used. However, completely numerical bound and continuum wavefunctions were obtained as solutions to a model potential. In comparison with the CDW-EIS model of Gulyás et al. [18], the CDW-EIS model of McSherry et al. [73] gives better agreement wih experiment for energies below 20 eV at all emission angles considered. It is noted that the results of Gulyás et al. [18] are in better accord than McSherry et al. [73] in two of the instances considered here—that is, at high electron energies for large emission angles of 90 and 125 degrees, respectively.

The question of describing the target states and whether or not the nonorthogonality of the target states in our present analytical solution, really makes a significant difference requires further research. Yet the qualitative agreement between the CDW-EIS model and experiment leads to the conclusion that it does represent the physics of the process correctly.

### B.  Historical Background of Saddle-Point Electron Emission

The discovery of the so-called saddle-point ionization mechanism was one of the most suprising and unexpected predictions within the theory of ion–atom colllsions. Using classical trajectory Monte Carlo (CTMC) calculations, Olson [32] was the first to predict the existence of such mechanisms by examining the single ionization of hydrogen by proton impact at energies greater than 25 keV. It is argued that when the electron is ejected with velocity $k$ relative to the target, $k$ is given by

$$k = \frac{v}{1 + (Z_P/Z_T)^{1/2}} \tag{78}$$

and it straddles the equi-force of the two-center Coulomb potential of the residual target and receding ion. The saddle point is therefore the top of this potential barrier. Olson [61] produced further CTMC predictions concerning saddle-point electrons in a study of $H^+ + H$ ionization in the energy range 40–200 keV. His results predicted a maximum in the forward direction of the electron spectra, which shifted from $k/v = \frac{1}{2}$ at 40 keV to $k/v \approx 0$ at 200 keV. Further studies by Olson et al. [31] concluded that saddle-point ionization was an important mechanism for $H^+ +$ He collisions for energies around 100 keV.

In the energy region of 1–15 keV where the dominant mechanism for target ionization is electron capture to a bound state of the projectile, one can understand the saddle-point ionization mechanism in terms of an adiabatic-energy curve-crossing model [62]. Pieksma et al. [63] claim experimental confirmation of this electron trapping near the classical saddle point for $H^+ + H$ at energies between 1 and 6 keV but suggest that at collision energies greater than 25 keV there is no conclusive evidence of any such mechanism. Such mechanisms have also been confirmed theoretically within the hidden-crossing model for $H^+ + H$ ionization at impact velocities of 0.4 a.u. [37,64,65]. Findings by Dörner et al. [66] on the velocity distribution of electrons emitted in 5- to 15-keV $H^+ +$ He collisions also report evidence of saddle-point trapping.

However, recent precise measurements of Shah et al. [67] for $H^+ + H$ total ionization over the energy range 1.25–9 keV contradict these findings with no observed saddle-point enhancement. McCartney [68] also suggested that there was no such evidence for such mechanisms within the energy range 10–500 keV.

### C.  Saddle-Point Ionization: Comparison Between CDW-EIS Theory and Experiment

In this section we consider measurements of doubly differential cross sections for electron emission at zero degrees arising from collisions of

1.  40-keV protons incident on molecular hydrogen and helium [38]
2.  100-keV protons incident on molecular hydrogen and helium [39]
3.  80-keV protons incident on neon [41]

Calculations using the CDW-EIS model [38] are shown to be in good accord with 40-keV protons incident on molecular hydrogen and helium, and at this energy both theory and experiment show no evidence of any saddle-point enhancement in the doubly differential cross sections. However, for collisions involving 100-keV protons incident on molecular hydrogen and helium the CDW-EIS calculations [39] predict the existence of the saddle-point mechanism, but this is not confirmed by experiment. Recent CDW-EIS calculations and measurement for 80-keV protons on Ne by McSherry et al. [41] find no evidence of the saddle-point electron emission for this collision.

### 1. $H^+$ Impact on He and $H_2$ at 40 keV

In Figs. 20 and 21 we compare the experimental and CDW-EIS results [38] for 40-keV $H^+$ projectiles incident on $H_2$ and He and for emission of electrons at 0° [doubly differential with respect to the electron polar angle of emission and the energy of the ejected electron $d^2\sigma/d\Omega dE_k$ ($10^{-16}$ cm$^2$ eV$^{-1}$ sr$^{-1}$)]. Both sets of results which did not require normalization are seen to be in very good accord, with the spectra being completely dominated by the ECC cusp.

In accordance with theory, the ECC peaks for the CDW-EIS calculations are found to extend much further than the experimental data. Such cusp-like

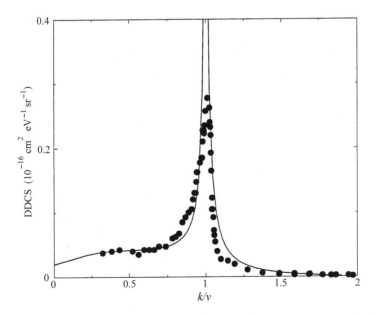

**Figure 20.** Electron emissions at $\theta = 0°$ for 40-keV $H^+$ ion impact in $H_2$. The double differential cross section (DDCS = $d^2\sigma/d\Omega dE_k$) is plotted against $k/v$, where $v$ is the impact velocity, $k$ is the ejected-electron momentum, and $d\Omega = 2\pi \sin\theta\, d\theta$. The filled circles represent the experimental data [38], and the CDW-EIS results are given by the solid line [38].

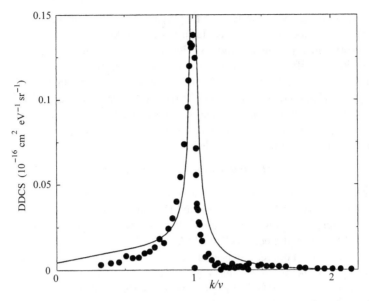

**Figure 21.**   Same as Fig. 20 but for electron emissions at 0° for 40-keV H$^+$ ion impact in He.

discontinuities are due to the Sommerfeld parameter $\zeta$ in Eq. (16) and N($\zeta$) in Eq. (17). The shape of the ECC cusp itself exhibits the usual asymmetry observed in the measurements of other work in the field [69–72] where the differential cross sections at emission energies above the peak fall off more steeply than those at energies below the peak.

The binary peak centered at $k = 2v$ is negligibly small at the present projectile energy because the projectile velocity is of the same magnitude as the velocity spread associated with the target electron. The existence of the structure associated with the saddle-point electrons is not confirmed at 40 keV either by the CDW-EIS calculations or experimentally. Previously, experimental evidence for the existence of saddle-point electrons has been retracted after further investigation [14]. In this case, Suárez et al. [14] attributed this artificial structure to the gas geometry that led to the enhancement of low-energy electrons as the target volume increased. In Figs. 22 and 23 we present three-dimensional surface plots for H$^+$ incident on He and H$_2$ at 40 keV. The graphs show the CDW-EIS doubly differential cross sections plotted as functions of the ratio $k/v$ and the polar angle of emission $\theta$. The angle of emission ranges from $-\pi$ to $+\pi$ radians. As expected, the most prominent feature is the electron capture to the continuum peak. There is, however, no evidence of the saddle-point mechanism.

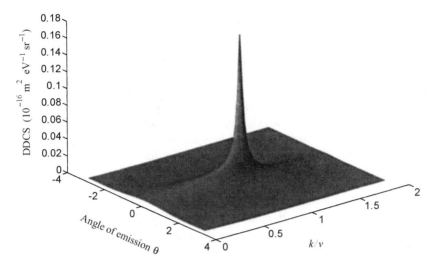

**Figure 22.** Electron emissions for 40-keV H$^+$ ion impact on He. A CDW-EIS surface plot [38] for the double differential cross section $d^2\sigma/d\Omega dE_k$ is plotted against $k/v$ (see the caption of Fig. 20).

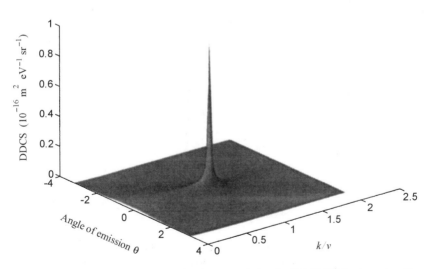

**Figure 23.** Same as Fig. 22 except for electron emission for 40-keV H$^+$ ion impact on H$_2$.

## 2.   $H^+$ Impact on He and $H_2$ at 100 keV

In Figs. 24 and 25 we show the measured double differential cross sections for electron emission at zero degrees in collisions of 100-keV protons with He and $H_2$ [39] compared to CDW-EIS predictions [39]. Uncertainties associated with the experimental results vary from $\pm 1\%$ near the electron capture to the continuum peak to about $\pm 15\%$ near the extreme wings of the distribution. These results have been scaled to provide a best fit with CDW-EIS calculations. In both cases there is satisfactory agreement between the CDW-EIS calculations and experiment, particularly with excellent agreement for electrons with velocities greater than $v$, where $v$ is the velocity of the projectile. For lower-energy electrons the eikonal description of the initial state may have its limitations, especially for lower-impact parameters.

The whole spectrum is dominated by the ECC peak, which, in accordance with theory, extends much higher than the experimental data. The cusp itself shows the usual asymmetry, with cross sections after the peak falling off more steeply than before it. The binary peak is negligibly small at this projectile energy because the projectile velocity is of the same magnitude as the velocity spread associated with the target electron.

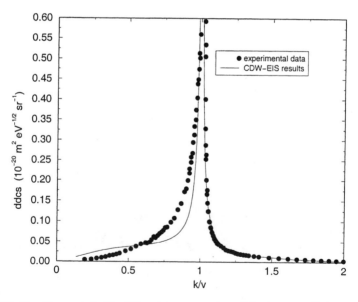

**Figure 24.**  Measured double differential cross sections [39] given by $d^2\sigma/d\Omega dk$ (where $d\Omega = 2\pi \sin\theta d\theta$) for electron emission at zero degrees in collisions of 100-keV protons with He) compared to CDW-EIS predictions [39].

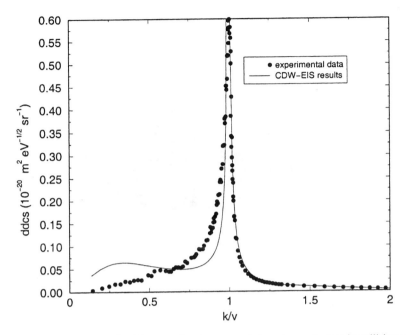

**Figure 25.** Same as Fig. 24 except for electron emission at zero degrees in collisions of 100-keV protons with $H_2$.

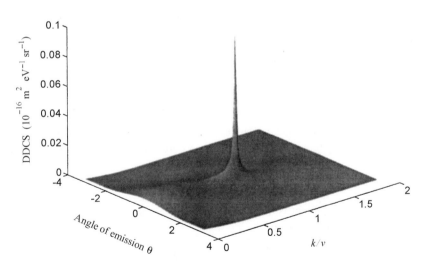

**Figure 26.** Electron emissions for 100-keV $H^+$ ion impact on He. The CDW-EIS calculations [38] now predict a ridge structure from $k/v = 0$ to $k/v \approx 0.6$ with a maximum at $k/v \approx 0.25$. The double differential cross sections are given by $d^2\sigma/d\Omega dE_k$, where $d\Omega = 2\pi \sin\theta\, d\theta$.

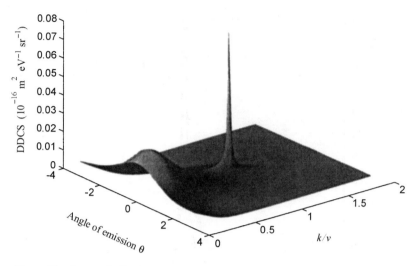

**Figure 27.** Same as in Fig. 26 except for electron emission of 100-keV H$^+$ ion impact on H$_2$.

We also present three-dimensional surface plots for H$^+$ incident on He and H$_2$ at 100 keV in Figs. 26 and 27. For the helium target the CDW-EIS calculations predict a broad ridge with its maximum at $k/v = 0.25$, which is consistent with the findings of Olson et al. [31]. Similarly for the H$_2$ target a saddle-point mechanism occurs again with a broad ridge with its maximum at $k/v = 0.25$. The peak is more pronounced in this case. Also from Figs. 26 and 27 it is noted that while the ECC peak diminishes rapidly as θ deviates from 0°, the saddle-point peak has a smooth, broad ridge.

### 3. H$^+$ Impact on Ne at 80 keV

In Fig. 28 the experimental double differential cross sections for electron emission at zero degrees of a neon target in collision with 80-keV protons are compared to theoretical predictions. Again we have quite good agreement between experimental results [41] and the CDW-EIS predictions [41].

The experimental data has been normalised to the theoretical predictions but the qualitative agreement is quite clear. Here we see that there is no suggestion of a saddle-point mechanism.

### 4. Further Research on Saddle-Point Ionization

To summarize this section on saddle-point ionization, we have calculated doubly differential cross sections for the single ionization of He, H$_2$, and Ne by proton

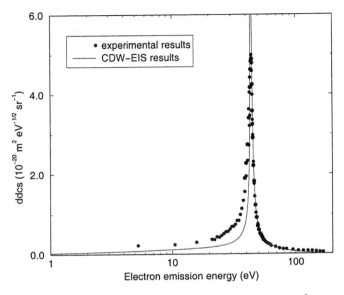

**Figure 28.** Measured double differential cross sections [41] given by $d^2\sigma/d\Omega\,dK$ (where $d\Omega = 2\pi\sin\theta\,d\theta$) for electron emission at zero degrees in collisions of 80-keV proton with Ne compared to CDW-EIS predictions [41].

impact. The observations are seen to be in good agreement with the CDW-EIS calculations. The existence of saddle-point electrons at collision energies of 40 keV is not confirmed, although the theoretical calculations at the higher energy of 100 keV predict such mechanisms. Further research is planned to produce accurate measurements for a wider range of energies. However, it would appear that the saddle-point mechanism is dwarfed by the ECC peak at small emission angles; moreover, we conclude that such processes make no significant contribution to the total cross sections for single ionization.

## VI.  SUMMARY

This review illustrates the complementary nature of recoil-ion momentum spectroscopy, projectile scattering measurements, and conventional electron emission spectroscopy in ion–atom ionizing collisions. We have examined recent applications of both the CDW and CDW-EIS approximations from this perspective. We have shown that both models provide a flexible and quite accurate theory of ionization in ion–atom collisions at intermediate and high energies and also allows simple physical analysis of the ionization process from the perspective of these different experimental techniques.

## Acknowledgments

D. M. McSherry acknowledges financial support from the Department of Education, Northern Ireland (DENI).

## References

1. R. Ali, V. Frohne, C. L. Cocke, M. Stöckli, and M. L. A. Raphaeliam, *Phys. Rev. Lett.* **69**, 2491 (1992).

2. R. Moshammer, J. Ullrich, M. Unverzagt, W. Schmidt, P. Jardin, R. E. Olson, R. Mann, R. Dörner, V. Mergel, U. Buck, and H. Schmidt-Böcking, *Phys. Rev. Lett.* **73**, 3371 (1994).

3. R. Dörner, V. Mergel, L. Zhaoyuan, J. Ullrich, L. Speilberger, R. E. Olson, and H. Schmidt-Böcking, *J. Phys. B: At. Mol. Opt. Phys.* **28**, 435 (1995).

4. R. Moshammer, J. Ullrich, H. Kollmus, W. Schmitt, M. Unverzagt, H. Schmidt-Böcking, C. J. Wood, and R. E. Olson, *Phys. Rev. A* **56**, 1351 (1997).

5. W. Schmitt, R. Moshammer, S. F. C. O'Rourke, H. Kollmus, L. Sarkadi, R. Mann, S. Hagmann, R. E. Olson, and J. Ullrich, *J. Phys. Rev. Lett.* **81**, 4337 (1998).

6. E. Y. Kamber, C. L. Cocke, S. Cheng, and S. L. Varghese, *Phys. Rev. Lett.* **60**, 2026 (1988).

7. F. G. Kristensen and E. Horsdal-Pedersen, *J. Phys. B: At. Mol. Opt. Phys.* **23**, 4129 (1990).

8. A. Gensmantel, J. Ullrich, R. Dörner, R. E. Olson, K. Ullmann, E. Forberich, S. Lencinas, and H. Schmidt-Böcking, *Phys. Rev. A* **45**, 4572 (1992).

9. G. B. Crooks and M. E. Rudd, *Phys. Rev. Lett.* **25**, 1599 (1970).

10. K. G. Harrison and M. Lucas, *Phys. Rev. Lett.* **33**, 149 (1970).

11. M. E. Rudd, L. H. Toburen, and N. Stolterfoht, *At. Data Nucl. Data. Tables* **18**, 413 (1976).

12. D. H. Lee, P. Richard, T. J. M. Zourus, J. M. Sanders, J. L. Shinpaugh, and H. Hidmi, *Phys. Rev. A* **41**, 4816 (1990).

13. M. E. Rudd, Y. K. Kim, D. H. Madison, and T. J. Gay, *Rev. Mod. Phys.* **64**, 441 (1992).

14. S. Suárez, C. Garibotti, W. Meckbach, and G. Bernardi, *Phys. Rev. Lett.* **70**, 418 (1993).

15. C. Liao, P. Richard, S. R. Grabbe, C. P. Bhalla, T. J. M. Zourus, and S. Hagmann, *Phys. Rev. A* **50**, 1328 (1994).

16. D. S. F. Crothers and J. F. McCann, *J. Phys. B: At. Mol. Phys.* **16**, 3229 (1983).

17. P. D. Fainstein, V. H. Ponce, and R. D. Rivarola, *J. Phys. B: At. Mol. Opt. Phys.* **21**, 287 (1988).

18. L. Gulyás, P. D. Fainstein, and A. Salin, *J. Phys. B: At. Mol. Opt. Phys.* **28**, 245 (1995).

19. S. F. C. O'Rourke, I. Shimamura, and D. S. F. Crothers, *Proc. R. Soc. Lond. A* **452**, 175 (1996).

20. S. F. C. O'Rourke and D. S. F. Crothers, *J. Phys. B: At. Mol. Opt. Phys.* **30**, 2443 (1997).

21. Dž. Belkič, *J. Phys. B: At. Mol. Phys.* **11**, 3529 (1978).

22. L. Gulyás, T. Kirchner, T. Shirai, and M. Horbatsch, *Phys. Rev. A* **63** (2000).

23. B. S. Nesbitt, S. F. C. O'Rourke, and D. S. F. Crothers, *Comput. Phys. Commun.* **114**, 385 (1998).

24. S. F. C. O'Rourke and D. S. F. Crothers, *Comput. Phys. Commun.* **109**, 184 (1998).

25. S. F. C. O'Rourke, D. M. McSherry, and D. S. F. Crothers, *Comput. Phys. Commun.* **131**, 129 (2000).

26. Kh. Khayyat, T. Weber, R. Dörner, M. Achler, V. Mergel, L. Spielberger, O. Jagutzki, U. Meyer, J. Ullrich, R. Moshammer, W. Schmitt, H. Knudsen, U. Mikkelsen, P. Aggerholm, E. Uggerhoej, S. P. Moeller, V. D. Rodriguez, S. F. C. O'Rourke, R. E. Olson, P. D. Fainstein, J. H. McGuire, and H. Schmidt-Böcking, *J. Phys. B: At. Mol. Opt. Phys.* **32**, L73 (1999).

27. V. D. Rodriguez, *J. Phys. B: At. Mol. Opt. Phys.* **29**, 275 (1996).

28. M. E. Rudd and T. Jorgensen Jr., *Phys. Rev.* **131**, 666 (1963).

29. M. E. Rudd, C. A. Sautter, and C. L. Bailey, *Phys. Rev* **151**, 20 (1966).

30. W. Meckbach, P. R. Focke, A. R. Goni, S. Suárez, J. Macek, and M. G. Menendez, *Phys. Rev. Lett.* **57**, 1587 (1986).

31. R. E. Olson, T. J. Gay, H. G. Berry, E. B. Hale, and V. D. Irby, *Phys. Rev. Lett.* **59**, 36 (1987).

32. R. E. Olson, *Phys. Rev. A* **27**, 1871 (1983).

33. T. J. Gay, M. W. Gealy, and M. E. Rudd, *J. Phys. B* **23**, L863 (1990).

34. W. Meckbach, S. Suárez, P. R. Focke, and G. Bernardi, *J. Phys. B: At. Mol. Phys.* **24**, 3763 (1991).

35. R. D. DuBois, *Phys. Rev. A* **48**, 1123 (1993).

36. G. Bernardi and W. Meckbach, *Phys. Rev. A* **51**, 1709 (1995).

37. M. Pieksma, *Nucl. Instrum. Methods B* **124**, 177 (1997).

38. B. S. Nesbitt, M. B. Shah, S. F. C. O'Rourke, C. McGrath, J. Geddes, and D. S. F. Crothers, *J. Phys. B: At. Mol. Opt. Phys.* **33**, 637 (2000).

39. C. McGrath, D. M. McSherry, M. B. Shah, S. F. C. O'Rourke, D. S. F. Crothers, G. Montgomery, H. B. Gilbody, C. Illescas, and A. Riera, *J. Phys. B: At. Mol. Phys.* **33**, 3693 (2000).

40. M. E. Rudd, L. H. Toburen, and N. Stolterfoht, *At. Data Nucl. Tables* **23**, 405 (1979).

41. D. M. McSherry, S. F. C. O'Rourke, C. McGrath, M. B. Shah, and D. S. F. Crothers, *Applications of Accelerators in Research and Industry*, J. L. Duggan and L. L. Morgan, eds., The American Institute of Physics, New York, 2001b, p. 168.

42. M. R. C. McDowell and J. P. Coleman, *Introduction to the Theory of Ion–Atom Collisions*, Amsterdam: North Holland, 1970.

43. B. H. Bransden and M. R. C. McDowell, *Charge Exchange and the Theory of Ion–Atom Collisions*, Clarendon, Oxford, 1990.

44. Dž. Belkič, R. Gayet, and A. Salin, *Phys. Rep.* **56**, 279 (1979).

45. E. Clementi and C. Roetti, *At. Data Nucl. Data Tables* **14**, 177 (1974).

46. I. N. Sneddon, *Fourier Transforms*. McGraw-Hill, New York, 1951.

47. E. Engel and S. H. Vosko, *Phys. Rev. A* **47**, 2800 (1993).

48. V. D. Rodriguez, Y. D. Wang, and C. D. Lin, *J. Phys. B: At. Mol. Opt. Phys.* **28**, L471 (1995).

49. S. F. C. O'Rourke, R. Moshammer, and J. Ullrich, *J. Phys. B: At. Mol. Opt. Phys.* **30**, 5281 (1997).

50. W. Schmitt, R. Moshammer, H. Kollmus, S. Hagmann, R. Mann, R. E. Olson, S. F. C. O'Rourke, and J. Ullrich, *Physica Scripta* **T80**, 335 (1999).

51. N. Stolterfoht, R. D. DuBois, and R. D. Rivarola, *Electron Emission in Heavy-Ion Atom Collisions*, Vol 20, J. P. Toennies, ed., Springer-Verlag, Berlin, 1997.

52. J. Ullrich, R. Moshammer, R. Dörner, O. Jagutzki, V. Mergel, H. Schmidt-Böcking, and L. Spielberger, *J. Phys. B: At. Mol. Opt. Phys.* **30**, 2917 (1997).

53. R. Moshammer, P. D. Fainstein, M. Schultz, W. Schmitt, H. Kollmus, R. Mann, S. Hagmann, and J. Ullrich, *J. Phys. Rev. Lett.* **83**, 4721 (1999).

54. J. M. Rost, *J. Phys. B: At. Mol. Opt. Phys.* **27**, 5923 (1994).

55. M. Kelley, W. T. Rogers, R. J. Celotta, and S. R. Mielczarek, *Phys. Rev. Lett.* **51**, 2191 (1983).

56. A. Salin, *J. Phys. B: At. Mol. Opt. Phys.* **22**, 3901 (1989).

57. S. F. C. O'Rourke, W. Schmitt, R. Moshammer, J. Ullrich, B. S. Nesbitt, and D. S. F. Crothers, *New Directions in Atomic Physics*, Whelan et al., eds., Kluwer Academic/Plenum Publishers, New York, 1999, p. 223.

58. P. D. Fainstein, V. H. Ponce, and R. D. Rivarola, *J. Phys. B: At. Mol. Opt. Phys.* **24**, 3091 (1991).

59. D. H. Madison, *Phys. Rev. A* **8**, 2449 (1973).

60. S. T. Manson, L. H. Toburen, D. H. Madison, and N. Stolterfoht, *Phys. Rev. A* **12**, 60 (1975).

61. R. E. Olson, *Phys. Rev. A* **33**, 4397 (1986).

62. M. Pieksma and S. Yu. Ovchinnikov, *J. Phys. B: At. Mol. Phys.* **27**, 4573 (1994).

63. M. Pieksma, S. Yu. Ovchinnikov, J. van Eck, W. B. Westerveld, and A. Niehaus, *Phys. Rev. Lett.* **73**, 46 (1994).

64. S. Yu. Ovchinnikov and J. H. Macek, *Phys. Rev. Lett.* **75**, 2474 (1995).

65. J. H. Macek, *Nucl. Instrum. Methods B* **124**, 191 (1997).

66. R. Dörner, H. Khemliche, M. H. Prior, C. L. Cocke, R. E. Olson, V. Mergel, J. Ullrich, and H. Schmidt-Bocking, *Phys. Rev. Lett.* **77**, 4520 (1996).

67. M. B. Shah, J. Geddes, B. M. McLaughlin, and H. B. Gilbody, *J. Phys. B: At. Mol. Opt. Phys.* **31**, L757 (1998).

68. M. McCartney, *Phys. Rev. A* **52**, 1213 (1995).

69. G. Bernardi, S. Suárez, P. D. Fainstein, C. R. Garibotti, W. Meckbach, and P. Focke, *Phys. Rev. A* **40** 6863 (1989).

70. P. Dahl, *J. Phys. B: At. Mol. Phys.* **18**, 1181 (1985).

71. D. K. Gibson and I. D. Reid, *J. Phys. B: At. Mol. Opt. Phys.* **19**, 3265 (1986).

72. L. H. Andersen, K. E. Jensen, and H. J. Knudsen, *J. Phys. B: At. Mol. Phys.* **19**, L161 (1986).

73. D. M. McSherry, S. F. C. O'Rourke, R. Moshammer, U. J. Ullrich, and D. S. F. Crothers, *Applications of Accelerators in Research and Industry*, J. L. Duggan and L. L. Morgan, eds., The American Institute of Physics, New York, 2001a, p. 133.

# EVOLUTION TIMES OF PROBABILITY DISTRIBUTIONS AND AVERAGES—EXACT SOLUTIONS OF THE KRAMERS' PROBLEM

ASKOLD N. MALAKHOV (DECEASED)

*Radiophysical Department, Nizhny Novgorod State University, Nizhny Novgorod, Russia*

ANDREY L. PANKRATOV

*Institute for Physics of Microstructures of RAS, Nizhny Novgorod, Russia*

## CONTENTS

*Advances in Chemical Physics, Volume 121,* Edited by I. Prigogine and Stuart A. Rice.
ISBN 0-471-20504-4  © 2002 John Wiley & Sons, Inc.

# I.  INTRODUCTION

The investigation of temporal scales (transition rates) of transition processes in various polystable systems driven by noise is a subject of great theoretical and practical importance in physics (semiconductor [1,2] and Josephson electronics [3], dynamics of magnetization of fine single-domain ferromagnetic particles [4–7]), chemistry and biology (transport of biomolecules in cell compartments and membranes [8], the motion of atoms and side groups in proteins [9], and the stochastic motion along the reaction coordinates of chemical and biochemical reactions [2,4,10–14]).

The first paper that was devoted to the escape problem in the context of the kinetics of chemical reactions and that presented approximate, but complete, analytic results was the paper by Kramers [11]. Kramers considered the mechanism of the transition process as noise-assisted reaction and used the Fokker–Planck equation for the probability density of Brownian particles to obtain several approximate expressions for the desired transition rates. The main approach of the Kramers' method is the assumption that the probability current over a potential barrier is small and thus constant. This condition is valid only if a potential barrier is sufficiently high in comparison with the noise intensity. For obtaining exact timescales and probability densities, it is necessary to solve the Fokker–Planck equation, which is the main difficulty of the problem of investigating diffusion transition processes.

The Fokker–Planck equation is a partial differential equation. In most cases, its time-dependent solution is not known analytically. Also, if the Fokker–Planck equation has more than one state variable, exact stationary solutions are very rare. That is why the most simple thing is to approximately obtain time characteristics when analyzing dynamics of diffusion transition processes.

Considering the one-dimensional Brownian diffusion (Brownian motion in the overdamped limit), we note that there are many different time characteristics, defined in different ways (see review [1] and books [2,15,16])—for example, decay time of metastable state or relaxation time to steady state. An often used method of eigenfunction analysis [2,15–18], when the timescale (the relaxation time) is supposed to be equal to an inverse minimal nonzero eigenvalue, is not applicable for the case of a large noise intensity because then higher eigenvalues should be also taken into account. In one-dimensional Fokker–Planck dynamics the moments of the first passage time (FPT) distribution can be calculated exactly, at least expressed by integrals [19]. But during the FPT approach, absorbing boundaries have additionally to be introduced. Both eigenfunction analysis and an FPT approach were widely used for describing different tasks in chemical physics [20–29].

However, most concrete tasks (see examples, listed above) are described by smooth potentials that do not have absorbing boundaries, and thus the moments of FPT may not give correct values of timescales in those cases.

The aim of this chapter is to describe approaches of obtaining exact time characteristics of diffusion stochastic processes (Markov processes) that are in fact a generalization of FPT approach and are based on the definition of characteristic timescale of evolution of an observable as integral relaxation time [5,6,30–41]. These approaches allow us to express the required timescales and to obtain almost exactly the evolution of probability and averages of stochastic processes in really wide range of parameters. We will not present the comparison of these methods because all of them lead to the same result due to the utilization of the same basic definition of the characteristic timescales, but we will describe these approaches in detail and outline their advantages in comparison with the FPT approach.

It should be noted that besides being widely used in the literature definition of characteristic timescale as integral relaxation time, recently "intrawell relaxation time" has been proposed [42] that represents some effective averaging of the MFPT over steady-state probability distribution and therefore gives the slowest timescale of a transition to a steady state, but a description of this approach is not within the scope of the present review.

## II. INTRODUCTION INTO THE BASIC THEORY OF RANDOM PROCESSES

### A. Continuous Markov Processes

This chapter describes methods of deriving the exact time characteristics of overdamped Brownian diffusion only, which in fact corresponds to continuous

Markov process. In the next few sections we will briefly introduce properties of Markov processes as well as equations describing Markov processes.

If we will consider arbitrary random process, then for this process the conditional probability density $W(x_n, t_n | x_1, t_1; \ldots; x_{n-1}, t_{n-1})$ depends on $x_1$, $x_2, \ldots, x_{n-1}$. This leads to definite "temporal connexity" of the process, to existence of strong aftereffect, and, finally, to more precise reflection of peculiarities of real smooth processes. However, mathematical analysis of such processes becomes significantly sophisticated, up to complete impossibility of their deep and detailed analysis. Because of this reason, some "tradeoff" models of random processes are of interest, which are simple in analysis and at the same time correctly and satisfactory describe real processes. Such processes, having wide dissemination and recognition, are Markov processes. Markov process is a mathematical idealization. It utilizes the assumption that noise affecting the system is white (i.e., has constant spectrum for all frequencies). Real processes may be substituted by a Markov process when the spectrum of real noise is much wider than all characteristic frequencies of the system.

A continuous Markov process (also known as a diffusive process) is characterized by the fact that during any small period of time $\Delta t$ some small (of the order of $\sqrt{\Delta t}$) variation of state takes place. The process $x(t)$ is called a Markov process if for any ordered $n$ moments of time $t_1 < \cdots < t < \cdots < t_n$, the $n$-dimensional conditional probability density depends only on the last fixed value:

$$W(x_n, t_n | x_1, t_1; \ldots; x_{n-1}, t_{n-1}) = W(x_n, t_n | x_{n-1}, t_{n-1}) \tag{2.1}$$

Markov processes are processes without aftereffect. Thus, the $n$-dimensional probability density of Markov process may be written as

$$W(x_1, t_1; \ldots; x_n, t_n) = W(x_1, t_1) \prod_{i=2}^{n} W(x_i, t_i | x_{i-1}, t_{i-1}) \tag{2.2}$$

Formula (2.2) contains only one-dimensional probability density $W(x_1, t_1)$ and the conditional probability density. The conditional probability density of Markov process is also called the "transition probability density" because the present state comprehensively determines the probabilities of next transitions. Characteristic property of Markov process is that the initial one-dimensional probability density and the transition probability density completely determine Markov random process. Therefore, in the following we will often call different temporal characteristics of Markov processes "the transition times," implying that these characteristics primarily describe change of the evolution of the Markov process from one state to another one.

The transition probability density satisfies the following conditions:

1. The transition probability density is a nonnegative and normalized quantity:

$$W(x, t|x_0, t_0) \geq 0, \qquad \int_{-\infty}^{+\infty} W(x, t|x_0, t_0)\, dx = 1$$

2. The transition probability density becomes Dirac delta function for coinciding moments of time (physically this means small variation of the state during small period of time):

$$\lim_{t \to t_0} W(x, t|x_0, t_0) = \delta(x - x_0)$$

3. The transition probability density fulfills the Chapman–Kolmogorov (or Smoluchowski) equation:

$$W(x_2, t_2|x_0, t_0) = \int_{-\infty}^{+\infty} W(x_2, t_2|x_1, t_1)W(x_1, t_1|x_0, t_0)\, dx_1 \qquad (2.3)$$

If the initial probability density $W(x_0, t_0)$ is known and the transition probability density $W(x, t|x_0, t_0)$ has been obtained, then one can easily get the one-dimensional probability density at arbitrary instant of time:

$$W(x, t) = \int_{-\infty}^{\infty} W(x_0, t_0)W(x, t|x_0, t_0)\, dx_0 \qquad (2.4)$$

## B. The Langevin and the Fokker–Planck Equations

In the most general case the diffusive Markov process (which in physical interpretation corresponds to Brownian motion in a field of force) is described by simple dynamic equation with noise source:

$$\frac{dx(t)}{dt} = -\frac{d\Phi(x, t)}{h\, dx} + \xi(t) \qquad (2.5)$$

where $\xi(t)$ may be treated as white Gaussian noise (Langevin force), $\langle \xi(t) \rangle = 0$, $\langle \xi(t)\xi(t + \tau) \rangle = D(x, t)\delta(\tau)$, $\Phi(x)$ is a potential profile, and $h$ is viscosity. The equation that has in addition the second time derivative of coordinate multiplied by the mass of a particle is also called the Langevin equation, but that one describes not the Markov process itself, but instead a set of two Markov processes: $x(t)$ and $dx(t)/dt$. Here we restrict our discussion by considering only Markov processes, and we will call Eq. (2.5) the Langevin equation, which in physical interpretation corresponds to overdamped Brownian motion. If the diffusion coefficient $D(x, t)$ does not depend on $x$, then Eq. (2.5) is called a Langevin equation with an additive noise source. For $D(x, t)$ depending on $x$, one speaks of a Langevin equation with multiplicative noise source. This distinction between additive and multiplicative noise may not be considered very significant because for the one-variable case (2.5), for time-independent drift and diffusion coefficients, and for $D(x, t) \neq 0$, the multiplicative noise always becomes an additive noise by a simple transformation of variables [2].

Equation (2.5) is a stochastic differential equation. Some required characteristics of stochastic process may be obtained even from this equation either by cumulant analysis technique [43] or by other methods, presented in detail in Ref. 15. But the most powerful methods of obtaining the required characteristics of stochastic processes are associated with the use of the Fokker–Planck equation for the transition probability density.

The transition probability density of continuous Markov process satisfies to the following partial differential equations ($W_{x_0}(x, t) \equiv W(x, t | x_0, t_0)$):

$$\frac{\partial W_{x_0}(x, t)}{\partial t} = -\frac{\partial}{\partial x}[a(x, t) W_{x_0}(x, t)] + \frac{\partial^2}{\partial x^2}\left[\frac{D(x, t)}{2} W_{x_0}(x, t)\right] \quad (2.6)$$

$$\frac{\partial W_{x_0}(x, t)}{\partial t_0} = -a(x_0, t_0)\frac{\partial}{\partial x_0} W_{x_0}(x, t) - \frac{D(x_0, t_0)}{2}\frac{\partial^2}{\partial x_0^2} W_{x_0}(x, t) \quad (2.7)$$

Equation (2.6) is called the Fokker–Planck equation (FPE) or forward Kolmogorov equation, because it contains time derivative of final moment of time $t > t_0$. This equation is also known as Smoluchowski equation. The second equation (2.7) is called the backward Kolmogorov equation, because it contains the time derivative of the initial moment of time $t_0 < t$. These names are associated with the fact that the first equation used Fokker (1914) [44] and Planck (1917) [45] for the description of Brownian motion, but Kolmogorov [46] was the first to give rigorous mathematical argumentation for Eq. (2.6) and he was first to derive Eq. (2.7). The derivation of the FPE may be found, for example, in textbooks [2,15,17,18].

The function $a(x, t)$ appearing in the FPE is called the drift coefficient, which, due to Stratonovich's definition of stochastic integral, has the form [2]

$$a(x, t) = -\frac{d\Phi(x, t)}{h\, dx} - \frac{1}{2}\frac{dD(x, t)}{dx}$$

where the first term is due to deterministic drift, while the second term is called the spurious drift or the noise-induced drift. It stems from the fact that during a change of $\xi(t)$, also the coordinate of Markov process $x(t)$ changes and therefore $\langle D(x(t), t)\xi(t)\rangle$ is no longer zero. In the case where the diffusion coefficient does not depend on the coordinate, the only deterministic drift term is present in the drift coefficient.

Both partial differential equations (2.6) and (2.7) are linear and of the parabolic type. The solution of these equations should be nonnegative and normalized to unity. Besides, this solution should satisfy the initial condition:

$$W(x, t \,|\, x_0, t_0) = \delta(x - x_0) \qquad (2.8)$$

For the solution of real tasks, depending on the concrete setup of the problem, either the forward or the backward Kolmogorov equation may be used. If the one-dimensional probability density with known initial distribution deserves needs to be determined, then it is natural to use the forward Kolmogorov equation. Contrariwise, if it is necessary to calculate the distribution of the mean first passage time as a function of initial state $x_0$, then one should use the backward Kolmogorov equation. Let us now focus at the time on Eq. (2.6) as much widely used than (2.7) and discuss boundary conditions and methods of solution of this equation.

The solution of Eq. (2.6) for infinite interval and delta-shaped initial distribution (2.8) is called the fundamental solution of Cauchy problem. If the initial value of the Markov process is not fixed, but distributed with the probability density $W_0(x)$, then this probability density should be taken as the initial condition:

$$W(x, t_0) = W_0(x) \qquad (2.9)$$

In this case the one-dimensional probability density $W(x, t)$ may be obtained in two different ways.

1. The first way is to obtain the transition probability density by the solution of Eq. (2.6) with the delta-shaped initial distribution and after that averaging it over the initial distribution $W_0(x)$ [see formula (2.4)].

2. The second way is to obtain the solution of Eq. (2.6) for one-dimensional probability density with the initial distribution (2.9). Indeed, multiplying (2.6) by $W(x_0, t_0)$ and integrating by $x_0$ while taking into account (2.4), we get the same Fokker–Planck equation (2.6).

Thus, the one-dimensional probability density of the Markov process fulfills the FPE and, for delta-shaped initial distribution, coincides with the transition probability density.

For obtaining the solution of the Fokker–Planck equation, besides the initial condition one should know boundary conditions. Boundary conditions may be quite diverse and determined by the essence of the task. The reader may find enough complete representation of boundary conditions in Ref. 15.

Let us discuss the four main types of boundary conditions: reflecting, absorbing, periodic, and the so-called natural boundary conditions that are much more widely used than others, especially for computer simulations.

First of all we should mention that the Fokker–Planck equation may be represented as a continuity equation:

$$\frac{\partial W(x,t)}{\partial t} + \frac{\partial G(x,t)}{\partial x} = 0 \tag{2.10}$$

Here $G(x,t)$ is the probability current:

$$G(x,t) = a(x,t)W(x,t) - \frac{1}{2}\frac{\partial}{\partial x}[D(x,t)W(x,t)] \tag{2.11}$$

*Reflecting Boundary.* The reflecting boundary may be represented as an infinitely high potential wall. Use of the reflecting boundary assumes that there is no probability current behind the boundary. Mathematically, the reflecting boundary condition is written as

$$G(d,t) = 0 \tag{2.12}$$

where $d$ is the boundary point. Any trajectory of random process is reflected when it contacts the boundary.

*Absorbing Boundary.* The absorbing boundary may be represented as an infinitely deep potential well just behind the boundary. Mathematically, the absorbing boundary condition is written as

$$W(d,t) = 0 \tag{2.13}$$

where $d$ is the boundary point. Any trajectory of random process is captured when it crosses the absorbing boundary and is not considered in the preboundary interval. If there are one reflecting boundary and one absorbing boundary, then eventually the whole probability will be captured by the absorbing boundary; and if we consider the probability density only in the interval between two boundaries, then the normalization condition is not fulfilled. If, however, we will think that the absorbing boundary is nothing else but an infinitely deep potential well and will take it into account, then total probability density (in preboundary region and behind it) will be normalized.

*Periodic Boundary Condition.* If one considers Markov process in periodic potential, then the condition of periodicity of the probability density may be treated as boundary condition:

$$W(x,t) = W(x+X,t) \tag{2.14}$$

where $X$ is the period of the potential. The use of this boundary condition is especially useful for computer simulations.

*Natural Boundary Conditions.* If the Markov process is considered in infinite interval, then boundary conditions at $\pm\infty$ are called natural. There are two possible situations. If the considered potential at $+\infty$ or $-\infty$ tends to $-\infty$ (infinitely deep potential well), then the absorbing boundary should be supposed at $+\infty$ or $-\infty$, respectively. If, however, the considered potential at $+\infty$ or $-\infty$ tends to $+\infty$, then it is natural to suppose the reflecting boundary at $+\infty$ or $-\infty$, respectively.

In conclusion, we can list several most widely used methods of solution of the FPE [1,2,15–18]:

1. Method of eigenfunction and eigenvalue analysis
2. Method of Laplace transformation
3. Method of characteristic function
4. Method of exchange of independent variables
5. Numerical methods

## III.   APPROXIMATE APPROACHES FOR ESCAPE TIME CALCULATION

### A.   The Kramers' Approach and Temperature Dependence of the Prefactor of the Kramers' Time

The original work of Kramers [11] stimulated research devoted to calculation of escape rates in different systems driven by noise. Now the problem of calculating escape rates is known as Kramers' problem [1,47].

Let us consider the potential $\Phi(x)$ describing a metastable state, depicted in Fig. 1.

Initially, an overdamped Brownian particle is located in the potential minimum, say somewhere between $x_1$ and $x_2$. Subjected to noise perturbations, the Brownian particle will, after some time, escape over the potential barrier of the height $\Delta\Phi$. It is necessary to obtain the mean decay time of metastable state [inverse of the mean decay time (escape time) is called the escape rate].

To calculate the mean escape time over a potential barrier, let us apply the Fokker–Planck equation, which, for a constant diffusion coefficient $D = 2kT/h$, may be also presented in the form

$$\frac{\partial W(x,t)}{\partial t} = \frac{\partial}{\partial x}\left\{\frac{kT}{h}e^{-\Phi(x)/kT}\frac{\partial}{\partial x}\left[e^{\Phi(x)/kT}W(x,t)\right]\right\} \qquad (3.1)$$

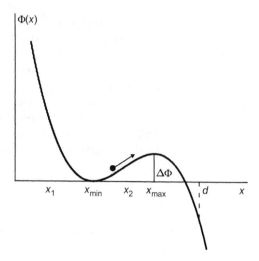

**Figure 1.** Potential describing metastable state.

where we substituted $a(x) = -\frac{d\Phi(x)}{h\,dx}$, where $k$ is the Boltzmann constant, $T$ is the temperature, and $h$ is viscosity.

Let us consider the case when the diffusion coefficient is small, or, more precisely, when the barrier height $\Delta\Phi$ is much larger than $kT$. As it turns out, one can obtain an analytic expression for the mean escape time in this limiting case, since then the probability current $G$ over the barrier top near $x_{max}$ is very small, so the probability density $W(x,t)$ almost does not vary in time, representing quasi-stationary distribution. For this quasi-stationary state the small probability current $G$ must be approximately independent of coordinate $x$ and can be presented in the form

$$G = -\left\{ \frac{kT}{h} e^{-\Phi(x)/kT} \frac{\partial}{\partial x} \left[ e^{\Phi(x)/kT} W(x,t) \right] \right\} \tag{3.2}$$

Integrating (3.2) between $x_{min}$ and $d$, we obtain

$$G \int_{x_{min}}^{d} e^{\Phi(x)/kT} \, dx = \frac{kT}{h} \left[ e^{\Phi(x_{min})/kT} W(x_{min}, t) - e^{\Phi(d)/kT} W(d, t) \right] \tag{3.3}$$

or if we assume that at $x = d$ the probability density is nearly zero (particles may for instance be taken away that corresponds to absorbing boundary), we can express the probability current by the probability density at $x = x_{min}$, that is,

$$G = \frac{kT}{h} e^{\Phi(x_{min})/kT} W(x_{min}, t) / \int_{x_{min}}^{d} e^{\Phi(x)/kT} \, dx \tag{3.4}$$

If the barrier is high, the probability density near $x_{min}$ will be given approximately by the stationary distribution:

$$W(x,t) \approx W(x_{min},t)e^{-[\Phi(x)-\Phi(x_{min})]/kT} \tag{3.5}$$

The probability $P$ to find the particle near $x_{min}$ is

$$P = \int_{x_1}^{x_2} W(x,t)\,dx \approx W(x_{min},t)e^{\Phi(x_{min})/kT}\int_{x_1}^{x_2} e^{-\Phi(x)/kT}\,dx \tag{3.6}$$

If $kT$ is small, the probability density becomes very small for $x$ values appreciably different from $x_{min}$, which means that $x_1$ and $x_2$ values need not be specified in detail.

The escape time is introduced as the probability $P$ divided by the probability current $G$. Then, using (3.4) and (3.6), we can obtain the following expression for the escape time:

$$\tau = \frac{h}{kT}\int_{x_1}^{x_2} e^{-\Phi(x)/kT}\,dx\int_{x_{min}}^{d} e^{\Phi(x)/kT}\,dx \tag{3.7}$$

Whereas the main contribution to the first integral stems from the region around $x_{min}$, the main contribution to the second integral stems from the region around $x_{max}$. We therefore expand $\Phi(x)$ for the first and the second integrals according to

$$\Phi(x) \approx \Phi(x_{min}) + \frac{1}{2}\Phi''(x_{min})(x-x_{min})^2 \tag{3.8}$$

$$\Phi(x) \approx \Phi(x_{max}) - \frac{1}{2}|\Phi''(x_{max})|(x-x_{max})^2 \tag{3.9}$$

We may then extend the integration boundaries in both integrals to $\pm\infty$ and thus obtain the well-known Kramers' escape time:

$$\tau = \frac{2\pi h}{\sqrt{\Phi''(x_{min})|\Phi''(x_{max})|}}e^{\Delta\Phi/kT} \tag{3.10}$$

where $\Delta\Phi = \Phi(x_{max}) - \Phi(x_{min})$. As shown by Edholm and Leimar [48], one can improve (3.10) by calculating the integrals (3.7) more accurately—for example, by using the expansion of the potential in (3.8) and (3.9) up to the fourth-order term. One can ask the question: What if the considered potential is such that either $\Phi''(x_{max}) = 0$ or $\Phi''(x_{min}) = 0$? You may see that Kramers' formula (3.10) does not work in this case. This difficulty may be easily overcome because we know how Kramers' formula has been derived: We may substitute the required

TABLE I
Temperature Dependence of the Prefactor of Escape Time

| $\tau_0(kT) \sim$ | | $\Phi_t(x) \sim$ | | | | | |
|---|---|---|---|---|---|---|---|
| | | $|x|^{1/2}$ | $|x|^{2/3}$ | $|x|$ | $x^2$ | $x^4$ | $x^\infty$ |
| $\Phi_b(x) \sim$ | $|x|^{1/2}$ | $(kT)^3$ | $(kT)^{5/2}$ | $(kT)^2$ | $(kT)^{3/2}$ | $(kT)^{5/4}$ | $(kT)^1$ |
| | $|x|^{2/3}$ | $(kT)^{5/2}$ | $(kT)^2$ | $(kT)^{3/2}$ | $(kT)^1$ | $(kT)^{3/4}$ | $(kT)^{1/2}$ |
| | $|x|$ | $(kT)^2$ | $(kT)^{3/2}$ | $(kT)^1$ | $(kT)^{1/2}$ | $(kT)^{1/4}$ | $(kT)^0$ |
| | $x^2$ | $(kT)^{3/2}$ | $(kT)^1$ | $(kT)^{1/2}$ | $(kT)^0$ | $(kT)^{-1/4}$ | $(kT)^{-1/2}$ |
| | $x^4$ | $(kT)^{5/4}$ | $(kT)^{3/4}$ | $(kT)^{1/4}$ | $(kT)^{-1/4}$ | $(kT)^{-1/2}$ | $(kT)^{-3/4}$ |
| | $|x|^\infty$ | $(kT)^1$ | $(kT)^{1/2}$ | $(kT)^0$ | $(kT)^{-1/2}$ | $(kT)^{-3/4}$ | $(kT)^{-1}$ |

potential into integrals in (3.7) and derive another formula, similar to Kramers' formula:

$$\tau = \tau_0(kT)e^{\Delta\Phi/kT} \qquad (3.11)$$

where the prefactor $\tau_0(kT)$ is a function of temperature and reflects particular shape of the potential. For example, one may easily obtain this formula for a piecewise potential of the fourth order. Formula (3.11) for $\tau_0(kT) = $ const is also known as the Arrhenius law.

Influence of the shape of potential well and barrier on escape times was studied in detail in paper by Agudov and Malakhov [49].

In Table I, the temperature dependencies of prefactor $\tau_0(kT)$ for potential barriers and wells of different shape are shown in the limiting case of small temperature (note, that $|x|^\infty$ means a rectangular potential profile). For the considered functions $\Phi_b(x)$ and $\Phi_t(x)$ the dependence $\tau_0(kT)$ vary from $\tau_0 \sim (kT)^3$ to $\tau_0 \sim (kT)^{-1}$. The functions $\Phi_b(x)$ and $\Phi_t(x)$ are, respectively, potentials at the bottom of the well and the top of the barrier. As follows from Table I, the Arrhenius law (3.11) [i.e. $\tau_0(kT) = $ const] occurs only for such forms of potential barrier and well that $1/p + 1/q = 1$. This will be the case for a parabolic well and a barrier ($p = 2$, $q = 2$), and also for a flat well ($p = \infty$) and a triangle barrier ($q = 1$), and, vice versa, for a triangle well ($p = 1$) and a flat barrier ($q = \infty$).

So, if one will compare the temperature dependence of the experimentally obtained escape times of some unknown system with the temperature dependence of Kramers' time presented in Table I, one can make conclusions about potential profile that describes the system.

## B. Eigenvalues as Transition Rates

Another widely used approximate approach for obtaining transition rates is the method of eigenfunction analysis. As an example, let us consider the symmetric bistable potential, depicted in Fig. 2.

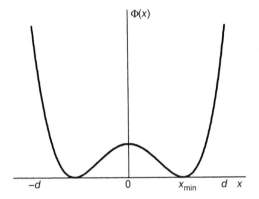

**Figure 2.** Bistable symmetric potential.

Let us calculate the relaxation time of particles in this potential (escape time over a barrier) which agrees with inverse of the lowest nonvanishing eigenvalue $\gamma_1$. Using the method of eigenfunction analysis as presented in detail in Refs. 2, 15, 17, and 18 we search for the solution of the Fokker–Planck equation in the form

$$W(x,t) = X(x) \cdot T(t) \tag{3.12}$$

where $X(x)$ and $T(t)$ are functions of coordinate and time, and we obtain the system of two equations for functions $X(x)$ and $T(t)$:

$$\frac{1}{T(t)}\frac{\partial T(t)}{\partial t} = -\gamma \tag{3.13}$$

$$\left\{\frac{\partial}{\partial x}\left[\frac{d\Phi(x)}{hdx}X(x)\right] + \frac{1}{2}\frac{\partial^2}{\partial x^2}[DX(x)]\right\} = -\gamma X(x) \tag{3.14}$$

where again for simplicity $D = 2kT/h$. Using the boundary conditions and a delta-shaped initial distribution, we can write the solution of the Fokker–Planck equation in the form

$$W(x,t) = \sum_{n=0}^{\infty} \frac{X_n(x)X_n(x_0)}{W_{st}(x_0)}e^{-\gamma_n(t-t_0)} \tag{3.15}$$

where $X_0(x) = W_{st}(x)$ and $\gamma_0 = 0$. Here we consider only the case where the steady-state probability distribution does exist: $W_{st}(x) \neq 0$, and thus we should suppose reflecting boundary conditions $G(\pm d) = 0$. Analyzing expression (3.15) and taking into account that the eigenvalues $\gamma_n$ represent a set such that $\gamma_1 < \gamma_2 < \cdots < \gamma_n$, we can see that the exponent with minimal eigenvalue will decay slower than the others and will thus reflect the largest timescale of decay which equals the inversed minimal nonzero eigenvalue.

So, Eq. (3.14) with boundary conditions is the equation for eigenfunction $X_n(x)$ of the $n$th order. For $X_0(x)$, Eq. (3.14) will be an equation for stationary probability distribution with zero eigenvalue $\gamma_0 = 0$, and for $X_1(x)$ the equation will have the following form:

$$\frac{\partial}{\partial x}\left\{\frac{kT}{h}e^{-\Phi(x)/kT}\frac{\partial}{\partial x}\left[e^{\Phi(x)/kT}X_1(x)\right]\right\} = -\gamma_1 X_1(x) \qquad (3.16)$$

Integrating Eq. (3.16) and taking into account the reflecting boundary conditions (probability current is equal to zero at the points $\pm d$), we get

$$\frac{kT}{h}\frac{\partial}{\partial x}e^{\Phi(x)/kT}X_1(x) = -\gamma_1 e^{\Phi(x)/kT}\int_x^d X_1(z)\,dz \qquad (3.17)$$

Integrating this equation once again, the following integral equation for eigenfunction $X_1(x)$ may be obtained:

$$X_1(x) = e^{-\Phi(x)/kT}\left[e^{\Phi(d)/kT}X_1(d) - \frac{h\gamma_1}{kT}\int_x^d e^{\Phi(y)/kT}\,dy\int_y^d X_1(z)\,dz\right] \qquad (3.18)$$

The eigenfunction $X_1(x)$ belonging to the lowest nonvanishing eigenvalue must be an odd function for the bistable potential, that is, $X_1(0) = 0$. The integral equation (3.18) together with reflecting boundary conditions determine the eigenfunction $X_1(x)$ and the eigenvalue $\gamma_1$. We may apply an iteration procedure that is based on the assumption that the noise intensity is small compared to the barrier height (this iteration procedure is described in the book by Risken [2]), and we obtain the following expression for the required eigenvalue in the first-order approximation:

$$\gamma_1 = (kT/h)\left/\int_0^d e^{\Phi(y)/kT}\,dy\int_y^d e^{-\Phi(z)/kT}\,dz\right. \qquad (3.19)$$

For a small noise intensity, the double integral may be evaluated analytically and finally we get the following expression for the escape time (inverse of the eigenvalue $\gamma_1$) of the considered bistable potential:

$$\tau_b = \frac{\pi h}{\sqrt{\Phi''(x_{\min})|\Phi''(0)|}}e^{\Delta\Phi/kT} \qquad (3.20)$$

The obtained escape time $\tau_b$ for the bistable potential is two times smaller than the Kramers' time (3.10): Because we have considered transition over the barrier top $x = 0$, we have obtained only a half.

## IV.   THE FIRST PASSAGE TIME APPROACH

The first approach to obtain exact time characteristics of Markov processes with nonlinear drift coefficients was proposed in 1933 by Pontryagin, Andronov, and Vitt [19]. This approach allows one to obtain exact values of moments of the first passage time for arbitrary time constant potentials and arbitrary noise intensity; moreover, the diffusion coefficient may be nonlinear function of coordinate. The only disadvantage of this method is that it requires an artificial introducing of absorbing boundaries, which change the process of diffusion in real smooth potentials.

### A.   Probability to Reach a Boundary by One-Dimensional Markov Processes

Let continuous one-dimensional Markov process $x(t)$ at initial instant of time $t = 0$ have a fixed value $x(0) = x_0$ within the interval $(c, d)$; that is, the initial probability density is the delta function:

$$W(x, 0) = \delta(x - x_0), \qquad x_0 \in (c, d)$$

It is necessary to find the probability $Q(t, x_0)$ that a random process, having initial value $x_0$, will reach during the time $t > 0$ the boundaries of the interval $(c, d)$; that is, it will reach either boundary $c$ or $d$: $Q(t, x_0) = \int_{-\infty}^{c} W(x, t)\, dx + \int_{d}^{+\infty} W(x, t)\, dx$.

Instead of the probability to reach boundaries, one can be interested in the probability

$$P(t, x_0) = 1 - Q(t, x_0)$$

of nonreaching the boundaries $c$ and $d$ by Markov process, having initial value $x_0$. In other words,

$$P(t, x_0) = P\{c < x(t) < d, 0 < t < T\}, \qquad x_0 \in (c, d)$$

where $T = T(c, x_0, d)$ is a random instant of the first passage time of boundaries $c$ or $d$.

We will not present here how to derive the first Pontryagin's equation for the probability $Q(t, x_0)$ or $P(t, x_0)$. The interested reader can see it in Ref. 19 or in Refs. 15 and 18. We only mention that the first Pontryagin's equation may be obtained either via transformation of the backward Kolmogorov equation (2.7) or by simple decomposition of the probability $P(t, x_0)$ into Taylor expansion in the vicinity of $x_0$ at different moments $\tau$ and $t + \tau$, some transformations and limiting transition to $\tau \to 0$ [18].

The first Pontryagin's equation looks like

$$\frac{\partial Q(t, x_0)}{\partial t} = a(x_0)\frac{\partial Q(t, x_0)}{\partial x_0} + \frac{D(x_0)}{2}\frac{\partial^2 Q(t, x_0)}{\partial x_0^2} \qquad (4.1)$$

Let us point out the initial and boundary conditions of Eq. (4.1). It is obvious that for all $x_0 \in (c, d)$ the probability to reach boundary at $t = 0$ is equal to zero:

$$Q(0, x_0) = 0, \qquad c < x_0 < d \qquad (4.2)$$

At the boundaries of the interval (i.e., for $x_0 = c$ and $x_0 = d$) the probability to reach boundaries for any instant of time $t$ is equal to unity:

$$Q(t, c) = Q(t, d) = 1 \qquad (4.3)$$

This means that for $x_0 = c, x_0 = d$ the boundary will be surely reached already at $t = 0$. Besides these conditions, usually one more condition must be fulfilled:

$$\lim_{t\to\infty} Q(t, x_0) = 1, \qquad c \le x_0 \le d$$

expressing the fact that the probability to pass boundaries somewhen for a long enough time is equal to unity.

The compulsory fulfillment of conditions (4.2) and (4.3) physically follows from the fact that a one-dimensional Markov process is nondifferentiable; that is, the derivative of Markov process has an infinite variance (instantaneous speed is an infinitely high). However, the particle with the probability equals unity drifts for the finite time to the finite distance. That is why the particle velocity changes its sign during the time, and the motion occurs in an opposite directions. If the particle is located at some finite distance from the boundary, it cannot reach the boundary in a trice—the condition (4.2). On the contrary, if the particle is located near a boundary, then it necessarily crosses the boundary— the condition (4.3).

Let us mention that we may analogically solve the tasks regarding the probability to cross either only the left boundary $c$ or the right one $d$ or regarding the probability to not leave the considered interval $[c, d]$. In this case, Eq. (4.1) is valid, and only boundary conditions should be changed.

Also, one can be interested in the probability of reaching the boundary by a Markov process, having random initial distribution. In this case, one should first solve the task with the fixed initial value $x_0$; and after that, averaging for all possible values of $x_0$ should be performed. If an initial value $x_0$ is distributed in the interval $(c_1, d_1) \supset (c, d)$ with the probability $W_0(x_0)$, then, following the theorem about the sum of probabilities, the complete probability to reach boundaries $c$ and $d$ is defined by the expression

$$Q(t) = \int_c^d Q(t, x_0) W_0(x_0) \, dx_0 + P\{c_1 < x_0 < c, t = 0\}$$
$$+ P\{d < x_0 < d_1, t = 0\} \tag{4.4}$$

## B.  Moments of the First Passage Time

One can obtain an exact analytic solution to the first Pontryagin equation only in a few simple cases. That is why in practice one is restricted by the calculation of moments of the first passage time of absorbing boundaries, and, in particular, by the mean and the variance of the first passage time.

If the probability density $w_T(t, x_0)$ of the first passage time of boundaries $c$ and $d$ exists, then by the definition [18] we obtain

$$w_T(t, x_0) = \frac{\partial}{\partial t} Q(t, x_0) = -\frac{\partial}{\partial t} P(t, x_0) \tag{4.5}$$

Taking a derivative from Eq. (4.1), we note that $w_T(t, x_0)$ fulfills the following equation:

$$\frac{\partial w_T(t, x_0)}{\partial t} = a(x_0) \frac{\partial w_T(t, x_0)}{\partial x_0} + \frac{D(x_0)}{2} \frac{\partial^2 w_T(t, x_0)}{\partial x_0{}^2} \tag{4.6}$$

with initial and boundary conditions

$$w_T(0, x_0) = 0, \qquad c < x_0 < d$$
$$w_T(t, c) = w_T(t, d) = \delta(t) \tag{4.7}$$

for the case of both absorbing boundaries and

$$w_T(t, d) = \delta(t), \qquad \left. \frac{\partial w_T(t, x_0)}{\partial x_0} \right|_{x_0 = c} = 0 \tag{4.8}$$

for the case of one absorbing (at the point $d$) and one reflecting (at the point $c$) boundaries.

The task to obtain the solution to Eq. (4.6) with the above-mentioned initial and boundary conditions is mathematically quite difficult even for simplest potentials $\Phi(x_0)$.

Moments of the first passage time may be expressed from the probability density $w_T(t, x_0)$ as

$$T_n = T_n(c, x_0, d) = \int_0^\infty t^n w_T(t, x_0) \, dt, \qquad n = 1, 2, 3, \ldots \tag{4.9}$$

Multiplying both sides of Eq. (4.6) by $e^{i\Omega t}$ and integrating it for $t$ going from 0 to $\infty$, we obtain the following differential equation for the characteristic function $\Theta(i\Omega, x_0)$:

$$-i\Omega\Theta(i\Omega, x_0) = a(x_0)\frac{\partial\Theta(i\Omega, x_0)}{\partial x_0} + \frac{D(x_0)}{2}\frac{\partial^2\Theta(i\Omega, x_0)}{\partial x_0^2} \qquad (4.10)$$

where $\Theta(i\Omega, x_0) = \int_0^\infty e^{i\Omega t}w_T(t, x_0)\, dt$.

Equation (4.10) allows to find one-dimensional moments of the first passage time. For this purpose let us use the well-known representation of the characteristic function as the set of moments:

$$\Theta(i\Omega, x_0) = 1 + \sum_{n=1}^\infty \frac{(i\Omega)^n}{n!} T_n(c, x_0, d) \qquad (4.11)$$

Substituting (4.11) and its derivatives into (4.10) and equating terms of the same order of $i\Omega$, we obtain the chain of linear differential equations of the second order with variable coefficients:

$$\frac{D(x_0)}{2}\frac{d^2 T_n(c, x_0, d)}{dx_0^2} + a(x_0)\frac{dT_n(c, x_0, d)}{dx_0} = -n \cdot T_{n-1}(c, x_0, d) \qquad (4.12)$$

Equations (4.11) allow us to sequentially find moments of the first passage time for $n = 1, 2, 3, \ldots$ ($T_0 = 1$). These equations should be solved at the corresponding boundary conditions, and by physical implication all moments $T_n(c, x_0, d)$ must have nonnegative values, $T_n(c, x_0, d) \geq 0$.

Boundary conditions for Eq. (4.12) may be obtained from the corresponding boundary conditions (4.7) and (4.8) of Eqs. (4.1) and (4.6). If boundaries $c$ and $d$ are absorbing, we obtain the following from Eq. (4.7):

$$T(c, c, d) = T(c, d, d) = 0 \qquad (4.13)$$

If one boundary, say $c$, is reflecting, then one can obtain the following from Eq. (4.8):

$$T(c, d, d) = 0, \qquad \frac{\partial T(c, x_0, d)}{\partial x_0}\bigg|_{x_0=c} = 0 \qquad (4.14)$$

If we start solving Eq. (4.12) from $n = 1$, then further moments $T_n(c, x_0, d)$ will be expressed from previous moments $T_m(c, x_0, d)$. In particular, for $n = 1, 2$ we obtain

$$\frac{D(x_0)}{2}\frac{d^2 T_1(c, x_0, d)}{dx_0^2} + a(x_0)\frac{dT_1(c, x_0, d)}{dx_0} + 1 = 0 \qquad (4.15)$$

$$\frac{D(x_0)}{2}\frac{d^2 T_2(c, x_0, d)}{dx_0^2} + a(x_0)\frac{dT_2(c, x_0, d)}{dx_0} + 2T_1(c, x_0, d) = 0 \qquad (4.16)$$

Equation (4.15) was first obtained by Pontryagin and is called the second Pontryagin equation.

The system of equations (4.12) may be easily solved. Indeed, making substitution $Z = dT_n(c, x_0, d)/dx_0$ each equation may be transformed in the first-order differential equation:

$$\frac{D(x_0)}{2} \frac{dZ}{dx_0} + a(x_0)Z = -n \cdot T_{n-1}(c, x_0, d) \tag{4.17}$$

The solution of (4.17) may be written by quadratures:

$$Z(x_0) = \frac{dT_n(c, x_0, d)}{dx_0} = e^{\varphi(x_0)} \left[ A - \int_c^{x_0} \frac{2nT_{n-1}(c, y, d)}{D(y)} e^{-\varphi(y)} \, dy \right] \tag{4.18}$$

where $\varphi(y) = \int \frac{2a(y)}{D(y)} \, dy$ and $A$ is an arbitrary constant, determined from boundary conditions.

When one boundary is reflecting (e.g., $c$) and another one is absorbing (e.g., $d$), then from (4.18) and boundary conditions (4.14) we obtain

$$T_n(c, x_0, d) = 2n \int_{x_0}^d e^{\varphi(x)} \int_c^x \frac{T_{n-1}(c, y, d)}{D(y)} e^{-\varphi(y)} \, dy \, dx \tag{4.19}$$

Because $dT_n(c, x_0, d)/dx_0 < 0$ for any $c < x_0 < d$ and $dT_n(c, x_0, d)/dx_0 = 0$ for $x_0 = c$, and, as follows from (4.12), $d^2T_n(c, x_0, d)/dx_0^2 < 0$ for $x_0 = c$, the maximal value of the function $T_n(c, x_0, d)$ is reached at $x_0 = c$.

For the case when both boundaries are absorbing, the required moments of the first passage time have more complicated form [18].

When the initial probability distribution is not a delta function, but some arbitrary function $W_0(x_0)$ where $x_0 \in (c, d)$, then it is possible to calculate moments of the first passage time, averaged over initial probability distribution:

$$T_n(c, d) = \int_c^d T_n(c, x_0, d)W_0(x_0) \, dx_0 \tag{4.20}$$

We note that recently the equivalence between the MFPT and Kramers' time was demonstrated in Ref. 50.

## V.   GENERALIZATIONS OF THE FIRST PASSAGE TIME APPROACH

### A.   Moments of Transition Time

As discussed in the previous section, the first passage time approach requires an artificial introduction of absorbing boundaries; therefore, the steady-state probability distribution in such systems does not exist, because eventually all particles will be absorbed by boundaries. But in the large number of real systems the steady-state distributions do exist, and in experiments there are usually measured stationary processes; thus, different steady-state character-istics, such as correlation functions, spectra, and different averages, are of interest.

The idea of calculating the characteristic timescale of the observable evolution as an integral under its curve (when the characteristic timescale is taken as the length of the rectangle with the equal square) was adopted a long time ago for calculation of correlation times and width of spectral densities (see, e.g., Ref. 51). This allowed to obtain analytic expressions of linewidths [51] of different types of oscillators that were in general not described by the Fokker–Planck equation. Later, this definition of timescales of different observables was widely used in the literature [5,6,14,24,30–41,52,53]. In the following we will refer to any such defined characteristic timescale as "integral relaxation time" [see Refs. 5 and 6], but considering concrete examples we will also specify the relation to the concrete observable (e.g., the correlation time).

However, mathematical evidence of such a definition of characteristic timescale has been understood only recently in connection with optimal estimates [54]. As an example we will consider evolution of the probability, but the consideration may be performed for any observable. We will speak about the transition time implying that it describes change of the evolution of the transition probability from one state to another one.

*The Transition Probability.* Suppose we have a Brownian particle located at an initial instant of time at the point $x_0$, which corresponds to initial delta-shaped probability distribution. It is necessary to find the probability $Q_{c,d}(t, x_0) = Q(t, x_0)$ of transition of the Brownian particle from the point $c \leq x_0 \leq d$ outside of the considered interval $(c, d)$ during the time $t > 0 : Q(t, x_0) = \int_{-\infty}^{c} W(x, t)\, dx + \int_{d}^{+\infty} W(x, t)\, dx$. The considered transition probability $Q(t, x_0)$ is different from the well-known probability to pass an absorbing boundary. Here we suppose that $c$ and $d$ are arbitrary chosen points of an arbitrary potential profile $\Phi(x)$, and boundary conditions at these points may be arbitrary: $W(c, t) \geq 0$, $W(d, t) \geq 0$.

The main distinction between the transition probability and the probability to pass the absorbing boundary is the possibility for a Brownian particle to come back in the considered interval $(c, d)$ after crossing boundary points (see, e.g., Ref. 55). This possibility may lead to a situation where despite the fact that a Brownian particle has already crossed points $c$ or $d$, at the time $t \to \infty$ this particle may be located within the interval $(c, d)$. Thus, the set of transition events may be not complete; that is, at the time $t \to \infty$ the probability $Q(t, x_0)$ may tend to the constant, smaller than unity: $\lim_{t\to\infty} Q(t, x_0) < 1$, as in the case where there is a steady-state distribution for the probability density $\lim_{t\to\infty} W(x,t) = W_{st}(x) \neq 0$. Alternatively, one can be interested in the probability of a Brownian particle to be found at the moment $t$ in the considered interval $(c,d)$ $P(t,x_0) = 1 - Q(t,x_0)$. In the following, for simplicity we will refer to $Q(t,x_0)$ as decay probability and will refer to $P(t,x_0)$ as survival probability.

*Moments of Transition Time.* Consider the probability $Q(t,x_0)$ of a Brownian particle, located at the point $x_0$ within the interval $(c,d)$, to be at the time $t > 0$ outside of the considered interval. We can decompose this probability to the set of moments. On the other hand, if we know all moments, we can in some cases construct a probability as the set of moments. Thus, analogically to moments of the first passage time we can introduce moments of transition time $\vartheta_n(c,x_0,d)$ taking into account that the set of transition events may be not complete, that is, $\lim_{t\to\infty} Q(t,x_0) < 1$:

$$\vartheta_n(c,x_0,d) = \langle t^n \rangle = \frac{\int_0^\infty t^n \frac{\partial Q(t,x_0)}{\partial t}\, dt}{\int_0^\infty \frac{\partial Q(t,x_0)}{\partial t}\, dt} = \frac{\int_0^\infty t^n \frac{\partial Q(t,x_0)}{\partial t}\, dt}{Q(\infty,x_0) - Q(0,x_0)} \qquad (5.1)$$

Here we can formally denote the derivative of the probability divided by the factor of normalization as $w_\tau(t,x_0)$ and thus introduce the probability density of transition time $w_{c,d}(t,x_0) = w_\tau(t,x_0)$ in the following way:

$$w_\tau(t,x_0) = \frac{\partial Q(t,x_0)}{\partial t} \frac{1}{[Q(\infty,x_0) - Q(0,x_0)]} \qquad (5.2)$$

It is easy to check that the normalization condition is satisfied at such a definition, $\int_0^\infty w_\tau(t,x_0)\, dt = 1$. The condition of nonnegativity of the probability density $w_\tau(t,x_0) \geq 0$ is, actually, the monotonic condition of the probability $Q(t,x_0)$. In the case where $c$ and $d$ are absorbing boundaries the probability density of transition time coincides with the probability density of the first passage time $w_T(t,x_0)$:

$$w_T(t,x_0) = \frac{\partial Q(t,x_0)}{\partial t} \qquad (5.3)$$

Here we distinguish $w_\tau(t,x_0)$ and $w_T(t,x_0)$ by different indexes $\tau$ and $T$ to note again that there are two different functions and $w_\tau(t,x_0) = w_T(t,x_0)$ in the case of absorbing boundaries only. In this context, the moments of the FPT are

$$T_n(c,x_0,d) = \langle t^n \rangle = \int_0^\infty t^n \frac{\partial Q(t,x_0)}{\partial t}\, dt = \int_0^\infty t^n w_T(t,x_0)\, dt$$

Integrating (5.1) by parts, one can obtain the expression for the mean transition time (MTT) $\vartheta_1(c, x_0, d) = \langle t \rangle$:

$$\vartheta_1(c, x_0, d) = \frac{\int_0^\infty [Q(\infty, x_0) - Q(t, x_0)] \, dt}{Q(\infty, x_0) - Q(0, x_0)} \qquad (5.4)$$

This definition completely coincides with the characteristic time of the probability evolution introduced in Ref. 32 from the geometrical consideration, when the characteristic scale of the evolution time was defined as the length of rectangle with the equal square, and the same definition was later used in Refs. 33–35. Similar ideology for the definition of the mean transition time was used in Ref. 30. Analogically to the MTT (5.4), the mean square $\vartheta_2(c, x_0, d) = \langle t^2 \rangle$ of the transition time may also be defined as

$$\vartheta_2(c, x_0, d) = 2 \frac{\int_0^\infty \left( \int_t^\infty [Q(\infty, x_0) - Q(\tau, x_0)] \, d\tau \right) dt}{Q(\infty, x_0) - Q(0, x_0)} \qquad (5.5)$$

Note that previously known time characteristics, such as moments of FPT, decay time of metastable state, or relaxation time to steady state, follow from moments of transition time if the concrete potential is assumed: a potential with an absorbing boundary, a potential describing a metastable state or a potential within which a nonzero steady-state distribution may exist, respectively. Besides, such a general representation of moments $\vartheta_n(c, x_0, d)$ (5.1) gives us an opportunity to apply the approach proposed by Malakhov [34,35] for obtaining the mean transition time and easily extend it to obtain any moments of transition time in arbitrary potentials, so $\vartheta_n(c, x_0, d)$ may be expressed by quadratures as it is known for moments of FPT.

Alternatively, the definition of the mean transition time (5.4) may be obtained on the basis of consideration of optimal estimates [54]. Let us define the transition time $\vartheta$ as the interval between moments of initial state of the system and abrupt change of the function, approximating the evolution of its probability $Q(t, x_0)$ with minimal error. As an approximation consider the following function: $\psi(t, x_0, \vartheta) = a_0(x_0) + a_1(x_0)[1(t) - 1(t - \vartheta(x_0))]$. In the following we will drop an argument of $a_0$, $a_1$, and the relaxation time $\vartheta$, assuming their dependence on coordinates of the considered interval $c$ and $d$ and on initial coordinate $x_0$. Optimal values of parameters of such approximating function satisfy the condition of minimum of functional:

$$U = \int_0^{t_N} [Q(t, x_0) - \psi(t, x_0, \vartheta)]^2 \, dt \qquad (5.6)$$

where $t_N$ is the observation time of the process. As it is known, a necessary condition of extremum of parameters $a_0$, $a_1$, and $\vartheta$ has the form

$$\frac{\partial U}{\partial a_0} = 0, \qquad \frac{\partial U}{\partial a_1} = 0, \qquad \frac{\partial U}{\partial \vartheta} = 0 \qquad (5.7)$$

It follows from the first condition that

$$\int_0^{t_N} \{Q(t, x_0) - a_0 - a_1 [1(t) - 1(t - \vartheta)]\} \, dt = 0$$

Transform this condition to the form

$$\int_0^{t_N} Q(t, x_0) \, dt = a_0 t_N + a_1 \vartheta \tag{5.8}$$

The condition of minimum of functional $U$ on $\vartheta$ may be written as

$$Q(\vartheta, x_0) = a_0 + a_1/2 \tag{5.9}$$

Analogically, the condition of minimum of functional $U$ on $a_1$ is

$$\int_0^{\vartheta} Q(t, x_0) \, dt = (a_0 + a_1)\vartheta \tag{5.10}$$

The presented estimate is nonlinear, but this does not lead to significant troubles in processing the results of experiments. An increase of the observation time $t_N$ allows us to adjust values of estimates, and slight changes of amplitudes $a_0$ and $a_1$ and a shift of the moment of abrupt change $\vartheta$ of the approximating function are observed.

When considering analytic description, asymptotically optimal estimates are of importance. Asymptotically optimal estimates assume infinite duration of the observation process for $t_N \to \infty$. For these estimates an additional condition for amplitude of a leap is superimposed: The amplitude is assumed to be equal to the difference between asymptotic and initial values of approximating function $a_1 = Q(0, x_0) - Q(\infty, x_0)$. The only moment of abrupt change of the function should be determined. In such an approach the required quantity may be obtained by the solution of a system of linear equations and represents a linear estimate of a parameter of the evolution of the process.

To get an analytical solution of the system of equations (5.8), (5.9), and (5.10), let us consider them in the asymptotic case $t_N \to \infty$. Here we should take into account that the limit of $a_0$ for $t_N \to \infty$ is $Q(\infty, x_0)$. In asymptotic form for $t_N \to \infty$, Eq. (5.8) is

$$\vartheta = \frac{\int_0^{\infty} [Q(\infty, x_0) - Q(t, x_0)] \, dt}{Q(\infty, x_0) - Q(0, x_0)} \tag{5.11}$$

Therefore, we have again arrived at (5.4), which, as follows from the above, is an asymptotically optimal estimate. From the expression (5.9), another well-known asymptotically optimal estimate immediately follows:

$$Q(\vartheta, x_0) = (Q(0, x_0) + Q(\infty, x_0))/2 \qquad (5.12)$$

but this estimate gives much less analytic expressions than the previous one. It should be noted, that asymptotically optimal estimates are correct only for monotonic evolutions of observables.

In many practical cases the MTT is a more adequate characteristic than the MFPT. As an example (for details see the end of Section V.E.5), if we consider the decay of a metastable state as a transition over a barrier top and we compare mean decay time obtained using the notion of integral relaxation time (case of a smooth potential without absorbing boundary) and the MFPT of the absorbing boundary located at the barrier top, we obtain a twofold difference between these time characteristics even in the case of a high potential barrier in comparison with the noise intensity (5.120). This is due to the fact that the MFPT does not take into account the backward probability current and therefore is sensitive to the location of an absorbing boundary. For the considered situation, if we will move the boundary point down from the barrier top, the MFPT will increase up to two times and tend to reach a value of the corresponding mean decay time which is less sensitive to the location of the boundary point over a barrier top. Such weak dependence of the mean decay time from the location of the boundary point at the barrier top or further is intuitively obvious: Much more time should be spent to reach the barrier top (activated escape) than to move down from the barrier top (dynamic motion). Another important example is noise delayed decay (NDD) of unstable states (see below Fig. 4, case $N$ without potential barrier). It was assumed before that the fluctuations can only accelerate the decay of unstable states [56]. However, in Refs. 57–69 it was found that there are systems that may drop out of these rules. In particular, in the systems considered in Refs. 57–69 the fluctuations can considerably increase the decay time of unstable and metastable states. This effect may be studied via MFPT (see, e.g., Ref. 64), but this characteristic significantly underestimates it [69]. As demonstrated in Ref. 69, the NDD phenomenon appears due to the action of two mechanisms. One of them is caused by the nonlinearity of the potential profile describing the unstable state within the considered interval. This mechanism is responsible for the resonant dependence of MFPT on the noise intensity. Another mechanism is caused by inverse probability current directed into the considered interval. The latter cannot be accounted for by the MFPT method. In Refs. 34 and 69, asymptotic expressions of the decay time of unstable states were obtained for small noise intensities, and it has been demonstrated that if the first derivative of the potential is negative (for the potential oriented as depicted in Fig. 4), the fluctuations acting in dynamic systems always increase the decay time of the unstable state in the limit of a small noise intensity.

Finally, for additional support of the correctness and practical usefulness of the above-presented definition of moments of transition time, we would like to mention the duality of MTT and MFPT. If one considers the symmetric potential, such that $\Phi(-\infty) = \Phi(+\infty) = +\infty$, and obtains moments of transition time over the point of symmetry, one will see that they absolutely coincide with the corresponding moments of the first passage time if the absorbing boundary is located at the point of symmetry as well (this is what we call "the principle of conformity" [70]). Therefore, it follows that the probability density (5.2) coincides with the probability density of the first passage time: $w_\tau(t, x_0) = w_T(t, x_0)$, but one can easily ensure that it is so, solving the FPE numerically. The proof of the principle of conformity is given in the appendix.

In the forthcoming sections we will consider several methods that have been used to derive different integral relaxation times for cases where both drift and diffusion coefficients do not depend on time, ranging from the considered mean transition time and to correlation times and time scales of evolution of different averages.

## B.  The Effective Eigenvalue and Correlation Time

In this section we consider the notion of an effective eigenvalue and the approach for calculation of correlation time by Risken and Jung [2,31]. A similar approach has been used for the calculation of integral relaxation time of magnetization by Garanin et al. [5,6].

Following Ref. 2 the correlation function of a stationary process $K(t)$ may be presented in the following form:

$$K(t) = K(0) \sum_{n=1}^{\infty} V_n \exp(-\lambda_n |t|) \tag{5.13}$$

where matrix elements $V_n$ are positive and their sum is one (for details see Ref. 2, Section 12.3):

$$\sum_{n=1}^{\infty} V_n = 1$$

The required correlation function (5.13) may be approximated by the single exponential function

$$K_{\text{eff}}(t) = K_2 \exp(-\lambda_{\text{eff}} |t|), \quad \frac{1}{\lambda_{\text{eff}}} = \sum_{n=1}^{\infty} \frac{V_n}{\lambda_n} \tag{5.14}$$

which has the same area and the same value at $t = 0$ as the exact expression. The same basic idea was used in Refs. 5 and 6 for the calculation of integral relaxation times of magnetization. The behavior of $\lambda_{\text{eff}}$ was studied in Refs. 71 and 72.

The correlation time, given by $1/\lambda_{\text{eff}}$, may be calculated exactly in the following way. Let us define the normalized correlation function of a stationary process by

$$\Psi(t) = K(t)/K(0)$$
$$K(t) = \langle \Delta r(x(t')) \Delta r(x(t' + t)) \rangle \qquad (5.15)$$
$$\Delta r(x(t')) = r(x(t')) - \langle r \rangle$$

The subtraction of the average $\langle r \rangle$ guarantees that the normalized correlation function $\Psi(t)$ vanishes for large times. Obviously, $\Psi(t)$ is normalized according to $\Psi(0) = 1$. A correlation time may be defined by

$$\tau_c = \int_0^\infty \Psi(t)\, dt \qquad (5.16)$$

For an exponential dependence we then have $\Psi(t) = \exp(-t/\tau_c)$. For the considered one-dimensional Markov process the correlation time may be found in the following way. Alternatively to (5.15) the correlation function may be written in the form

$$K(t) = \int \Delta r(x) \tilde{W}(x, t)\, dx \qquad (5.17)$$

where $\tilde{W}(x, t)$ obeys the FPE (2.6) with the initial condition

$$\tilde{W}(x, 0) = \Delta r(x) W_{st}(x) \qquad (5.18)$$

where $W_{st}(x) = \frac{N}{D(x)} \exp\{\int \frac{2a(x)}{D(x)}\, dx\}$ is the stationary probability distribution. Introducing

$$\rho(x) = \int_0^\infty \tilde{W}(x, t)\, dt \qquad (5.19)$$

Eq. (5.16) takes the form

$$\tau_c = \frac{1}{K(0)} \int_{-\infty}^\infty \Delta r(x) \rho(x)\, dx \qquad (5.20)$$

Due to initial condition (5.18), $\rho(x)$ must obey

$$-\Delta r(x)W_{st}(x) = \left\{ -\frac{d}{dx}a(x) + \frac{d^2}{dx^2}\frac{D(x)}{2} \right\}\rho(x) \tag{5.21}$$

This equation may be integrated leading to

$$\rho(x) = W_{st}(x) \int_{-\infty}^{x} \frac{2f(x')}{D(x')W_{st}(x')}\,dx' \tag{5.22}$$

with $f(x)$ given by

$$f(x) = -\int_{-\infty}^{x} \Delta r(x)W_{st}(x)\,dx \tag{5.23}$$

Inserting (5.22) into (5.20) we find, after integration by parts, the following analytical expression for the correlation time:

$$\tau_c = \frac{1}{K(0)} \int_{-\infty}^{\infty} \frac{2f^2(x)}{D(x)W_{st}(x)}\,dx \tag{5.24}$$

### C.  Generalized Moment Expansion for Relaxation Processes

To our knowledge, the first paper devoted to obtaining characteristic time scales of different observables governed by the Fokker–Planck equation in systems having steady states was written by Nadler and Schulten [30]. Their approach is based on the generalized moment expansion of observables and, thus, called the "generalized moment approximation" (GMA).

The observables considered are of the type

$$M(t) = \int_{c}^{d} \int_{c}^{d} f(x)W(x,t\,|\,x_0)g(x_0)\,dx_0\,dx \tag{5.25}$$

where $W(x,t\,|\,x_0)$ is the transition probability density governed by the Fokker–Planck equation

$$\frac{\partial W(x,t)}{\partial t} = \frac{\partial}{\partial x}\left\{ W_{st}(x)\frac{\partial}{\partial x}\left[ \frac{D(x)}{2W_{st}(x)} \right] \right\} \tag{5.26}$$

$g(x_0)$ is initial probability distribution and $f(x)$ is some test function that monitors the distribution at the time $t$. The reflecting boundary conditions at points $c$ and $d$ are supposed, which leads to the existence of steady-state probability distribution $W_{st}(x)$:

$$W_{st}(x) = \frac{C}{D(x)}\exp\left[ \int_{x'}^{x} \frac{2a(x)}{D(x)}\,dx \right] \tag{5.27}$$

where $C$ is the normalization constant.

The observable has initial value $M(0) = \langle f(x)g_0(x) \rangle$ and relaxes asymptotically to $M(\infty) = \langle f(x) \rangle \langle g_0(x) \rangle$. Here $g_0(x) = g(x)/W_{st}(x)$. Because the time development of $M(t)$ is solely due to the relaxation process, one needs to consider only $\Delta M(t) = M(t) - M(\infty)$.

The starting point of the generalized moment approximation (GMA) is the Laplace transformation of an observable:

$$\Delta M(s) = \int_0^\infty \Delta M(t) e^{-st} \, dt \tag{5.28}$$

$\Delta M(s)$ may be expanded for low and high frequencies:

$$\Delta M(s) \sim_{s \to 0} \sum_{n=0}^\infty \mu_{-(n+1)}(-s)^n \tag{5.29}$$

$$\Delta M(s) \sim_{s \to \infty} \sum_{n=0}^\infty \mu_n(-1/s)^n \tag{5.30}$$

where the expansion coefficients $\mu_n$, the "generalized moments," are given by

$$\mu_n = (-1)^n \int_c^d g(x) \{(L^+(x))^n\}_b f(x) \, dx \tag{5.31}$$

where $\{ \ \}_b$ denotes operation in a space of functions that obey the adjoint reflecting boundary conditions, and $L^+(x)$ is the adjoint Fokker–Planck operator:

$$L^+(x) = -\left\{ a(x)\frac{\partial}{\partial x} + D(x)\frac{\partial^2}{\partial x^2} \right\} \tag{5.32}$$

In view of expansions (5.29) and (5.30), we will refer to $\mu_n$, $n \geq 0$, as the high-frequency moments and to $\mu_n$, $n < 0$, as the low-frequency moments. The moment $\mu_0$ is identical to the initial value $\Delta M(t)$ and assumes the simple form:

$$\mu_0 = \langle f(x)g_0(x) \rangle - \langle f(x) \rangle \langle g_0(x) \rangle \tag{5.33}$$

For negative $n$ (see Ref. 30), the following recurrent expressions for the moments $\mu_{-n}$ may be obtained:

$$\mu_{-n} = \int_c^d \frac{4\,dx}{W_{st}(x)} \int_c^x \frac{W_{st}(y)}{D(y)} \mu_{-(n-1)}(y) \, dy \int_c^y \frac{W_{st}(z)}{D(z)} (g_0(z) - \langle g_0(z) \rangle) \, dz \tag{5.34}$$

where

$$\mu_{-n}(x) = C - \int_c^x \frac{dy}{W_{st}(y)} \int_c^y \frac{2W_{st}(z)}{D(z)} \mu_{-(n-1)}(z) \, dz \tag{5.35}$$

where $C$ is an integration constant, chosen to satisfy the orthogonality property. For $n = 1$

$$\mu_{-1} = \int_c^d \frac{4dx}{W_{st}(x)} \int_c^x \frac{W_{st}(y)}{D(y)} (f(y) - \langle f(y) \rangle) \, dy \int_c^y \frac{W_{st}(z)}{D(z)} (g_0(z) - \langle g_0(z) \rangle) \, dz$$

(5.36)

holds. Moments with negative index, which account for the low-frequency behavior of observables in relaxation processes, can be evaluated by means of simple quadratures. Let us consider now how the moments $\mu_n$ may be employed to approximate the observable $\Delta M(t)$.

We want to approximate $\Delta M(s)$ by a Padé approximant $\Delta m(s)$. The functional form of $\Delta m(s)$ should be such that the corresponding time-dependent function $\Delta m(t)$ is a series of $N$ exponentials describing the relation of $\Delta M(t)$ to $\Delta M(\infty) = 0$. This implies that $\Delta m(s)$ is an $[N - 1, N]$-Padé approximant that can be written in the form

$$\Delta m(s) = \sum_{n=1}^{N} a_n / (\lambda_n + s)$$

(5.37)

or, correspondingly,

$$\Delta m(t) = \sum_{n=1}^{N} a_n \exp(-\lambda_n t)$$

(5.38)

The function $\Delta m(s)$ should describe the low- and high-frequency behavior of $\Delta M(s)$ to a desired degree. We require that $\Delta m(s)$ reproduces $N_h$ high- and $N_l$ low-frequency moments. Because $\Delta m(s)$ is determined by an even number of constants $a_n$ and $\lambda_n$, one needs to choose $N_h + N_l = 2N$. We refer to the resulting description as the $(N_h, N_l)$-generalized-moment approximation (GMA). The description represents a two-sided Padé approximation. The moments determine the parameters $a_n$ and $\lambda_n$ through the relations

$$\sum_{n=1}^{N} a_n \lambda_n^m = \mu_m$$

(5.39)

where $m = -N_l, -N_l + 1, \ldots, N_h - 1$.

Algebraic solution of Eq. (5.39) is feasible only for $N = 1, 2$. For $N > 2$ the numerical solution of (5.39) is possible by means of an equivalent eigenvalue problem (for references see Ref. 30).

The most simple GMA is the $(1, 1)$ approximation which reproduces the moments $\mu_0$ and $\mu_1$. In this case, the relaxation of $\Delta M(t)$ is approximated by a single exponential

$$\Delta M(t) = \mu_0 \exp(-t/\tau)$$

(5.40)

where $\tau = \mu_{-1}/\mu_0$ is the mean relaxation time. As has been demonstrated in Ref. 30 for a particular example of rectangular barrierless potential well, this simple one-exponential approximation is often satisfactory and describes the required observables with a good precision. We should note that this is indeed so as will be demonstrated below.

### D.  Differential Recurrence Relation and Floquet Approach

#### 1.  Differential Recurrence Relations

A quite different approach from all other presented in this review has been recently proposed by Coffey [41]. This approach allows both the MFPT and the integral relaxation time to be exactly calculated irrespective of the number of degrees of freedom from the differential recurrence relations generated by the Floquet representation of the FPE.

In order to achieve the most simple presentation of the calculations, we shall restrict ourselves to a one-dimensional state space in the case of constant diffusion coefficient $D = 2kT/h$ and consider the MFPT (the extension of the method to a multidimensional state space is given in the Appendix of Ref. 41). Thus the underlying probability density diffusion equation is again the Fokker–Planck equation (2.6) that for the case of constant diffusion coefficient we present in the form:

$$\frac{\partial W(x,t)}{\partial t} = \frac{1}{B}\left\{\frac{\partial}{\partial x}\left[\frac{d\varphi(x)}{dx}W(x,t)\right] + \frac{\partial^2 W(x,t)}{\partial x^2}\right\} \tag{5.41}$$

where $B = 2/D = h/kT$ and $\varphi(x) = \Phi(x)/kT$ is the dimensionless potential. Furthermore, we shall suppose that $\Phi(x)$ is the symmetric bistable potential (rotator with two equivalent sites)

$$\Phi(x) = U\sin^2(x) \tag{5.42}$$

Because the solution of Eq. (5.41) must be periodic in $x$ that is $W(x + 2\pi) = W(x)$, we may assume that it has the form of the Fourier series

$$W(x,t) = \sum_{p=-\infty}^{\infty} a_p(t)e^{ipx} \tag{5.43}$$

where for convenience (noting that the potential has minima at $0$, $\pi$ and a central maximum at $\pi/2$) the range of $x$ is taken as $-\pi/2 < x < 3\pi/2$.

On substituting Eq. (5.43) into Eq. (5.41) we have, using the orthogonality properties of the circular functions,

$$\dot{a}_p(t) + \frac{p^2}{B}a_p(t) = \frac{\sigma p}{B}[a_{p-2}(t) - a_{p+2}(t)] \tag{5.44}$$

where $2\sigma = U/kT$. The differential-recurrence relation for $a_{-p}(t)$ is from Eq. (5.44)

$$\dot{a}_{-p}(t) + \frac{p^2}{B} a_{-p}(t) = \frac{\sigma p}{B} [a_{-(p-2)}(t) - a_{-(p+2)}(t)] \tag{5.45}$$

which is useful in the calculation that follows because the Fourier coefficients of the Fourier cosine and sine series, namely,

$$W(x,t) = \frac{f_0}{2} + \sum_{p=1}^{\infty} f_p(t) \cos(px) + \sum_{p=1}^{\infty} g_p(t) \sin(px) \tag{5.46}$$

corresponding to the complex series (5.43) are by virtue of Eqs. (5.44) and (5.45)

$$f_{-p}(t) = f_p(t) = \frac{1}{\pi} \int_{-\pi/2}^{3\pi/2} W(x,t) \cos(px) \, dx$$

$$g_{-p}(t) = -g_p(t) = -\frac{1}{\pi} \int_{-\pi/2}^{3\pi/2} W(x,t) \sin(px) \, dx \tag{5.47}$$

Thus Eq. (5.44) need only be solved for positive $p$.

We also remark that Eq. (5.44) may be decomposed into separate sets of equations for the odd and even $a_p(t)$ which are decoupled from each other. Essentially similar differential recurrence relations for a variety of relaxation problems may be derived as described in Refs. 4, 36, and 73–76, where the frequency response and correlation times were determined exactly using scalar or matrix continued fraction methods. Our purpose now is to demonstrate how such differential recurrence relations may be used to calculate mean first passage times by referring to the particular case of Eq. (5.44).

### 2. Calculation of Mean First Passage Times from Differential Recurrence Relations

In order to illustrate how the Floquet representation of the FPE, Eq. (5.44), may be used to calculate first passage times, we first take the Laplace transform of Eq. (5.41) for the probability density $(Y(x,s) = \int_0^\infty W(x,t) e^{-st} \, dt,\ \hat{a}_p(s) = \int_0^\infty a_p(t) e^{-st} \, dt)$ which for a delta function initial distribution at $x_0$ becomes

$$\frac{d^2 Y(x,s)}{dx^2} + \frac{d}{dx}\left[ \frac{d\varphi(x)}{dx} Y(x,s) \right] - sBY(x,s) = -B\delta(x - x_0) \tag{5.48}$$

The corresponding Fourier coefficients satisfy the differential recurrence relations

$$s\hat{a}_p(s) - \frac{\exp(-ipx_0)}{2\pi} = \frac{\sigma p}{B} [\hat{a}_{p-2}(s) - \hat{a}_{p+2}(s)] \tag{5.49}$$

The use of the differential recurrence relations to calculate the mean first passage time is based on the observation that if in Eq. (5.48) one ignores the term $sY(x,s)$ (which is tantamount to assuming that the process is quasi-stationary, i.e., all characteristic frequencies associated with it are very small), then one has

$$\frac{d^2 Y(x,s)}{dx^2} + \frac{d}{dx}\left[\frac{d\varphi(x)}{dx} Y(x,s)\right] = -B\delta(x - x_0) \qquad (5.50)$$

which is precisely the differential equation given by Risken for the stationary probability if at $x_0$ a unit rate of probability is injected into the system; integrating Eq. (5.50) from $x_0 - \epsilon$ to $x_0 + \epsilon$ leads to $G(x_0 + \epsilon) - G(x_0 - \epsilon) = 1$, where $G$ is the probability current given by Eq. (2.11).

The mean first passage time at which the random variable $\xi(t)$ specifying the angular position of the rotator first leaves a domain $L$ defined by the absorbing boundaries $x_1$ and $x_2$ where $W(x, t | x_0, 0)$ and consequently $Y(x, s = 0 | x_0) \equiv Y(x, x_0)$ vanishes may now be calculated because [see Eq. (8.5) of Ref. 2]

$$T(x_0) = \int_{x_1}^{x_2} Y(x, x_0)\, dx \qquad (5.51)$$

Here we are interested in escape out of the domain $L$ specified by a single cycle of the potential that is out of a domain of length $\pi$ that is the domain of the well. Because the bistable potential of Eq. (5.42) has a maximum at $x = \pi/2$ and minima at $x = 0$, $x = \pi$, it will be convenient to take our domain as the interval $-\pi/2 < x < \pi/2$. Thus we will impose absorbing boundaries at $x = -\pi/2$, $x = \pi/2$. Next we shall impose a second condition that all particles are initially located at the bottom of the potential well so that $x_0 = 0$. The first boundary condition (absorbing barriers at $-\pi/2$, $\pi/2$) implies that only odd terms in $p$ in the Fourier series will contribute to $Y(x)$. While the second ensures that only the cosine terms in the series will contribute because there is a null set of initial values for the sine terms. Hence

$$T(x_0) = T(0)$$

is given by

$$T(0) = \int_{-\pi/2}^{\pi/2} \sum_{p=0}^{\infty} \hat{f}_{2p+1}(0) \cos(2p + 1)x\, dx \qquad (5.52)$$

and the MFPT to go from $-\pi/2$ to $3\pi/2$ which is the escape time is

$$\tau_e = 2T(0) = 4 \sum_{p=0}^{\infty} \frac{(-1)^p}{2p + 1} \hat{f}_{2p+1}(0) \qquad (5.53)$$

In the high barrier limit in this particular problem the inverse escape time is the Kramers' escape rate.

### 3.   Calculation of $\tau$ by a Continued Fraction Method

The differential recurrence relations for $f_{2p+1}(t)$ are [4,77]

$$\dot{f}_{2p+1}(t) + \frac{(2p+1)^2}{B}f_{2p+1}(t) = \frac{\sigma(2p+1)}{B}[f_{2p-1}(t) - f_{2p+3}(t)] \qquad (5.54)$$

the solution of which for $\hat{f}_1(s)$ is [77]

$$\hat{f}_1(s) = \frac{B}{sB + (1-\sigma) + \sigma\hat{S}_3(s)}$$
$$\times \left[ f_1(0) + \sum_{p=1}^{\infty} \frac{(-1)^p}{2p+1}f_{2p+1}(0) \prod_{k=1}^{p} \hat{S}_{2k+1}(s) \right] \qquad (5.55)$$

where the continued fraction $\hat{S}_p(s)$ is

$$\hat{S}_p(s) = \frac{\sigma p}{sB + p^2 + \sigma p\hat{S}_{p+2}(s)} \qquad (5.56)$$

and successive $\hat{f}_p(s)$, $p > 1$, are determined from [4,77]

$$\hat{f}_p = \hat{S}_p\hat{f}_{p-2} + q_p \qquad (5.57)$$

where

$$q_p = \frac{f_p(0)B - \sigma p q_{p+2}}{sB + p^2 + \sigma p\hat{S}_{p+2}} \qquad (5.58)$$

whence

$$\hat{f}_3(s) = \frac{B\hat{S}_3(s)}{sB + (1-\sigma) + \sigma\hat{S}_3(s)} \left[ f_1(0) + \sum_{p=1}^{\infty} \frac{(-1)^p}{2p+1}f_{2p+1}(0) \prod_{k=1}^{p} \hat{S}_{2k+1}(s) \right]$$
$$- \frac{B}{\sigma} \left[ \sum_{p=1}^{\infty} \frac{(-1)^p}{2p+1}f_{2p+1}(0) \prod_{k=1}^{p} \hat{S}_{2k+1}(s) \right] \qquad (5.59)$$

where of course the $f_{2p+1}(0)$ are the initial values of $f_{2p+1}(t)$. Since $x_0$ is not an end point in the domain of integration the initial values are

$$f_{2p+1}(0) = \frac{1}{\pi}\cos(2p+1)x_0 = \frac{1}{\pi} \qquad (5.60)$$

because we suppose that $x_0 = 0$. All $\hat{f}_{2p+1}(s)$ can now be written down in the manner of $\hat{f}_1(s)$ and $\hat{f}_2(s)$ using Eqs. (5.57) and (5.58), with the calculations being much simpler than those of the correlation time in Ref. 77 on account of the rather simple initial condition of Eq. (5.60). This is unnecessary, however, as we shall demonstrate. First we recall that for the purpose of the calculation of $Y(x, x_0)$ the relevant quantity is the $s = 0$ value of $\hat{f}_{2p+1}$. Furthermore, the $s = 0$ value of the continued fraction $\hat{S}_{2k+1}(s)$ is [77]

$$\hat{S}_{2k+1}(0) = \frac{\sigma/(2k+1)}{1 + [\sigma/(2k+1)]\hat{S}_{2k+3}(0)} \tag{5.61}$$

the solution of which, which is finite at $\sigma = 0$, is

$$\hat{S}_{2k+1}(0) = \frac{I_{k+(1/2)}(\sigma)}{I_{k-(1/2)}(\sigma)} \tag{5.62}$$

where $I_{k+(1/2)}(\sigma)$ are the modified Bessel functions of the first kind of order $k + \frac{1}{2}$. Thus just as in Ref. 77, $\hat{f}_1(0)$ may be written as

$$\hat{f}_1(0) = \frac{B}{2\sigma}(e^{2\sigma} - 1) \sum_{p=0}^{\infty} \frac{(-1)^p}{2p+1} \frac{1}{\pi} L_p(\sigma) \tag{5.63}$$

where the $L_p(\sigma)$ satisfy the recurrence relation

$$\begin{aligned} L_{p-1}(\sigma) - L_{p+1}(\sigma) &= \frac{2p+1}{\sigma} L_p(\sigma) \\ L_p(\sigma) &= \frac{I_{p+(1/2)}(\sigma)}{I_{(1/2)}(\sigma)} \end{aligned} \tag{5.64}$$

The leading term in Eq. (5.63) arises because

$$\begin{aligned} \frac{B}{(1-\sigma) + \sigma\hat{S}_3(0)} &= \frac{B}{(1-\sigma) + \sigma(I_{(3/2)}(\sigma)/I_{(1/2)}(\sigma))} \\ &= \frac{B}{2\sigma}(e^{2\sigma} - 1) \end{aligned} \tag{5.65}$$

In like manner we have

$$\hat{f}_3(0) = \frac{B\hat{S}_3(0)}{(1-\sigma) + \sigma\hat{S}_3(0)} \frac{1}{\pi} \sum_{p=0}^{\infty} \frac{(-1)^p}{2p+1} L_p(\sigma) - \frac{B}{\sigma\pi} \sum_{p=1}^{\infty} \frac{(-1)^p}{2p+1} L_p(\sigma) \tag{5.66}$$

The large $\sigma$ limit (high barrier limit, $2\sigma = U/kT$) of both Eqs. (5.63) and (5.66) is the same, namely,

$$\hat{f}_1(0) = \hat{f}_3(0) \approx \frac{B}{2\sigma}\frac{e^{2\sigma}}{\pi}\sum_{p=0}^{\infty}\frac{(-1)^p}{2p+1} = \frac{B}{8\sigma}e^{2\sigma} \tag{5.67}$$

by the properties of the Riemann zeta function [78] [Eq. (5.67)] because

$$\lim_{\sigma\to\infty} L_p(\sigma) = 1$$

and because the second term in Eq. (5.66) decays as $\sigma^{-1}$ for large $\sigma$. One may also deduce from Eqs. (5.57) and (5.58) that the behavior of all $\hat{f}_{2p+1}(0)$ for large $\sigma$ is the same. Thus we may write down $Y(x, x_0)$ in the high barrier limit as

$$Y(x, 0) \approx \frac{B}{8\sigma}e^{2\sigma}\sum_{p=1}^{\infty}\cos(2p+1)x \tag{5.68}$$

which with Eq. (5.52) yields the following for the MFPT from $x = 0$ to $x = \pi/2$ or $-\pi/2$:

$$T_{\mathrm{MFPT}} \approx \frac{B}{8\sigma}e^{2\sigma}\sum_{p=0}^{\infty}\frac{2(-1)^p}{2p+1} \approx \frac{\pi}{2}\frac{B}{8\sigma}e^{2\sigma} \tag{5.69}$$

The escape time $\tau_e$ out of the well that is the domain $-\pi/2 < x < \pi/2$ is

$$\tau_e \approx 2T_{\mathrm{MFPT}} \approx \frac{\pi}{4}\frac{B}{2\sigma}e^{2\sigma} \tag{5.70}$$

in agreement with the results of the Kramers theory [Eq. (102) of Ref. 77] and the asymptotic expression for the smallest nonvanishing eigenvalue [79]. Equivalently, the escape time in this case is the MFPT from $-\pi/2$ to $3\pi/2$. In the opposite limit when the potential barrier is zero, Eq. (5.53) becomes (again using the properties of the Riemann zeta and Langevin functions)

$$\tau_e = \frac{\pi^2 B}{8} \tag{5.71}$$

in agreement with the predictions of the Einstein theory of the Brownian movement as adapted to the calculation of MFPTs by Klein [80] and in contrast to the zero barrier value of the correlation time which is $B$.

### E.   The Approach by Malakhov and Its Further Development

Another approach for computing the transition times had been proposed by Malakhov [34,35]. This approach also utilizes the definition of the desired timescale as integral relaxation time and is based on obtaining the solution of the Laplace transformed FPE as a set with respect to small Laplace parameter $s$ and allows us to obtain a wide variety of temporal characteristics of Markov processes with time constant drift and diffusion coefficients and for a wide variety of boundary conditions ranging from natural to periodic.

In this section we will consider this approach in detail for different types of potential profiles $\varphi(x) = \Phi(x)/kT$, and to avoid cumbersome calculations we present the analysis for the constant diffusion coefficient $D = 2kT/h$, but the results, of course, may be easily generalized for any $D(x) \neq 0$.

### 1.    Statement of the Problem

It is convenient to present the Fokker–Planck equation in the following dimensionless form:

$$\frac{\partial W(x,t)}{\partial t} = -\frac{\partial G(x,t)}{\partial x} = \frac{1}{B}\left\{\frac{\partial}{\partial x}\left[\frac{d\varphi(x)}{dx}W(x,t)\right] + \frac{\partial^2 W(x,t)}{\partial x^2}\right\} \qquad (5.72)$$

where $B = 2/D = h/kT$, $G(x,t)$ is the probability current, and $\varphi(x) = 2\Phi(x)/hD = \Phi(x)/kT$ is the dimensionless potential profile.

We suppose that at initial instant $t = 0$ all Brownian particles are located at the point $x = x_0$, which corresponds to the initial condition $W(x,0) = \delta(x - x_0)$. The initial delta-shaped probability distribution spreads with time, and its later evolution strongly depends on the form of the potential profile $\varphi(x)$. We shall consider the problem for the three archetypal potential profiles that are sketched in Figs. 3–5.

If a potential profile is of the type I (see Fig. 3) when $\varphi(x)$ goes to plus infinity fast enough at $x \to \pm\infty$, there is the steady-state probability distribution

$$W(x,\infty) = Ae^{-\varphi(x)}, \qquad A = 1/\int_{-\infty}^{+\infty} \exp[-\varphi(x)]\, dx > 0 \qquad (5.73)$$

In this case our aim is to determine *the relaxation time* $\Theta$ that is the timescale of the probability density evolution from the initial $W(x,0)$ to the final value $W(x,\infty)$ for any $x \neq x_0$. On the other hand we may consider the probability

$$P(t,x_0) = P(t) = \int_c^d W(x,t)\, dx \qquad (5.74)$$

in a given interval $[c,d]$, which in the following we shall call *the decision interval*, and seek the relaxation time of this probability $P(t)$ which changes from the initial value $P(0) = \int_c^d W(x,0)\, dx$ to the final one $P(\infty) = \int_c^d W(x,\infty)\, dx$. A potential

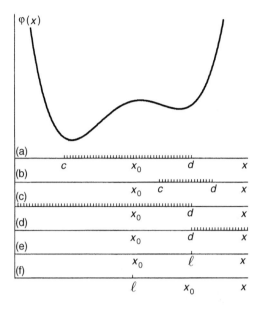

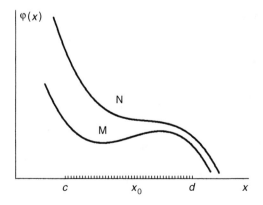

**Figure 3.** A sketch of a potential profile of type I. The $x$ axes (a)–(f) represent various dispositions of decision intervals $[c, d]$, $[-\infty, d]$, $[d, +\infty]$ and points of observation $\ell$ with respect to the $x_0$ coordinate of the initial delta-shaped probability distribution.

profile of type II (Fig. 4) tends fast enough to plus infinity at $x \to -\infty$ and to minus infinity at $x \to +\infty$. A potential profile of type III (Fig. 5) drops to minus infinity at $x \to \pm\infty$. Let us note that if there is a potential profile that increases to plus infinity at $x \to +\infty$ and decreases to minus infinity at $x \to -\infty$, then it may be reduced to the profile of type II by reversing of the $x$ axis.

For potential profiles of types II and III there are no nonzero stationary probability densities because all diffusing particles in the course of time will leave the initial interval and go toward the regions where the potential $\varphi(x)$ tends to minus infinity. In these situations we may pose two questions:

**Figure 4.** Sketches of potential profiles of type II with a metastable ($M$) and a nonstable ($N$) state.

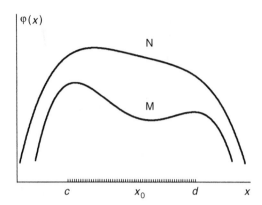

**Figure 5.** Sketches of potential profiles of type III with a metastable (*M*) and a nonstable (*N*) state.

1. How long may a metastable state *M* exist in decision interval $[c, d]$?
2. How long may a nonstable state *N* remain in decision interval $[c, d]$?

In the first position it is reasonable to define *the decay time* $\tau$ (or *the lifetime* $\tau$) of the metastable state as the timescale of probability (5.74) evolution from the initial value $P(0) = 1$ to the zeroth final value $P(\infty) = 0$.

In the second position we denote the timescale of probability (5.74) evolution from $P(0) = 1$ to $P(\infty) = 0$ as *the decay time* $\tau$ of the nonstable state.

Instead of containing natural boundaries at $x \to \pm\infty$, all potential profiles shown in Figs. 3–5 may contain, of course, reflecting and/or absorbing boundaries whose coordinates we shall denote as $\lambda_1, \lambda_2 : \lambda_1 \le c < d \le \lambda_2$. If we consider cases with absorbing boundaries, then we arrive at an MFPT that has the same meaning: It is the timescale of the probability evolution; that is, the MFPT is actually nothing other than the decay time of a metastable or nonstable state.

Now let us define the timescales stated above in a general case as

$$\vartheta = \frac{\int_0^\infty [P(t) - P(\infty)]\, dt}{P(0) - P(\infty)} = \frac{\int_0^\infty [P(\infty) - P(t)]\, dt}{P(\infty) - P(0)} \tag{5.75}$$

for the characteristic time of probability evolution and

$$\Theta(\ell) = \frac{\int_0^\infty [W(\ell, \infty) - W(\ell, t)]\, dt}{W(\ell, \infty)} \tag{5.76}$$

for the relaxation time in a single point $x = \ell \neq x_0$. These definitions are legitimate, if variations of $P(t)$ or $W(\ell, t)$ are sufficiently fast so that integrals (5.75), (5.76) converge and if $P(t)$ and $W(\ell, t)$ during their process of tendency to $P(\infty)$ and $W(\ell, \infty)$ do not intersect the final values $P(\infty)$ and $W(\ell, \infty)$. For the fulfillment of the last condition, monotonic variations of $P(t)$ and $W(\ell, t)$ are sufficient.

Our problem is to obtain the above-stated timescales—the relaxation time, the decay time of a metastable state, and the decay time of a nonstable state—in such a way that the desired results may be expressed directly in terms of the given potential profiles $\varphi(x)$.

Below we provide the full resolution of this problem for potential profiles of types I, II, and III separately.

## 2.   Method of Attack

We introduce into consideration the Laplace transformations of the probability density and the probability current

$$Y(x, s) = \int_0^\infty W(x, t) e^{-st}\, dt, \qquad \hat{G}(x, s) = \int_0^\infty G(x, t) e^{-st}\, dt \qquad (5.77)$$

Then the FPE, according to (5.72) and (5.77) and to the initial condition, may be rewritten as

$$\frac{d^2 Y(x, s)}{dx^2} + \frac{d}{dx}\left[\frac{d\varphi(x)}{dx}\, Y(x, s)\right] - sBY(x, s) = -B\delta(x - x_0) \qquad (5.78)$$

The Laplace transformed probability current is

$$\hat{G}(x, s) = -\frac{1}{B}\left[\frac{d\varphi(x)}{dx}\, Y(x, s) + \frac{dY(x, s)}{dx}\right] \qquad (5.79)$$

In Laplace transform terms the timescale (5.75) has the form

$$\vartheta = \lim_{s \to 0} \frac{s\hat{P}(s) - P(\infty)}{s[P(0) - P(\infty)]} \qquad (5.80)$$

where according to (5.74) we have

$$\hat{P}(s) = \int_0^\infty P(t) e^{-st}\, dt = \int_c^d Y(x, s)\, dx$$

Integration of Eq. (5.72) with respect to $x$ between the limits $c$ and $d$ leads to

$$\frac{dP(t)}{dt} = -\int_c^d \frac{\partial G(x, t)}{\partial x}\, dx = G(c, t) - G(d, t)$$

In the Laplace transform terms we get

$$s\hat{P}(s) - P(0) = \hat{G}(c, s) - \hat{G}(d, s) \qquad (5.81)$$

Substituting (5.81) into (5.80), one obtains

$$\vartheta = \lim_{s \to 0} \frac{[P(0) - P(\infty)] - [\hat{G}(d,s) - \hat{G}(c,s)]}{s[P(0) - P(\infty)]} \tag{5.82}$$

For the relaxation time (5.76) in a single point we find in a similar manner

$$\Theta(\ell) = \lim_{s \to 0} \frac{W(\ell, \infty) - sY(\ell, s)}{sW(\ell, \infty)} \tag{5.83}$$

Thus, to obtain the timescales $\vartheta$ and $\Theta(\ell)$ we need to find the solution $Y(x,s)$ of Eq. (5.78) at appropriate boundary conditions and evaluate limits (5.82) and (5.83) for $s \to 0$. It is precisely that way by which the various timescales for piecewise-linear and piecewise-parabolic potential profiles were derived in Refs. 32 and 33.

But now we have an arbitrary potential profile $\varphi(x)$ in Eq. (5.78). Nobody knows a solution of this equation for any $\varphi(x)$. That is why we shall use another approach.

Let us note that it is not necessary to know the solution $Y(x,s)$ as a whole, but its behavior at $s \to 0$ only. For this reason we expand $Y(x,s)$ and $\hat{G}(x,s)$ in power series in $s$:

$$\begin{aligned} Z(x,s) &\equiv sY(x,s) = Z_0(x) + sZ_1(x) + s^2Z_2(x) + \cdots \\ H(x,s) &\equiv s\hat{G}(x,s) = H_0(x) + sH_1(x) + s^2H_2(x) + \cdots \end{aligned} \tag{5.84}$$

In accordance with the limit theorems of the Laplace transformation (see, e.g., Ref. 81), we obtain

$$\begin{aligned} Z_0(x) &= \lim_{s \to 0} sY(x,s) = W(x, \infty) = W_{st}(x) \\ H_0(x) &= \lim_{s \to 0} s\hat{G}(x,s) = G(x, \infty) = G_{st}(x) \end{aligned}$$

It is obvious that the steady-state quantities of the probability density $W_{st}(x)$ and the probability current $G_{st}(x)$ may be obtained without any difficulties at appropriate boundary conditions directly from Eq. (5.72).

Inserting (5.84) into (5.82), we obtain

$$\begin{aligned} \vartheta = \lim_{s \to 0} \Bigg\{ &\frac{[P(0) - P(\infty)] - [H_1(d) - H_1(c)]}{s[P(0) - P(\infty)]} \\ &- \frac{[H_2(d) - H_2(c)]}{[P(0) - P(\infty)]} - s\frac{[H_3(d) - H_3(c)]}{[P(0) - P(\infty)]} - \cdots \Bigg\} \end{aligned}$$

Here it is taken into account that for all profiles in question the steady-state probability current $H_0(x)$ equals 0 for any finite $x$.

As will be demonstrated below for all profiles the following condition takes place

$$H_1(d) - H_1(c) = P(0) - P(\infty) \tag{5.85}$$

Consequently, the desired timescale reads as follows:

$$\vartheta = \frac{H_2(c) - H_2(d)}{P(0) - P(\infty)} \tag{5.86}$$

Substituting (5.84) into (5.83), we find the relaxation time in a single point:

$$\Theta(\ell) = -\frac{Z_1(\ell)}{W(\ell, \infty)} = -\frac{Z_1(\ell)}{Z_0(\ell)} \tag{5.87}$$

Hence, to attain our aim it is necessary to calculate functions $H_2(x)$ and $Z_1(x)$. Inserting (5.84) into (5.78) and (5.79), it is easy to write the following equations:

$$\frac{d}{dx}\left(\frac{dZ_0}{dx} + \varphi' Z_0\right) = 0 \qquad \text{(a)}$$

$$\frac{d}{dx}\left(\frac{dZ_1}{dx} + \varphi' Z_1\right) = BZ_0 - B\delta(x - x_0) \qquad \text{(b)} \tag{5.88}$$

$$\frac{d}{dx}\left(\frac{dZ_2}{dx} + \varphi' Z_2\right) = BZ_1 \qquad \text{(c)}$$

$$\vdots \qquad \vdots$$

$$H_0 = -\frac{1}{B}\left[\frac{dZ_0}{dx} + \varphi' Z_0\right] \qquad \text{(a)}$$

$$H_1 = -\frac{1}{B}\left[\frac{dZ_1}{dx} + \varphi' Z_1\right] \qquad \text{(b)} \tag{5.89}$$

$$H_2 = -\frac{1}{B}\left[\frac{dZ_2}{dx} + \varphi' Z_2\right] \qquad \text{(c)}$$

$$\vdots \qquad \vdots$$

From (5.88) and (5.89) we obtain

$$\frac{dH_0}{dx} = 0 \qquad \text{(a)}$$

$$\frac{dH_1}{dx} = -Z_0 + \delta(x - x_0) \qquad \text{(b)} \tag{5.90}$$

$$\frac{dH_2}{dx} = -Z_1 \qquad \text{(c)}$$

$$\vdots \qquad \vdots$$

Combining Eqs. (5.88)–(5.90), one finds the closed equations for $H(x)$:

$$\frac{dH_0}{dx} = 0 \qquad \text{(a)}$$

$$\frac{d^2H_1}{dx^2} + \varphi'\frac{dH_1}{dx} = BH_0 + \left(\frac{d}{dx} + \varphi'\right)\delta(x - x_0) \qquad \text{(b)}$$

$$\frac{d^2H_2}{dx^2} + \varphi'\frac{dH_2}{dx} = BH_1 \qquad \text{(c)}$$

$$\vdots \qquad \vdots$$

(5.91)

When solving these equations it is necessary to take into account the boundary conditions that are different for different types of potential profiles.

### 3.  Basic Results Relating to Relaxation Times

Let us begin our consideration in detail from the derivation of the relaxation time which implies that the potential profile $\varphi(x)$ of type I tends fast enough to plus infinity at $x \to \pm\infty$. In this case the boundary conditions are $G(\pm\infty, t) = 0$; that is, all functions $H_k(x)$ must be zero at $x = \pm\infty$. According to (5.73) the steady-state distribution equals

$$W_{st}(x) = Z_0(x) = Ae^{-\varphi(x)}, \qquad A = 1\bigg/ \int_{-\infty}^{+\infty} e^{-\varphi(x)}\,dx > 0$$

This situation is depicted in Fig. 3, where the decision interval $[c, d]$ is chosen in accordance with the concrete stated task and may involve or not involve the initial distribution $W(x, 0) = \delta(x - x_0)$.

It is clear that the probability $P(t) = \int_c^d W(x, t)\,dx$ varies from $P(0) = 1$ or $P(0) = 0$ to $P(\infty) = A\int_c^d e^{-\varphi(v)}\,dv$.

From Eq. (5.90b) it follows that

$$H_1(x) = -A\int_{-\infty}^{x} e^{-\varphi(v)}\,dv + 1(x - x_0) + C$$

where $C$ is arbitrary constant and the unit step function $1(x)$ is defined as

$$1(x) = \begin{cases} 0, & x < 0 \\ 1/2, & x = 0 \\ 1, & x > 0 \end{cases}$$

The boundary conditions $H_1(\pm\infty) = 0$ lead to $C = 0$. Hence

$$H_1(x) = -F(x) + 1(x - x_0) \qquad (5.92)$$

where

$$F(x) = A \int_{-\infty}^{x} e^{-\varphi(v)} \, dv \qquad (5.93)$$

is the integral distribution function for the probability density $W_{st}(x)$.

It is easy to see that

$$H_1(d) - H_1(c) = \begin{cases} -A \int_c^d e^{-\varphi(v)} dv + 1 = P(0) - P(\infty), & c < x_0 < d \\ -A \int_c^d e^{-\varphi(v)} \, dv = P(0) - P(\infty), & x_0 < c < d \end{cases}$$

Thus, condition (5.85) is true for potential profiles of type I, and therefore the relaxation time $\Theta$ is really found according to (5.86).

The function $H_2(x)$ may be found from (5.91c) taking into account (5.92) and (5.93):

$$H_2 = C_1 F(x) + B \int_{-\infty}^{x} e^{-\varphi(u)} \, du \int_{-\infty}^{u} e^{\varphi(v)} [1(v - x_0) - F(v)] \, dv + C_2$$

where $C_1$ and $C_2$ are arbitrary constants, which must be defined from boundary conditions $H_2(\pm\infty) = 0$. As the result, we have $C_2 = 0$ and

$$C_1 = -B \int_{-\infty}^{+\infty} e^{-\varphi(u)} \, du \int_{-\infty}^{u} e^{\varphi(v)} [1(v - x_0) - F(v)] \, dv$$

Consequently,

$$H_2(x) = B \left\{ \int_{-\infty}^{x} e^{-\varphi(u)} \, du \int_{-\infty}^{u} e^{\varphi(v)} [1(v - x_0) - F(v)] \, dv \right.$$
$$\left. -F(x) \int_{-\infty}^{+\infty} e^{-\varphi(u)} \, du \int_{-\infty}^{u} e^{\varphi(v)} [1(v - x_0) - F(v)] \, dv \right\} \qquad (5.94)$$

By changing the integration order, one obtains

$$H_2(c) - H_2(d) = \frac{B}{A} \left\{ P(\infty) \int_{-\infty}^{+\infty} e^{\varphi(v)} [1(v - x_0) - F(v)][1 - F(v)] \, dv \right.$$
$$- P(\infty) \int_{-\infty}^{c} e^{\varphi(v)} [1(v - x_0) - F(v)] \, dv$$
$$\left. - \int_{c}^{d} e^{\varphi(v)} [1(v - x_0) - F(v)][F(d) - F(v)] \, dv \right\} \qquad (5.95)$$

*The Initial Probability Distribution Is Within the Decision Interval.* After some transformations taking into account that $c < x_0 < d$, we obtain finally from (5.95) and from (see (5.86))

$$\Theta = \frac{H_2(c) - H_2(d)}{1 - P(\infty)}$$

the following quantity of the relaxation time for the situation depicted in Fig. 3(a):

$$\Theta = \frac{B}{A(1 - P_\infty)} \left\{ (1 - P_\infty) \int_c^d e^{\varphi(v)} F(v)[1 - F(v)] \, dv \right.$$

$$+ P_\infty \int_{-\infty}^c e^{\varphi(v)} F^2(v) \, dv + P_\infty \int_d^\infty e^{\varphi(v)} [1 - F(v)]^2 \, dv$$

$$\left. - (1 - F_2) \int_c^{x_0} e^{\varphi(v)} F(v) \, dv - F_1 \int_{x_0}^d e^{\varphi(v)} [1 - F(v)] \, dv \right\} \quad (5.96)$$

where $F_1 = F(c)$, $F_2 = F(d)$, and $P_\infty \equiv P(\infty)$.

In the particular case where $c = -\infty$ ($F(d) = P_\infty$, $F(c) = 0$) the relaxation time is given by

$$\Theta = \frac{B}{A} \left\{ \int_{x_0}^d e^{\varphi(v)} F(v)[1 - F(v)] \, dv + \frac{P_\infty}{(1 - P_\infty)} \int_d^\infty e^{\varphi(v)} [1 - F(v)]^2 \, dv \right.$$

$$\left. - \int_{-\infty}^{x_0} e^{\varphi(v)} F^2(v) \, dv \right\} \quad (5.97)$$

If, on the other hand, $d = \infty$ ($F(d) = 1$, $F(c) = 1 - P_\infty$) from (5.96), one finds

$$\Theta = \frac{B}{A} \left\{ \int_c^{x_0} e^{\varphi(v)} F(v)[1 - F(v)] \, dv + \frac{P_\infty}{(1 - P_\infty)} \int_{-\infty}^c e^{\varphi(v)} F^2(v) \, dv \right.$$

$$\left. - \int_{x_0}^\infty e^{\varphi(v)} [1 - F(v)]^2 \, dv \right\} \quad (5.98)$$

In the case of the symmetry of the potential profile with respect to the coordinate of the initial distribution $x_0$ (which may be taken as $x_0 = 0$) and of the symmetrical decision interval $[-d, +d]$ from (5.96), one can obtain the following expression for the relaxation time:

$$\Theta = \frac{B}{A_1} \left\{ \int_0^d e^{\varphi(v)} f(v)[1 - f(v)] \, dv + \frac{P_\infty}{(1 - P_\infty)} \int_d^\infty e^{\varphi(v)} [1 - f(v)]^2 \, dv \right\} \quad (5.99)$$

where a new function is introduced:

$$f(x) = A_1 \int_0^x e^{-\varphi(v)}\, dv, \qquad A_1^{-1} = \int_0^\infty e^{-\varphi(v)}\, dv, \qquad P_\infty = f(d)$$

As may be shown, the same relaxation time (5.99) will take place if the initial distribution is located in an immediate vicinity of the origin (i.e., when $x_0 = +0$) and when the reflecting wall is sited at the point $x = 0$.

In the situation where there are two reflecting walls at the points $\lambda_1 = 0$ and $\lambda_2 = \lambda$ and the decision interval is $[c = 0, \, d]$, so that $0 < x_0 < d < \lambda$, expression (5.96) becomes ($F_1 = 0$, $F_2 = P_\infty$)

$$\Theta = \frac{B}{A_2} \left\{ \int_0^d e^{\varphi(v)} \psi(v)[1 - \psi(v)]\, dv + \frac{P_\infty}{(1 - P_\infty)} \int_d^\lambda e^{\varphi(v)}[1 - \psi(v)]^2\, dv \right.$$

$$\left. - \int_0^{x_0} e^{\varphi(v)} \psi(v)\, dv \right\} \tag{5.100}$$

where

$$\psi(x) = A_2 \int_0^x e^{-\varphi(v)}\, dv, \qquad A_2^{-1} = \int_0^\lambda e^{-\varphi(v)}\, dv, \qquad P_\infty = \psi(d)$$

*The Initial Probability Distribution Lies Outside the Decision Interval.* In this case, which is depicted in Fig. 3(b), the chosen points $c$ and $d$ satisfy the inequalities $x_0 < c < d$. Because of $P(0) = 0$ the relaxation time is now [see (5.86)]

$$\Theta = \frac{H_2(d) - H_2(c)}{P_\infty} \tag{5.101}$$

The numerator of this fraction is represented by (5.95) as before, but due to the above-mentioned inequalities the relaxation time is expressed by another formula ($F_1 = F(c)$, $F_2 = F(d)$):

$$\Theta = \frac{B}{A} \left\{ \int_{-\infty}^{+\infty} e^{\varphi(v)} F(v)[1 - F(v)]\, dv - \int_{-\infty}^{x_0} e^{\varphi(v)} F(v)\, dv - \int_d^\infty e^{\varphi(v)}[1 - F(v)]\, dv \right.$$

$$\left. - \frac{1}{F_2 - F_1} \int_c^d e^{\varphi(v)}[1 - F(v)][F(v) - F_1]\, dv \right\} \tag{5.102}$$

In particular case where $d = \infty$ ($F_2 = 1$, $P_\infty = 1 - F_1$)

$$
\Theta = \frac{B}{A} \left\{ \int_{x_0}^{c} e^{\varphi(v)} F(v)[1 - F(v)] \, dv + \frac{F_1}{1 - F_1} \int_{c}^{\infty} e^{\varphi(v)} [1 - F(v)]^2 \, dv \right.
$$
$$
\left. - \int_{-\infty}^{x_0} e^{\varphi(v)} F(v)^2 \, dv \right\} \tag{5.103}
$$

*The Relaxation Time in a Single Point.* We turn now to formula (5.87). The function $Z_1(x)$ may be obtained in two ways: First, one can solve Eq. (5.88b), second, one can use Eq. (5.90c) keeping in mind that the function $H_2(x)$ is already obtained [see formula (5.94)] and in addition it satisfies the required boundary conditions ($H_2(\pm\infty) = 0$).

The second way is shorter. From (5.90c) and (5.94) we have

$$
-Z_1(x) = \frac{dH_2(x)}{dx} = B e^{-\varphi(x)} \left\{ \int_{-\infty}^{x} e^{-\varphi(v)} [1(v - x_0) - F(v)] \, dv \right.
$$
$$
\left. - A \int_{-\infty}^{+\infty} e^{-\varphi(u)} \, du \int_{-\infty}^{u} e^{\varphi(v)} [1(v - x_0) - F(v)] \, dv \right\}
$$

Hence, according to (5.87) and by changing the order of integration in the last integral, the relaxation time may be expressed as

$$
\Theta(\ell) = \frac{B}{A} \left\{ \int_{-\infty}^{\ell} e^{\varphi(v)} [1(v - x_0) - F(v)] \, dv - \int_{-\infty}^{+\infty} e^{\varphi(v)} [1(v - x_0) \right.
$$
$$
\left. - F(v)][1 - F(v)] \, dv \right\} \tag{5.104}
$$

Considering the relaxation time at the point $x = \ell > x_0$ [Fig. 3(e)], one obtains the following after some transformations:

$$
\Theta(\ell) = \frac{B}{A} \left\{ \int_{-\infty}^{+\infty} e^{\varphi(v)} F(v)[1 - F(v)] \, dv - \int_{-\infty}^{x_0} e^{\varphi(v)} F(v) \, dv \right.
$$
$$
\left. - \int_{\ell}^{\infty} e^{\varphi(v)} [1 - F(v)] \, dv \right\} \tag{5.105}
$$

It is easy to check that the same relaxation time may be derived from expression (5.102) for $c \le \ell \le d$ at $c \to \ell$, $d \to \ell$.

For the case where $\ell < x_0$ [Fig. 3(f)] from (5.104) after some transformations, it follows that the relaxation time takes the form

$$
\Theta(\ell) = \frac{B}{A} \left\{ \int_{-\infty}^{+\infty} e^{\varphi(v)} F(v)[1 - F(v)] \, dv - \int_{-\infty}^{\ell} e^{\varphi(v)} F(v) \, dv \right.
$$
$$
\left. - \int_{x_0}^{\infty} e^{\varphi(v)}[1 - F(v)] \, dv \right\} \tag{5.106}
$$

When comparing (5.106) with (5.105), it becomes evident that these expressions coincide to make the interchange $x_0 \leftrightarrow \ell$. This fact demonstrates the so-called *reciprocity principle*: In any linear system, some effect does not vary if the source (at position $x = x_0$) and the observation point ($x = \ell$) will be interchanged. The linearity of our system is represented by the linearity of the Fokker–Planck equation (5.72).

### 4.  Basic Results Relating to Decay Times

*Potential Profiles of Type II.* Consider now the potential profiles of type II, which may represent the occurrence of the metastable state or the nonstable state (Fig. 4) and which tend to plus infinity at $x \to -\infty$ and to minus infinity at $x \to +\infty$. The boundary conditions are now $G(-\infty, t) = 0$ and $W(+\infty, t) = 0$. For this reason all functions $H_k(x)$ must be zero at $x = -\infty$ and all functions $Z_k(x)$ must be zero at $x = +\infty$. For such potential profiles the nonzero steady-state distributions do not exist and consequently $Z_0(x) \equiv 0$.

As for probability $P(t) = \int_c^d W(x, t) \, dx$, where $c < x_0 < d$, it is evident that $P(0) = 1$ and $P(\infty) = 0$. From (5.86) it follows that (denoting $\tau$ as the decay time of the metastable state or the decay time of the unstable state)

$$
\tau = H_2(c) - H_2(d) \tag{5.107}
$$

From Eq. (5.90b) we get $H_1(x) = 1(x - x_0)$ and consequently condition (5.85) is fulfilled. From Eq. (5.91c) one can find

$$
\frac{dH_2(x)}{dx} = e^{-\varphi(x)} \left[ C_1 + B \int_{-\infty}^{x} e^{\varphi(v)} 1(v - x_0) \, dv \right] \tag{5.108}
$$

where $C_1$ is an arbitrary constant.

Because $Z_1(x) = -dH_2(x)/dx$ [see Eq. (5.90c)] must be zero at $x = +\infty$, we obtain

$$
C_1 = -B \int_{-\infty}^{+\infty} e^{\varphi(v)} 1(v - x_0) \, dv = -B \int_{x_0}^{+\infty} e^{\varphi(v)} \, dv
$$

Consequently, the integration of Eq. (5.108) gives

$$H_2(x) = B \left\{ \int_{-\infty}^{x} e^{-\varphi(u)} \, du \int_{-\infty}^{u} e^{\varphi(v)} 1(v - x_0) \, dv \right.$$
$$\left. - \int_{x_0}^{+\infty} e^{\varphi(v)} \, dv \cdot \int_{-\infty}^{x} e^{-\varphi(v)} \, dv \right\} + C_2$$

Because of $H_2(-\infty) = 0$, an arbitrary constant $C_2$ must be zero.
After some transformations we have

$$H_2(x) = B \begin{cases} \int_{x_0}^{x} e^{\varphi(v)} \, dv \int_{v}^{x} e^{-\varphi(u)} \, du - \int_{x_0}^{\infty} e^{\varphi(v)} \, dv \cdot \int_{-\infty}^{x} e^{-\varphi(u)} \, du, & x > x_0 \\ - \int_{x_0}^{\infty} e^{\varphi(v)} \, dv \cdot \int_{-\infty}^{x} e^{-\varphi(u)} \, du, & x < x_0 \end{cases}$$

Hence, according to (5.107), taking into account that $c < x_0 < d$, we finally arrive at the exact expression of the escape time of the Brownian particles from decision interval $[c, d]$ for an arbitrary potential profile $\varphi(x)$ of type II (Fig. 4):

$$\tau = B \left\{ \int_{x_0}^{d} e^{\varphi(v)} \, dv \int_{c}^{v} e^{-\varphi(u)} \, du + \int_{d}^{\infty} e^{\varphi(v)} \, dv \cdot \int_{c}^{d} e^{-\varphi(u)} \, du \right\} \qquad (5.109)$$

In the case where the absorbing boundary is at $x = \lambda_2 > d$, expression (5.109) takes the form

$$\tau = B \left\{ \int_{x_0}^{d} e^{\varphi(v)} \, dv \int_{c}^{v} e^{-\varphi(u)} \, du + \int_{d}^{\lambda_2} e^{\varphi(v)} \, dv \cdot \int_{c}^{d} e^{-\varphi(u)} \, du \right\} \qquad (5.110)$$

For the decision interval extended to the absorbing boundary ($d = \lambda_2$) from (5.110), one obtains

$$\tau = B \int_{x_0}^{\lambda_2} e^{\varphi(v)} \, dv \int_{c}^{v} e^{-\varphi(u)} \, du \qquad (5.111)$$

If, in addition, $c = -\infty$, we have come to the MFPT

$$\tau = B \int_{x_0}^{\lambda_2} e^{\varphi(v)} \, dv \int_{-\infty}^{v} e^{-\varphi(u)} \, du \equiv T(x_0, \lambda_2) \qquad (5.112)$$

that we have already mentioned. In other words, the MFPT coincides with the decay time of the state existing in decision interval $[-\infty, \lambda_2]$.

For the decision interval $[-\infty, d]$, escape time (5.109) reads

$$\tau = T(x_0, d) + B\left\{ \int_d^\infty e^{\varphi(v)}\, dv \cdot \int_{-\infty}^d e^{-\varphi(u)}\, du \right\} \qquad (5.113)$$

We obtain $\tau > T(x_0, d)$. Such, indeed, is the case, because the escape of the Brownian particles from the decision interval is more rapid when the absorbing wall is at $x = d$ than it is in the presence of the comparatively slowly dropping potential profile (see Fig. 4).

In the case where the reflecting wall and the absorbing wall are at the ends of the decision interval $[c, d]$, the decay time

$$\tau = B \int_{x_0}^d e^{\varphi(v)}\, dv \int_c^v e^{-\varphi(u)}\, du \qquad (5.114)$$

obtained from (5.110) coincides with the MFPT.

We now call attention to an interesting paradoxical fact: For two different situations we have obtained the same decay times represented by formulas (5.111) and (5.114). It is obvious that the processes of changing the probability $P(t) = \int_c^d W(x, t)\, dx$ from $P(0) = 1$ to $P(\infty) = 0$ in the two above-mentioned situations must be distinct because of various potential profiles $\varphi(x)$ for $x < c$. What is the reason for this coincidence of the decay times? This occurrence is due to integral properties of the timescales defined by (5.75). Different behaviors of $P(t)$ may lead to the same $\vartheta$.

Let us consider, for example, two simplest cases where this fact takes place (Fig. 6). There are two different potential profiles (a) and (b) with $x_0 = +0$ and with the same decision interval $[0, d]$. The decay time of the metastable state arranged in the decision interval according to (5.111) and (5.114) is the same and equals $\tau = d^2/D$. At the same time, it is clear that evolutions of $P(t)$ are different. In case (b), decreasing of the probability is precisely faster because of the fastest outcome of the initial distribution from the decision interval to the left.

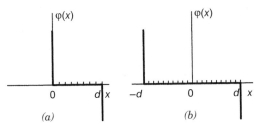

**Figure 6.**  The simplest example of two different [(a) and (b)] potential profiles with the same lifetime $\tau$.

*Potential Profiles of Type III.* We now direct our attention to the potential profiles of type III (Fig. 5). The boundary conditions are now $W(\pm\infty) = 0$. Consequently, all functions $Z_k(x)$ must be zero at $x = \pm\infty$. As before, $Z_0(x) \equiv 0$, $P(t) = \int_c^d W(x,t)\,dx$, $c < x_0 < d$, $P(0) = 1$, $P(\infty) = 0$, and

$$\tau = H_2(c) - H_2(d)$$

The calculation of $H_1(x)$ from Eq. (5.90b) gives

$$H_1(x) = 1(x - x_0) + C_1$$

where an arbitrary constant $C_1$ remains unknown because we do not know boundary conditions on the functions $H_k(x)$. At the same time, condition (5.85) is fulfilled.

The calculation of $dH_2(x)/dx$ from Eq. (5.91c) results in

$$\frac{dH_2(x)}{dx} = e^{-\varphi(x)}\left[C_2 + B \int_{-\infty}^{x} e^{\varphi(v)}[1(v - x_0) + C_1]\,dv\right]$$

Taking into account Eq. (5.90c), we can determine arbitrary constants $C_1$ and $C_2$ from the boundary conditions: $-Z_1(\pm\infty) = \frac{dH_2(x)}{dx}\big|_{x=\pm\infty} = 0$. This leads to

$$C_1 = -\int_{x_0}^{+\infty} e^{\varphi(v)}\,dv \bigg/ \int_{-\infty}^{+\infty} e^{\varphi(v)}\,dv, \qquad C_2 = -B\int_{-\infty}^{+\infty} e^{\varphi(v)}[1(v - x_0) + C_1]\,dv$$

and to

$$H_2(x) = -B\left\{\int_{-\infty}^{x} e^{-\varphi(u)}\,du \int_{u}^{+\infty} e^{\varphi(v)}[1(v - x_0) + C_1]\,dv\right\} + C_3$$

Therefore, independently of the value of an arbitrary constant $C_3$, we obtain

$$\tau = B\int_{c}^{d} e^{-\varphi(u)}\,du \int_{u}^{\infty} e^{\varphi(v)}[1(v - x_0) + C_1]\,dv$$

By reversing the order of the integrals, we arrive at the net expression for the decay time for the situation depicted in Fig. 5:

$$\tau = B\left\{f_1 \cdot \int_{x_0}^{d} e^{\varphi(v)}\,dv \int_{c}^{v} e^{-\varphi(u)}\,du - f_2 \cdot \int_{c}^{x_0} e^{\varphi(v)}\,dv \int_{c}^{v} e^{-\varphi(u)}\,du \right.$$
$$\left. + f_1 \cdot \int_{d}^{\infty} e^{\varphi(v)}\,dv \cdot \int_{c}^{d} e^{-\varphi(u)}\,du\right\} \tag{5.115}$$

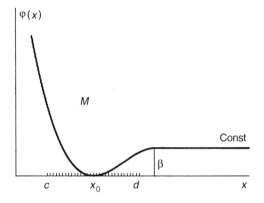

**Figure 7.** An example of the potential profile with the metastable state ($M$) for which the convergence condition is disrupt.

where

$$f_1 = \int_{-\infty}^{x_0} e^{\varphi(v)}\, dv/f_3, \quad f_2 = \int_{x_0}^{\infty} e^{\varphi(v)}\, dv/f_3, \quad f_3 = \int_{-\infty}^{\infty} e^{\varphi(v)}\, dv \qquad (5.116)$$

### 5. Some Comments and Interdependence of Relaxation and Decay Times

*Convergence Condition.* Let us return to the convergence condition of integrals (5.75) and (5.76). Consider, for example, the potential profile depicted in Fig. 7 and let us try to calculate the decay time of the metastable state $M$ in accordance with (5.75). It is easy to verify that for this potential profile $\vartheta = \infty$, that is, the integral in (5.75) diverges, though the probability $P(t)$ varies from $P(0) = 1$ to $P(\infty) = 0$. The important matter is that this probability changes from the initial value to the final one slowly enough. To determine the time characteristics for such profiles it is necessary to use another approach (see, e.g., Refs. 82–84), defining the relaxation time at the level $1/2$ (5.12).

It is very interesting to note that in this case the factor $e^{2\beta}$ arises in the escape time instead of the Kramers' factor $e^{\beta}$. This circumstance is associated with the good possibility for the Brownian particles to diffuse back to the potential well from a flat part of the potential profile, resulting in strong increasing of the escape time from the well (see, e.g., Ref. 83).

*Monotony Condition.* Let us turn back to the monotony condition of the variations of $P(t)$ or $W(\ell, t)$. If, for example, the point $\ell$ is arranged near $x_0$, where the initial probability distribution $W(x, 0) = \delta(x - x_0)$ is located, the probability density $W(\ell, t)$ early in the evolution may noticeably exceed the final value $W(\ell, \infty)$. For such a situation the relaxation time $\Theta(\ell)$ according to (5.76) may take not only a zero value, but also a negative one. In other words,

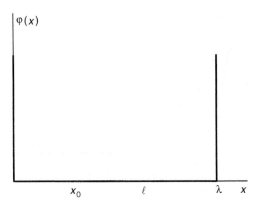

**Figure 8.** A rectangular potential well.

nonmonotony of $P(t)$ or $W(\ell, t)$ leads in accordance with (5.75) and (5.76) to defective results. We analyze this situation with aid of a simple example.

Consider the potential profile depicted in Fig. 8. At the points $x = 0$ and $x = \lambda$ the reflecting walls are arranged. The relaxation time at the point $\ell$ may be calculated with the aid of (5.105), taking into account that $\varphi(x) = 0$, $x \in [0, \lambda]$:

$$\Theta(\ell) = \frac{B}{A}\left\{\int_0^\lambda F(v)[1 - F(v)]\,dv - \int_0^{x_0} F(v)\,dv - \int_\ell^\lambda [1 - F(v)]\,dv\right\} \quad (5.117)$$

Here

$$A^{-1} = \int_0^\lambda dv = \lambda, \qquad F(v) = A\int_0^v d\xi = v/\lambda$$

As a result we have ($B = 2/D$)

$$\Theta(\ell) = \frac{\lambda^2}{3D}\left\{1 - 3\left(\frac{x_0}{\lambda}\right)^2 - 3\left(1 - \frac{\ell}{\lambda}\right)^2\right\} \quad (5.118)$$

Three terms in (5.117) correspond to three terms in (5.118).

We have obtained the relaxation time—that is, the time of attainment of the equilibrium state or, in other words, the transition time to the stationary distribution $W(\ell, \infty)$ at the point $\ell$ in the rectangular potential profile. This time depends on the delta-function position $x_0$ and on the observation point location $\ell$.

The relaxation time is maximal for $x_0 = 0$ and $\ell = \lambda$—that is, when $x_0$ and $\ell$ are widely spaced. If the distance between $x_0$ and $\ell$ decreases, the relaxation time goes down.

But if the distance between $x_0$ and $\ell$ is too small, expression (5.118) may give $\Theta(\ell) = 0$ or even $\Theta(\ell) < 0$. What does it mean? This signals that the condition of monotonic variation of $W(\ell, t)$ from zero to $W(\ell, \infty)$ is broken and that the obtained values of the relaxation times are false.

The special investigation has shown that, in particular, for $x_0 = 0$ the relaxation time defined by (5.118) is true for $\ell/\lambda \geq 0.5$. At the same time it would be reasonable to take the first term in (5.118)—that is, $\Theta = \lambda^2/3D$—as an upper bound for the relaxation time in the rectangular well to avoid the difficulties with the possible nonmonotonic behavior of the $W(\ell, t)$ regardless of the values $x_0$ and $\ell$.

The same reasoning may be extended on all above-obtained results regarding relaxation times, but with some care especially when there are no any potential barriers between $x_0$ and the observation point.

*Some Common Interdependence of Relaxation Times.* Consider an arbitrary disposition of the initial distribution point $x_0$ and the boundary point $d$ [see Fig. 3(c,d)]. We may take into account two decision intervals $I_1 = [-\infty, d]$ and $I_2 = [d, +\infty]$ and define two probabilities:

$$P_1(t) = \int_{-\infty}^{d} W(x, t)\, dx, \qquad P_2(t) = \int_{d}^{\infty} W(x, t)\, dx$$

In line with these probabilities, we may introduce into consideration two relaxation times $\Theta_1$ and $\Theta_2$ according to a common definition (5.75). From the evident equality $P_1(t) + P_2(t) = 1$, it follows that $\Theta_1 = \Theta_2$ for both $d > x_0$ and $d < x_0$.

If there is the symmetrical potential profile $\varphi(x) = \varphi(-x)$ and the initial distribution is located at the origin (i.e., $x_0 = 0$), then all results concerning the relaxation times will be the same as for the potential profile in which the reflecting wall is at $x = 0$ and at $x_0 = +0$. This coincidence of the relaxation times may be proven in a common case taking into account that the probability current at $x = 0$ is equal to zero at any instant of time.

*Interdependence Between Relaxation and Decay Times.* Consider the potential profile of type I which is symmetrical with respect to $x = d > 0$ and where the initial distribution is located at $x_0 = 0$ and the decision interval is $[-\infty, d]$ (see Appendix). The relaxation time that follows from (5.97) is

$$\Theta = \frac{B}{A} \left\{ \int_0^d e^{\varphi(v)} F(v)[1 - F(v)]\, dv - \int_{-\infty}^0 e^{\varphi(v)} F^2(v)\, dv \right.$$
$$\left. + \frac{F(d)}{1 - F(d)} \int_d^\infty e^{\varphi(v)} [1 - F(v)]^2\, dv \right\}$$

This formula may be transformed into

$$\Theta = \frac{B}{A}\left\{\int_0^d e^{\varphi(v)}F(v)\,dv - \int_{-\infty}^d e^{\varphi(v)}F^2(v)\,dv\right.$$
$$\left. + \frac{F(d)}{1-F(d)}\int_d^\infty e^{\varphi(v)}[1-F(v)]^2\,dv\right\}$$

Because of the potential profile symmetry we have $F(d) = 1/2$, $F(d-z) = 1 - F(d+z)$ and the second integral goes to $\int_d^\infty e^{\varphi(v)}[1-F(v)]^2\,dv$. As a result, we have

$$\Theta = \frac{B}{A}\int_0^d e^{\varphi(v)}F(v)\,dv = B\int_0^d e^{\varphi(v)}\,dv\int_{-\infty}^v e^{-\varphi(u)}\,du = T(0,d) \qquad (5.119)$$

Thus, we have proven the so-called *principle of conformity* which was demonstrated for an arbitrary symmetrical potential profile of type I by another more complicated way in Refs. 70 and 85.

Consider the bistable one-level system with symmetrical potential profile $\varphi(x) = \varphi(-x)$ with respect to the origin, where the potential barrier of the height $\beta$ is located (Fig. 9). According to the principle of conformity and (5.119) the relaxation time $\Theta$ of this bistable system (the decision interval is $[-\infty, 0]$) is equal to

$$\Theta = B\int_{-d}^0 e^{\varphi(v)}\,dv\int_{-\infty}^v e^{-\varphi(u)}\,du = T(-d,0)$$

If we will locate the absorbing wall at the point $x = d$ (the dashed line in Fig. 9), the MFPT $T(-d,d)$ reads

$$T(-d,d) = B\int_{-d}^d e^{\varphi(v)}\,dv\int_{-\infty}^v e^{-\varphi(u)}\,du$$

After some transformations using the symmetrical properties of the potential profile, one finds

$$T(-d,d) = 2T(-d,0) + \tau_0$$

where

$$\tau_0 = 2B\int_0^d e^{\varphi(v)}\,dv\int_0^v e^{-\varphi(u)}\,du$$

It may be shown that for the high enough potential barrier $\beta \gg 1$ the quantity $\tau_0$ is significantly smaller than $2T(-d, 0)$ and we arrive at

$$\Theta = \frac{1}{2} T(-d, +d) \tag{5.120}$$

Expression (5.120) means that the relaxation time in an arbitrary symmetrical one-level bistable system is two times smaller than the MFPT $T(-d, +d)$—that is, two times smaller than the decay time of the metastable state—shown with the dashed line taken into account in Fig. 9.

This is concerned with the fact that in the case of the relaxation time, roughly speaking only *half of all* Brownian particles should leave the initial potential minimum to reach the equilibrium state, while for the profile of the decay time case *all* particles should leave the initial minimum. Expression (5.120), of course, is true only in the case of the sufficiently large potential barrier, separating the stable states of the bistable system, when the inverse probability current from the second minimum to the initial one may be neglected (see Ref. 33).

The comparison with known results (e.g., Refs. 2, 20, 21, 86, and 87) for variety of examples of metastable, bistable and periodic potentials was done in Refs. 32–35 and 85.

## 6. *Derivation of Moments of Transition Time*

In this section we restrict our consideration by potentials of type II only (in this case moments of transition time are moments of decay time). The approach, considered above for derivation of the mean transition time may also be used to obtain higher moments, given by Eq. (5.1).

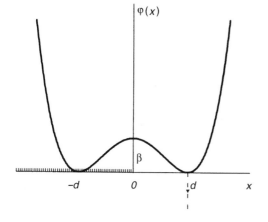

**Figure 9.** A potential profile of type I representing a one-level bistable system. The dashed line shows the absorbing wall.

Let us write in the explicit form the expressions for functions $H_n(x)$ (we remind that $H(x, s) = s\hat{G}(x, s) = H_0(x) + sH_1(x) + s^2H_2(x) + \cdots$) using the boundary conditions $W(+\infty, t) = 0$ and $G(-\infty, t) = 0$ ($H_1(x) = 1(x - x_0)$):

$$H_2(x) = -B \int_{-\infty}^{x} e^{-\varphi(v)} \int_{v}^{\infty} e^{\varphi(y)} 1(y - x_0) \, dy \, dv$$

$$H_n(x) = -B \int_{-\infty}^{x} e^{-\varphi(v)} \int_{v}^{\infty} e^{\varphi(y)} H_{n-1}(y) \, dy \, dv, \qquad n = 3, 4, 5, \ldots$$

(5.121)

As one can check, from formula (5.1) (taking the integral by parts and Laplace transforming it using the property $P(x_0, 0) - s\hat{P}(x_0, s) = \hat{G}(d, s) - \hat{G}(c, s)$ together with the expansion of $H(x, s)$ via $H_n(x)$, one can obtain the following expressions for moments of decay time [88]:

$$\begin{aligned}
\tau_1(c, x_0, d) &= -(H_2(d) - H_2(c)) \\
\tau_2(c, x_0, d) &= 2(H_3(d) - H_3(c)) \\
\tau_3(c, x_0, d) &= -2 \cdot 3(H_4(d) - H_4(c)), \ldots \\
\tau_n(c, x_0, d) &= (-1)^n_n!(H_{n+1}(d) - H_{n+1}(c))
\end{aligned}$$

(5.122)

From these recurrent relations, one obtains the following expression for the second moment in the case $c = -\infty$ ($c < x_0 < d$):

$$\begin{aligned}
\tau_2(-\infty, x_0, d) = 2B^2 \Bigg\{ &[\tau_1(-\infty, x_0, d)]^2 \\
&+ \int_{-\infty}^{d} e^{-\varphi(x)} dx \cdot \int_{x_0}^{\infty} e^{\varphi(v)} \int_{d}^{v} e^{-\varphi(u)} \int_{u}^{\infty} e^{\varphi(z)} \, dz \, du \, dv \\
&- \int_{x_0}^{d} e^{-\varphi(x)} \int_{x_0}^{x} e^{\varphi(v)} \int_{d}^{v} e^{-\varphi(u)} \int_{u}^{\infty} e^{\varphi(z)} \, dz \, du \, dv \, dx \Bigg\}
\end{aligned}$$

(5.123)

where $\tau_1(-\infty, x_0, d)$ is the first moment:

$$\begin{aligned}
\tau_1(-\infty, x_0, d) = B \Bigg\{ &\int_{-\infty}^{d} e^{-\varphi(x)} dx \cdot \int_{x_0}^{\infty} e^{\varphi(v)} \, dv \\
&- \int_{x_0}^{d} e^{-\varphi(x)} \int_{x_0}^{x} e^{\varphi(v)} \, dv \, dx \Bigg\}
\end{aligned}$$

(5.124)

Instead of analyzing the structure and the properties of the second and higher moments, in the next section, we will perform analysis of temporal evolution of the survival probability.

### 7.  Timescales of Evolution of Averages and Correlation Time

Analogically, one can apply the considered approach to derive timescales of evolution of averages [89] and correlation time [90,91]. We will search for the average $m_f(t)$ in the form

$$m_f(t) = \langle f(x) \rangle = \int_{-\infty}^{+\infty} f(x) W(x,t)\, dx \qquad (5.125)$$

As we have mentioned above, the FPE (5.72) is a continuity equation. To obtain necessary average $m_f(t)$ (5.125), let us multiply it by the function $f(x)$ and integrate with respect to $x$ from $-\infty$ to $+\infty$. Then we get

$$\frac{dm_f(t)}{dt} = -\int_{-\infty}^{+\infty} f(x)\frac{\partial G(x,t)}{\partial x}\, dx \qquad (5.126)$$

This is already an ordinary differential equation of the first order ($f(x)$ is a known deterministic function), but nobody knows how to find $G(x,t)$.

Let us define the characteristic scale of time evolution of the average $m_f(t)$ as an integral relaxation time:

$$\tau_f(x_0) = \frac{\int_0^\infty \left[ m_f(t) - m_f(\infty) \right] dt}{m_f(0) - m_f(\infty)} \qquad (5.127)$$

Definition (5.127) as before is general in the sense that it is valid for any initial condition. But here we restrict ourselves by the delta-shaped initial distribution and consider $\tau_f(x_0)$ as a function of $x_0$. For arbitrary initial distribution the required timescale may be obtained from $\tau_f(x_0)$ by separate averaging of both numerator and denominator $m_f(0) - m_f(\infty)$ over initial distribution, because $m_f(0)$ is also a function of $x_0$.

The restrictions of the definition (5.127) are the same as before: It gives correct results only for monotonically evolving functions $m_f(t)$ and $m_f(t)$ should fastly enough approach its steady-state value $m_f(\infty)$ for convergence of the integral in (5.127).

Performing Laplace transform of formula (5.127), Eq. (5.126) (Laplace transformation of (5.126) gives: $sm_f(s) - m_f(0) = -\int_{-\infty}^{+\infty} f(x)[\partial \hat{G}(x,s)/\partial x]\, dx$, where $m_f(s) = \int_0^\infty m_f(t)e^{-st}dt$) and combining them, we obtain

$$\tau_f(x_0) = \lim_{s \to 0} \frac{sm_f(s) - m_f(\infty)}{s[m_f(0) - m_f(\infty)]}$$

$$= \lim_{s \to 0} \frac{m_f(0) - m_f(\infty) - \int_{-\infty}^{+\infty} f(x)[\partial \hat{G}(x,s)/\partial x]\, dx}{s[m_f(0) - m_f(\infty)]} \qquad (5.128)$$

where $\hat{G}(x,s)$ is the Laplace transformation of the probability current $\hat{G}(x,s) = \int_0^\infty G(x,t)e^{-st}\, dt$.

Introducing the function $H(x,s) = s\hat{G}(x,s)$ and expanding it into power series in $s$, we restrict our computation by obtaining $H_2(x)$ only, since it can be demonstrated that

$$\tau_f(x_0) = -\frac{\int_{-\infty}^{+\infty} f(x)\, dH_2(x)}{[m_f(0) - m_f(\infty)]} \tag{5.129}$$

Substituting the concrete form of $H_2(x)$ (5.94) into formula (5.129), one can obtain the characteristic scale of time evolution of any average $m_f(t)$ for arbitrary potential such that $\varphi(\pm\infty) = \infty$:

$$\begin{aligned}
\tau_f(x_0) = \frac{B}{m_f(0) - m_f(\infty)} \Bigg\{ &\int_{-\infty}^{\infty} f(x)e^{-\varphi(x)} \int_{x_0}^{x} e^{\varphi(u)} F(u)\, du\, dx \\
&- A\int_{-\infty}^{\infty} f(x)e^{-\varphi(x)} dx \int_{-\infty}^{\infty} e^{-\varphi(x)} \int_{x_0}^{x} e^{\varphi(u)} F(u)\, du\, dx \\
&+ A\int_{-\infty}^{\infty} f(x)e^{-\varphi(x)} dx \int_{x_0}^{\infty} e^{-\varphi(x)} \int_{x_0}^{x} e^{\varphi(u)}\, du\, dx \\
&- \int_{x_0}^{\infty} f(x)e^{-\varphi(x)} \int_{x_0}^{x} e^{\varphi(u)}\, du\, dx \Bigg\}
\end{aligned} \tag{5.130}$$

where $F(x)$ is expressed by

$$F(u) = \int_{-\infty}^{u} e^{-\varphi(v)}\, dv \Big/ \int_{-\infty}^{+\infty} e^{-\varphi(v)}\, dv \tag{5.131}$$

and $A = 1/\int_{-\infty}^{+\infty} e^{-\varphi(x)} dx$.

Analogically, the correlation time as evolution time of the correlation function $K_x(t) = \langle x(t')x(t'+t)\rangle$ [90] or the correlation time of more general function $K_f(t) = \langle f(x(t'))f(x(t'+t))\rangle$ (in Ref. 91 the correlation time of $\sin(x(t))$ has been computed) may be obtained. Here we present the correlation time of $K_x(t) = \langle x(t')x(t'+t)\rangle$ [90] defined as

$$\tau_c = \frac{1}{\sigma^2} \int_0^{\infty} [K_x(t) - \langle x \rangle^2]\, dt \tag{5.132}$$

where $\sigma^2 = \langle x^2 \rangle - \langle x \rangle^2$. As can be demonstrated [90], the correlation time is given by

$$\tau_c = \frac{B}{A\sigma^2} \int_{-\infty}^{+\infty} e^{\varphi(u)} [H(u) - \langle x \rangle F(u)]^2\, du \tag{5.133}$$

where

$$H(x) = \int_{-\infty}^{x} uW(u)\,du, \qquad \langle x \rangle = H(\infty), \qquad \sigma^2 = \int_{-\infty}^{+\infty} (u - \langle u \rangle)^2 W(u)\,du$$

$$W(x) = Ae^{-\varphi(x)}, \qquad\qquad A = 1 \Big/ \int_{-\infty}^{+\infty} e^{-\varphi(x)}\,dx$$

One can check that this result coincides with the result by Risken and Jung (5.24) for the case of constant diffusion coefficient.

## VI.    TIME EVOLUTION OF OBSERVABLES

### A.    Time Constant Potentials

It is known that when the transition of an overdamped Brownian particle occurs over a potential barrier high enough in comparison with noise intensity $\Delta\Phi/kT \gg 1$, time evolution of many observables (e.g., the probability of transition or the correlation function) is a simple exponent $\sim \exp(-t/\tau)$ [2,15], where $\tau$ is the corresponding timescale (the mean transition time or the correlation time). Such representation of an observable is evident, for example, from the method of eigenfunction analysis. In this case the corresponding timescale (mean transition time) gives complete information about the probability evolution. The boundaries of validity of exponential approximation of the probability were studied in Refs. 24 and 30. In Ref. 24 the authors extended the mean first passage time to the case of "radiation" boundary condition and for two barrierless examples demonstrated good coincidence between exponential approximation and numerically obtained probability. In a more general case the exponential behavior of observables was demonstrated in Ref. 30 for relaxation processes in systems having steady states. Using the approach of "generalized moment approximation," the authors of Ref. 30 obtained the exact mean relaxation time to steady state (see Section V.C), and for particular example of a rectangular potential well they demonstrated good coincidence of exponential approximation with numerically obtained observables. They considered in Ref. 30 an example where the rectangular well did not have a potential barrier, and the authors of that paper supposed that their approach (and the corresponding formulas) should also give good approximation in tasks with diffusive barrier crossing for different examples of potentials and in a wide range of noise intensity.

In this section we will analyze the validity of exponential approximation of observables in wide range of noise intensity [88,89,91,92].

### 1.    Time Evolution of Survival Probability

Let us perform the analysis of temporal evolution of the probability $P(t, x_0)$ of a Brownian particle, located at the point $x_0$ ($t = 0$) within the interval $(c, d)$, to be at the time $t > 0$ inside the considered interval: $P(t, x_0) = \int_c^d W(x, t)\,dx$ (survival probability). We suppose that $c$ and $d$ are arbitrary chosen points of an arbitrary potential profile $\varphi(x)$, and boundary conditions at these points may be arbitrary: $W(c, t) \geq 0$ and $W(d, t) \geq 0$, and we restrict our consideration by profiles of type II (Fig. 4).

The corresponding moments of decay time were obtained above (5.122). Let us analyze in more detail their structure. One can represent the $n$th moment in the following form:

$$\tau_n(c, x_0, d) = n!\tau_1^n(c, x_0, d) + r_n(c, x_0, d) \tag{6.1}$$

This is a natural representation of $\tau_n(c, x_0, d)$ due to the structure of recurrent formulas (5.121), which is seen from the particular form of the first and the second moments given by (5.123) and (5.124). Using the approach, applied in the paper by Shenoy and Agarwal [93] for analysis of moments of the first passage time, it can be demonstrated that in the limit of a high barrier $\Delta\varphi \gg 1$ ($\Delta\varphi = \Delta\Phi/kT$ is the dimensionless barrier height) the remainders $r_n(c, x_0, d)$ in formula (6.1) may be neglected. For $\Delta\varphi \approx 1$, however, a rigorous analysis should be performed for estimation of $r_n(c, x_0, d)$. Let us suppose that the remainders $r_n(c, x_0, d)$ may be neglected in a wide range of parameters, and below we will check numerically when our assumption is valid.

The cumulants [2,43] of decay time $\mathscr{æ}_n$ are much more useful for our purpose to construct the probability $P(t, x_0)$—that is, the integral transformation of the introduced probability density of decay time $w_\tau(t, x_0)$ (5.2). Unlike the representation via moments, the Fourier transformation of the probability density (5.2)—the characteristic function—decomposed into the set of cumulants may be inversely transformed into the probability density.

Analogically to the representation for moments (6.1), a similar representation can be obtained for cumulants $\mathscr{æ}_n$:

$$\mathscr{æ}_n(c, x_0, d) = (n - 1)!\mathscr{æ}_1^n(c, x_0, d) + R_n(c, x_0, d) \tag{6.2}$$

It is known that the characteristic function $\Theta(\omega, x_0) = \int_0^\infty w_\tau(t, x_0)e^{j\omega t}\,dt$ ($j = \sqrt{-1}$) can be represented as the set of cumulants ($w_\tau(t, x_0) = 0$ for $t < 0$):

$$\Theta(\omega, x_0) = \exp\left[\sum_{n=1}^\infty \frac{\mathscr{æ}_n(c, x_0, d)}{n!}(j\omega)^n\right] \tag{6.3}$$

In the case where the remainders $R_n(c, x_0, d)$ in (6.2) (or $r_n(c, x_0, d)$ in (6.1)) may be neglected, the set (6.3) may be summarized and inverse Fourier transformed:

$$w_\tau(t, x_0) = \frac{e^{-t/\tau}}{\tau} \tag{6.4}$$

where $\tau$ is the mean decay time [34,35] ($\tau(c, x_0, d) \equiv \tau_1 \equiv \ae_1$):

$$\tau(c, x_0, d) = B \left\{ \int_{x_0}^{d} e^{\varphi(x)} \int_{c}^{x} e^{-\varphi(v)} \, dv \, dx + \int_{d}^{\infty} e^{\varphi(x)} dx \int_{c}^{d} e^{-\varphi(v)} \, dv \right\} \quad (6.5)$$

This expression is a direct transformation of formula (5.124), where now $c$ is arbitrary, such that $c < x_0 < d$.

Probably, a similar procedure was previously used (see Refs. 1 and 93–95) for summation of the set of moments of the first passage time, when exponential distribution of the first passage time probability density was demonstrated for the case of a high potential barrier in comparison with noise intensity.

Integrating probability density (6.4), taking into account definition (5.2), we get the following expression for the survival probability $P(t, x_0)$ ($P(0, x_0) = 1$, $P(\infty, x_0) = 0$):

$$P(t, x_0) = \exp(-t/\tau) \quad (6.6)$$

where mean decay time $\tau$ is expressed by (6.5). Probability (6.6) represents a well-known exponential decay of a metastable state with a high potential barrier [15]. Where is the boundary of validity of formula (6.6) and when can we neglect $r_n$ and $R_n$ in formulas (6.1) and (6.2)? To answer this question, we have considered three examples of potentials having metastable states and have compared numerically obtained survival probability $P(t, x_0) = \int_{c}^{d} W(x, t) \, dx$ with its exponential approximation (6.6). We used the usual explicit difference scheme to solve the FPE (5.72), supposing the reflecting boundary condition $G(c_b, t) = 0$ ($c_b < c$) far above the potential minimum and the absorbing one $W(d_b, t) = 0$ ($d_b > d$) far below the potential maximum, instead of boundary conditions at $\pm\infty$, such that the influence of phantom boundaries at $c_b$ and $d_b$ on the process of diffusion was negligible. The first considered system is described by the potential $\Phi(x) = ax^2 - bx^3$. We have taken the following particular parameters: $a = 2$ and $b = 1$, which lead to the barrier height $\Delta\Phi \approx 1.2$, $c = -2$, $d = 2a/3b$, and $kT = 0.5; 1; 3$ (in computer simulations we set the viscosity $h = 1$). The corresponding curves of the numerically simulated probability and its exponential approximation are presented in Fig. 10. In the worse case when $kT = 1$ the maximal difference between the corresponding curves is 3.2%. For comparison, there is also presented a curve of exponential approximation with the mean first passage time (MFPT) of the point $d$ for $kT = 1$ (dashed line). One can see that in the latter case the error is significantly larger.

The second considered system is described by the potential $\Phi(x) = ax^4 - bx^5$. We have taken the following particular parameters: $a = 1$ and $b = 0.5$, which lead to the barrier height $\Delta\Phi \approx 1.3$, $c = -1.5$, $d = 4a/5b$, and $kT = 0.5; 1; 3$. The corresponding curves of the numerically simulated probability and its

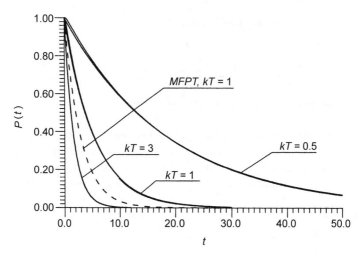

**Figure 10.** Evolution of the survival probability for the potential $\Phi(x) = ax^2 - bx^3$ for different values of noise intensity; the dashed curve denoted as MFPT (mean first passage time) represents exponential approximation with MFPT substituted into the factor of exponent.

exponential approximation are presented in Fig. 11. In the worse case ($kT = 1$) the maximal difference between the corresponding curves is 3.4%.

The third considered system is described by the potential $\Phi(x) = 1 - \cos(x) - ax$. This potential is multistable. We have considered it in the interval $[-10, 10]$, taking into account three neighboring minima. We have taken

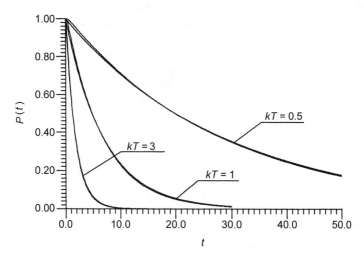

**Figure 11.** Evolution of the survival probability for the potential $\Phi(x) = ax^4 - bx^5$ for different values of noise intensity.

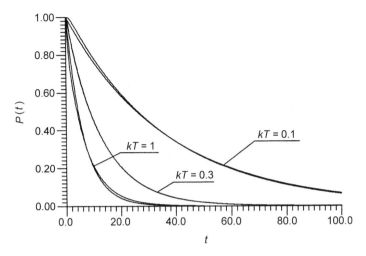

**Figure 12.** Evolution of the survival probability for the potential $\Phi(x) = 1 - \cos(x) - ax$ for different values of noise intensity.

$a = 0.85$, which leads to the barrier height $\Delta\Phi \approx 0.1$, $c = -\pi - \arcsin(a)$, $d = \pi - \arcsin(a)$, $x_0 = \arcsin(a)$, and $kT = 0.1; 0.3; 1$. The corresponding curves of the numerically simulated probability and its exponential approximation are presented in Fig. 12. In difference with two previous examples, this potential was considered in essentially longer interval and with smaller barrier. The difference between curves of the numerically simulated probability and its exponential approximation is larger. Nevertheless, the qualitative coincidence is good enough.

Finally, we have considered an example of metastable state without potential barrier: $\Phi(x) = -bx^3$, where $b = 1$, $x_0 = -1$, $d = 0$, $c = -3$, and $kT = 0.1$; $1; 5$. A dashed curve is used to present an exponential approximation with the MFPT of the point $d$ for $kT = 1$ (Fig. 13). It is seen that even for such an example the exponential approximation [with the mean decay time (6.5)] gives an adequate description of the probability evolution and that this approximation works better for larger noise intensity.

## 2.  Temporal Evolution of Averages

Once we know the required timescale of the evolution of an average, we can present the required average in the form

$$m_f(t) = (m_f(0) - m_f(\infty)) \exp(-t/\tau_f(x_0)) + m_f(\infty) \tag{6.7}$$

The applicability of this formula for several examples of the time evolution of the mean coordinate $m(t) = \langle x(t) \rangle$ will be checked below.

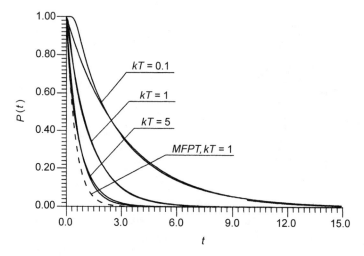

**Figure 13.** Evolution of the survival probability for the potential $\Phi(x) = -bx^3$ for different values of noise intensity; the dashed curve denoted as MFPT (mean first passage time) represents exponential approximation with MFPT substituted into the factor of exponent.

As an example of the description presented above, let us consider the time evolution of a mean coordinate of the Markov process:

$$m(t) = \langle x(t) \rangle = \int_{-\infty}^{+\infty} x W(x, t)\, dx \qquad (6.8)$$

The characteristic timescale of evolution of the mean coordinate in the general case may be easily obtained from (5.130) by substituting $x$ for $f(x)$. But for symmetric potentials $\varphi(x) = \varphi(-x)$ the expression of timescale of mean coordinate evolution may be significantly simplified ($m(\infty) = 0$):

$$\tau_m(x_0) = \frac{B}{x_0} \left\{ \int_0^{+\infty} x e^{-\varphi(x)}\, dx \cdot \int_0^{x_0} e^{\varphi(u)}\, du + \int_0^{x_0} x e^{-\varphi(x)} \int_{x_0}^{x} e^{\varphi(u)}\, du\, dx \right\} \qquad (6.9)$$

If $x_0 = 0$, it is not difficult to check that $\tau_m(x_0) = 0$.

Let us consider now some examples of symmetric potentials and check the applicability of exponential approximation:

$$m(t) = \langle x(t) \rangle = x_0 \exp(-t/\tau_m(x_0)) \qquad (6.10)$$

First, we should consider the time evolution of the mean coordinate in the monostable parabolic potential $\varphi(x) = ax^2/2$ (linear system), because for this case the time evolution of the mean is known:

$$m_{\text{par}}(t) = x_0 \exp\left(-at/B\right) \tag{6.11}$$

where $a = a'/kT$ and $B = h/kT$, so $\tau_{\text{par}} = B/a$ for the linear system and does not depend on noise intensity and the coordinate of initial distribution $x_0$. On the other hand, $\tau_m(x_0)$ is expressed by formula (6.9). Substituting parabolic potential $\varphi(x) = ax^2/2$ in formula (6.9), making simple evaluations and changing the order of integrals, it can be easily demonstrated that $\tau_m(x_0) = B/a = h/a' = \tau_{\text{par}}$, so for purely parabolic potential the time of mean evolution (6.9) is independent of both noise intensity and $x_0$, which is required. This fact proves the correctness of the used approach.

The second considered example is described by the monostable potential of the fourth order: $\Phi(x) = ax^4/4$. In this nonlinear case the applicability of exponential approximation significantly depends on the location of initial distribution and the noise intensity. Nevertheless, the exponential approximation of time evolution of the mean gives qualitatively correct results and may be used as first estimation in wide range of noise intensity (see Fig. 14, $a = 1$). Moreover, if we will increase noise intensity further, we will see that the error of our approximation decreases and for $kT = 50$ we obtain that the exponential approximation and the results of computer simulation coincide (see Fig. 15, plotted in the logarithmic scale, $a = 1$, $x_0 = 3$). From this plot we can conclude that the nonlinear system is "linearized" by a strong noise, an effect which is qualitatively obvious but which should be investigated further by the analysis of variance and higher cumulants.

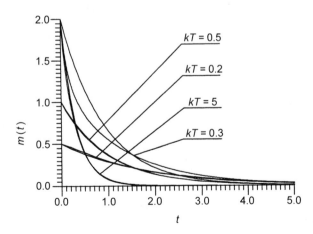

**Figure 14.** Evolution of the mean coordinate in the potential $\Phi(x) = x^4/4$ for different values of noise intensity.

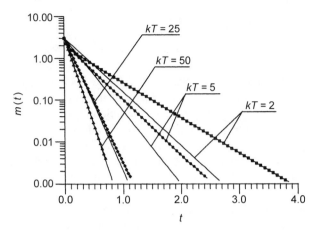

**Figure 15.** Evolution of the mean coordinate in the potential $\Phi(x) = x^4/4$ for different values of noise intensity (logarithmic scale).

The third considered example is described by the bistable potential—the so-called "quartic" potential: $\Phi(x) = ax^4/4 - bx^2/2$. In this case the applicability of exponential approximation also significantly depends on the coordinate of initial distribution. If $x_0$ is far from the potential minimum, then there exist two characteristic timescales: fast dynamic transition to potential minimum and slow noise-induced escape over potential barrier. In this case the exponential approximation gives a not-so-adequate description of temporal dynamics of the mean; however, it may be used as a first estimation. But if $x_0$ coincides with the potential minimum, then the exponential approximation of the mean coordinate differs only a few percent from results of computer simulation even in the case when noise intensity is significantly larger than the potential barrier height (strongly nonequilibrium case) (see Fig. 16, $a = 1$, $b = 2$, $x_0 = 1.414$). If, however, we consider the case where the initial distribution $x_0$ is far from the potential minimum, but the noise intensity is large, we will see again as in the previous example that essential nonlinearity of the potential is suppressed by strong fluctuations and the evolution of the mean coordinate becomes exponential (see Fig. 17, plotted in the logarithmic scale, $a = 1$, $b = 2$, $x_0 = 2.5$).

### 3. Discussion of Applicability of Single Exponential Approximation

Temporal behavior of the correlation function was studied in Ref. 91 using a particular example of the correlation function of $\sin x(t)$ in a periodic potential with periodic boundary conditions. In that case the use of single exponential approximation had also given a rather adequate description. The considered

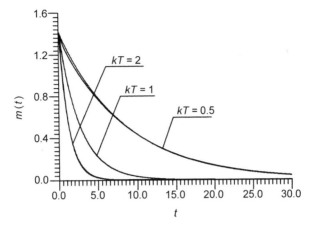

**Figure 16.** Evolution of the mean coordinate in the potential $\Phi(x) = x^4/4 - x^2$ for different values of noise intensity with initial distribution located in the potential minimum.

examples of observables lead to the following conclusions about the possibility to use single exponential approximation:

1. The single exponential approximation works especially well for observables that are less sensitive to the location of initial distribution, such as transition probabilities and correlation functions.
2. In all other cases it is usually enough to apply a double exponential approximation to obtain the required observable with a good precision,

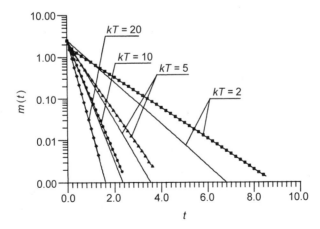

**Figure 17.** Evolution of the mean coordinate in the potential $\Phi(x) = x^4/4 - x^2$ for different values of noise intensity with initial distribution located far from a potential minimum (logarithmic scale).

and one can have recourse to a two-sided Padé approximation as suggested in Ref. 30.

3. The exponential approximation may lead to a significant error in the case when the noise intensity is small, the potential is tilted, and the barrier is absent (purely dynamical motion slightly modulated by noise perturbations). But, to the contrary, as it has been observed for all considered examples, the single exponential approximation is more adequate for a noise-assisted process: either (a) a noise-induced escape over a barrier or (b) motion under intensive fluctuations.

## B.  Time Periodic Potentials: Resonant Activation and Suprathreshold Stochastic Resonance

Investigation of nonlinear dynamical systems driven by noise and periodic signal is an important topic in many areas of physics. In the past decade several interesting phenomena, such as resonant activation [96], stochastic resonance [97], ratchet effect [98,99], noise-enhanced stability of unstable systems [61], and field-induced stabilization of activation processes [100], have been observed in these systems. In particular, for underdamped systems driven by periodic signals, the effect of resonant activation was reported by many authors (see Ref. 96 and references therein). The phenomenon consists of a decrease of the decay time of metastable state (or, equivalently, increase of decay rate) at certain frequencies of the periodic signal. For overdamped systems the resonant activation was first observed in the case of stochastically driven barriers fluctuating according to some probability distribution [101–105] and, recently, for barriers that are either (a) deterministically flipping between two states [105] or (b) continuously (sinusoidally) driven [100]. In the deterministic continuous case, however, the study was limited to the case of small driving amplitudes.

The application of methods described above with the help of adiabatic approximation allows us to study different characteristics of Markov processes subjected to driving signals. The use of exact mean transition times (instead of Kramers' time) helps to obtain an analytical description of probability evolution for arbitrary noise intensity and arbitrary amplitude of the driving signal, and such approximate description provides good coincidence with computer simulation results up to driving frequencies of the order of cutoff frequency of the system and even allows us to predict the effect of resonant activation for the case of a strong periodic driving (we call the "strong" or "suprathreshold" driving the case when the driving amplitude exceeds the static threshold amplitude).

As an example, let us consider again a system with a metastable state described by the potential of type II (Fig. 4) [106] (an example with the potential of type I was considered in Ref. 107):

$$\varphi(x,t) = (-bx^3 + ax^2 + Ax\cos(\omega t + \psi))/kT \qquad (6.12)$$

with $\psi$ an arbitrary phase. The particle is initially located in the potential well near the minimum. In the course of time the potential barrier moves up and down and the probability to find a particle in the minimum tends to zero. Again we are interested in the evolution of the survival probability:

$$P(t) = \int_{-\infty}^{d} W(x,t)\,dx \tag{6.13}$$

where $d$ is the coordinate of the barrier top at the instant of time when the barrier height has its maximal value.

For the analysis of the resonant activation effect in the considered system it is suitable to use as monitoring characteristic the first moment of the expansion of the survival probability (6.13). If one decomposes the probability to the set of moments as was done above for time-constant potentials, it can be demonstrated that $\tau(\omega)$ defined as

$$\tau(\omega) = \frac{\int_0^{\infty}[P(t) - P(\infty)]\,dt}{[P(0) - P(\infty)]} \tag{6.14}$$

is the first moment of such expansion [the mean decay time (MDT)]. In the considered case, $\tau(\omega)$ correctly describes the probability evolution (a comparison of some characteristic scale of the probability evolution, e.g., decrease of the probability $e$ times, and the MDT provides rather good coincidence both for the case of zero initial phase of driving and for the phase averaged characteristics).

In computer simulations the probability $P(t)$ is obtained by solving the FPE with the potential (6.12) for the following parameter values $d = (a + \sqrt{a^2 + 3Ab})/3b$, $a = 1, b = 1, A = 1, 0.3, 0.1$. For convenience we take the initial distribution located at $x_0 = 0 \neq x_{\min}$ (the phenomenon is quite insensitive to the location of the initial condition). Let us consider first the case of zero phase $\psi = 0$. With this choice the potential barrier at the initial instant of time has maximal height and is decreasing during the first half of the period. The probability evolution for $\psi = 0$ is depicted in Fig. 18 for $kT = 0.1$, $A = 1$ (strong driving) and for different values of the frequency, from which it is clear that the decay of metastable state occurs earlier for $\omega \approx 1$ than for other values. This manifestation of the resonant activation is also seen in Fig. 19 where the mean decay time for different values of the noise intensity is presented. We see that the resonant activation is almost absent at large noise intensities (for $kT = 1$ the effect has the order of error) and becomes stronger when the noise intensity is decreased.

The complete curve $\tau(\omega)$ in Fig. 19 is difficult to describe analytically, but one can have recourse to the adiabatic approximation. This approximation has been used in the context of stochastic resonance by many authors [96,108]. Here

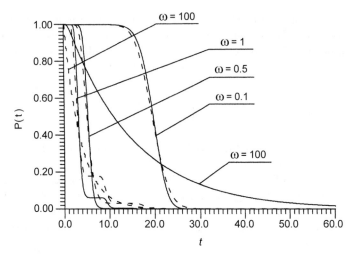

**Figure 18.** Evolution of the survival probability for different values of frequency of the driving signal, $kT = 0.1$, $A = 1$. Solid lines represent results of computer simulation, and dashed lines represent an adiabatic approximation (6.15).

we remark that the part of the curve $\tau(\omega)$ for $0 \leq \omega \leq 1$ may be well-described by a modified adiabatic approximation that allows us to extend the usual analysis to arbitrary driving amplitudes and noise intensities. To this end the probability to find a particle at the time $t$ in the potential minimum takes the

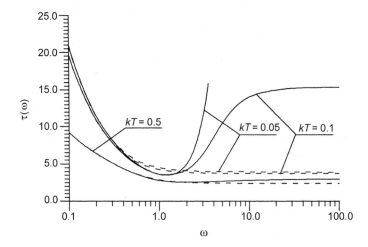

**Figure 19.** The mean decay time as a function of frequency of the driving signal for different values of noise intensity, $kT = 0.5, 0.1, 0.05$, $A = 1$. The phase is equal to zero. Solid lines represent results of computer simulation, and dashed lines represent an adiabatic approximation (6.15).

form

$$P(x_0, t) = \exp\left\{-\int_0^t \frac{1}{\tau_p(x_0, t')}\, dt'\right\} \tag{6.15}$$

where $\tau_p(x_0, t')$ is the exact mean decay time [34,35] obtained for the corresponding time-constant potential:

$$\tau_p(x_0) = B\left\{\int_{x_0}^d e^{\varphi(y)} \int_{-\infty}^y e^{-\varphi(x)}\, dx\, dy + \int_d^\infty e^{\varphi(y)}\, dy \int_{-\infty}^d e^{-\varphi(x)}\, dx\right\} \tag{6.16}$$

[Note that with respect to the usual adiabatic analysis we have *ad hoc* substituted the approximate Kramers' time by the exact one, Eq. (6.16), and found a surprisingly good agreement of this approximate expression with the computer simulation results in a rather broad range of parameters.]

The corresponding curves derived from Eqs. (6.15) and (6.16) are reported in Figs. 18 and 19 as dashed lines, from which we see that there is a good agreement between the modified adiabatic approximation and the numerical results up to $\omega \sim 1$. Moreover, the approximation improves with the increase of the noise intensity. This could be due to the fact that the adiabatic approximation [96,108] is based on the concept of instantaneous escape, and for higher noise intensity the escape becomes faster.

In the opposite limit, $\omega \gg 1$, $\tau(\omega)$ can be described by formula (6.16), with the potential (6.12) averaged over the period of the driving signal. Therefore we can obtain the following empirical expressions for the "amplitude" of the resonant activation effect for $\omega = 0$, $\omega = \infty$:

$$\tau_0 = \frac{\tau_p(x_0)}{\tau_a(x_0, \infty)}, \qquad \tau_\infty = \frac{\tau_p(x_0, \bar\varphi(x, t))}{\tau_a(x_0, \infty)} \tag{6.17}$$

where $\tau_a(x_0, \infty)$ denotes the minimal value of $\tau(\omega)$ which is approximately equal to the value given by adiabatic approximation at $\omega \to \infty$ and $\bar\varphi(x, t)$ denotes potential (6.12) averaged over the period. It is important to note that the resonant activation effect can be predicted on the basis of asymptotic consideration in the ranges $0 < \omega < 1$ and $\omega \gg 1$ without having to rely on computer simulations because the following relations take place: $\tau(\omega \approx 1) < \tau(\omega = 0)$, $\tau(\omega \approx 1) < \tau(\omega = \infty)$.

Similar analysis of the MDT may be performed for arbitrary initial phase $\psi \neq 0$. We note that, depending on the initial phase, $\tau(\omega)$ may vary significantly (especially in the low-frequency limit). This is due to the fact that the height of the potential barrier at initial instant of time has a large variation (from zero to some maximal value). Because in real experiments the initial phase is usually

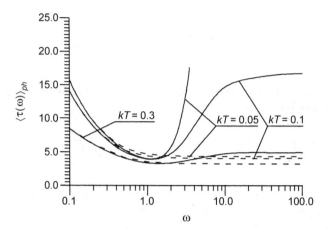

**Figure 20.** The phase-averaged mean decay time as a function of frequency of the driving signal for different values of noise intensity, $kT = 0.3, 0.1, 0.05, A = 1$. Solid lines represent results of computer simulation, and dashed lines represent adiabatic approximation (6.15).

not accessible, one has to consider it as an uniformly distributed random variable, so that it is natural to average the mean decay time upon the initial phase distribution $\langle \tau(\omega) \rangle_{ph}$. This is done in Fig. 20, from which we see that $\langle \tau(\omega) \rangle_{ph}$ has the same qualitative behavior as for $\psi = 0$ (see Fig. 19) and therefore the observed effect is rather independent of the phase.

It is worth to remark, that the curves of the MDT for different values of noise intensity ($kT < 0.1$) actually coincide with the one for $kT = 0$ in the frequency range, corresponding to the minimum of the curve $\tau(\omega)$ ($\omega \sim 1$) (see Figs. 19 and 20). This means that in this region of parameters the fluctuations are suppressed by the strong external signal, and, therefore, tuning a real device in this regime may significantly decrease noise-induced errors.

We have also investigated the dependence of the phase-averaged mean decay time on the amplitude of the driving signal. This is shown in Fig. 21 where the phase-averaged MDT is reported for different values of the driving amplitude. From this figure we see that the phenomenon exists also for relatively small values of the amplitude ($A = 0.3, 0.1$) for which no transitions occur in the absence of noise. As has been demonstrated in Ref. 106, the location of the minimum $\omega_{min}$ of $\langle \tau(\omega) \rangle_{ph}$, as well as the value of the minimum of $\langle \tau(\omega) \rangle_{ph}$, significantly depends on $kT$. On the other hand, for very small amplitudes the resonant activation is significantly reduced and the corresponding frequency where the minimum of $\tau(\omega)$ occurs decreases toward $\omega \sim 0.5$. In this region, however, a description of the phenomenon can be done in terms of the theory developed in Refs. 109—112 and the results coincide with the conclusions, reported in Ref. 96.

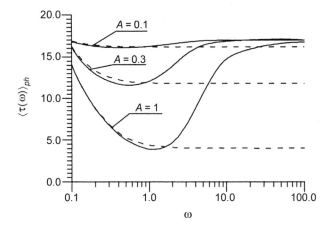

**Figure 21.** The phase-averaged mean decay time as a function of frequency of the driving signal for different values of the amplitude, $A = 1, 0.3, 0.1, kT = 0.1$. Solid lines represent results of computer simulation, and dashed lines represent an adiabatic approximation (6.15).

It is intuitively obvious that this phenomenon should also exist in systems having steady states (e.g., in a system described by "quartic" potential that has been intensively studied in the context of stochastic resonance), but it is more natural to investigate the resonant properties of signal-to-noise ratio (SNR) in those cases [113].

Consider a process of Brownian diffusion in a potential profile $\varphi(x, t) = \Phi(x, t)/kT$:

$$\varphi(x, t) = (bx^4 - ax^2 + xA\sin(\omega t + \psi))/kT \qquad (6.18)$$

where $\psi$ is initial phase. The quantity of our interest is the SNR. In accordance with Ref. 97 we denote SNR as

$$\mathrm{SNR} = \frac{1}{S_N(\omega)} \lim_{\Delta\Omega \to 0} \int_{\omega - \Delta\Omega}^{\omega + \Delta\Omega} S(\Omega) \, d\Omega \qquad (6.19)$$

where

$$S(\Omega) = \int_{-\infty}^{+\infty} e^{-i\Omega\tau} K[t + \tau, t] \, d\tau \qquad (6.20)$$

is the spectral density, $S_N(\omega)$ is noisy pedestal at the driving frequency $\omega$, and $K[t + \tau, t]$ is the correlation function:

$$K[t + \tau, t] = \langle\langle x(t + \tau)x(t)\rangle\rangle \qquad (6.21)$$

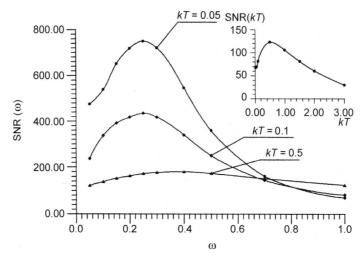

**Figure 22.**   Signal-to-noise ratio as a function of driving frequency. *Inset*: SNR as a function of $kT$.

where the inner brackets denote the ensemble average and the outer brackets indicate the average over initial phase $\psi$.

In computer simulations we had chosen the following parameters of the potential: $b = 1$, $a = 2$. With such a choice the coordinates of minima equal $x_{min} = \pm 1$, the barrier height in the absence of driving is $\Delta\Phi = 1$, the critical amplitude $A_c$ is around 1.5, and we have chosen $A = 2$ to be far enough from $A_c$. In order to obtain the correlation function $K[t + \tau, t]$ we solved the FPE (2.6) numerically, using the Crank–Nicholson scheme.

In order to study the resonant behavior of spectral density, let us plot the SNR as function of driving frequency $\omega$. From Fig. 22 one can see, that SNR as function of $\omega$ has strongly pronounced maximum. The location of this maximum at $\omega = \omega_{max}$ approximately corresponds to the timescale matching condition: $\omega_{max} \approx \pi/\tau_{min}$, where $\tau_{min}$ is the minimal transition time from one state to another one.

When the driving frequency is higher than the cutoff frequency of the system, $\omega \geq \omega_c$, noise helps the particle to move to another state and the conventional stochastic resonance may be observed (see the inset of Fig. 22 for $\omega = 1$). Therefore, for the case of strong periodic driving, the signal-to-noise ratio of the bistable system as well as the mean decay time of a metastable state demonstrate resonant behavior as function of frequency of the driving signal, which reflects the same origin of these phenomena.

In conclusion, we note that the practical application of phenomena of resonant activation and suprathreshold stochastic resonance provide an intriguing possibility to tune a concrete device in a regime with minimal noise-induced error.

## VII. CONCLUSIONS

In the frame of the present review, we discussed different approaches for description of an overdamped Brownian motion based on the notion of integral relaxation time. As we have demonstrated, these approaches allow one to analytically derive exact time characteristics of one-dimensional Brownian diffusion for the case of time constant drift and diffusion coefficients in arbitrary potentials and for arbitrary noise intensity. The advantage of the use of integral relaxation times is that on one hand they may be calculated for a wide variety of desirable characteristics, such as transition probabilities, correlation functions, and different averages, and, on the other hand, they are naturally accessible from experiments.

Another important thing is that in many situations for the considered diffusion processes these characteristic timescales give a rather good description of observables via utilization of single exponential approximation. The exponential approximation works especially well for observables that are less sensitive to the location of initial distribution, such as transition probabilities and correlation functions. In all other cases it is usually enough to apply double exponential approximation to obtain the required observable with a good precision, and one can have recourse to the two-sided Padé approximation as suggested in Ref. 30. The exponential approximation may lead to a significant error in the case where the noise intensity is small, the potential is tilted, and the barrier is absent (purely dynamical motion slightly modulated by noise perturbations). To the contrary, as has been observed for all considered examples, the single exponential approximation is more adequate for a noise-assisted process: either (a) noise-induced escape over a barrier or (b) motion under intensive fluctuations. Moreover, these temporal characteristics are useful for the description of diffusion processes in time-dependent potentials, where one can have recourse to adiabatic approximation and obtain an adequate description of an observable up to the cutoff frequency of the considered system.

Finally we note that the presented approaches may be easily generalized for the case of multidimensional systems with axial symmetry. The generalization for arbitrary multidimensional potentials had been discussed in Refs. 41 and 114.

### Acknowledgments

A. L. Pankratov wishes to thank the Department of Physics "E.R.Caianello" of the University of Salerno, and he wishes to personally thank Professor M. Salerno for the offered position where part of this work has been done. This work has been supported by the Russian Foundation for Basic

Research (Project N 00-02-16528, Project N 99-02-17544, and Project N 00-15-96620) and by the Ministry of High Education of Russian Federation (Project N 992874).

## APPENDIX: THE PRINCIPLE OF CONFORMITY

We consider the process of Brownian diffusion in a potential $\varphi(x)$. The probability density of a Brownian particle is governed by the FPE (5.72) with delta-function initial condition. The moments of transition time are given by (5.1).

Let us formulate the principle of conformity:

The moments of transition time of a dynamical system driven by noise, described by arbitrary potential $\varphi(x)$ such that $\varphi(\pm\infty) = \infty$, symmetric relatively to some point $x = d$, with initial delta-shaped distribution, located at the point $x_0 < d$ [Fig. A1(a)], coincides with the corresponding moments of the first passage time for the same potential, having an absorbing boundary at the point of symmetry of the original potential profile [Fig. A1(b)].

Let us prove that formula (4.19) (in this particular case for $c = -\infty$) not only gives values of moments of FPT of the absorbing boundary, but also expresses moments of transition time of the system with noise, described by an arbitrary symmetric with respect to the point $d$ potential profile.

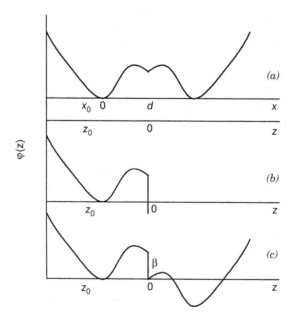

**Figure A1.**   Potential profiles illustrating the principle of conformity.

Superpose the point of symmetry of the dimensionless potential profile with the origin of the coordinate $z = x - d$ such that $\varphi(-z) = \varphi(z)$ and put $z_0 = x_0 - d < 0$. For the Laplace-transformed probability density $Y(z, s) = \int_0^\infty W(z, t)e^{-st}\, dt$ from Eq. (5.72) we may write the following equation:

$$\frac{d^2Y(z, s)}{dz^2} + \frac{d}{dz}\left[\frac{d\varphi(z)}{dz}\, Y(z, s)\right] - sBY(z, s) = -B\delta(z - z_0) \tag{9.1}$$

Note that the probability current in Laplace transform terms is

$$\hat{G}(z, s) = \int_0^\infty G(z, t)e^{-st}\, dt = -\frac{1}{B}\left[\frac{d\varphi(z)}{dz}\, Y(z, s) + \frac{dY(z, s)}{dz}\right] \tag{9.2}$$

Suppose that we know two linearly independent solutions $U(z) = U(z, s)$ and $V(z) = V(z, s)$ of the homogeneous equation corresponding to (9.1) (i.e., when the right-hand side of Eq. (9.1) is equal to zero), such that $U(z) \to 0$ at $z \to +\infty$ and $V(z) \to 0$ at $z \to -\infty$. Because of symmetry of the function $\varphi(z)$, these independent solutions may be also chosen as symmetrical, such that $U(-z) = V(z)$, $U(0) = V(0)$, $[dU(z)/dz]_{z=0} = -[dV(z)/dz]_{z=0} < 0$. In this case the general solution of Eq. (9.1) may be represented as follows:

$$Y(z, s) = \begin{cases} Y_1(z) + y^-(z), & z \le z_0 \\ Y_1(z) + y^+(z), & z_0 \le z \le 0 \\ Y_2(z), & z \ge 0 \end{cases} \tag{9.3}$$

where

$$Y_1(z) = C_1 V(z), \qquad\qquad Y_2(z) = C_2 U(z)$$

$$y^-(z) = \frac{B}{W[z_0]}\, U(z_0)V(z), \qquad y^+(z) = \frac{B}{W[z_0]}\, V(z_0)U(z)$$

Here $W[z] = U(z)\frac{dV(z)}{dz} - V(z)\frac{dU(z)}{dz}$ is Wronskian, and $C_1$ and $C_2$ are arbitrary constants that may be found from the continuity condition of the probability density and the probability current at the origin:

$$Y_1(0) + y^+(0) = Y_2(0), \qquad \hat{G}(z = -0, s) = \hat{G}(z = +0, s) \tag{9.4}$$

Calculating from (9.4) the values of arbitrary constants and putting them into (9.3), one can obtain the following value for the probability current Laplace transform $\hat{G}(z, s)$ (9.2) at the point of symmetry $z = 0$:

$$\hat{G}(0, s) = \frac{V(z_0)}{W[z_0]}\left[\frac{dV(z)}{dz}\right]_{z=0} \tag{9.5}$$

Actually, we can prove the principle of conformity step-by-step for all moments of transition time (5.4),(5.5), and so on, but it is more simple to prove it for the probability density of transition time $w_\tau(t, z_0)$ (5.2). Taking the Laplace transform from the expression (5.2) and noting that $s\hat{Q}(s, z_0) - Q(0, z_0) = \hat{G}(0, s)$, one can obtain the following formula for $w_\tau(s, z_0)$:

$$w_\tau(s, z_0) = \frac{s\hat{Q}(s, z_0) - Q(0, z_0)}{Q(\infty, z_0) - Q(0, z_0)} = \frac{\hat{G}(0, s)}{Q(\infty, z_0) - Q(0, z_0)} \qquad (9.6)$$

where $\hat{Q}(s, z_0) = \int_0^\infty Y(z, s)\, dz$ is the Laplace-transformed decay probability. In our particular case $Q(0, z_0) = 0$, $Q(\infty, z_0) = 1/2$, because the steady-state probability density $W(z, \infty)$ will spread symmetrically from both sides of the point of symmetry $z = 0$. Thus, combining (9.6) and (9.5) we obtain the following formula for the Laplace-transformed probability density of transition time:

$$w_\tau(s, z_0) = \frac{2V(z_0)}{W[z_0]} \left[ \frac{dV(z)}{dz} \right]_{z=0} \qquad (9.7)$$

Before finding the Laplace-transformed probability density $w_T(s, z_0)$ of FPT for the potential, depicted in Fig. A1(b), let us obtain the Laplace-transformed probability density $w_\tau(s, z_0)$ of transition time for the system whose potential is depicted in Fig. A1(c). This potential is transformed from the original profile [Fig. A1(a)] by the vertical shift of the right-hand part of the profile by step $\beta$ which is arbitrary in value and sign. So far as in this case the derivative $d\varphi(z)/dz$ in Eq. (9.1) is the same for all points except $z = 0$, we can use again linear-independent solutions $U(z)$ and $V(z)$, and the potential jump that equals $\beta$ at the point $z = 0$ may be taken into account by the new joint condition at $z = 0$. The probability current at this point is continuous as before, but the probability density $W(z, t)$ has now the step, so the second condition of (9.4) is the same, but instead of the first one we should write $Y_1(0) + y^+(0) = Y_2(0)e^{-\beta}$. It gives new values of arbitrary constants $C_1$ and $C_2$ and a new value of the probability current at the point $z = 0$. Now the Laplace transformation of the probability current is

$$\hat{G}(0, s) = \frac{2V(z_0)}{W[z_0](1 + e^{-\beta})} \left[ \frac{dV(z)}{dz} \right]_{z=0} \qquad (9.8)$$

One can find that for the potential depicted in Fig. A1(c) the quantity $Q(\infty, z_0)$ has been also changed and now equals $Q(\infty, z_0) = 1/(1 + e^{-\beta})$, while the quantity of $Q(0, z_0)$ certainly, as before, is $Q(0, z_0) = 0$. It is easy to check that the substitution of new values of $\hat{G}(0, s)$ and $Q(\infty, z_0)$ into formula (9.6) gives the same formula (9.7) for $w_\tau(s, z_0)$. Putting now $\beta = \infty$—that is, locating the absorbing boundary at the point $z = 0$ $(x = d)$—we obtain the same formula (9.7), not for the probability density of the transition time but for the probability density of FPT $w_T(s, z_0)$. It is known that if the Laplace transformations of two functions coincide, then their origins coincide too. So, if we substitute the coinciding probability densities $w_\tau(t, z_0)$ and $w_T(t, z_0)$ into formula (5.1) [see formulas (5.2) and (5.3)] for the cases of the symmetric potential profile and the profile with the absorbing boundary at the point of symmetry, we should obtain equal values for the moments of transition time and FPT.

Thus, we have proved the principle of conformity for both probability densities and moments of the transition time of symmetrical potential profile and FPT of the absorbing boundary located at the point of symmetry.

It is obvious that moments of FPT to the point 0 are the same for the profiles depicted in Fig. A1. For moments of transition time, this coincidence is not so easily understandable, and the fact that the principle of conformity is valid for the potential depicted in Fig. A1(c) leads to an unusual conclusion: Neither the quantity nor the sign of the potential step $\beta$ influences the moments of the transition time.

For the mean transition time, this fact may be explained in the following way: If the transition process is going from up to down, then the probability current is large, but it is necessary to fill the lower minimum by the larger part of the probability to reach the steady state; if the transition process is going from down to up, then the probability current is small, and it is necessary to fill the upper minimum by the smaller part of the probability to reach the steady state.

The difference in the quantities of currents is completely compensated by quantities of final probabilities of reaching the steady state.

An interested reader can easily check the principle of conformity numerically and see that if the probability of the FPT $Q_T(t, x_0)$ [Fig. A1(b)] is known, then the decay probability $Q_\tau(t, x_0)$ [Fig. A1(a)] is expressed as $Q_\tau(t, x_0) = [1 + Q_T(t, x_0)]/2$.

Note finally that the principle of conformity, proved for delta-shaped initial distribution of the probability density, may also be extended to arbitrary initial distributions located within the considered interval $W(x, 0) = W_{in}(x), x \in (c, d)$.

## References

1. P. Hanggi, P. Talkner, and M. Borkovec, *Rev. Mod. Phys.* **62**, 251 (1990).

2. H. Risken, *The Fokker–Planck Equation*, 2nd ed., Springer Verlag, Berlin, 1989.

3. K. K. Likharev, *Dynamics of Josephson Junctions and Circuits*, Gordon and Breach, New York, 1986.

4. W. T. Coffey, Yu. P. Kalmykov, and J. T. Waldron, *The Langevin Equation*, World Scientific, Singapore, 1996.

5. D. A. Garanin, V. V. Ishchenko, and L. V. Panina, *Teor. Mat. Fiz.* **82**, 242 (1990) [*Teor. Mat. Phys.* **82**, 169 (1990)].

6. D. A. Garanin, *Phys. Rev. E* **54**, 3250 (1996).

7. W. T. Coffey, D. S. F. Crothers, Yu. P. Kalmykov, and J. T. Waldron, *Phys. Rev. B* **51**, 15947 (1995).

8. F. W. Weigel, *Phys. Rep.* **95**, 283 (1983).

9. J. A. McCammon, *Rep. Prog. Phys.* **47**, 1 (1984).

10. R. S. Berry, S. A. Rice, and J. Ross, *Physical Chemistry*, Wiley, New York, 1980.

11. H. Kramers, *Physica* **7**, 284 (1940).

12. S. H. Northrup and J. T. Hynes, *J. Chem. Phys.* **69**, 5264 (1978).

13. G. van der Zwan and J. T. Hynes, *J. Chem. Phys.* **77**, 1295 (1982).

14. N. Agmon and J. J. Hopfield, *J. Chem. Phys.* **78**, 6947 (1983).

15. C. W. Gardiner, *Handbook of Stochastic Methods*, Springer-Verlag, 1985.

16. N. G. van Kempen, *Stochastic Process in Physics and Chemistry*, 2nd ed., North-Holland, Amsterdam, 1992.

17. R. L. Stratonovich, *Topics of the Theory of Random Noise*, Gordon and Breach, New York, 1963.

18. V. I. Tikhonov and M. A. Mironov, *Markovian Processes*, Sovetskoe Radio, Moscow, 1978, in Russian.

19. L. A. Pontryagin, A. A. Andronov, and A. A. Vitt, *Zh. Eksp. Teor. Fiz.* **3**, 165 (1933) [translated by J. B. Barbour and reproduced in *Noise in Nonlinear Dynamics*, Vol. 1, F. Moss and P. V. E. McClintock, eds., Cambridge University Press, Cambridge, 1989, p. 329.

20. R. S. Larson and M. D. Kostin, *J. Chem. Phys.* **69**, 4821 (1978).

21. R. S. Larson, *J. Chem. Phys.* **81**, 1731 (1984).

22. G. H. Weiss, *Adv. Chem. Phys.* **13**, 1 (1967).

23. I. Oppenheim, K. E. Shuler, and G. H. Weiss, *Stochastic Processes in Chemical Physics*, MIT, Cambridge, 1977.

24. A. Szabo, K. Schulten, and Z. Schulten, *J. Chem. Phys.* **72**, 4350 (1980).

25. K. Schulten, Z. Schulten, and A. Szabo, *J. Chem. Phys.* **74**, 4426 (1981).

26. V. Seshadri, B. J. West, and K. Lindenberg, *J. Chem. Phys.* **72**, 1145 (1980).

27. B. J. West, K. Lindenberg, and V. Seshadri, *J. Chem. Phys.* **72**, 1151 (1980).

28. J. Deutch, *J. Chem. Phys.* **73**, 4700 (1980).

29. B. Carmeli and A. Nitzan, *J. Chem. Phys.* **76**, 5321 (1982).

30. W. Nadler and K. Schulten, *J. Chem. Phys.* **82**, 151 (1985).

31. P. Jung and H. Risken, *Z. Phys. B* **59**, 469 (1985).

32. N. V. Agudov and A. N. Malakhov, *Radiophys. Quantum Electron.* **36**, 97 (1993).

33. A. N. Malakhov and A. L. Pankratov, *Physica A* **229**, 109 (1996).

34. A. N. Malakhov and A. L. Pankratov, *Physica C* **269**, 46 (1996).

35. A. N. Malakhov, *CHAOS* **7**, 488 (1997).

36. W. T. Coffey, *Adv. Chem. Phys.* **103**, 259 (1998).

37. W. T. Coffey and D. S. F. Crothers, *Phys. Rev. E* **54**, 4768 (1996).

38. Yu. P. Kalmykov, W. T. Coffey, and J. T. Waldron, *J. Chem. Phys.* **105**, 2112 (1996).

39. Yu. P. Kalmykov, J. L. Dejardin, and W. T. Coffey, *Phys. Rev. E* **55**, 2509 (1997).

40. W. T. Coffey, D. S. F. Crothers, and Yu. P. Kalmykov, *Phys. Rev. E* **55**, 4812 (1997).

41. W. T. Coffey, *J. Chem. Phys.* **111**, 8350 (1999).

42. I. Derenyi and R. D. Astumian, *Phys. Rev. Lett.* **82**, 2623 (1999).

43. A. N. Malakhov, *Cumulant Analysis of Random Non-Gaussian Processes and Its Transformations*, Sovetskoe Radio, Moscow, 1978, in Russian.

44. A. D. Fokker, *Ann. Phys. (Leipzig)* **43**, 310 (1915).

45. M. Planck, *Sitzungsber. Preuss. Acad. Wiss. Phys. Math. Kl.* **325** (1917).

46. A. N. Kolmogorov, *Math. Ann.* **104**, 415 (1931).

47. V. I. Melnikov, *Phys. Rep.* **209**, 1 (1991).

48. O. Edholm, O. Leimar, *Physica A* **98**, 313 (1979).

49. N. V. Agudov and A. N. Malakhov, *Appl. Nonlinear Dyn.* **3**, N3, 75 (1995) (in Russian).

50. P. Reimann, G. J. Schmid, and P. Hanggi, *Phys. Rev. E* **60**, R1 (1999).

51. A. N. Malakhov, *Fluctuations in Self-Oscillating Systems*, Science, Moscow, 1968, in Russian.

52. M. San Miguel, L. Pesquera, M. A. Rodriguez, and A. Hernandez-Machado, *Phys. Rev. A* **35**, 208 (1987).

53. W. Coffey, D. S. F. Crothers, Yu. P. Kalmykov, E. S. Massawe, and J. T. Waldron, *Phys. Rev. E* **49**, 1869 (1994).

54. A. N. Malakhov and E. L. Pankratov, *Radiophysics and Quantum Electonics*, **44**, 339 (2001).

55. Jose Luis Garsia-Palacios and Francisco J. Lazaro, *Phys. Rev. B* **58**, 14937 (1998).

56. P. Colet, M. San Miguel, J. Casademunt, and J. M. Sancho, *Phys. Rev. A* **39**, 149 (1989).

57. J. E. Hirsch, B. A. Huberman, and D. V. Scalapino, *Phys. Rev. A* **25**, 519 (1982).

58. I. Dayan, M. Gitterman, and G. H. Weiss, *Phys. Rev. A* **46**, 757 (1992).

59. M. Gitterman and G. H. Weiss, *J. Stat. Phys.* **70**, 107 (1993).

60. J. M. Casado and M. Morillo, *Phys. Rev. E* **49**, 1136 (1994).

61. R. N. Mantegna and B. Spagnolo, *Phys. Rev. Lett.* **76**, 563 (1996).

62. R. N. Mantegna and B. Spagnolo, *Int. J. Bifurcation Chaos Appl. Sci. Eng.* **8**, 783 (1998).

63. N. V. Agudov and A. N. Malakhov, *Int. J. Bifurcation Chaos Appl. Sci. Eng.* **5**, 531 (1995).

64. N. V. Agudov, *Phys. Rev. E* **57**, 2618 (1998).

65. M. Frankowicz and G. Nicolis, *J. Stat. Phys.* **33**, 595 (1983).

66. M. Frankowicz, M. Malek Mansour, and G. Nicolis, *Physica A* **125**, 237 (1984).

67. E. Arimondo, D. Dangoisse, and L. Fronzoni, *Europhys. Lett.* **4**, 287 (1987).

68. J. Iwaniszewski, *Phys. Rev. A* **45**, 8436 (1992).

69. N. V. Agudov and A. N. Malakhov, *Phys. Rev. E* **60**, 6333 (1999).

70. A. L. Pankratov, *Phys. Lett. A* **234**, 329 (1997).

71. F. T. Arecci, M. Giglio, and A. Sona, *Phys. Lett. A* **25**, 341 (1967).

72. S. Chopra and L. Mandel, *IEEE J.* **QE-8**, 324 (1972).

73. W. T. Coffey, D. S. F. Crothers, J. L. Dormann, L. G. Geoghegan, and E. C. Kennedy, *Phys. Rev. B* **58**, 3249 (1998).

74. Yu. P. Kalmykov, S. V. Titov, and W. T. Coffey, *Phys. Rev. B* **58**, 3267 (1998).

75. Yu. P. Kalmykov and S. V. Titov, *Phys. Rev. Lett.* **82**, 2967 (1999).

76. W. T. Coffey, *J. Chem. Phys.* **93**, 724 (1990).

77. W. T. Coffey, Yu. P. Kalmykov, E. S. Massawe, and J. T. Waldron, *J. Chem. Phys.* **99**, 4011 (1993).

78. M. Abramowitz and I. Stegun, *Handbook of Mathematical Functions*, Dover, New York, 1964.

79. J. I. Lauritzen, Jr., and R. W. Zwanzig, Jr., *Adv. Mol. Relax. Processes* **5**, 339 (1973).

80. G. Klein, *Proc. R. Soc. London, Ser. A* **211**, 431 (1952).

81. G. Korn and T.Korn, *Mathematical Handbook for Scientists and Engineers*, McGraw-Hill, New York, 1961.

82. A. N. Malakhov, *Radiophys. Quantum Electron.* **34**, 451 (1991).

83. A. N. Malakhov, *Radiophys. Quantum Electron.* **34**, 571 (1991).

84. Z. Bilkadi, J. D. Parsons, J. A. Mann, and R. D. Neumann, *J. Chem. Phys.* **72**, 960 (1980).

85. A. N. Malakhov and A. L. Pankratov, *Radiophys. Quantum Electron.* **38**, 167 (1995).

86. H. L. Frisch, V. Privman, C. Nicolis, and G. Nicolis, *J. Phys. A: Math. Gen.* **23**, L1147 (1990).

87. M. Mörsch, H. Risken, and H. Vollmer, *Z. Phys. B* **32**, 245 (1979).

88. S. P. Nikitenkova and A. L. Pankratov, Preprint N adap-org/9811004 at http://xxx.lanl.gov.

89. A. L. Pankratov, *Phys. Lett. A* **255**, 17 (1999).

90. A. A. Dubkov, A. N. Malakhov, and A. I. Saichev, *Radiophys. Quantum Electron.* **43**, 335 (2000).

91. A. A. Antonov, A. L. Pankratov, A. V. Yulin, and J. Mygind, *Phys. Rev. B* **61**, 9809 (2000).

92. S. P. Nikitenkova and A. L. Pankratov, *Phys. Rev. E* **58**, 6964 (1998).

93. S. R. Shenoy and G. S. Agarwal, *Phys. Rev. A* **29**, 1315 (1984).

94. R. Roy, R. Short, J. Durnin, and L. Mandel, *Phys. Rev. Lett.* **45**, 1486 (1980).

95. K. Lindenberg and B. West, *J. Stat. Phys.* **42**, 201 (1986).

96. P. Jung, *Physics Rep.* **234**, 175 (1993).

97. For a review on stochastic resonance see L. Gammaitoni, P. Hanggi, P. Jung, and F. Marchesoni, *Rev. Mod. Phys.* **70**, 223 (1998) and references therein.

98. F. Julicher, A. Ajdari, and J. Prost, *Rev. Mod. Phys.* **69**, 1269 (1997).

99. C. R. Doering, *Physica A* **254**, 1 (1998).

100. N. G. Stocks and R. Manella, *Phys. Rev. Lett.* **80**, 4835 (1998).

101. C. R. Doering and J. Godoua, *Phys. Rev. Lett.* **69**, 2318 (1992).

102. M. Bier and R. Dean Astumian, *Phys. Rev. Lett.* **71**, 1649 (1993).

103. P. Pechukas and P. Hanggi, *Phys. Rev. Lett.* **73**, 2772 (1994).

104. P. Reimann and P. Hanggi, in *Lectures on Stochastic Dynamics*, Springer Series LNP 484, Springer, New York, 1997, p. 127.

105. M. Boguna, J. M. Porra, J. Masoliver, and K. Lindenberg, *Phys. Rev. E* **57**, 3990 (1998).

106. A. L. Pankratov and M. Salerno, *Phys. Lett. A* **273**, 162 (2000).

107. A. L. Pankratov and M. Salerno, *Phys. Rev. E* **61**, 1206 (2000).

108. T. Zhou, F. Moss, and P. Jung, *Phys. Rev. A* **42**, 3161 (1990).

109. V. N. Smelyanskiy, M. I. Dykman, H. Rabitz, and B. E. Vugmeister, *Phys. Rev. Lett.* **79**, 3113 (1997).

110. V. N. Smelyanskiy, M. I. Dykman, and B. Golding, *Phys. Rev. Lett.* **82**, 3193 (1999).

111. D. G. Luchinsky, R. Mannella, P. V. E. McClintock, M. I. Dykman, and V. N. Smelyanskiy, *J. Phys. A: Math. Gen.* **32**, L321 (1999).

112. J. Lehmann, P. Reimann, and P. Hanggi, *Phys. Rev. Lett.* **84**, 1639 (2000).

113. A. L. Pankratov, *Phys. Rev. E*, in press, cond-mat/0103448.

114. W. Nadler and K. Schulten, *J. Chem. Phys.* **84**, 4015 (1986).

# *AB INITIO* QUANTUM MOLECULAR DYNAMICS

MICHAL BEN-NUN and TODD. J. MARTÍNEZ

*Department of Chemistry and the Beckman Institute, University of Illinois, Urbana, Illinois, U.S.A.*

## CONTENTS

## I.  INTRODUCTION

The introduction of the Born–Oppenheimer approximation (BOA) set the stage for the development of electronic structure theory and molecular dynamics as separate disciplines. Certainly this separation has been fruitful and has in large measure fostered the rapid development of the fields. However, it is also clear that a comprehensive approach to chemistry must remain cognizant of the interplay between electronic structure and nuclear dynamics. Inferring dynamical behavior

*Advances in Chemical Physics, Volume 121*, Edited by I. Prigogine and Stuart A. Rice.
ISBN 0-471-20504-4  © 2002 John Wiley & Sons, Inc.

from potential energy surfaces (PESs) can be deceptive, especially when energy flow is restricted on the relevant timescales. The subpicosecond lifetimes often observed for excited electronic state dynamics imply that this caution will be especially valid in photochemistry. On the other hand, fitting of multidimensional PESs to empirical functional forms involves a tradeoff between accuracy and human effort. Erasing the boundary between electronic structure and molecular dynamics is the goal of *ab initio* molecular dynamics (AIMD) methods, which are just entering mainstream chemical physics because of recent methodological and computational advances. Simultaneous solution of the electronic Schrödinger equation and Newton's equations reemphasizes the close relationship between the electrons that govern the form of the PES and the nuclear dynamics that occurs on this PES. From a practical standpoint, arbitrary bond rearrangements can be described without difficult and often impossible fitting procedures. The first AIMD calculation [1] was performed in the 1970s, with the quantum chemical part of the calculation treated largely as a complicated and computationally expensive force routine. The introduction of the Car-Parrinello (CP) method [2] fueled the rapid increase in popularity of AIMD methods over the last 15 years [3–7], although we hasten to add that not all AIMD methods use the extended Lagrangian scheme which is the hallmark of the CP method [8–13]. Interestingly, the core of the CP method is a blurring of the line separating electronic structure and molecular dynamics. The coefficients that comprise the unknowns in the quantum chemical problem are endowed with a fictitious dynamics designed to mimic complete solution of the electronic structure at every time step. We believe that this trend will continue, and there is hope that it will lead to a better understanding of the interplay between electronic and nuclear motion.

Most of the AIMD simulations described in the literature have assumed that Newtonian dynamics was sufficient for the nuclei. While this is often justified, there are important cases where the quantum mechanical nature of the nuclei is crucial for even a qualitative understanding. For example, tunneling is intrinsically quantum mechanical and can be important in chemistry involving proton transfer. A second area where nuclei must be described quantum mechanically is when the BOA breaks down, as is always the case when multiple coupled electronic states participate in chemistry. In particular, photochemical processes are often dominated by conical intersections [14,15], where two electronic states are exactly degenerate and the BOA fails. In this chapter, we discuss our recent development of the *ab initio* multiple spawning (AIMS) method which solves the elecronic and nuclear Schrödinger equations simultaneously; this makes AIMD approaches applicable for problems where quantum mechanical effects of both electrons and nuclei are important. We present an overview of what has been achieved, and make a special effort to point out areas where further improvements can be made. Theoretical aspects of the AIMS method are discussed, including both the electronic and nuclear parts of the problem. Several applications to fundamental problems in the chemistry of excited electronic states are presented, and we conclude with our thoughts on future interesting directions.

## II. THEORY

The development of an *ab initio* quantum molecular dynamics method is guided by the need to overcome two main obstacles. First, one needs to develop an efficient, yet accurate, method for solving the electronic Schrödinger equation for both ground and excited electronic states. Second, the quantum mechanical character of the nuclear dynamics must be addressed. (This is necessary for the description of photochemical and tunneling processes.) This section provides a detailed discussion of the approaches we have taken to solve these two problems.

### A. Electronic Structure for Multiple Electronic States

A first-principles treatment of photochemistry requires repeated solution of the electronic Schrödinger equation for multiple electronic states, including the nonadiabatic coupling matrix elements that induce transitions between electronic states. These requirements make computational efficiency paramount, even more so than for traditional time-independent quantum chemistry. At the same time, accuracy must be maintained because there is no point in developing a first-principles approach if the underlying PESs are not (at least) qualitatively accurate. The conflicting requirements of accuracy and efficiency are already present in time-independent quantum chemistry, but they are made more severe in the time-dependent case because of both the large number of PES evaluations and the requirement of global accuracy in the PESs.

Single-reference methods, such as single-excitation configuration interaction (CIS) [16,17], are computationally attractive and in certain circumstances are capable of describing both ground and excited electronic states. These methods often provide reasonable vertical excitation energies for the excited states that dominate electronic absorption spectra. However, the accuracy, and hence utility, of single-reference methods in general, and single-reference/single-excitation methods in particular, diminishes rapidly as the excited-state trajectory/wavefunction leaves the Franck–Condon region. This is due to two problems. First, while electronic states with doubly excited character are usually optically forbidden (and hence less important in the electronic absorption spectrum), they can play a significant role in photochemistry and cannot be modeled accurately with CIS. Second, avoided crossings and conical intersections are often ubiquitous in the manifold of excited states. Even if the wavefunctions for each of the interacting states are reasonably described by a single configuration outside the crossing/intersection region, a multireference description becomes necessary in these regions because the character of the wavefunctions changes rapidly. For example, CIS fails to correctly predict the global minimum on the lowest-valence adiabatic excited-state surface of ethylene [18]. Despite their computational advantages, most single-reference methods are not appropriate for our needs because of their inability to predict the correct shape of the excited potential energy surface(s). It remains to be seen whether these criticisms apply to time-dependent formulations of density functional theory (TDDFT) [19–21], which are superficially similar to single-reference/single-excitation methods.

A further problem in *ab initio* photochemistry is the need to avoid variational bias to the ground electronic state; that is, the quality of the ground- and excited-state wavefunctions should be similar. A widely used procedure that alleviates this problem is state-averaging [22–25]. Here the orbitals are determined to minimize a weighted average of the ground-state and one or more excited-state energies. The resulting orbitals are often called a "best-compromise" set because they are *not* optimal for *any* single target electronic state. Once the orbitals have been determined, state-dependent orbital relaxation must be incorporated, typically via the inclusion of single excitations in a configuration interaction wavefunction. In this approach, single excitations are taken from the same set of reference configurations that was used to determine the orbitals in the state-averaged multiconfiguration SCF (MCSCF). Variants of this technique, such as the first-order CI method of Schaefer [26] and the POL-CI method of Goddard [27], have been successfully used in past treatments of excited states.

A simpler computational procedure with similar aims is occupation averaging. Here, the orbitals that are of variable occupancy (comprising the "active space") in the ground and excited electronic states are equally populated with electrons. (For example, in the case of $\pi \rightarrow \pi^*$ excitation of ethylene or cyclobutene, the $\pi$ and $\pi^*$ orbitals would each be singly occupied.) There are several ways to accomplish this averaging, differing in the treatment of electronic spin coupling. The active space orbitals may be determined by an SCF procedure where they are high-spin coupled in a single determinant. This approach is related to "half-electron" semiempirical theories [28], and similarly motivated approaches have recently been investigated in coupled cluster methods [29]. For ethylene with a two-orbital active space, this corresponds to using the SCF orbitals from a triplet single-determinant wavefunction. This procedure has been tested for ethylene, where the lowest triplet Hartree–Fock wavefunction is less prone than the ground-state singlet wavefunction to overemphasize Rydberg character in the orbitals. It has therefore been argued that the triplet coupled orbitals provide a better starting point for CI expansions [30]. A second approach for the specific case of ethylene determines the orbitals using the framework of a generalized valence bond (GVB) wavefunction [31] where the covalent and ionic states are constrained to have equal weights. For example, the GVB wavefunction for ethylene is

$$\psi_{GVB} = c_{cov}\hat{A}[\psi_{core}(\chi_{Cp,r}\chi_{Cp,l} + \chi_{Cp,l}\chi_{Cp,r})(\alpha\beta - \beta\alpha)]$$
$$+ c_{ion}\hat{A}[\psi_{core}(\chi_{Cp,r}\chi_{Cp,r} + \chi_{Cp,l}\chi_{Cp,l})(\alpha\beta - \beta\alpha)] \quad (2.1)$$

where $\hat{A}$ is the antisymmetrizing operator, $\psi_{core}$ represents all the $\sigma$ framework electrons, and $\chi_{Cp,r}$ and $\chi_{Cp,l}$ denote the nonorthogonal GVB orbitals on the right and left carbon atoms (respectively), which are dominated by contributions from the carbon $2p$ atomic orbitals. The GVB wavefunction, Eq. (2.1), can be written in an equivalent form using the orthogonal molecular orbitals:

$$\psi_{GVB} = c_b\hat{A}[\psi_{core}\phi_\pi\phi_\pi(\alpha\beta - \beta\alpha)] + c_a\hat{A}[\psi_{core}\phi_{\pi^*}\phi_{\pi^*}(\alpha\beta - \beta\alpha)] \quad (2.2)$$

In the usual GVB procedure, both the orbitals and the coefficients $c_{cov}, c_{ion}$ or, equivalently, $c_b, c_a$ would be optimized. The occupation-averaged orbitals appropriate for the GVB wavefunction are defined to minimize the average energy of the individual terms in Eq. (2.2):

$$E_{average} = 1/2[E(\psi_{core}\phi_\pi\phi_\pi) + E(\psi_{core}\phi_{\pi^*}\phi_{\pi^*})] \quad (2.3)$$

Theoretically, this is somewhat more appealing than the use of triplet orbitals, because the orbitals in this procedure are derived from a wavefunction averaged over states with the desired singlet spin coupling. In our studies on ethylene, we have found little difference between the two approaches. For example, the global features of the PESs are qualitatively unchanged for these two choices of starting orbitals, and even the vertical excitation energies are within 0.1 of each other. This is apparently due to the subsequent CI expansion, which is sufficiently flexible to correct the shape of the orbitals in either case. In our *ab initio* molecular dynamics of ethylene we have used the GVB-occupation-averaged (GVB-OA) orbitals [32], but in benchmark calculations we have used the simpler Hartree–Fock-occupation-averaged (HF-OA) orbitals (with high spin coupling) [33]. In either case, the set of reference configurations from which single excitations are drawn in the subsequent CI expansion is of the complete active space (CAS) type, allowing all possible configurations of the active electrons in the occupation-averaged orbitals that are consistent with the Pauli exclusion principle. We refer to this form of wavefunction as HF-OA-CAS($n/m$)*S or GVB-OA-CAS($n/m$)*S, where $n$ and $m$ denote the number of electrons and orbitals in the active space which defines the reference configurations. The S indicates that single excitations are taken from the CAS($n/m$) reference configurations.

The accuracy of the potential energy surfaces is determined also by the size of the electronic basis set. We have used double-$\zeta$ quality basis sets, which are the minimum that can be expected to describe both ground and excited electronic states. For small organic molecules near their equilibrium configuration, one often finds Rydberg states among the low-lying excited electronic states. This implies that the inclusion of Rydberg basis functions would be desirable. At present, computational considerations render this impractical for quantum *ab initio* molecular dynamics. However, one can assess the accuracy of the dynamics using benchmark calculations and large basis sets. In Section III.B, we discuss the accuracy of the electronic wavefunction ansatz and remark on the importance of valence-Rydberg mixing (and its neglect in the dynamical calculations).

A final electronic structure issue is the nonadiabatic coupling between electronic states. The form of the coupling depends on whether an adiabatic or diabatic representation has been chosen. The adiabatic representation diagonalizes the electronic potential energy, while the diabatic representation minimizes the change in electronic character due to the nuclear perturbations and hence approximately diagonalizes the nuclear kinetic energy. The diabatic representation leads to smoother potential energy surfaces, and it is therefore often the preferred representation for time-dependent studies (i.e., molecular dynamics). However, without information about the electronic wavefunction at different geometries it is difficult to obtain a unique and path-independent set of diabatic states [34,35]. Hence, we (and others) prefer to use the unique adiabatic electronic states, in which case the form of the interstate coupling is

$$(\mathbf{d}^{IJ}) = \langle \phi_I(\mathbf{r}; \mathbf{R}) | \frac{\partial}{\partial \mathbf{R}} | \phi_J(\mathbf{r}; \mathbf{R}) \rangle_{\mathbf{r}} \tag{2.4}$$

In Eq. (2.4) the parametric dependence of the electronic wavefunction on the nuclear coordinates ($\mathbf{R}$) is denoted by the semicolon, and the integration is only over the electronic coordinates ($\mathbf{r}$). For our wavefunction ansatz, both the orbitals and the CI coefficients depend on the nuclear geometry and therefore both contribute to the derivative in Eq. (2.4). Because the orbitals are (state or occupation) averaged, the contribution from the CI coefficients is usually dominant [36,37] and consequently we have neglected the orbital contribution to the nonadiabatic coupling. Currently, we find the derivatives of the CI coefficients using numerical differentiation, but the required theory for analytic evaluation has been published [25]. In either case (numerical or analytic) the evaluation of the nonadiabatic coupling requires some care because one needs to ensure continuity. This can only be achieved if a consistent phase convention is adopted with respect to the electronic wavefunction. Failure to ensure this results in a wildly oscillating nonadiabatic coupling function.

## B. Adaptive Time-Dependent Nuclear Basis Set

The AIMS method treats both the electrons and the nuclei quantum mechanically. The previous section dealt with the quantum nature of the electrons, and here we discuss our strategy for incorporating nuclear quantum effects—the full multiple spawning (FMS) method that forms the nuclear dynamics component of AIMS. Quantum mechanical effects of the nuclei are necessary for proper modeling of photochemical processes because the excited-state lifetime of the molecule is usually short, and its (radiationless) decay back to the ground electronic state is often mediated by conical intersections—points of true degeneracy between two electronic states. Implicit in classical mechanics is the assumption that a single potential energy surface governs the dynamics. Hence, multielectronic state dynamics cannot be described using classical mechanics. At least formally, quantum mechanics can treat multielectronic states straightforwardly.

Within the context of traditional quantum chemistry, the interface between a quantum mechanical treatment of the electrons and the nuclei is quite problematic because the locality of quantum chemistry conflicts with the global nature of the nuclear Schrödinger equation. In principle, quantum nuclear dynamics requires the entire PES at each time step, but quantum chemistry only provides local information—given a nuclear geometry, it can return the potential and its derivatives. This immediate conflict imposes a stringent limitation on the method for the nuclear dynamics—compatibility with conventional quantum chemistry requires some form of localization of the nuclear Schrödinger equation. Furthermore, given the computational expense of quantum chemistry, the more local the method, the better; that is, at each time step we wish to require as few as possible PES and gradient evaluations. The complete locality of classical mechanics represents the ideal in this regard, leading us to develop a method that retains a classical flavor, reducing to pure classical mechanics in one limit, while allowing for quantum effects and converging to exact numerical solution of the Schrödinger equation in another limit.

The multiple-spawning method solves the nuclear Schrödinger equation using an adaptive time-dependent basis set that is generated using classical mechanics [38–41]. Individual basis functions are of the frozen Gaussian form. In the mid-1970s, Heller [42] pioneered the "frozen Gaussian approximation" (FGA); since that time, Gaussian wavepacket methods, along with Fourier methods [43–45], played a key role in popularizing time-dependent approaches to the nuclear Schrödinger equation. Gaussian wavepacket methods have successfully described a number of short time processes (see, for example, the original work of Heller and co-workers [46–49] as well as more recent publications [50–56]). Unlike the original FGA method, we take full account of the nonorthogonal nature of Gaussian basis sets by inverting the time-dependent overlap matrix at each time step and coupling the complex coefficients of the nuclear basis functions. The multiconfigurational nuclear wavefunction we use is of the form

$$\Psi = \sum_I \chi_I(\mathbf{R}; t)\phi_I(\mathbf{r}; \mathbf{R}) \qquad (2.5)$$

where the index $I$ denotes the electronic state, $\phi_I(\mathbf{r}; \mathbf{R})$ is an electronic wavefunction (which is allowed to depend parametrically on the nuclear coordinates), and $\chi_I(\mathbf{R}; t)$ is the time-dependent nuclear wavefunction associated with the $I$th electronic state. An arbitrary number of electronic states and nuclear degrees of freedom is permitted in Eq. (2.5), and we use bold letters to denote vectors or matrices. Unlike mean-field based methods [57–60], the AIMS wavefunction ansatz associates a unique nuclear wavefunction with each electronic state, thereby allowing for qualitatively different nuclear dynamics on different electronic states (e.g., bound vs. dissociative). The nuclear wavefunction on each electronic state is represented as a superposition of multidimensional traveling frozen Gaussian basis functions with time-dependent coefficients:

$$\chi_I(\mathbf{R}; t) = \sum_{j=1}^{N_I(t)} C_j^I(t) \chi_j^I(\mathbf{R}; \bar{\mathbf{R}}_j^I(t), \bar{\mathbf{P}}_j^I(t), \bar{\gamma}_j^I(t), \boldsymbol{\alpha}_j^I) \qquad (2.6)$$

where the index $j$ labels nuclear basis functions on electronic state $I$, $N_I(t)$ is the number of nuclear basis functions on electronic state $I$ at time $t$ (this number is allowed to change during the propagation), and we have explicitly denoted the time-dependent parameters of the individual basis functions. Individual, multidimensional, nuclear basis functions are expressed as a product of one-dimensional Gaussian basis functions

$$\chi_j^I(\mathbf{R}; \bar{\mathbf{R}}_j^I(t), \bar{\mathbf{P}}_j^I(t), \bar{\gamma}_j^I(t), \boldsymbol{\alpha}_j^I) = e^{i\bar{\gamma}_j^I(t)t} \prod_{\rho=1}^{3N} \chi_{\rho_j}^I(R; \bar{R}_{\rho_j}^I(t), \bar{P}_{\rho_j}^I(t), \alpha_{\rho_j}^I) \qquad (2.7a)$$

$$\chi_{\rho_j}^I(R; \bar{R}_{\rho_j}^I(t), \bar{P}_{\rho_j}^I(t), \alpha_{\rho_j}^I) = \left(\frac{2\alpha_{\rho_j}^I}{\pi}\right)^{1/4} \prod_{\rho=1}^{3N} \exp[-\alpha_{\rho_j}^I(R_{\rho_j} - \bar{R}_{\rho_j}^I(t))^2$$
$$+ i\bar{P}_{\rho_j}^I(t)(R_{\rho_j} - \bar{R}_{\rho_j}^I(t))] \qquad (2.7b)$$

where the index $\rho$ enumerates the 3N coordinates of the molecule, typically chosen to be Cartesian coordinates. The frozen Gaussian basis functions are parameterized with a time-independent width ($\alpha_{\rho_j}^I$) and time-dependent position, momentum, and nuclear phase [$\bar{R}_{\rho_j}^I(t), \bar{P}_{\rho_j}^I(t), \bar{\gamma}_j^I(t)$, respectively]. We have chosen to propagate the time-dependent position and momentum parameters using Hamilton's equations of motion

$$\frac{\partial \bar{R}_{\rho_j}^I}{\partial t} = \frac{\bar{P}_{\rho_j}^I}{m_\rho}$$
$$\frac{\partial \bar{P}_{\rho_j}^I}{\partial t} = -\frac{\partial V_{II}(\mathbf{R})}{\partial R_{\rho_j}}\bigg|_{\bar{R}_{\rho_j}^I(t)} \qquad (2.8)$$

and the nuclear phase is propagated according to the usual semiclassical prescription—that is, as the integral of the Lagrangian:

$$\frac{\partial \bar{\gamma}_j^I}{\partial t} = -V_{II}(\bar{\mathbf{R}}_j^I(t)) + \sum_{\rho=1}^{3N} \frac{(\bar{P}_{\rho_j}^I(t))^2}{2m_\rho} \qquad (2.9)$$

In Eqs. (2.8) and (2.9), $V_{II}(\mathbf{R})$ is the potential energy of the $I$th electronic state, and $m_\rho$ is the mass of the $\rho$th degree of freedom.

Implicit in the choice of classical propagator for the position and momentum parameters is the assumption that classical mechanics provides a reasonable description, on average, of the quantum dynamics. In favorable cases, a relatively small number of nuclear basis functions may be required to obtain an accurate description of the wavefunction. However, one must realize that this will not always be true, leading one to consider means of incorporating quantum effects directly into the time-dependence of the nuclear basis function parameters. One such alternative propagates the position and momentum parameters using a time-dependent variational principle (TDVP) [61,62], but we find this choice less desirable for two reasons. The first is conceptual: such a variational propagator does not reduce to the well-understood classical limit in the extreme case of a single nuclear basis function. The second reason is practical: A variational propagator requires the calculation of more matrix elements and their derivatives than otherwise required. This conflicts with the locality of quantum chemistry and is thus undesirable for *ab initio* quantum nuclear dynamics. Numerical considerations also guide the use of a fixed width parameter ($\alpha_{\rho_j}^I$): Propagation of this parameter requires the calculation of second derivatives of the potential energy surfaces which can be extremely tedious if the PESs are determined by direct solution of the electronic Schrödinger equation. It is reasonable to assume that these two choices result in a penalty: Compared to fully variational treatment of a Gaussian basis set [55,63,64], we will require a larger nuclear basis set to achieve numerical convergence. The finite size of the nuclear basis set is the first approximation we make; and, as usual, larger basis sets generally lead to improved results. The issue of basis set size and convergence is discussed in more detail in Section III.B. Here we only note that at each point in time the size of the nuclear basis set ($N_{\text{nuc}}(t)$) is given by the sum of nuclear basis functions on all electronic states included in the calculation:

$$N_{\text{nuc}}(t) = \sum_I N_I(t) \qquad (2.10)$$

In general, the optimal choice for the time-independent width is not known *a priori* (the only exception is a separable harmonic PES with one nuclear basis function in which case [42] $\alpha^I_{p_j} = m_p \omega^I_p / 2$, where $\omega^I_p$ is the harmonic vibrational frequency of the pth coordinate on electronic state $I$). We have always chosen the width to be independent of both the electronic state index and the nuclear basis function index (i.e., $\alpha^I_{p_j} = \alpha_p$ for all $I$ and $j$). Specific values were chosen by requiring the results to be stable with respect to the width of the basis functions. Our experience is that as long as the basis functions are not very wide, the results are relatively independent of the precise choice of the width parameters. Nevertheless, since they are chosen empirically, we must view them as parameters that characterize the nuclear basis set.

A set of coupled equations for the evolution of the basis function coefficients is obtained by substituting the wavefunction ansatz of Eqs. (2.5)–(2.7) into the nuclear Schrödinger equation

$$\frac{d\mathbf{C}^I(t)}{dt} = -i(\mathbf{S}_{II}^{-1}) \left\{ [\mathbf{H}_{II} - i\dot{\mathbf{S}}_{II}]\mathbf{C}^I + \sum_{J \neq I} \mathbf{H}_{IJ}\mathbf{C}^J \right\} \qquad (2.11)$$

For compactness and clarity, Eq. (2.11) is written in matrix notation. It is similar to the more familiar case of a time-independent basis set expansion but with two important differences: The AIMS basis is time-dependent and nonorthogonal. As a consequence, the proper propagation of the coefficients requires the inverse of the (time-dependent) nuclear overlap matrix

$$(S_{II})_{kl} = \langle \chi^I_k | \chi^I_l \rangle_{\mathbf{R}} \qquad (2.12)$$

as well as its right acting time derivative

$$(\dot{S}_{II})_{kl} = \left\langle \chi^I_k \left| \frac{\partial}{\partial t} \chi^I_l \right\rangle_{\mathbf{R}} \qquad (2.13)$$

Both of these matrix elements are readily computed analytically (the subscript $\mathbf{R}$ denotes integration over the nuclear coordinates and by definition $\mathbf{S}_{IJ}$ and $\dot{\mathbf{S}}_{IJ}$ vanish for $I \neq J$). In Eq. (2.11), $\mathbf{H}_{IJ}$ is the full Hamiltonian matrix including both electronic and nuclear terms. Each matrix element of $\mathbf{H}$ is written as the sum of the nuclear kinetic energy ($\hat{T}_{\mathbf{R}}$) and the electronic Hamiltonian ($\hat{H}_e$)

$$H_{IkJk'} = \langle \chi^I_k \phi_I | \hat{H}_e + \hat{T}_{\mathbf{R}} | \phi_J \chi^J_{k'} \rangle \qquad (2.14)$$

and the integration is over both the electronic and nuclear coordinates. The electronic Hamiltonian includes all of the Coulomb interactions as well as the electronic kinetic energy

$$\langle \chi^I_k \phi_I | \hat{H}_e | \phi_J \chi^J_{k'} \rangle = \langle \chi^I_k | H^e_{IJ} | \chi^J_{k'} \rangle + 2D_{IkJk'} + G_{IkJk'} \qquad (2.15)$$

where

$$H_{IJ}^e = \langle \phi_I | H_e | \phi_J \rangle_{\mathbf{r}} \tag{2.16a}$$

$$D_{Ik,Jk'} = \langle \chi_k^I | \sum_{\rho=1}^{3N} d_\rho^{IJ} \frac{1}{2m_\rho} \frac{\partial}{\partial R_\rho} | \chi_{k'}^J \rangle \tag{2.16b}$$

$$G_{Ik,Jk'} = \langle \chi_k^I | \sum_{\rho=1}^{3N} \frac{1}{2m_\rho} \langle \phi_I | \frac{\partial^2}{\partial R_\rho^2} | \phi_J \rangle | \chi_{k'}^J \rangle \tag{2.16c}$$

In the previous section, we discussed the calculation of the PESs needed in Eq. (2.16a) as well as the nonadiabatic coupling terms of Eqs. (2.16b) and (2.16c). We have noted that in the diabatic representation the off-diagonal elements of Eq. (2.16a) are responsible for the coupling between electronic states while $D_{IJ}$ and $G_{IJ}$ vanish. In the adiabatic representation the opposite is true: The off-diagonal elements of Eq. (2.16a) vanish while $D_{IJ}$ and $G_{IJ}$ do not. In this representation, our calculation of the nonadiabatic coupling is approximate because we assume that $G_{IJ}$ is negligible and we make an approximation in the calculation of $D_{IJ}$. (See end of Section II.A for more details.)

The obstacle to simultaneous quantum chemistry and quantum nuclear dynamics is apparent in Eqs. (2.16a)–(2.16c). At each time step, the propagation of the complex coefficients, Eq. (2.11), requires the calculation of diagonal and off-diagonal matrix elements of the Hamiltonian. These matrix elements are to be calculated for each pair of nuclear basis functions. In the case of *ab initio* quantum dynamics, the potential energy surfaces are known only locally, and therefore the calculation of these matrix elements (even for a single pair of basis functions) poses a numerical difficulty, and severe approximations have to be made. These approximations are discussed in detail in Section II.D. In the case of analytic PESs it is sometimes possible to evaluate these multidimensional integrals analytically. In either case (analytic or *ab initio*) the matrix elements of the nuclear kinetic energy

$$\langle \chi_k^I | \hat{T}_{nuc} | \chi_{k'}^J \rangle = \langle \chi_k^I | \sum_{\rho=1}^{3N} \frac{\partial^2}{2m_\rho \partial R_\rho^2} | \chi_{k'}^J \rangle \delta_{IJ} \tag{2.17}$$

are computed analytically.

## C. Selection of Initial Conditions, Propagation, and Spawning

From the very beginning we have emphasized that the AIMS method uses an adaptive basis set. This property is at the core of the AIMS method, and much of this subsection will be devoted to this topic. However, before we elaborate on it, we discuss the selection of initial conditions and some topics regarding the propagation.

A complete description of the method requires a procedure for selecting the initial conditions. At $t = 0$, initial values for the complex basis set coefficients and the parameters that define the nuclear basis set (position, momentum, and nuclear phase) must be provided. Typically at the beginning of the simulation only one electronic state is populated, and the wavefunction on this state is modeled as a sum over discrete trajectories. The size of initial basis set ($N_I(t = 0)$) is clearly important, and this point will be discussed later. Once the initial basis set size is chosen, the parameters of each nuclear basis function must be chosen. In most of our calculations, these parameters were drawn randomly from the appropriate Wigner distribution [65], but the earliest work used a quasi-classical procedure [39,66,67]. At this point, the complex amplitudes are determined by projection of the AIMS wavefunction on the target initial state ($\Psi_{t=0}^{\text{exact}}$)

$$C_k^I(0) = \sum_{k'=1}^{N_I(0)} S_{Iklk'}^{-1} \langle \chi_k^I(t=0) | \Psi_{t=0}^{\text{exact}} \rangle \qquad (2.18)$$

We have used various integrators (e.g., Runga–Kutta, velocity verlet, midpoint) to propagate the coupled set of first-order differential equations: Eqs. (2.8) and (2.9) for the parameters of the Gaussian basis functions and Eq. (2.11) for the complex amplitudes. The specific choice is guided by the complexity of the problem and/or the stiffness of the differential equations.

As the calculation progresses, the size of the basis set is adjusted in a physically motivated way. This basis set expansion is aimed only at describing quantum mechanical effects associated with electronic nonadiabaticity and *not* at correcting the underlying classical dynamics of individual nuclear basis functions. Hence, a new basis function may be spawned whenever an existing basis function passes through a region of significant electronic coupling to another electronic state. In general, the new function is spawned on a different electronic state than the existing basis function that generates it. More recently we have allowed for spawning on the same electronic state—for example, to allow for tunneling effects—and this extension is discussed in Section II.F. By allowing the basis set to grow only at specific regions (or instances), the growth of basis set size with time can be controlled. Although one can envision cases where the basis set size grows to an unmanageable level, so far this has rarely been a problem in practice because nonadiabatic events are very short in most chemical problems and often infrequent. The practical implementation of the spawning algorithm that expands the basis set is as follows. First, we define two parameters $\lambda_0$ and $\lambda_f$ that signal when a region of nonadiabatic coupling has been entered and exited, respectively. (The values of these parameters are chosen by running test calculations and monitoring the magnitude of an effective nonadiabatic coupling; see below.) At each time step and for each nuclear basis function in the wavepacket, we calculate an effective (dimensionless) coupling

$$
\Lambda_{IJ}^{\text{eff}} = 
\begin{cases}
\left| \frac{V_{IJ}(\mathbf{R})}{V_{II}(\mathbf{R}) - V_{JJ}(\mathbf{R})} \right|, & \text{diabatic} \\
|\dot{\mathbf{R}} \cdot \mathbf{d}_{IJ}|, & \text{adiabatic}
\end{cases}
\tag{2.19}
$$

where the overdot indicates the time derivative. If the value of $\Lambda_{IJ}^{\text{eff}}$ at a given time step (and for a particular basis function) is larger than the parameter $\lambda_0$, this signals that a region of significant electronic coupling has been entered, and that basis function has the opportunity to spawn a new basis function(s). This point in time is labeled $t_0$. The parent basis function is now propagated forward in time according to Hamilton's equations of motion. This propagation is uncoupled from the propagation of the wavepacket; that is, we do not propagate the complex amplitudes of Eq. (2.11). We stop the uncoupled propagation when the magnitude of $\Lambda_{IJ}^{\text{eff}}$ drops below $\lambda_f$ because this indicates that the region of nonadiabatic coupling has ended. This time is labeled $t_f$. Within this region of electronic coupling ($t_f - t_0$), a predetermined number of basis functions may be spawned and we label this parameter MULTISPAWN. The first basis function is spawned at the point in time in which $\Lambda_{IJ}^{\text{eff}}$ reaches its maximum value (this time is labeled $t_{\max}$), and the rest of the basis functions are spread out evenly in time before and after $t_{\max}$. For example, if MULTISPAWN = 3, one basis function is spawned at $t_0 + 0.5(t_{\max} - t_0)$, a second at $t_{\max}$, and a third at $t_{\max} + 0.5(t_f - t_{\max})$.

A pictorial description of the spawning algorithm is given in Fig. 1 using a collinear $A + BC \rightarrow AB + C$ reaction. The upper and lower set of panels correspond to two diabatic potential energy surfaces (represented by contour lines), correlating to $A + BC$ and $AB + C$ respectively. The two diabatic surfaces are coupled via a constant potential energy term, and they are plotted in Jacobi coordinates: the A to BC center-of-mass distance, $R$, and the BC distance, $r$. The nuclear wavefunctions are superimposed on the contour lines. The calculation begins with population on a single diabatic PES, uppermost left panel. As the basis functions approach the nonadiabatic region, new basis functions are created (i.e., "spawned") on the other diabatic state. The locations of individual basis functions are indicated by the triangles. Initially (middle panels), the parent basis functions overlap the ones they spawned yet the subsequent dynamics (right panels) are very different: the parent wavefunction corresponds to an $A + BC$ arrangement and the spawned wavefunction to an $AB + C$ arrangement.

The immediate question is where (in phase space) to place the newly spawned basis functions. The optimal choice will maximize the absolute value of the coupling matrix element between the existing basis function (i.e., the

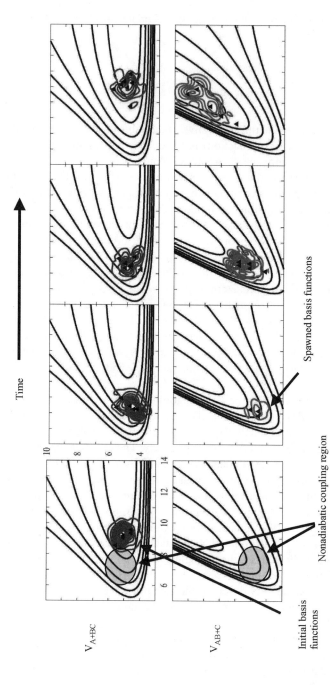

**Figure 1.** Schematic illustration of the spawning algorithm for a collinear A + BC → AB + C reaction. The top and bottom panels correspond to two diabatic potential energy surfaces, correlating to A + BC and AB + C, respectively. The potential energy surfaces are represented in Jacobi coordinates (the A to BC center-of-mass distance and the B–C distance). Superimposed on the PESs are the nuclear wavefunctions in the diabatic representation. The calculation begins with population on a single diabatic potential energy surface (upper left panel); and as basis functions traverse the nonadiabatic region (roughly indicated by the shaded circle), new basis functions are created on the other diabatic state (second, third, and fourth lower panels). The locations of individual Gaussian basis functions are indicated by the triangles. (Figure adapted from Ref. 40.)

Time

$V_{A+BC}$

$V_{AB+C}$

Initial basis functions

Nonadiabatic coupling region

Spawned basis functions

parent) and the newly spawned basis function: $|\langle\chi_k^I|H_{IJ}|\chi_{k'}^J\rangle|$. The connection with classical mechanics suggests also requiring that the classical energy of the spawned basis function be the same as its parent. This requirement is in direct contradiction with short-time first-order perturbation theory analysis that predicts that the spawned function should be proportional to the parent basis function multiplied by the nonadiabatic coupling function. Hence, for example in the case of constant nonadiabatic coupling, first-order perturbation theory predicts that the spawned basis function will have the same momentum and position as its parent (and therefore a different classical energy whenever the parent basis function does not lie exactly at either the crossing seam of the diabats or a conical intersection in the adiabatic representation). Nevertheless, the equal energy constraint is justified from a long-time analysis (e.g., through the state-to-state form of Fermi's Golden Rule [68]), and one may therefore expect the expectation value of the energy (in an exact quantum mechanical calculation) to smoothly interpolate from the short-time (first-order perturbation theory) limit to the long-time limit. Figure 2 demonstrates this for a simple (yet realistic) case of a one-dimensional avoided crossing model. In this example, only one electronic state is populated at time zero; but as the wavepacket traverses the nonadiabatic region, it bifurcates and a new component is generated on the other electronic state. Apart from a phase relation, these two components propagate separately. After the two components of the wavepacket have left the region of nonadiabatic coupling, the expectation value of the energy can be defined for each electronic state individually; that is, there are no interference terms between the wavepackets on each electronic state:

$$\langle\Psi(\mathbf{R};t)|\hat{H}|\Psi(\mathbf{R};t)\rangle = N_1(t)\langle\chi_1(\mathbf{R};t)|H_{11}|\chi_1(\mathbf{R};t)\rangle$$
$$+ N_2(t)\langle\chi_2(\mathbf{R};t)|H_{22}|\chi_2(\mathbf{R};t)\rangle \qquad (2.20)$$

The energy-conserving nature of the transition (which is strictly valid only in the limit that the nonadiabatic coupling region is traversed infinitely slowly) implies that each of these energy expectation values be the same, where we stress that the above expression and the preceding statement are only valid outside of the nonadiabatic coupling region. Classically, the equality of these two quantum mechanical expectation values implies that the average energy of the basis functions representing the wavepacket on each electronic state should be the same. Because the phase space centers of the basis functions follow Hamilton's equations, it is necessarily true that the classical energy chosen for the spawned basis function will be conserved throughout the propagation. Therefore, the proper asymptotic mean energy should be imposed at the time of spawning, and the requirement of equality between the average classical energy of the wavepackets on different electronic states is met by requiring the

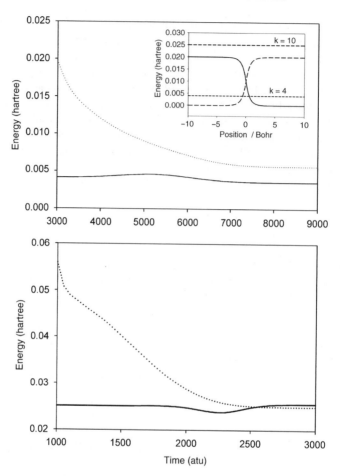

**Figure 2.** Expectation value of energy as a function of time (in atomic units) on the ground and excited electronic states for Tully's simple one-dimensional avoided crossing model [96]. The inset in the upper panel depicts the two diabatic states, indicated by full and dashed lines. Initially only one electronic state (dashed line) is populated and the wavepacket is a Gaussian centered at $x = -10$ bohr with a width of 0.5 bohr$^{-2}$. As the wavepacket traverses the nonadiabatic region, population is transferred to the other (full line) electronic state. The horizontal lines in the inset indicate the energy of the wavepacket for the two cases shown in the upper and lower panels. Long dashed line: the wavevector $k = 10$, yielding sufficient energy to populate both electronic states asymptotically. Short dashed line: the wavevector $k = 4$, yielding insufficient energy to cross the barrier on the ground adiabatic state (not shown). In both panels, the full line denotes the expectation value of the energy of the initially populated state, and the dotted one denotes this expectation value on the newly populated state. Upper panel: $k = 4$; lower panel: $k = 10$. In both cases, the expectation value of energy on the newly populated state (dotted line) starts as predicted by short-time first-order perturbation theory, but at the end of the nonadiabatic event it is nearly energy-conserving as would be predicted for an infinitely long nonadiabatic event. The mass used is 2000 atomic units, similar to the mass of a hydrogen atom.

parent and spawned basis functions to have the same classical energy. We reemphasize at this point that the Schrödinger equation is solved for the time-dependent coefficients of all basis functions. Therefore, our only duty is to justify that the restricted variational space we sample with the spawning procedure indeed covers the physically relevant space. To the extent that we sample only what is necessary, computational efficiency will be optimal. However, if the spawned functions under-sample the relevant Hilbert space, the accuracy of the method will suffer.

When we include the constraint that the newly spawned basis function will have the same classical energy as the parent basis function, we have an over-complete set of equations:

$$\frac{\partial}{\partial \bar{R}^J_{\rho k'}} \langle \chi^I_k | H_{IJ} | \chi^J_{k'} \rangle = 0, \qquad \rho = 1, \ldots, 3N$$

$$\frac{\partial}{\partial \bar{P}^J_{\rho k'}} \langle \chi^I_k | H_{IJ} | \chi^J_{k'} \rangle = 0, \qquad \rho = 1, \ldots, 3N \qquad (2.21)$$

$$V_{II}\left(\bar{\mathbf{R}}^I_k\right) + \sum_{\rho=1}^{3N} \frac{\bar{P}^{I\,2}_k}{2m_\rho} = V_{JJ}\left(\bar{\mathbf{R}}^J_{k'}\right) + \sum_{\rho=1}^{3N} \frac{\bar{P}^{J\,2}_{\rho k'}}{2m_\rho}$$

These equations can be solved in a least-squares sense, but in general they do not have a unique solution. The finite phase space width of the basis functions tends to dampen the sensitivity of the results, especially branching ratios, to the particular solution that is chosen. This sensitivity is further reduced when convergence with respect to MULTISPAWN is demonstrated.

Nevertheless, it is instructive to consider the simplest solutions that arise in a concrete example. Assuming two diabatic states in one dimension with a constant interstate coupling, maximizing the Hamiltonian coupling matrix element is equivalent to maximizing an overlap integral. This integral has the form of a product of two Gaussian functions whose arguments are the differences in momentum and position between the parent and spawned basis function. Two simple approximate solutions are "position-preserving" and "momentum-preserving" spawns, where the spawned basis function is placed at either the same position or momentum as its parent. The conjugate variable must then be adjusted to satisfy the constraint equation. In a one-dimensional problem, there is no ambiguity concerning the adjustment made to the conjugate variable, and the position-preserving spawn gives rise to exactly the momentum adjustment which Tully has used in the surface hopping method [69] and which was later justified by Herman via a semiclassical argument [70]. Interestingly, the second type of solution, the momentum-preserving spawn, can give rise to

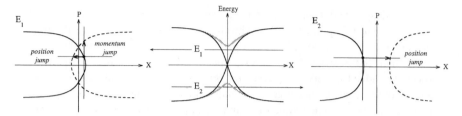

**Figure 3.** Schematic illustration of position- and momentum-preserving spawns for the one-dimensional avoided crossing model of Fig. 2. The two diabatic (black) and adiabatic (gray) potentials are depicted in the middle panel, and the two horizontal lines indicate the energies of the trajectories shown on the right and left panels, $E_2$ and $E_1$, respectively. Left panel: Phase space plot of a diabatic trajectory at energy $E_1$ that is above the ground-state adiabatic barrier. The full line indicates the trajectory traversed by the populated basis function, and the dashed line indicates the trajectory traversed by the unpopulated basis function (at the same energy $E_1$). The horizontal and perpendicular arrows show momentum- and position-preserving spawns where the spawned basis function is placed at either the same momentum or position as its parent. Right panel: Same as left panel but for energy $E_2$ that is below the ground-state adiabatic barrier. At energies below the adiabatic barrier, only momentum-preserving spawns are possible. For this specific example, the momentum-preserving spawn describes tunneling. (Figure adapted from Ref. 40.)

tunneling effects as demonstrated in Fig. 3. The two types of spawns for a simple one-dimensional avoided crossing model are depicted in this figure. Note that at energies below the adiabatic barrier (rightmost panel in Fig. 3), there is no solution corresponding to a position-preserving spawn, and the momentum-preserving spawn describes tunneling. However, this is a very special type of tunneling, contingent on the multielectronic state representation of the problem. For example, in the model shown in Fig. 3, both types of spawns are forbidden if the adiabatic representation of the electronic states is used.

We have obtained further empirical information on the optimal choice of the position and momentum of the new basis function by studying the Wigner transform of the wavepacket calculations. In Fig. 4 we show snapshots of the ground- (black contours) and excited-state (gray contours) Wigner distributions for the one-dimensional problem of Figs. 2 and 3. The distributions are shown at three instances of time, namely, the onset of population transfer (left panels), the peak of population transfer (middle panels), and long after the population transfer is completed (right panels). The upper and lower set of panels are for relative kinetic energies corresponding to wavevectors of $k = 4$ and 10, respectively (see inset in upper panel of Fig. 2.) Along with the Wigner distributions, we plot the expectation value of position versus momentum for the wavepacket traveling on the initially populated electronic state (black line), a mirror image of this path (gray line), and the actual expectation value of the position versus

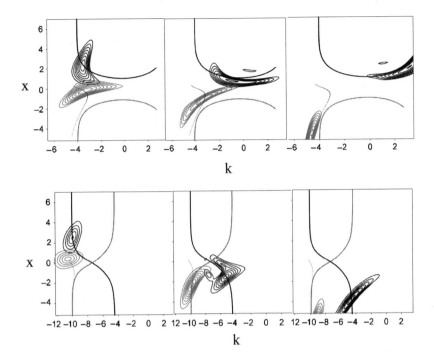

**Figure 4.** Phase space evolution of Wigner distributions for the two-state, one-dimensional model of Figs. 2 and 3. The upper and lower panels are for different initial relative kinetic energies ($k = 4$ and 10, respectively; see inset in upper panel of Fig. 2). The black and gray contour lines denote the Wigner distributions (computed using the exact quantum mechanical wavefunction) on the initially and newly populated electronic states, respectively. In all panels, the thick black line denotes the expectation value (position vs. momentum for the entire simulation) of the wavefunction, on the initially populated state, and the thick gray line is a mirror image of the thick black line, analogous to Fig. 3. The dotted line is the actual expectation value (position vs. momentum for the entire simulation) on the newly populated state. In both the upper and lower panels, snapshots of the Wigner distributions are shown at the onset of the nonadiabatic event (left panels), the peak of the nonadiabatic event (middle panels), and immediately after the completion of the nonadiabatic event (right panels). For clarity, the population on each state has been scaled to unity prior to calculating the Wigner transform and therefore the contours do not reflect the population of the states. At low energy (upper set of panels) the nonadiabatic event is not well-described in terms of the idealized limit of position- and/or momentum-preserving spawns. The event begins with a momentum-preserving (i.e., position-jump) population transfer (left panel) and continues with a mixture of position- and momentum-preserving population transfer (middle panel). After the nonadiabatic event is completed (right panel), the expectation value of position and momentum on the newly populated state (dotted line) is similar (but not identical) to that of the idealized spawned trajectory (gray line). Lower set of panels: Same as above, but for an energy that is above the barrier on the ground adiabatic state. Here, the actual expectation values of position and momentum on the newly populated state (dotted line) do coincide with those of the idealized spawned trajectory (gray line) and the position-preserving nature of the "spawn" is evident in the middle panel.

momentum on the newly populated state (dotted line). This figure demonstrates a few points. First, in both cases (upper and lower panels) the population transfer begins with a momentum preserving event (leftmost panel). Second, the peak of the nonadiabatic event is characterized by a mixture of position- and momentum-preserving "spawns." Third, when the nonadiabatic event is completed, the mirror phase space path (gray line) is either an excellent (lower right panel) or reasonable (upper right panel) approximation to the exact path of the wavepacket (dotted line).

In practice we take the position-preserving spawn, and the momentum of the new basis function is calculated by

$$\mathbf{P}_{new} = \mathbf{P}_{old} - D\hat{\mathbf{h}}_{IJ} \qquad (2.22)$$

where $\hat{\mathbf{h}}$ is a unit vector along the nonadiabatic coupling direction

$$\hat{\mathbf{h}}_{IJ} = \frac{\mathbf{d}_{IJ}}{|\mathbf{d}_{IJ}|} \qquad (2.23)$$

and $D$ is a scalar variable whose value is determined by conservation of energy. In some cases, it may happen that the surface to which a spawn should occur is not classically energetically accessible, or that there is sufficient energy to spawn but there is not enough momentum along the direction used to adjust the energy during spawning. In these instances, we use a steepest descent quenching of the position-preserving spawn until the constraint equation is satisfied. In one dimension, this leads to a momentum-preserving spawn, but in multiple dimensions this can give rise to spawned functions that do not preserve either position or momentum.

Certain additional numerical considerations should be satisfied before a spawning attempt is successful. First, in order to avoid unnecessary basis set expansion, we require that the parent of a spawned basis function have a population greater than or equal to $P_{min}$, where the population of the $k$th basis function on electronic state $I$ is defined as

$$p_k^I = |C_k^I|^2 \qquad (2.24)$$

This threshold prevents basis functions with small population (which are only negligibly contributing to the nuclear wavefunction in any case) from giving rise to new basis functions. The ideal of $P_{min} = 0$ is usually computationally wasteful, leading to many unpopulated basis functions. However, it is also important to note that the uncertainty in branching ratios incurred by finite $P_{min}$ is dependent on the average population of a basis function in the wavepacket. Second, it is pointless to create new basis functions that are redundant with existing basis functions. We therefore enforce a maximum overlap, $S_{max}$, that any new basis function is allowed to have with any existing basis function in the wavepacket. Any potentially newly spawned basis function that has been back-propagated to $t_0$ and has an overlap greater than $S_{max}$ with any other basis function on the same electronic state in the same wavepacket at time $t_0$ is not spawned. Finally, we note that we do allow for back-spawning; that is, a newly spawned basis function on a given electronic state is allowed to transfer population back to the other electronic state.

Once we have determined which basis functions will be spawned (and their initial positions and momenta), they are back-propagated in time to $t_0$, where they are added to the set of basis functions. The coupled propagation of all of the basis functions is then continued, beginning again from time $t_0$; that is, Eq. (2.11) is integrated again along with Hamilton's equations of motion. The final-state analysis of the results is discussed in Section II.D, following a review of various approximations that can, and often should, be made.

## D. Approximations and Analysis of Results

So far, the only approximation in our description of the FMS method has been the use of a finite basis set. When we test for numerical convergence (small model systems and empirical PESs), we often do not make any other approximations; but for large systems and/or *ab initio*-determined PESs (AIMS), additional approximations have to be made. These approximations are discussed in this subsection in chronological order (i.e., we begin with the initial basis set and proceed with propagation and analysis of the results).

In Section II.C we discussed the selection of initial conditions for the nuclear basis set and its propagation. After selecting the position and momentum parameters for each nuclear basis function, the complex amplitudes were determined by a least-squares fitting procedure [cf. Eq. (2.18)] and then propagated simultaneously. If this procedure of coupled propagation is followed, then one is attempting a particular form of wavepacket propagation with classical mechanics as a guide for basis set selection, expansion (i.e., spawning), and propagation. In this mode, convergence to exact quantum mechanical results is ensured for a sufficiently large number of basis functions. When quantal aspects of the single-surface dynamics are important, this option should be used (as well as a coherent analysis of the final results). The first approximation in FMS (beyond the use of a finite basis set) assigns a unit initial amplitude to each of the initial basis functions and propagates each one independently; however, each basis function can spawn new basis functions in nonadiabatic regions, and these in turn can spawn additional basis functions, and these multiple descendents of a single parent are fully coupled (on all potential energy surfaces) by the FMS algorithm. This "independent-first-generation" (IFG) approximation assumes that a properly chosen swarm of classical trajectories will suffice to describe the dynamics occurring on a single electronic state. While this approximation allows one to carry out a very complete sampling of initial-state phase space without introducing an unmanageable amount of coupling (between basis functions), it may not be sufficiently accurate when quantum interference effects on a single electronic state are important. In most of our AIMS calculations, we have used the IFG approximation because of computational limitations; but when predetermined PESs were available, we have usually preferred coupled propagation.

The next set of approximations concerns the propagator and the evaluation of the matrix elements that are required for Eq. (2.11). At each time step and for each nuclear basis function, numerical integration of Eq. (2.11) requires diagonal and off-diagonal terms of the Hamiltonian matrix elements ($\langle \chi_k^I | H_{II}^e | \chi_{k'}^I \rangle$ and $\langle \chi_k^I | H_{IJ}^e | \chi_{k'}^J \rangle$, respectively). The need to evaluate these integrals is the greatest drawback of the FMS method. Given an analytic representation of the PESs and their couplings, it is sometimes possible to evaluate these multidimensional integrals analytically; but when the PESs (and their couplings) are known only locally, as in AIMD and AIMS, this is not possible. By expanding the wavefunction in a (traveling) Gaussian basis set, we have localized individual components of the wavefunction and thus any integrals involving the semilocal nuclear basis functions. Motivated by this localization, we evaluate the required integrals using a first-order saddle-point (SP) approximation [71]:

$$\langle \chi_k^I | H_{IJ}^e | \chi_{k'}^J \rangle = H_{IJ}^e(\tilde{\mathbf{R}}) \langle x_k^I | x_{k'}^J \rangle \qquad (2.25)$$

where $\tilde{\mathbf{R}}$ is the location of the centroid of the product of the basis functions $\chi_k$ and $\chi_{k'}$. This approximation is applied to both diagonal ($I = J$) and off-diagonal ($I \neq J$) elements of the Hamiltonian, and it involves only the potential energy operator (the part involving kinetic-energy operator can be evaluated analytically for Gaussians basis functions). The SP approximation resembles the Mulliken–Ruedenberg and related approximations that have been used in electronic structure theory for the approximate evaluation of multicenter two-electron integrals [72]. This approximation has been tested for simple one-dimensional model problems [38,73,74] with favorable results. Nevertheless, it is a severe approximation whose quality should be tested more rigorously and, whenever possible, improved. Two obvious possible improvements involve the use of higher-order SP approximations—for example, second-order

$$\langle \chi_k^I | H_{IJ}^e | \chi_{k'}^J \rangle = H_{IJ}^e(\tilde{\mathbf{R}}) \langle x_k^I | x_{k'}^J \rangle + \langle \chi_k^I | \mathbf{R} - \tilde{\mathbf{R}} | \chi_{k'}^J \rangle \frac{d}{d\mathbf{R}} H_{IJ}^e(\mathbf{R}) \Big|_{\tilde{\mathbf{R}}} \qquad (2.26)$$

or various forms of numerical quadrature. Although these approximations require more computational effort and are currently beyond our capabilities, they could greatly improve the accuracy of the integrals. A less obvious improvement, which does not require any additional calculations, is to incorporate more of the available information (about the PESs and their couplings) into the calculation of individual matrix elements. Although our knowledge of the PESs in AIMD is only local, it does increase with time. For example, for a single classical trajectory the potential and its derivative are only known at one point (geometry) at the first time step (assuming a simple first-order integrator). After $n$ steps they are known at $n$ geometries. In the AIMS method, we actually know the potential at many more points (geometries) at each instance of time. To see this, assume for simplicity that only one electronic state ($I$) is populated and that at time $t$ there are $N_I(t)$ nuclear basis functions. In such a case, at time $t$ the potential is known at $N_I(t)[N_I(t) + 1]/2$ points (geometries) even if we use the simple first-order SP approximation. As in the case of a classical propagation, this information grows linearly with time (even when the size of the nuclear basis set does not change). Equation (2.25) does not incorporate any of this additional information. Because the additional information (from other basis functions and from previous points in time) is irregular, it is not obvious how to incorporate (or weight) it. However, it is also obvious that not using it is quite wasteful, and future developments of the method should address this point. Our first attempt to exploit the temporal nonlocality of the Schrödinger equation is discussed in Section II.F.

The use of the SP approximation solves the problem of matrix element evaluation. Still, at each time step, $O(N_{nuc}^2(t))$ evaluations of potential energies and couplings are required. For *ab initio*-determined PESs, this can be quite tedious. Furthermore, the dynamics of the complex amplitude coefficients are governed by the Schrödinger equation even in the absence of intersurface coupling. Hence, the trajectory amplitudes will attempt to correct as much as possible the inadequacies of classical mechanics by exchanging population between trajectories with the same electronic state label. This behavior has no analog in classical mechanics, and it can obscure the classical interpretation of the results. In many cases this is desirable, but if the objective is to obtain a method that is "classical-like," then one should separate the dynamics occurring on single and multiple electronic states. By invoking the idea of operator splitting, this separation can be achieved and the number of potential energy evaluations can be further reduced. For a two-electronic-state problem we write the Hamiltonian operator as a sum of single-state (SS) and interstate (IS) terms:

$$\hat{H} = \hat{H}_{SS} + \hat{H}_{IS} = \begin{pmatrix} \hat{H}_{11} & \\ & \hat{H}_{22} \end{pmatrix} + \begin{pmatrix} & \hat{H}_{12} \\ \hat{H}_{21} & \end{pmatrix} \qquad (2.27)$$

Invoking a Trotter factorization [75], we write the time evolution operator as

$$\exp(-i\hat{H}t) \approx \exp(-i\hat{H}_{SS}t/2)\exp(-i\hat{H}_{IS})\exp(-i\hat{H}_{SS}t/2) \tag{2.28}$$

The nonclassical evolution of the trajectory amplitudes can now be avoided by forbidding the interaction of basis functions during the single-surface propagation. This is accomplished by substituting the frozen Gaussian propagator [46] for $\exp(-i\hat{H}_{ss}t/2)$. At each time step, Eq. (2.28) with the frozen Gaussian propagator requires only $N_{nuc}(t)$ PES evaluations, whereas a coupled single-state propagation requires

$$\sum_{I=1}^{N\,\text{states}} N_I(t)[N_I(t) + 1]/2 = 1/2\left[N_{nuc}(t) + \sum_{I=1}^{N\,\text{states}} N_I^2(t)\right] \tag{2.29}$$

PES evaluations. In either case (coupled or split-operator with frozen Gaussian propagator)

$$\sum_{I=1}^{N\,\text{states}} \sum_{J>I}^{N\,\text{states}} N_I(t)N_J(t) \tag{2.30}$$

evaluations of the off-diagonal matrix elements are required at each time step. The number of intrastate matrix elements needed is reduced from $O(N^2)$ to $O(N)$ by using the frozen Gaussian propagator in Eq. (2.28). This more favorable scaling can significantly reduce the computational expense of the AIMS method. However, care should be taken when using the frozen Gaussian propagator because it does not conserve the normalization of the wavefunction. Left unchecked, this would destroy the ability to predict branching ratios. Yet, strict single-surface propagation cannot change the relative fraction of nuclear populations on different electronic states. We therefore renormalize the amplitudes of the nuclear basis functions after the single-surface propagation so that this requirement is obeyed. From a quantum mechanical perspective, this is a crude approximation (recall that the exact single-surface propagator does exchange population between basis functions with the same electronic state label). It is, however, an adequate approximation when an ensemble of classical trajectories can provide an adequate description of the exact, quantum mechanical, dynamics. Although this may seem to be an extreme requirement, the basic underlying assumption of the AIMS method is that classical mechanics provides a good zeroth-order propagator in the absence of specific temporally localized quantum mechanical events. Because the quantum mechanical intrastate interactions between trajectories primarily alter the relative phases between trajectories, it is reasonable to assume that it will be difficult to model phase interference effects (e.g. Stueckelberg oscillations [76,77]) with this approach. The neglect of these interactions may result in errors in the relative phases and therefore in the amplitudes and positions of the oscillations. More research is needed to determine the range of applicability of this approximation.

The saddle-point and split-operator approximations are compatible. However, we see no compelling reason to use the operator splitting without approximating the single-surface propagator. In what follows we refer to the AIMS method with saddle-point approximation as AIMS-SP and refer to the AIMS method with both saddle-point and split-operator factorization (with an approximate renormalized frozen Gaussian single-surface propagator) as AIMS-SP-SO.

Because the AIMS method associates a unique nuclear wavefunction with each electronic state, one has direct access to dynamical quantities on individual states. This is unlike mean-field based approaches that use only one nuclear wavefunction for all electronic states [59]. One can therefore calculate branching ratios

$$n_I(t) = \sum_{k,k'}^{N_I(t)} C_k^{I^*} \mathbf{S}_{IkIk'} C_{k'}^I \tag{2.31}$$

and coherent expectation values

$$\langle O(t) \rangle_I = \frac{\sum_{k,k'}^{N_I(t)} C_k^{I^*} C_{k'}^I \langle \chi_k^I | \hat{O} | \chi_{k'}^I \rangle}{\sum_{k,k'}^{N_I(t)} C_k^{I^*} \mathbf{S}_{IkIk'} C_{k'}^I} \tag{2.32}$$

where $\hat{O}$ can be any operator (e.g., position or momentum). The coherent analysis of the results is compatible with a coupled propagation of the nuclear basis functions. When the AIMS method is used with no approximations beyond the use of a finite basis set, the initial amplitudes are propagated simultaneously and the analysis of the results includes the coherence terms, then the results are guaranteed to converge to the exact quantum mechanical results, given a large enough initial basis set and a robust spawning algorithm. An incoherent analysis of the results

$$\langle O(t) \rangle_I = \frac{\sum_k^{N_I(t)} C_k^{I^*} C_k^I \langle \chi_k^I | \hat{O} | \chi_k^I \rangle}{\sum_k^{N_I(t)} C_k^{I^*} \mathbf{S}_{IkIk} C_k^I} \tag{2.33}$$

is compatible with the IFG approximation. If the IFG approximation and incoherent analysis are used together with the saddle-point and split-operator frozen Gaussian propagator (i.e., AIMS-SP-SO), and a single electronic state is included in the dynamics, then the method is identical to classical mechanics. Thus, as we have required at the very beginning the AIMS method reduces to quantum mechanics in one limit and to classical mechanics in the other. Heller's frozen Gaussian approximation is obtained when the IFG approximation is used with AIMS-SP-SO, and the final results are analyzed coherently.

## E.   Comparison to Other Methods

Recently, several other groups have been working to include nuclear quantum effects in AIMD simulations. Path integral approaches [78–87] are attractive because they exhibit the local behavior that is needed for the quantum chemistry/ quantum dynamics interface. However, real-time dynamics with path integrals usually requires many more paths than can be practically computed when the PESs are generated by solving the quantum chemical problem directly. Imaginary-time path integral methods converge much more rapidly and provide equilibrium properties including quantum nuclear effects. The first imaginary-time path integral methods using AIMD techniques have appeared [88,89] and been applied to investigate proton tunneling in water [6,90–93]. The centroid molecular dynamics (CMD) method can follow real-time dynamics with a proper treatment of dispersion and zero point effects. Voth and co-workers [94] have recently presented an AIMD implementation of CMD that seems very promising.

It is also possible to merge semiclassical methods with AIMD—for example, the trajectory surface-hopping (TSH) method of Tully [95,96] for nonadiabatic dynamics and a similarly motivated method for tunneling developed by Makri and Miller [97,98]. There have been a few attempts along these lines. The original Tully–Preston surface-hopping method [95] has been used to investigate photoinduced isomerization in a retinal protonated Schiff base analog [99]. The more recent "fewest-switches" variant of TSH [96] has been used to investigate dynamics around conical intersections in the $Na_3F_2$ cluster [100]. We have used a surface hopping formulation of tunneling [97,98] in conjunction with hybrid DFT electronic structure methods in order to model the proton transfer tunneling splitting in malonaldehyde, obtaining quantitative agreement with experiment [101].

One can also ask about the relationship of the FMS method, as opposed to AIMS, with other wavepacket and semiclassical nonadiabatic dynamics methods. We first compare FMS to previous methods in cases where there is no spawning, and then proceed to compare with previous methods for nonadiabatic dynamics. We stress that we have always allowed for spawning in our applications of the method, and indeed the whole point of the FMS method is to address problems where localized nuclear quantum mechanical effects are important. Nevertheless, it is useful to place the method in context by asking how it relates to previous methods in the absence of its adaptive basis set character. There have been many attempts to use Gaussian basis functions in wavepacket dynamics, and we cannot mention all of these. Instead, we limit ourselves to those methods that we feel are most closely related to FMS, with apologies to those that are not included. A nice review that covers some of the earlier methods in more detail, with attention also to nonadiabatic dynamics methods, has appeared in the literature [102].

When spawning is not allowed, the FMS method becomes very similar to Heller's FGA approach [46]. The differences depend on the particular variant of the FMS method that is used. As mentioned above, FMS without spawning and using the IFG and SP approximations with coherent analysis of the final results is exactly the FGA method. If one further analyzes the results incoherently, it becomes exactly classical mechanics. Without the IFG approximation—that is, when all the initial wavepackets are fully coupled—the FMS method becomes a form of variational basis set wavepacket propagation. The use of a variational principle in the context of a Gaussian basis set was first suggested, but not implemented, by Heller [63]. Several implementations appeared later [55,64,103–106], and Metiu and co-workers spent considerable effort trying to interpret the nonclassical propagation of position and momentum parameters which results when the TDVP is applied [55,107,108]. In the FMS method, we intentionally avoid the variational propagation of the width, position, and momentum parameters in order to ensure a well-defined classical limit in the absence of spawning. The methods proposed by Skodje and Truhlar [64] are thus the closest analogs, since most of the other variational approaches have applied the TDVP to all of the basis set parameters.

Recently, there has been much effort [50,109,110] devoted to understanding and improving the Herman–Kluk propagator [70,111], which provides a means for determining the time evolution of the coefficients in the FGA without directly coupling trajectory basis functions. These ideas have been interfaced with the semiclassical initial value representation (IVR) with considerable success [112]. It would be difficult to incorporate these methods directly in the context of AIMS because they require second derivatives of the PES, and the number of trajectories required to obtain stable results often exceeds $10^4$ even in one- and two-dimensional problems [50,110,113]. Although it seems that there should be a formal connection between the Herman–Kluk formula for the basis function coefficients (derived with stationary phase approximations) and the expression obtained through application of the TDVP in the limit of an infinite basis set, no attempt to make a rigorous connection has yet been reported.

When spawning is allowed, which is invariably the case in our applications, the closest analog to FMS in the context of Gaussian wavepacket methods are the curve-crossing variants of Metiu's minimum error method (CC-MEM) dynamics [56,114,115]. To the best of our knowledge, these were the first attempts to add basis functions to the basis set during the propagation of Gaussian wavepackets. However, the methods were only explored in one dimension, and therefore the question of where and how to place new basis functions was not very important. In high-dimensional problems, this question becomes critical and the spawning algorithm of FMS has been designed with such problems in mind. A further difference between CC-MEM and FMS is that CC-MEM, like most other variational Gaussian wavepacket methods, applies the TDVP to all basis set parameters, including position, momentum, and width.

Coalson has recently introduced a wavepacket-path integral method [53,116] for nonadiabatic problems which has similarities to FMS. Time is discretized around the nonadiabatic event, as in FMS; but all possible paths with respect to an electronic surface index are then directly enumerated, and the results of propagation with all possible sets of impulsive hops within the nonadiabatic region are summed coherently in a path integral-like expression. An approximation is introduced to maintain the Gaussian shape of the wavepackets as they propagate through the nonadiabatic region. It remains to be seen whether this method can be implemented successfully in problems with more than two nuclear degrees of freedom or more than one nonadiabatic event.

A number of methods for nonadiabatic dynamics formulated directly in phase space—that is, solving the Liouville equation—have been proposed [117–121]. These also have some similarity to FMS, with the possible advantage of avoiding the averaging over wavepackets that is necessary to model finite temperature effects in a Schrödinger equation-based approach. Kapral and Ciccotti have stressed that it may be more appropriate to mix quantum and classical treatments of different nuclei within the context of the quantum mechanical Liouville equation [120]. Within the FMS approach, the mixed quantum-classical nuclear problem is conceptually sidestepped by treating all degrees of freedom quantum mechanically. In practice, it is not clear whether this avoids the problem when small basis sets are used. These Liouville methods are promising, but, again, it remains to be seen how they will fare in high-dimensional cases with multiple nonadiabatic events.

Perhaps the most popular method for nonadiabatic dynamics has been Tully's "surface-hopping" method (TSH). Of course, the whole idea of spawning has some similarity to the surface hops in TSH. Nevertheless, there are key differences between the methods. The FMS method in all of its variants possesses a well-defined nuclear wavefunction, and therefore it is straightforward to extract correlation functions and spectra, as we have shown in past work [122]. However, even when one restricts the analysis to branching ratios and momentum distributions, there remain differences. We have applied [38] the FMS method directly to the one-dimensional problems first used to test the "fewest-switches" form of TSH [96], and the results point out two possible problems with TSH which can be resolved in FMS. First, phase interference effects in FMS come about through the interaction of different basis functions, that is, trajectories. In the TSH method, these interferences are modeled through phase evolution on different electronic states along a single trajectory. Complete averaging of these interferences would lead to the disappearance of nonadiabatic effects—Prezhdo and Rossky [123] have referred to this as the "quantum

Zeno paradox." On the other hand, recovering these interferences from a single path leads to excessive correlation, as evidenced by the highly oscillatory results obtained with TSH for Tully's third, "extended coupling with reflection," model. This is remedied effortlessly in FMS, and one may speculate that FMS will tend to the opposite behavior: Interferences that are truly present will tend to be damped if insufficient basis functions are available. This is probably preferable to the behavior seen in TSH, where there is a tendency to accentuate phase interferences and it is often unclear whether the interference effects are treated correctly. This last point can be seen in the results of the second, "dual avoided crossing," model, where the TSH results exhibit oscillation, but with the wrong structure at low energies. The correct behavior can be reproduced by the FMS calculations with only ten basis functions [38].

A much more detailed comparison of TSH and FMS-M ("minimal" FMS, employing the IFG and SP approximations and incoherent analysis of final results) has been carried out [124] for a six-dimensional problem where exact results are available. Although the FMS-M method generally improves on TSH results, there are residual errors whose origin remains undetermined. We refer the reader to this paper for a detailed comparison of the TSH and FMS-M methods as applied to a particular problem. Here, we only want to point out the major advantage of the FMS methods over TSH, which is that FMS forms a hierarchy of methods, culminating in the exact solution of the nuclear Schrö-dinger equation. Even though one may never reach the ideal of converged quantum nuclear dynamics in high-dimensional problems, one can carry out test calculations to assess the effects of improving the calculation. For example, one can increase the size of the initial basis set, decrease spawning thresholds, increase MULTISPAWN, and use alternative methods for evaluating the required integrals—for example, higher-order SP approximations or numerical quadrature. In contrast, it is difficult or impossible to assess the accuracy of the approximations in TSH without recourse to an exact calculation for comparison.

### F. Advanced Topics

The previous subsections defined the AIMS method, the various approximations that one could employ, and the resulting different limits: classical mechanics, Heller's frozen Gaussian approximation, and exact quantum mechanics. As emphasized throughout the derivation, the method can be computationally costly, and this is one of the reasons for developing and investigating the accuracy of various approximations. Alternatively, and often in addition, one could try to develop algorithms that reduce the computational cost of the method without compromising its accuracy. In this subsection we discuss two such extensions. Each of these developments has been extensively discussed in a publication, and interested readers should additionally consult the relevant papers (Refs. 125 and 41, respectively). We conclude this subsection with a discussion of the first steps that we have taken to treat tunneling effects within the framework of the AIMS method [126].

### 1.  Interpolated Potential Energy Surfaces

The two primary advantages of *ab initio*-based molecular dynamics methods (flexibility in the PESs and avoidance of tedious fitting to empirical functional forms) come at a challenging computational cost, especially for photochemical problems where electronic excited states and nonadiabatic couplings are required. At the same time, information about the PESs obtained during the propagation is typically discarded, and therefore one does not benefit from the fact that the nuclei may often visit similar regions in configuration space. In particular, the dynamics of certain highly harmonic modes (e.g., C–H stretches) may be quite trivial.

One way to address this issue is to combine first-principles dynamics with automatic interpolation of PESs. For nonadiabatic dynamics, interpolation alone is problematic because the adiabatic PESs often have sharp features near the region of strong interstate coupling. Any smooth interpolating function will fail to reproduce these sharp features in the PESs as well as the often strongly varying nonadiabatic coupling. While both difficulties can be avoided by using a diabatic representation of the electronic states, we insist on using the adiabatic representation in the AIMS method for two reasons. First, the diabatic representation is not unique and, without information about the electronic wavefunction at different geometries, is often path-dependent. Second, for chemical applications the nuclear kinetic energy is typically low (i.e., thermal), making the adiabatic representation more natural. In practice, this means that fewer basis functions will need to be spawned. By using *ab initio* methods for configurations near conical intersections while interpolating the PESs (and not their couplings) elsewhere, we can overcome the nonsmoothness problems.

We have chosen to use the modified Shepard interpolation scheme introduced by Collins and co-workers [127–131]. At any nuclear configuration $\mathbf{R}$, the potential energy, $V(\mathbf{R})$, is represented as a sum of Taylor expansions in reciprocal bond lengths, $\mathbf{Z}$. Each of these Taylor expansions is centered at a different data point, $Z(i)$, and is truncated at second order, although higher derivatives may in principle be included if desired. The choice of internal coordinates minimizes the number of data points required for the interpolation. As Collins and co-workers have shown [132], the correct identical particle symmetry may be enforced on the PESs by extending the data set to include all possible permutations of each data point. For simplicity we ignore this symmetry issue and hence write the interpolated potential energy as

$$V(\mathbf{R}) = \sum_i w_i(\mathbf{Z})T_i(\mathbf{Z}) \tag{2.34}$$

where $w_i$ is a normalized weight function which ensures that the patching of the Taylor polynomials, $T_i$, is smooth:

$$w_i(\mathbf{Z}) = \frac{|\mathbf{Z} - \mathbf{Z}(i)|^{-p}}{\sum_j |\mathbf{Z} - \mathbf{Z}(j)|^{-p}} \tag{2.35}$$

where $p$ must be taken [128] such that $p > (3N - 6) + q$, where $q$ is the order of the Taylor expansion. This purely geometric requirement ensures that the weight function decreases faster than the increase of the Taylor expansion. The way in which one chooses where in configuration space to place the data points is crucial, and Collins et al. [127–130] have developed an iterative algorithm for choosing these points. Briefly, new data points are selected from configurations visited by $n_t$ classical trajectories calculated on the PES interpolated from the data set of the previous iteration. The process begins by calculating a few tens of data points along some minimum energy path from reactants to products, and it is deemed converged when an observable of interest ceases to change with the addition of more data.

Not all the data points from the $n_t$ trajectories are used in the interpolation. The $n_{\text{sel}}$ new data points are selected using the "$h$ weight" function [133] that balances the desire to place new points as far as possible from the existing $n_d$ data points with the need to have a higher data density in dynamically important regions. In particular, the relative importance of a candidate data point $\mathbf{Z}^k$ ($k$ denotes the trajectory) is given by

$$h[\mathbf{Z}^k] = \frac{\sum_{j \neq k}^{n_d} |\mathbf{Z}^k - \mathbf{Z}^j|^{-p}}{\sum_{i=1}^{n_d} |\mathbf{Z}^k - \mathbf{Z}(i)|^{-p}} \tag{2.36}$$

This criterion is based on pure geometric considerations. When the second derivatives must be computed by finite differences of analytic gradients, as is often the case in practice for highly correlated electronic wavefunctions, it can be improved by a "prescreening" procedure. Before calculating the energy and gradients at small (positive and negative) displacements of a proposed new data point, $\mathbf{R}_{\text{new}}$, the values of the interpolated and *ab initio* potential energies are compared. If they agree within the target accuracy of the PES, the point is not added to the data set and the second derivative matrix is not evaluated. This results in considerable computational savings, with little or no loss of accuracy, because many calculations are required to obtain the second derivative information whereas only one is required for the energy.

The interpolation method was extended to include multiple electronic states by requiring that the same data points be used to interpolate all electronic states. These points were chosen (by the prescreened $h$-weight procedure) from classical trajectories that run alternately on each of the electronic states.

Because the concept of minimum energy path is not well-defined when multiple electronic states are involved, the initial data set is simply taken as the union of points which one considers important on each of the electronic states—for example, local minima on each electronic state. The weights of each data point, $w_i$ in Eq. (2.34), were taken to be the same on all electronic states because they only depend on the location of the data points. Hence, the difference between electronic states ($V^I_{\text{Shepard}}(\mathbf{R})$) is manifested only in the parameters of each of the Taylor expansions:

$$V^I_{\text{Shepard}}(\mathbf{R}) = \sum_i w_i(\mathbf{Z})T^I_i(\mathbf{Z}) \tag{2.37}$$

Because the Shepard interpolation method is global, it is affected by cusped regions in the PES. To avoid this we must define a crossing or nonadiabatic region where the Shepard interpolation will likely fail. Because we do not interpolate the nonadiabatic coupling, we define this region using a threshold energy gap (and not a threshold effective nonadiabatic coupling), $\Delta E_{\text{min}}$. Attempts to fit cusped regions are then avoided by rejecting all data points for which the energy gap (between any pair of electronic states) is less than $\Delta E_{\text{min}}$. In these surface-crossing regions, the interpolation function is replaced by direct *ab initio* evaluation of the PESs (and their couplings). As with the spawning threshold previously discussed, the precise value of this parameter is somewhat arbitrary. If it is too small, the interpolation will converge slowly when conical intersections are present. If it is too large, large regions of the PES will be modeled by direct *ab initio* evaluation, making the procedure more costly than necessary.

The *ab initio* and interpolated potential functions are coupled using a smooth switching function, written in terms of the energy difference between the electronic states:

$$V^I(\mathbf{R}) = w(\mathbf{R})V^I_{ab\ initio}(\mathbf{R}) + (1 - w(\mathbf{R}))V^I_{\text{Shepard}}(\mathbf{R}) \tag{2.38}$$

where

$$\tilde{w}(\mathbf{R}) = \frac{1}{2}\left[1 + \tanh\left(\frac{\Delta E(\mathbf{R}) - \Delta E_c}{\Delta E_w}\right)\right] \tag{2.39}$$

and

$$w(\mathbf{R}) = \begin{cases} 0, & \tilde{w}(\mathbf{R}) < w_{\text{threshold}} \\ 1, & \tilde{w}(\mathbf{R}) > 1 - w_{\text{threshold}} \\ \tilde{w}(\mathbf{R}), & \text{otherwise} \end{cases} \tag{2.40}$$

In Eq. (2.39), $\Delta E(\mathbf{R})$ is the minimum (over all pairs of electronic states considered) of the absolute value of the energy gap at the given nuclear geometry, $w_{threshold}$ is a numerical parameter taken small enough ($\approx 10^{-5}$) that it has no effect on the dynamics, and $\Delta E_c$ and $\Delta E_w$ are the center and range of the switching function, respectively.

This hybrid approach can significantly extend the domain of applicability of the AIMS method. The use of interpolation significantly reduces the computational effort associated with the dynamics over most of the timescale of interest, while regions where the PESs are difficult to interpolate are treated by direct solution of the electronic Schrödinger equation during the dynamics. The applicability and accuracy of the method was tested using a triatomic model: collisional quenching of $Li(p)$ by $H_2$ [125], which is discussed in Section III.A below.

## 2. Time-Displaced Basis Set

We have investigated another procedure for reducing the computational expense of the AIMS method, which capitalizes on the temporal nonlocality of the Schrödinger equation and the deterministic aspect of the AIMS method. Recall that apart from the Monte Carlo procedure that we employ for selecting initial conditions, the prescription for basis set propagation and expansion is deterministic. We emphasize the deterministic aspect because the time-displaced procedure relies on this property.

The time-displaced basis (TDB) method incorporates information about the PES in the region spanned by the wavefunction at previous times without resorting to interpolation. It can significantly reduce the number of matrix element evaluations (normally scaling quadratically with the number of basis functions) without compromising the quality of the results. As we will show, under certain assumptions it is possible to reduce the scaling of the procedure from quadratic to linear in the number of basis functions. The basic idea is as follows: Instead of choosing $N_{init}$ independent initial basis functions (using a MC procedure), choose only $N_s$ independent initial basis functions, where $N_s < N_{init}$. The basis set is then enlarged to $N_{init}$ by adding $N_{init} - N_s$ basis functions that are displaced forward and backward in time with respect to the $N_s$ independent basis functions. This is motivated by the observation that one can cover a given energy shell in phase space either by generating many independent trajectories or by following a single trajectory for a long time. This is not meant to imply that one will be able to achieve anything approaching complete coverage of the accessible phase space for multi-dimensional problems. Rather, we point out that there are two independent ways of expanding the basis set: multiple independent classical trajectories and time displacement along these trajectories. To better understand this idea, consider the simple example of $N_{init} = 3$ and $N_s = 1$. Instead of choosing three independent basis functions, we choose only one and refer to it as the "seed." (The position and momentum that define a seed basis function are chosen randomly from the appropriate Wigner distribution at $t = 0$.) Two additional TD basis functions are associated with the seed and are related to it by forward and backward propagation in time. Because classical mechanics guides the propagation of the nuclear basis functions, a single distinct path is traveled by the three basis functions. This path is defined by the basis function that at $t = 0$ is most forward displaced in time, and all other basis functions belonging to the same seed follow this path. Because the spawning procedure is deterministic, this property carries over to the next generation of basis functions: The ones spawned by the "head" basis function define the path that is traversed by all the other spawned basis functions that are children of basis functions belonging to the same seed.

To quantify the resulting reduction in computational cost, we compare the number of matrix element evaluations for $N_{init}$ independent basis functions to that required when only $N_s$ of these $N_{init}$ are independent. The total number of basis functions is $N_{init}$ in both cases. For simplicity, we ignore the symmetry of the Hamiltonian matrix—we assume that all $N_{init}^2$ elements are evaluated even though $\mathbf{H}$ is Hermitian. When all $N_{init}$ basis functions are independent, traveling on $N_{init}$ distinct paths in phase space, $N_{init}^2$ matrix element evaluations are required at each time step. When the total number of basis functions is $N_{init}$, but only $N_s$ distinct paths are traversed by these $N_{init}$ basis functions, the number of matrix element evaluations reduces to $O(N_s^2 + N_s N_{init})$. The number of matrix elements is no longer quadratic in $N_{init}$, but rather linear. This result is illustrated in Fig. 5. Here, the total number of basis functions is six, and there are two seeds (i.e., $N_s = 2$). For clarity the two seeds are spatially separated (in reality, basis functions that belong to different seeds are not so neatly separated), and within each seed the "leading" (in time) basis function is shaded. The basis set is shown at two time points, $t$ and $t + \Delta t$. At each point in time, any matrix element that involves a shaded basis function must be evaluated, and these are denoted by arrows connecting basis functions. This is to be compared with the case where all basis functions are independent. In this specific example, 11 matrix elements are required for the TDB set while 21 are required if all basis functions are chosen independently.

The key question is how many independent initial conditions are required for a basis set of size $N_{init}$. While one would like to minimize the number of seeds, the quality of the results is bound to deteriorate if too few are used. There is no general answer to this question, and therefore this number should be viewed as another parameter that defines the basis set and governs the numerical convergence of the method. For each class of problems, numerical tests should be used to determine $N_s$. Our experience with this procedure is limited, and therefore we make only one general comment. The number of independent initial conditions must be large enough so that its energy spectrum faithfully mimics the energy spectrum of the initial wavefunction $\Psi_{t=0}^{exact}$. As a

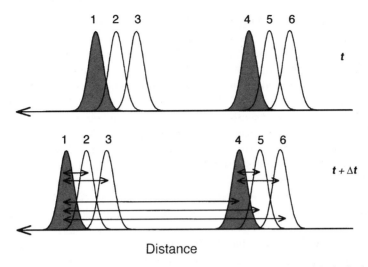

**Figure 5.** Schematic illustration of the time evolution of a time-displaced basis. Basis states 1, 2, and 3 belong to one seed while 4, 5, and 6 belong to another. The basis set is shown at two time points, and the leading basis functions are shaded in gray. The arrows connecting basis functions indicate required new matrix elements at time $t + \Delta t$. For this specific example, 11 new matrix elements are evaluated at each point in time, compared to 21 if all basis functions had been chosen independently. (Figure adapted from Ref. 40.)

consequence of the uncertainty principle, the quantum mechanical "phase space" accessed by the wavefunction is not restricted to lie on a classical energy shell, but rather will have a finite energy width which can never be modeled with a single classical trajectory, time displaced or not. Therefore, the limit of a single seed should not be expected to be sufficient.

The TDB approach can be applied alone or combined with the hybrid approach discussed in the previous section. We have tested it using a two-dimensional analytical model, and some of the results are presented in Section III.A. A more extensive discussion can be found in Ref. 41.

### 3. Tunneling

So far our discussion of the AIMS method has only emphasized quantum effects that are due to breakdown of the Born–Oppenheimer approximation. There are other instances when classical mechanics fails to describe the dynamics, and in this subsection we discuss an extension of the AIMS method that incorporates tunneling effects. Unlike the previous two subsections, this extension does not attempt to reduce the computational cost of the method but rather to expand its scope. This extension to the AIMS method affects only the spawning algorithm.

All other aspects of the method (e.g., selection of initial conditions, propagation, and analysis of the results) remain the same.

Because the development of the AIMS method relied on the concept of nonadiabatic regions, significant modifications/additions are needed if tunneling effects are to be modeled. In the original prescription, the basis set was allowed to expand only during nonadiabatic events and only between different electronic states. Within this framework, tunneling effects can occur only if they are mediated by a second electronic state—that is, when they accompany an electronically nonadiabatic event. Under thermal conditions the Born–Oppenheimer approximation works well and therefore one would like to incorporate tunneling effects even when the dynamics is electronically adiabatic. Because tunneling is intrinsically nonclassical, some form of spawning procedure will be required—this time on the same electronic state. To determine where and when spawning is required, an analog to the concept of "nonadiabatic event" will be introduced. As for the multielectronic state case, the development of the algorithm was dictated by the requirement of compatibility with quantum chemistry. In particular we have assumed that information about the PES is limited and local; that is, the PES is known only at a few geometries and is generated "on the fly."

The algorithm that we have developed requires labeling of $N_T$ tunneling particles and $N_{DA}$ donor/acceptor particles. The former includes all the particles that will be allowed to tunnel during the simulation (e.g., protons), and the latter describes all the particles that can be covalently attached to a tunneling particle. The distinction between a donor and acceptor simply denotes whether or not the particle is currently covalently attached to a tunneling particle, and a distance-based criterion is used to determine this. During the simulation, donor and acceptor particles can change their identity, but both they and the tunneling particles have to be identified at the beginning of the calculation. The balance between basis set size and accuracy is kept by two numerical parameters, namely, the tunneling threshold ($R_T$) and the number of basis functions that will be spawned per tunneling event (MULTISPAWN). The tunneling threshold is analogous to the spawning threshold, $\lambda_0$, and it defines the minimum extension of a covalent bond involving a tunneling particle that suffices to indicate that a tunneling event should be considered. The magnitude of $R_T$ controls the propensity to add new basis functions (the smaller the value of $R_T$, the greater the number of basis functions that are likely to be added). As in the multielectronic state case, higher accuracy is achieved by increasing MULTI-SPAWN and decreasing $R_T$.

At each time step, and for a tunneling particle within each nuclear basis function, we check whether a tunneling event might be occurring. We first determine its current donor particle, defined as the current donor/acceptor particle to which it is closest. If this distance is shorter than $R_T$, this particle does not need to be considered further, until the next time step. On the other hand, if this distance exceeds $R_T$, the particle should be allowed to tunnel. In such a case we displace the tunneling particle along the tunneling vector(s), $R^i_{\text{Tunnel}}$. There are $N_{DA} - 1$ tunneling vectors per tunneling particle. To determine them, we place the tunneling particle next to each acceptor and quench the system to its local minimum. From the resulting $N_{DA} - 1$ local minima, only those with potential energy less than the current total energy are considered further. $R^i_{\text{Tunnel}}$ is defined as the vector connecting (a) the local minimum where the tunneling particle is bound to its current donor to (b) the one where it is bound to the $i$th acceptor. For each local minimum that remains under consideration, we propagate the basis function until a classical turning point along the tunneling vector is reached (at time $t_{tp}$) or until the tunneling particle is closer to the new minimum than the old one. In the latter case, the acceptor can be reached classically, and therefore there is no need to consider this path/tunneling event. If a turning point was reached, we displace the basis function along $R^i_{\text{Tunnel}}$ until a new classically allowed region is encountered and record the position and momentum of all the particles at that point. The basis function is then back-propagated for time $t_{bp} = t_{tp} - t_0$, where $t_0$ labels the beginning of the tunneling event. This brings the parent basis function back to the point where its uncoupled (classical) propagation began. Within the back-propagation time, $t_{bp}$, MULTISPAWN new basis functions are spawned with zero population. The initial conditions for the MULTISPAWN new basis functions are obtained by back-propagation of a trajectory using the positions and momenta recorded when the new classically allowed region was reached. This procedure is repeated for all tunneling particles and for all basis functions. When it is completed, the coupled propagation of the basis functions and their coefficients proceeds. A flow chart of the algorithm is provided in Fig. 6.

Whereas the multi- and single-surface spawning algorithms are different (the former allows only for spawning on a different electronic state, and the latter allows only for spawning on the same electronic state), the physical motivation for both is the same: We first attempt to identify and discretize the quantum mechanical events and then determine the initial conditions for the newly spawned basis functions. The spawning events are isolated by identifying the physical property that triggers basis set expansion: nonadiabatic coupling in the multistate problem and distance in the single state (tunneling) problem. Initial conditions are obtained by propagating the parent basis function through the tunneling or nonadiabatic region. This propagation is straightforward in the multistate case, but it poses a problem in the tunneling case because the selection of the direction of the tunneling path (that determines the position of the new basis functions) is not obvious. The optimal choice of this path has been discussed in many papers [134–137]. We have proposed to use a simple, straight-line path (in the full multidimensional coordinate space). We believe

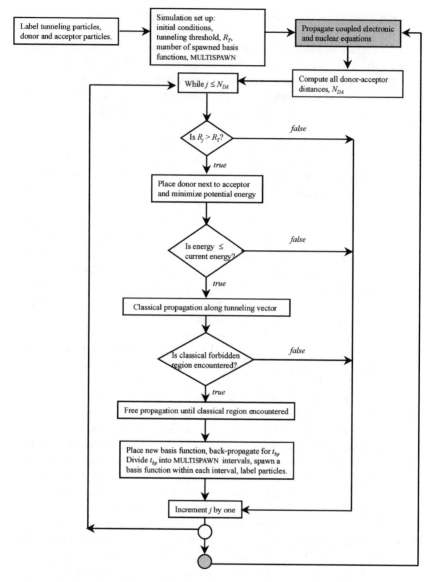

**Figure 6.**    Flow chart of the multiple-spawning code and the spawning algorithm for the case of single-state spawning. The spawning algorithm is executed after each propagation time step (upper-rightmost rectangle). When the execution of the algorithm is completed, the program returns to the propagator (gray circle at bottom of sketch.) For clarity, only a single tunneling particle is considered in this sketch. (Figure adapted from Ref. 126.)

that in the absence of detailed global information about the PES, this is the only viable choice.

This algorithm attempts to be general and is therefore rather complicated— certainly more complicated than the multisurface spawning algorithm. Our experience with it is limited to two-dimensional model problems [126], and therefore we cannot be certain about its generality. Because it requires a preassignment of tunneling particles and donors and acceptors and also because the choice of the tunneling vector is not obvious, it will be harder to demonstrate stability of the final results than it has been for the multisurface algorithm. The results will depend on the placement of new basis functions, and careful investigations will be required in order to optimize the placement of new basis functions and test the validity of the procedure.

## III. APPLICATIONS AND NUMERICAL CONVERGENCE

In this section we discuss the accuracy of the AIMS method and review some applications. Considerably larger molecules can be studied with the FMS method if one is willing to use empirical functional forms; for example, we have carried out calculations of photochemistry in retinal protonated Schiff base [138] and bacteriorhodopsin [139]. The nuclear wavefunction in the latter calculation includes more than 11,000 degrees of freedom. In the following, we restrict our discussion to AIMS applications that solve the electronic Schrödinger equation simultaneously with the nuclear dynamics. However, when discussing numerical convergence, we will alternate between AIMS and FMS depending on whether it is convergence of the electronic structure or nuclear dynamics which is being discussed.

### A. Applications

We review three systems to which the AIMS method has been applied. The first is the collision-induced electronic quenching of $Li(2p)$ by an $H_2$ molecule [140]. This is perhaps the simplest realistic example of electronic to vibrational/ translational energy transfer and can therefore serve as a useful paradigm for understanding electronic quenching. The electronic states relevant to the collision-induced quenching process can be labeled as $Li(2s) + H_2$ and $Li(2p) + H_2$, where the $p$ orbital in the second state is aligned parallel to the $H_2$ molecular axis. An OA-HF-CAS(2,2)*S electronic wavefunction (and a double-$\zeta$ quality basis set [141,142]) was used in the calculations and the initial state of the system is nonstationary. The Li atom is far (8 Å) from the center of mass of the $H_2$ molecule that is in a coherent rotational–vibrational state chosen according to the quasi-classical prescription to model $v = 0, j = 0$. At this level of theory, the system exhibits a conical intersection at an atom–diatom distance of 2.86 bohr. The form of the potential energy surfaces in the vicinity of the

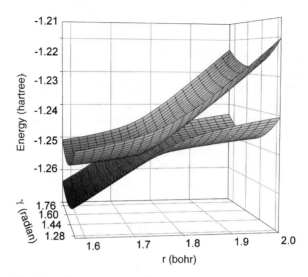

**Figure 7.** Two-dimensional cut of the ground- and excited-state adiabatic potential energy surfaces of $Li + H_2$ in the vicinity of the conical intersection. The $Li–H_2$ distance is fixed at 2.8 bohr, and the ground and excited states correspond to $Li(2s) + H_2$ and $Li(2p) + H_2$, where the $p$ orbital in the latter is aligned parallel to the $H_2$ molecular axis. $\gamma$ is the angle between the H–H internuclear distance, $r$, and the Li-to-$H_2$ center-of-mass distance. Note the sloped nature of the intersection as a function of the H–H distance, $r$, which occurs because the intersection is located on the repulsive wall. (Figure adapted from Ref. 140.)

conical intersection is shown in Fig. 7, where the $Li–H_2$ distance is fixed to $R = 2.86$ bohr and $\gamma$ is the angle between $R$ and the $H_2$ internuclear distance, $r$. Accessing the conical intersection requires stretching of the $H_2$ bond beyond the usual outer classical turning point at the zero point energy. Because the intersection is located on the repulsive walls of the PESs, it is strikingly sloped, deviating quite strongly from the model conical form that is often discussed. Of course, in a small enough neighborhood around the intersection a conical form is guaranteed, although it may be tilted [143,144].

Commonly, it is asserted that upward transitions from the lower adiabat to the upper one should be less likely than downward transitions because of the "funneling" property of the intersection [144,145]. This is clearly seen in the usual model conical intersection—as seen, for example, in Fig. 1 of Ref. 146, where there is (a) a well, or "funnel," in the upper adiabat which guides the wavepacket to the intersection and (b) a peak on the lower adiabat which tends to guide the wavepacket away from the intersection. The potential energy surfaces shown in Fig. 7 differ from this canonical picture, and in particular it is not at all clear that the wavepacket on the lower adiabatic state will be funneled away from the intersection. For the conditions chosen in our calculations, we

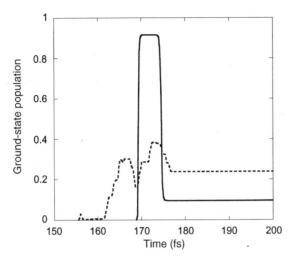

**Figure 8.** $Li + H_2$: Ground-state population as a function of time for a representative initial basis function (solid line) and the average over 25 (different) initial basis functions sampled (using a quasi-classical Monte Carlo procedure) from the $Li(2p) + H_2(v = 0, j = 0)$ initial state at an impact parameter of 2 bohr. Individual nonadiabatic events for each basis function are completed in less than a femtosecond (solid line); and due to the sloped nature of the conical intersection (see Fig. 7), there is considerable "up-funneling" (i.e., back-transfer) of population from the ground to the excited electronic state. (Figure adapted from Ref. 140.)

found that the nuclear population on the lower adiabatic state is not funneled away from the intersection sufficiently quickly. Consequently, there is efficient recrossing (back to the upper adiabatic state) and therefore inefficient quenching. This "up-funneling" phenomenon is demonstrated in Fig. 8, which shows the ground-state population as a function of time for a particular (arbitrarily chosen) set of initial conditions (solid line). As expected for quenching of a localized wavepacket via a conical intersection, the population transfer is extremely fast (individual nonadiabatic events are completed in less than a femtosecond). More interestingly, the ground state is populated and then immediately depopulated. The first transition corresponds to a downward transition from the upper to the lower adiabatic electronic state and is fully expected. The second transition, however, is an upward transition. This behavior is in contrast to the common view that conical intersections should serve as "funnels" directing population toward the lower adiabatic state. Because the conical nature of an intersection is operative over some lengthscale, which (depending on the nuclear velocity) may or may not be relevant to the nuclear dynamics, the assertion that downward transitions dominate at a conical intersection can be called into question. For the $Li + H_2$ system (and our particular choice of initial conditions) this assertion is incorrect, and in general the veracity of such statements will depend on the topography of the conical intersection and the nuclear velocities. A recent study by Yarkony [147] investigates funneling and up-funneling behavior in a two-dimensional model conical intersection, supporting these conclusions.

One expects the timescale of the nonadiabatic transition to broaden for a stationary initial state, where the nuclear wavepacket will be less localized. To mimic the case of a stationary initial state, we have averaged the results of 25 nonstationary initial conditions and the resulting ground-state population is shown as the dashed line in Fig. 8. The expected broadening is seen, but the nonadiabatic events are still close to the impulsive limit. Additional averaging of the results would further smooth the dashed line.

The $Li + H_2$ problem was used to test the *ab initio*/interpolated extension of the AIMS method. As described in Section II.F [125], the interpolation function is called on to represent the smooth regions of the potential energy surfaces, while *ab initio* quantum chemistry, in this case full configuration interaction, represents the regions near conical intersections and the nonadiabatic coupling between electronic states. The accuracy of the hybrid approach was investigated by (1) comparing various cuts of the potential energy surfaces, (2) comparing time traces of individual trajectories/basis functions, and (3) comparing averaged expectation values. In Fig. 9 the hybrid and fully *ab initio* excited-state potential energy surfaces are compared (the system is constrained to $C_{2v}$ symmetry). Explicit comparison of the *ab initio* and hybrid interpolated/*ab initio* surfaces shows RMS errors of 0.63 and 0.69 kcal/mol on the ground- and excited-state potential energy surfaces, respectively. The ground- and excited-state PESs vary by 3.0 and 1.5 eV over this region. The final (averaged) branching ratios were also in good agreement: 51% quenching for the hybrid method and 54% for the full *ab initio* dynamics. (For a comparison of individual time traces as well as averaged ones see the original paper, Ref. 125.) The computational effort associated with the hybrid method was over an order of magnitude less than that for the full *ab initio* dynamics, an impressive reduction given the high accuracy of the interpolated surfaces. The results are encouraging, but it remains to be seen whether similar accuracy and reduction in computational effort can be obtained for larger systems (10–20 atoms).

A second molecule that we have studied using the AIMS method is ethylene. The photochemistry of ethylene is interesting as a paradigm for *cis–trans* isomerization in unsaturated hydrocarbons. Unsaturated alkanes pose a challenge to quantum chemistry because the description of their lowest excited electronic states requires careful treatment of electron correlation. For example, the ordering of the lowest-lying singly and doubly excited electronic states ($B_u$ and $A_g$, respectively) is sensitive to the details of the wavefunction used. In the case of butadiene, this ordering has been the topic of a long controversy [148].

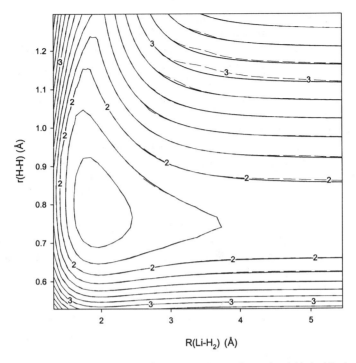

r(H-H) (Å)

R(Li-H$_2$) (Å)

**Figure 9.** Comparison of *ab initio* (full line) and *ab initio*/interpolated (dashed line) potential energy surfaces for the first electronically excited state of Li + H$_2$ system restricted to $C_{2v}$ geometry. Contours are labeled in eV. (Figure adapted from Ref. 125.)

Because ethylene is the shortest unsaturated hydrocarbon, its photochemistry is special in some respects. Simple particle-in-a-box considerations suggest (and theory and experiment confirm) that ethylene will have the largest excitation energy (as compared to longer conjugated systems). At the same time, ethylene also has a small number of internal modes. Consequently, photoexcitation of ethylene leads to fragmentation in addition to the isomerization that is the hallmark of longer polyenes [149–151]. This added complexity is in addition to some unresolved issues regarding the absorption and resonance Raman spectrum of ethylene, which are partially due to our incomplete knowledge of the character of the manifold of excited electronic states. Nevertheless, certain crude statements about the singly excited state of ethylene can be made. Upon absorption of a photon, an electron is promoted from a bonding $\pi$ molecular orbital (MO) to an antibonding $\pi^*$ MO. The ground electronic state of ethylene is planar and stable with respect to twisting, whereas the $\pi\pi^*$ state favors a twisted $D_{2d}$ geometry that minimizes both the kinetic energy associated with the antibonding $\pi^*$ orbital and the Coulomb repulsion between the $p$ electrons of the two carbon atoms. Consequently, the electronic excitation results in geometric relaxation toward a stretched (formally the bond order is reduced from two to one) and twisted geometry. The conventional picture of photoisomerization indeed identifies torsion about the C=C bond as the reaction coordinate and concentrates on the $\pi\pi$ and $\pi\pi^*$ electronic states. Computation of the ground and singly excited potential energy surfaces along this coordinate shows that this view is considerably oversimplified (regardless of the detailed way in which the bond and angle coordinates are allowed to vary). In particular, the minimal $S_0$–$S_1$ energy gap that results is approximately 60 kcal/mol, implying a long excited-state lifetime and significant fluorescence. Yet there is no detectable fluorescence from the excited state of ethylene. The absence of fluorescence suggests a short excited-state lifetime, which has recently been investigated using femtosecond pump-probe spectroscopy [152].

The broad and diffuse absorption spectrum of ethylene has been assigned by Wilkinson and Mulliken [153] to the $\pi \rightarrow \pi^*$ valence (V) and Rydberg (R) states. Based on an isoelectronic analogy between $O_2$ and $C_2H_4$, they assigned the single progression in the V state band to the C=C stretching motion [153]. Later investigators questioned this assignment and suggested a purely torsional progression [154]. Next, based on a spectral study of ethylene isotopomers, Foo and Innes [155] agreed with the reassignment but suggested a mixture of stretching and torsion. When theoretical investigation [156] predicted that the change in the C=C bond length on the V state was less than 0.1 Å, Mulliken himself became convinced that the torsion dominated the spectrum [157,158]. The accepted assignment of mixed torsion and stretching was challenged by Siebrand and co-workers [159] who presented theoretical evidence that there is no visible stretching activity in the spectrum. Consequently, there have been few challenges to the torsional assignment of the spectrum, but recently the very identity of the bands has been questioned [160]. These uncertainties regarding the excited-state motion were exacerbated when the possible role of the pyramidalization coordinate was suggested.

The pyramidalization coordinate of ethylene (and longer polyenes) has been first studied in the context of the concept of "sudden polarization." Salem and co-workers [161–163] noted that mono-pyramidalization of ethylene (keeping the molecule in $C_s$ symmetry) results in a large dipole moment. The onset of this phenomenon is quite sudden; that is, small distortions result in a large change in the dipole moment (which is zero by symmetry at both the planar and twisted geometries). This arises due to an avoided crossing between the valence and ionic states (V and Z in Mulliken's notation) near the twisted (nonpyramidalized) geometry. The original speculation that the electrical signal thus generated might trigger conformational changes in the visual pigment proteins [162] was abandoned when it was shown that proton transfer was required for vision [164]. Irrespective of its importance in the retinal protonated Schiff base chromophore of visual pigments, and in spite of the crude electronic structure treatment that led to the original observation, both experiment and theory suggest that the pyramidalization coordinate does participate in the excited-state dynamics. The theoretical evidence is based on restricted excited-state optimizations that found mono-pyramidalization to be a stabilizing distortion [165,166]. Resonance Raman experiments find overtone activity in both out-of-plane wagging and rocking vibrations [167], supporting a role for pyramidalization in the initial motion of ethylene following the electronic excitation.

Using a GVB-OA-CAS(2/2)*S wavefunction and a double-$\zeta$ quality basis set, we have carried out AIMS simulations of the photodynamics upon $\pi \rightarrow \pi^*$ excitation. In the following we limit our discussion to the photochemical mechanism of *cis–trans* isomerization and fragmentation dynamics, but we have also used AIMS to compute the electronic absorption and resonance Raman spectra [122]. The AIMS simulations treat the excitation as being instantaneous and centered at the absorption maximum. Hence, the initial state nuclear basis functions are sampled from the ground-state Wigner distribution in the harmonic approximation. Ten basis functions are used to describe the initial state and we employ the IFG, SP, and split-operator approximations, that is, AIMS-SP-SO. Overall approximately 100 basis functions are spawned during the dynamics, and we follow the dynamics up to 0.5 ps (using a time step of 0.25 fs).

The first dynamical question to ask concerns the excited-state lifetime. In Fig. 10, we show the AIMS results for the excited-state lifetime as a function of time following photoexcitation at $t = 0$. Both the raw data (solid line) and a Gaussian fit (dashed line) are shown. The inferred time constant of 180 fs is in agreement with the expected subpicosecond dynamics. The decay is clearly nonexponential and its onset is delayed: Appreciable decay from the excited state does not begin until approximately 50 fs after the excitation. The results shown in Fig. 10 have been directly compared to recent pump-probe experiments of Radloff and co-workers [152], where the probe pulse induces ionization of the excited state. Assuming exponential decay of the ionizable excited state, Radloff and co-workers obtained a lifetime of $30 \pm 15$ fs. If the AIMS simulations are correct, this is much too short to be considered as an excited-state lifetime and the possibility of a dark form of the excited state must be considered. Using the AIMS data and assuming that the molecule ionizes with 100% efficiency provided that the ionization potential is lower than the threshold given by the probe pulse, we have simulated the experiment of Radloff et al. The results are shown in Fig. 11 along with an exponential fit. The excited-state lifetime extracted from this fit ($35 \pm 2$ fs) agrees well with the experimental [152] one ($30 \pm 15$ fs). This agreement is encouraging and important, but it also suggests that the experiment is probing the excited state only within a limited window around the Franck–Condon region, thus providing a lower bound on the excited-state lifetime. In this scenario, most of the

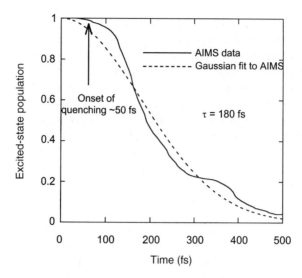

**Figure 10.** Excited-state population of ethylene as a function of time in femtoseconds (full line). (Results are averaged over 10 initial basis functions selected from the Wigner distribution for the ground state in the harmonic approximation.) Quenching to the ground electronic state begins approximately 50 fs after the electronic excitation, and a Gaussian fit to the AIMS data (dashed line) predicts an excited-state lifetime of 180 fs. (Figure adapted from Ref. 214.)

excited-state dynamics, after significant twisting, which is discussed below, is invisible to the experiment. Furthermore, the AIMS results are clearly poorly modeled by an exponential decay, questioning the wisdom of exponential fitting of the experimental data.

By following the centroids of the dominant nuclear basis functions along with their electronic wavefunction, detailed information about the excited-state (nuclear and electronic) dynamics can be extracted. Following excitation of the planar molecule, the C=C bond stretches and then the molecule begins to twist. After ~50 fs the molecule is typically twisted, but excited-state quenching does not even begin until 50 fs. Therefore torsion is not the sole coordinate responsible for the return to the ground electronic state. Significant quenching to the ground electronic state requires pyramidalization of one of the methylene units. Attempts at hydrogen migration are also observed, but only rarely have we seen the molecule decay to the ground electronic state via a hydrogen-migration conical intersection. The corresponding electronic dynamics involves three electronic states: N, V, and Z in Mulliken notation. The excited-state dynamics can be characterized as consisting of electron transfer between the two methylene units. The intramolecular electron transfer dynamics is punctuated by quenching back to the ground electronic state each time that the excited-state molecule reaches one of the excited-state minima, since these

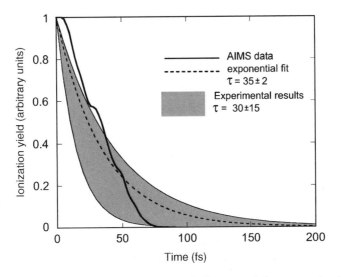

**Figure 11.** $C_2H_4$ ion yield as a function of time in femtoseconds for a pump-photoionization probe experiment. Heavy line: Predicted ion yield using the AIMS data and assuming an ionization threshold of 3.5 eV. Dashed line: Exponential fit to the AIMS ion yield predicting an excited state lifetime of 35 fs. Gray shaded area: Reported ion yield [152] obtained using an exponential fit to the experimental data predicting an excited state lifetime of $30 \pm 15$ fs. (Figure adapted from Ref. 214.)

are in close proximity to a conical intersection. This picture of the ethylene photochemistry that involves pyramidalization and torsional motions, and both the V and Z states, is quite different from the conventional one centering on the torsional coordinate and the role of the V state.

One of the advantages of AIMS is the ability to describe bond rearrangement. In Fig. 12 we show a sample of the kinds of reactive events that are observed in the simulations. Both panels show the time evolution of individual basis functions that are used to construct the time-dependent basis set. (Note that the time evolution of individual basis functions is not the same as the time evolution of expectation values. The two are equivalent only when only a single basis function is populated. In general, expectation values can be evaluated using the AIMS wavefunction and the appropriate operator.) The right and left panels originate (i.e., are spawned) from the same parent basis function (traveling on the excited electronic state), and this is a representative example in that one usually observes more than one reactive outcome from a particular parent basis function. In each panel, the time traces of the four C–H bond distances are shown, beginning when the depicted basis function is spawned from the excited state. The final products shown (in both panels) are transient, and they are expected to further decompose to acetylene. However, this decomposition occurs on a longer timescale, and only rarely have we observed

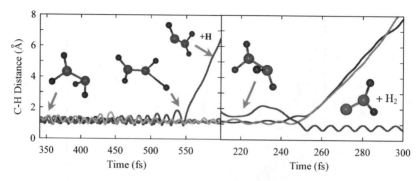

**Figure 12.** Sample of reactive outcomes of ethylene photochemistry. Right and left panels represent different basis functions (traveling on the ground electronic state) spawned from the same parent basis function (traveling on the excited electronic state) at different points in time. The final populations of the two basis functions are 56% and 15% (right and left panels, respectively). The time traces denote the four C–H bond distances as a function of time for the two (different) ground-state basis functions. Left panel: Formation of vinyl radical and atomic hydrogen. Right panel: Concerted elimination of molecular hydrogen and formation of vinylidene. Both products (vinyl radical and vinylidene) are expected to be transient, ultimately decomposing into acetylene. (Figure adapted from Ref. 215.)

acetylene formation within the simulation time window of 0.5 ps. This is not surprising, because there are significant barriers to be overcome from any of the transient photoproducts that we observe: vinyl radical, ethylidene, and vinily-dene. Both calculation [168] and experiment [169] agree on a barrier of 1.66 eV for decomposition of vinyl radical to acetylene and atomic hydrogen. The barrier heights for decomposition of ethylidene and vinylidene are calculated [168,170,171] to be in the ranges 1.46–1.52 and 0.06–0.13 eV, respectively. Typically, bond rearrangement in our simulations occurs on the ground electronic state, but it is also observed much more rarely on the $S_1$ state. Experimental branching ratios between atomic and molecular hydrogen products, as well as molecular hydrogen internal and translational energy distributions, have been published [172–175]. Direct comparison to these experiments will require either longer simulation times or the use of statistical assumptions to extrapolate to $t = \infty$.

As a third and last example, we discuss the photodynamics of cyclobutene. The electrocyclic ring-opening reaction of cyclobutene (to 1,3-butadiene) is a classic example of a pericyclic rearrangement whose outcome is predicted by the Woodward–Hoffmann [176] (WH) rules and complementary theories [177–179]. The WH rules predict that for reactants containing $4n + 2$ $\pi$ electrons the photochemical ring-opening reaction will proceed with disrotatory stereochemistry, while for reactants with $4n$ $\pi$ electrons the photochemical ring opening will proceed in conrotatory fashion. For the case of cyclohexadiene, this prediction (conrotatory) has been experimentally confirmed using both stereochemical analysis of the products in alkyl-substituted reactants [180–182] and resonance Raman spectroscopy [183,184]. Surprisingly, the results for the smaller cyclobutene (CB) molecule are less conclusive. Alkyl substitutions of CB have produced mixtures of allowed (disrotatory) and forbidden (conrotatory) photoproducts [185] and, because of controversies in the assignment of some bands, [186,187] the interpretation of resonance Raman experiments [188,189] (which provide a more direct probe of the excited-state dynamics) has been questioned. On the basis of the original assignment of the vibrational force field of CB [190], and a later assignment by Craig et al. [186], the resonance Raman spectrum [188] has been interpreted to show traces of a disrotatory ring opening [188,189]. However, based on their own assignment of the vibrational force field, and on their interpretation of the assignment of Wiberg and Rosenberg [187], Negri et al. [191] concluded that the resonance Raman spectrum does not show "any positive hint of the activity of distrotatory ring-opening motion."

The AIMS calculations used a HF-OA-CAS(4/4)$^*$S wavefunction where the four active orbitals are those that become the two $\pi$ orbitals in the butadiene photoproduct. Diffuse electronic basis functions were not included in the calculations. Therefore, we do not expect to reliably model those (long-time) features of the photochemistry and formation of ethylene, acetylene, and methylene-cyclopropane photoproducts [185,192] that are thought to arise from excitation to a low-lying $\pi \rightarrow 3s$ Rydberg state. Instead, our treatment focuses on the features that arise due to the state with the strongest oscillator strength: the $\pi \rightarrow \pi^*$ excited state. As in the case of ethylene, the excitation is assumed to be instantaneous and centered at the absorption maximum. Only four basis functions are used to describe the initial state, and we again use the AIMS-SP-SO method. Only short-time dynamics has been investigated (up to 50 fs), and during this short propagation time (and small set of initial conditions) we did not observe any nonadiabatic effects.

Figure 13 depicts a few snapshots of a typical excited-state trajectory, and we discuss the observed geometrical changes in chronological order. Following the electronic excitation the first motion is a stretching of the C=C bond. (Within 15 fs, the C=C bond extends by $\sim$0.4 Å.) This is expected for a $\pi \rightarrow \pi^*$ transition and is in agreement with the resonance Raman spectrum [188] that is dominated by the totally symmetric C=C stretching mode. An impulsive change in the hybridization of the methylenic carbons is also observed. As the $CH_2$–$CH_2$ bond breaks, the HCH angle changes from $\sim 109°$ ($sp^3$) to $\sim 120°$ ($sp^2$). The change in hybridization begins almost immediately after the electronic excitation and is completed within 50 fs. In the resonance Raman spectrum [188], this motion is manifested by a pronounced activity of the totally symmetric $CH_2$ scissors mode. Finally, and most importantly, Fig. 13

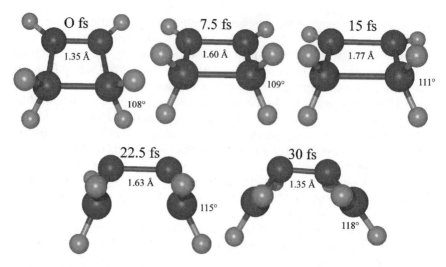

**Figure 13.** Snapshots of a typical excited state trajectory of cyclobutene. Values of the C–C bond distance and HCH hybridization angle are indicated. Immediately after the electronic excitation (at $t = 0$) the C–C bond begins to stretch. This is followed by a change in hybridization of the methylene carbons (from $sp^3$ to $sp^2$) and a pronounced disrotatory motion. (Figure adapted from Ref. 214.)

unambiguously shows that the excited-state dynamics are directed along the WH-predicted disrotatory reaction coordinate. In fact, this is the most pronounced feature observed in Fig. 13. In Fig. 14 we plot the absolute value of the average disrotatory angle as a function of time. The left inset of this panel defines the disrotatory coordinate, a coordinate that is *not* a normal mode of the molecule. (We plot the absolute value of the disrotatory angle because this motion can occur in two equivalent directions: Hydrogen atoms above/below the carbon skeleton rotate inward/outward or outward/inward.) The disrotatory motion begins shortly (approximately 10 fs) after the electronic excitation, and its amplitude is large (120°). No significant motion along the conrotatory reaction coordinate is observed. The right inset in the upper panel of Fig. 14 explains these observations. Here we show (solid line) a one-dimensional cut of the excited-state potential energy along the disrotatory and conrotatory coordinates (all other coordinates are kept at their ground-state equilibrium value). The Franck–Condon point on the excited electronic state remains a minimum with respect to conrotatory motion, but becomes unstable with respect to disrotatory motion. The dynamical conclusions we draw concerning the rapid onset of disrotatory motion arise largely because of the form of the excited state PES shown in the inset. Thus, we have also computed this cut (dashed line) using the MOLPRO program [193] with a larger cc-pVTZ basis set [194] and a

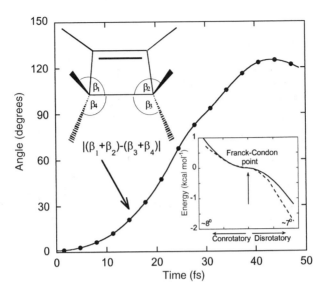

**Figure 14.** The absolute value of the average disrotatory angle as a function of time in femtoseconds. (The disrotatory angle is defined in the upper left inset.) Lower inset: A one-dimensional cut of the excited-state potential energy surface along the disrotatory and conrotatory coordinates. All other coordinates are kept at their ground-state equilibrium value, and the full and dashed lines correspond to two levels of electronic structure theory (see text for details). (Figure adapted from Ref. 216.)

more sophisticated electronic wavefunction (SA-3-CAS(4/4)*SD). The same qualitative behavior along the conrotatory and disrotatory coordinates is observed with this more accurate treatment of the electronic structure.

We now turn to the electronic structure during the dynamics. In Fig. 15 we show an isosurface rendition of the two frontier excited-state natural orbitals in the CI wavefunction. (At the beginning of the calculation these orbitals correspond most closely to $\pi$ and $\pi^*$ orbitals.) These are the natural orbitals of the basis function trajectory shown in Fig. 13. At early times (leftmost frames at 7.5 fs), before the onset of disrotatory motion, the excited-state wavefunction can be described using a single determinant with one electron in a $\pi$-like orbital ($\phi_a$) and one in a $\pi^*$-like orbital ($\phi_b$). During the disrotatory motion the character of the natural orbitals is more complicated and two determinants are required to describe the electronic wavefunction. In one determinant both electrons are in the $\phi_a$ orbital, and in the other they are both in the $\phi_b$ orbital. Both orbitals ($\phi_a$ and $\phi_b$) show significant $\sigma-\pi$ mixing which is expected due to the significant disrotatory motion. After the disrotatory motion is completed (rightmost frames at 22.5 fs), the excited-state wavefunction is again described

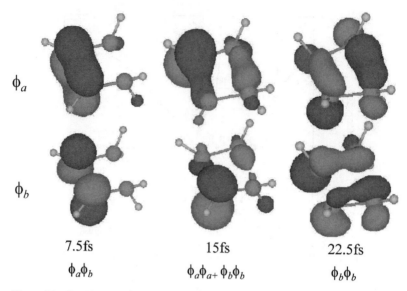

$\phi_a$

$\phi_b$

7.5fs       15fs       22.5fs

$\phi_a\phi_b$     $\phi_a\phi_{a+}\phi_b\phi_b$     $\phi_b\phi_b$

**Figure 15.** Snapshots of the two frontier excited-state natural orbitals (computed using the HF-OA-CAS(4/4)*S wavefunction) of the excited-state trajectory of cyclobutene shown in Fig. 13. Left panels: Before the onset of disrotatory motion, the excited-state wavefunction can be described using a single determinant with one electron in a $\pi$-like orbital ($\phi_a$) and one in a $\pi^*$-like orbital ($\phi_b$). Middle panels: During the disrotatory motion the simplest description of the electronic wavefunction requires two determinants. In one determinant both electrons are in the $\phi_a$ orbital, and in the other they are both in the $\phi_b$ orbital. Both orbitals ($\phi_a$ and $\phi_b$) show significant $\sigma-\pi$ mixing, which is a consequence of the significant disrotatory motion. Right panels: When the disrotatory motion is completed, the excited-state wavefunction is described by a single determinant in which both electrons are in the $\phi_b$ orbital. Note how the shape of the orbitals changes as the initial bonds are broken and the two new $\pi$ bonds are formed.

using a single determinant. This time both electrons are in the $\phi_b$ orbital, which reflects the formation of the two new $\pi$ bonds.

The interpretation of resonance Raman spectra is critically dependent on the assignment of the observed vibrational frequencies in the normal mode approximation. It is therefore crucial that the coordinate of interest correspond to a single normal mode and that the frequency of this normal mode be correctly assigned. In the case of CB, this has resulted in some controversy. The disrotatory motion is a non-totally symmetric coordinate ($b_1$ symmetry), projecting only on non-totally symmetric normal modes that may appear as odd overtones in the experimental spectra. Mathies and co-workers [188] observed an overtone of the 1075 cm$^{-1}$ mode at 2150 cm$^{-1}$. Using the normal mode assignment of previous workers [186,190], they concluded that this mode is a non-totally symmetric CH$_2$ twist, projecting directly onto the disrotatory twist of the CH$_2$ groups. In 1995, this interpretation was challenged by Negri et al. [191]. They revisited the normal mode analysis of CB and concluded, in agreement with a previous assignment by Wiberg and Rosenberg [187], that the 1075 cm$^{-1}$ mode is mainly a $CH_2$ deformation (and not twist), while the lower-frequency 848 cm$^{-1}$ mode is a $CH_2$ twist mode with a dominant disrotatory component. (Calculations of vibrational frequencies of $d_2$- and $d_4$-CB isotopomers provided further confirmation of their analysis.) This conflicting assignment of the CB normal modes led to the conclusion that the overtone at 2150 cm$^{-1}$ does not provide any proof of disrotatory motion. Instead, Negri argued that the overtone of the 848 cm$^{-1}$ mode would provide such proof. Unfortunately, one cannot determine if this overtone is present because of the intense band at 1650 cm$^{-1}$, which is attributed to butadiene photoproduct [188].

We have analyzed the character of the normal modes and agree with Negri et al. [191] that the assignment [186,190] on which Mathies [188] based his interpretation is incorrect. Although there is some component of disrotatory motion in the 1075 cm$^{-1}$ normal mode, it is dominantly a $CH_2$ wag. Nevertheless, by following the excited-state dynamics we can conclude that the WH tendency is established during the first femtoseconds of the ring-opening. This suggests a role for impulsive character and kinematic effects on the efficacy of the WH rules for photochemical reactions. Indeed, one might then expect that classification of the cyclobutene and substituted cyclobutene ring-opening reactions that do and do not lead to the WH-predicted stereochemistry could be correlated with the effective mass of the substituents: The heavier the substituents, the more likely that the initial WH-directed impulse could be overcome by the detailed landscape of the excited-state potential energy surface. Further calculations and experiments—using, for example, deuterated cyclobutene—are needed to make progress in formulating such a theory.

## B.  Numerical Convergence

When judging the quality of the results produced by the AIMS method, it is important to distinguish the nuclear and electronic aspects; that is, one should test the accuracy of each one separately. Ideally one would like to demonstrate the convergence with a series of AIMS calculations following the usual quantum chemical hierarchy for the electronic structure part of the problem. However, computational limitations make this impractical at present. On the other hand, the more traditional time-independent approach of finding and characterizing special points on the PES will always require fewer PES evaluations than a time-dependent dynamical study. Hence, a higher level of electronic structure theory may be used in the time-independent case, and one can verify that the important features of the PESs remain consistent with the dynamical conclusions from an AIMS study. This has been done for the ethylene molecule, as discussed below. One of the main reasons for these extensive studies of the excited states of ethylene is the need to assess the role and importance of Rydberg states. Because of computational limitations, the AIMS simulations were carried out with a double-$\zeta$ basis set. However, ethylene is known to possess many low-lying electronic states of Rydberg character; these cannot be reproduced without a more extensive basis set containing diffuse functions, and possibly also without a more detailed treatment of electron correlation.

Our high-level ethylene electronic structure calculations have used extended basis sets (aug-cc-pVDZ [195,196]) and state-averaged CASSCF wavefunctions. In some cases the orbitals determined from the SA-CASSCF have been used in an internally contracted multireference single and double-excitation configuration interaction treatment (MRSDCI) [197,198]. All the calculations used the MOLPRO [193] electronic structure code, and symmetry was never imposed on the electronic wavefunction. The calculations have followed the time ordering observed in the AIMS simulations, where one finds that immediately following the excitation the C–C bond stretches and subsequently the molecule begins to twist and pyramidalize. To investigate the extent of C–C bond stretching on the valence excited state (V state in Mulliken's notation), we have optimized the molecule under the constraint of a planar $D_{2h}$ geometry. The formal reduction in the C–C bond order (the electronic excitation promotes an electron from a $\pi$ orbital to a $\pi^*$ orbital) is expected to result in stretching and twisting of this bond. The extent of this stretching is known to be sensitive to the amount of Rydberg-valence mixing in the V state. If this mixing is large, the extension is small because the delocalized nature of the Rydberg electrons leads these states to prefer only minor extension of the C–C bond. Previous theoretical studies [156] have found very slightly ($\approx 0.05$ Å) extended C–C bond lengths (for planar V-state ethylene). In contrast we find (using a SA-6-CAS(2,6)$^*$SD wavefunction) significant stretching of the C–C bond by 0.139 Å. Furthermore, as the C–C bond is stretched, the ordering of the excited electronic states changes and the V state, which at the Franck–Condon geometry is the fourth excited state ($S_4$), becomes the third excited state ($S_3$) for $R_{CC} > 1.45$ Å. This significant stretching was also seen in the AIMS simulations that used considerably smaller basis sets (and less accurate wavefunctions).

The two panels in Fig. 16 show two-dimensional cuts of the PESs calculated using the OA-GVB-CAS(2/2)$^*$S and SA-3-CAS(2/6)$^*$SD wavefunctions (in the latter case the aug-cc-pVDZ basis set was used). At both levels of theory the twisted geometry of ethylene ($D_{2d}$ symmetry) is not a true minimum, but rather a saddle point. This is at variance with previous studies at lower levels of theory [18], but in agreement with our AIMS results. The lowest points on the $S_1$ PES are conical intersections, not true minima. Of these, we have found two nearly isoenergetic intersections: (a) the twisted/pyramidalized geometry that the AIMS simulations also highlighted and (b) another that corresponds to ethylidene. A third type of intersection involving hydrogen migration is similar to the

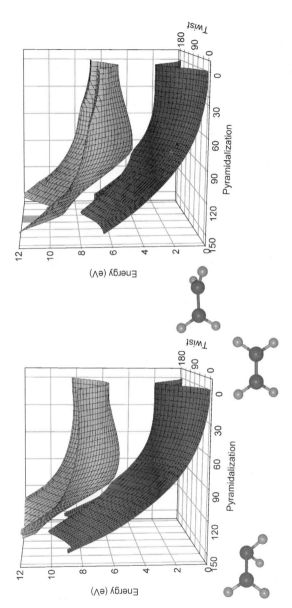

**Figure 16.** Left panel: The ground and first excited electronic states of ethylene, computed using the OA-GVB-CAS(2/2)*S wavefunction, as a function of the pyramidalization and twist angles. (All other coordinates, except the C–C bond distance, are kept at their ground-state equilibrium value and the C–C bond distance is stretched to 1.426 Å.) On the ground electronic state the molecule is planar (twist angle 0° or 180°), and on the lowest excited state the twisted geometry (twist angle 90°) is a saddle point. Accessing the conical intersection requires pyramidalization of one of the methylene fragments. Right panel: Same as left panel but computed using the aug-cc-pVDZ basis set and a three-state-averaged CAS(2/6) wavefunction augmented with single and double excitations, that is, SA-3-CAS(2/6)*SD. The form of the excited-state PES is in agreement with the simpler model of the left panel; in particular, the lowest energy point on the lowest excited state remains pyramidalized. (Figure adapted from Ref. 214.)

493

one found by Ohmine [199]. This intersection is not a minimal energy point on the intersection seam but is energetically accessible after photoexcitation. Our AIMS simulations find that it does not usually lead to efficient $S_1 \rightarrow S_0$ quenching, possibly because of PES topography (cf. the $Li + H_2$ example discussed above.) An intersection between the valence and ionic states (V and Z) was also identified. This intersection was not seen in the AIMS simulations, most of which were restricted to include only $S_0$ and $S_1$ states.

Near the equilibrium geometry of ground-state ethylene, the first three low-lying excited states are Rydberg states, while at the photochemically important geometries (e.g., twisted, pyramidalized) the lowest excited state has purely valence character and the Rydberg states are higher in energy. As a consequence, the excited molecule must first descend through the manifold of Rydberg states before it can decay back to the ground electronic state. The torsion angle can facilitate this by stabilizing the valence states and destabilizing the Rydberg states. We have investigated this cascade of conical intersections using a SA-2-CAS(2,6) wavefunction. First, we located the intersection between the valence state ($S_4$) and the adjacent Rydberg state ($S_3$). Once this intersection was located (in the proximity of the planar geometry), we computed a one-dimensional cut along the twist coordinate, keeping all other coordinates at their values at the $S_4/S_3$ crossing. In Fig. 17 we show the resulting cascade of valence/Rydberg conical intersections. As the molecule is twisted, the energy of the Rydberg states increases while that of the valence state decreases. The molecule must twist by $\sim 60°$ to leave the manifold of Rydberg states. The energy of the ionic (Z) state begins to decrease at larger twist angles, and at the $90°$-twisted geometry, this state is strongly coupled to the valence state (as evidenced by the strong avoided crossing at $\sim 80°$).

Overall the results of high-level electronic structure studies confirm the features of the excited state PESs that lead to the photodynamics observed in the AIMS simulations. However, they also highlight the valence/Rydberg conical intersections that exist in ethylene. It would be interesting to characterize the dynamics as the molecule descends through the manifold of Rydberg states (which were not included in our dynamical simulations). A priori, it is not clear to what extent the dynamics will be diabatic—that is, to what extent the Rydberg states are just spectators in the dynamics. More detailed AIMS simulations that treat the Rydberg states properly and allow for coupling of up to six electronic states are planned to address this issue.

When the FMS method was first introduced, a series of test calculations were performed using analytical PESs. These calculations tested the numerical convergence with respect to the parameters that define the nuclear basis set (number of basis functions and their width) and the spawning algorithm (e.g., $\lambda_0$ and MULTISPAWN). These studies were used to validate the method, and therefore we refrained from making any approximations beyond the use of a

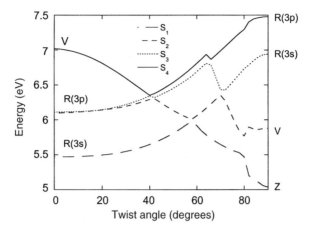

**Figure 17.** The four lowest excited electronic states of ethylene, computed using the SA-6-CAS(2/6) wavefunction (and the aug-cc-pVDZ basis set), as a function of the twist angle. All other coordinates are kept at their values at the first crossing between $S_4$ (full line) and $S_3$ (dotted line): $R_{CC} = 1.3389$ Å, $R_{CH} = 1.128$ Å, and $\angle HCH = 112.1°$. The four states are labeled using both adiabatic ($S_1$, $S_2$, $S_3$, and $S_4$) and diabatic [V, Z, R($3s$) and R($3p$)] notations. At the planar geometry (twist angle is $0°$) the first three excited states are Rydberg states and $S_4$ is the V state. As the molecule is twisted, the Rydberg states are destabilized while the V state is stabilized. This results in a cascade of V/R conical intersections. A twist angle of $\sim 60°$ is required to leave the manifold of Rydberg states completely. As the molecule is twisted further, the energy of the ionic state (Z) decreases, and this state is strongly coupled to the valence (V) state at the $90°$-twisted geometry. (Figure adapted from Ref. 33.)

finite basis set. Using a set of one-dimensional problems introduced by Tully [96], we first showed that the method is capable of reproducing exact (numerically converged) quantum results with a small number of basis functions [38]. These one-dimensional tests may seem trivial (to us as well as to the readers), but the important conclusion is not the ability to reproduce exact results but rather the ability to do this with a small number of basis functions—typically less than 10 for these one-dimensional cases. This provided the first empirical evidence for the underlying assumption of the AIMS method, namely that the basis set expansion could converge quickly if classical mechanics provided a good zeroth-order picture. The next set of tests used a two-dimensional, two-electronic-state problem [40,41]. We intentionally chose a model that avoided harmonic and/or separable PESs. In this two-dimensional avoided crossing model, shown in Fig. 1, each of the two diabatic surfaces describes a collinear nonreactive collision between an atom and a diatomic molecule, and the interstate coupling is set to a constant. One diabatic potential corresponds to an $A + BC$ arrangement, and the other corresponds to $AB + C$. The corresponding ground- and excited-state adiabatic potentials describe a reactive atom exchange reaction and a bound linear triatomic molecule, respectively. We used a Morse oscillator for the $AB$ ($BC$) vibration and an exponentially repulsive form for the $A-B$ ($B-C$) interaction. All three atoms were taken to be identical; and force constants, anharmonicities, and masses were chosen according to spectroscopic information for $Li_2$. The PESs are highly anharmonic, and both degrees of freedom are strongly coupled. All the calculations were carried out in the diabatic representation, and the required matrix elements were evaluated analytically. We studied a broad range of relative kinetic energies (atom with respect to center of mass of diatomic molecule) beginning at the ground-state adiabatic barrier and ending at an energy above the bottom of the excited state well. The initial nuclear wavefunction was centered in the asymptotic region (atom far from the molecule which is in the ground vibrational state), and 30 basis functions were used to represent the initial state. Using this initial basis set, quantitative agreement in expectation values and branching ratios for this broad range of energies was demonstrated. Convergence with respect to the initial size of the basis set was investigated by repeating the calculations using 10 and 20 initial basis functions (all other parameters—e.g., $\lambda_0$ and MULTI-SPAWN—were kept the same). The final branching ratios were found to gradually converge to the exact value as the size of the basis increased from 10 to 20 and then 30, and the overall time dependence of the transmission probability was already reproduced with only 10 initial basis functions.

The same two-dimensional two-electronic-state model was used to test the TDB algorithm. All the parameters were kept as before, but instead of using 30 independent initial conditions we chose 10 independent initial conditions ($N_s = 10$) and augmented the basis set to 30 by displacing each initial condition forward and backward in time. The displacements should not be too small (basis functions become linearly dependent) or too large (basis functions become uncoupled). Given these considerations, we did not specify the amount of time for displacement, but rather displaced the basis functions so that the overlap between any two neighboring basis functions that belong to the same seed is ~0.7. For an initial state that is represented by 10 independent (and 20 time-displaced) basis functions versus 30 independent basis functions, the reduction in computational cost is almost a factor of two. Achieving this reduction in computational effort does carry a penalty: Matrix elements that will be reused must be stored. For each seed (10 in this case), all the matrix elements at each point in time between the forward-most and backward-most displaced basis functions should be stored. The total number of matrix elements to be stored is therefore given by the product of the number of new matrix element evaluations (255 in this example) and the number of time points between the forward-most and backward-most basis functions, $N_t$. In the present case, which we expect to be quite typical, the average value of $N_t$ was approximately 20, and therefore the memory requirement was modest. In general, the extra storage requirement

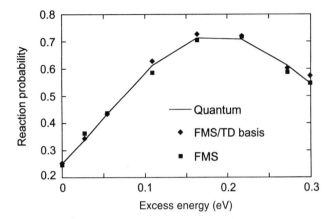

**Figure 18.** Diabatic reaction probability as a function of excess energy in eV for the collinear A + BC → AB + C model shown in Fig. 1. (Excess energy is measured relative to the barrier height of the ground-state adiabatic potential energy surface.) Full line: Exact quantum mechanical results. Diamonds: Multiple-spawning with a time-displaced basis set. Squares: multiple-spawning with a regular basis set. All three calculations are in the diabatic representation, and the range of energies shown begins at the ground-state adiabatic potential energy barrier and extends to an energy above the bottom of the excited-state adiabatic well. (Figure adapted from Ref. 41.)

scales as $O(N_s^2 + N_s N)$, which is insignificant compared to the Hamiltonian matrix itself.

Figure 18 depicts the diabatic reaction probability ($P_{1\rightarrow2} = A + BC \rightarrow AB + C$) as a function of the excess energy for the TDB (diamonds), regular basis set (squares), and numerically exact (full line) methods. (The excess energy is defined as the difference between the initial relative kinetic energy and the barrier height in the adiabatic ground-state potential energy surface; cf. Fig. 4 in Ref. 40.) The agreement between the TDB results and the numerically exact results is very good throughout the energy range shown, and, more importantly, the results of the multiple-spawning method do not change when the TDB is used. Similar agreement is also obtained for expectation values as a function of time.

For this and other problems, the spread in the results depends on the initial size of the basis set. For small basis sets, the final branching ratios depend on the initial conditions—position and momentum parameters that define each basis function. Because in the TDB only 10 (out of 30) initial conditions were chosen independently, it is instructive to compare its results to the ones obtained when only 10 independent basis functions are used to represent the initial wavefunction. The purpose of this comparison is to examine the dependence of the branching ratios on the initial conditions and not to demonstrate an improvement in the results. Such an improvement is expected because the

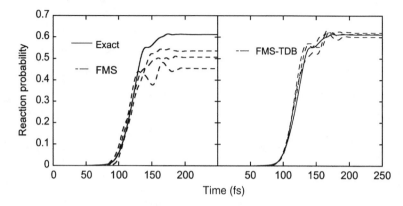

**Figure 19.** Diabatic reaction probability as a function of time (in femtoseconds) for the collinear $A + BC \rightarrow AB + C$ model of Fig. 1, at an excess energy of 0.109 eV. In both panels, the full line designates the exact quantum mechanical results and the dashed lines are multiple spawning results with different initial conditions. Right panel: TDB using an initial basis set with 30 basis functions and 10 seeds. Left panel: Regular basis set using an initial basis set with 10 basis functions. (Figure adapted from Ref. 41.)

computational effort increases when 10 independent basis functions are augmented with 20 TDB functions. This comparison is shown in Fig. 19. In both panels, the full line designates the numerically exact results and the dashed lines designate three different runs of the regular multiple spawning (left panel) and the TDB version (right panel). The TDB results are clearly converged, whereas the ones without the TDB functions are not (i.e., they depend on the initial conditions).

The results of this test of the TDB-FMS method are encouraging, and we expect the gain in efficiency to be more significant for larger molecules and/or longer time evolutions. Furthermore, as noted briefly before, the approximate evaluation of matrix elements of the Hamiltonian may be improved if we can further exploit the temporal nonlocality of the Schrödinger equation.

In the past decade, vibronic coupling models have been used extensively and successfully to explain the short-time excited-state dynamics of small to medium-sized molecules [200–202]. In many cases, these models were used in conjunction with the MCTDH method [203–207] and the comparison to experimental data (typically electronic absorption spectra) validated both the MCTDH method and the model potentials, which were obtained by fitting high-level quantum chemistry calculations. In certain cases the *ab initio*-determined parameters were modified to agree with experimental results (e.g., excitation energies). The MCTDH method assumes the existence of factorizable parameterized PESs and is thus very different from AIMS. However, it does scale more favorably with system size than other numerically exact quantum dynamics techniques (e.g., pseudospectral Fourier methods). Therefore, the availability of the MCTDH method and its application to vibronic coupling models provide a unique opportunity to compare FMS and numerically converged quantum dynamics for problems of large dimensionality (e.g., up to 24 degrees of freedom). Vibronic coupling model Hamiltonians are constructed using the mass-weighted ground-state normal modes, $Q$, and the diabatic representation. They typically contain (i) a term that describes the ground-state Hamiltonian in the diabatic representation, (ii) a matrix of state energies, (iii) linear on-diagonal expansion terms, (iv) linear off-diagonal coupling terms, (v) quadratic and bilinear diagonal terms, and (vi) bilinear off-diagonal terms. The first four terms appear in all vibronic coupling models, while the last two occur only when the Hamiltonian is expanded to second order. As a test case for the FMS method, we have chosen the electronic absorption spectrum of pyrazine. Various model Hamiltonians have been developed for this system, enabling tests of the FMS method using three- [208], four- [209], and 24-dimensional [202] model Hamiltonians. Both the three- and four- dimensional Hamiltonians used here include only linear coupling terms, while the 24-dimensional one includes also quadratic on- and off-diagonal coupling terms. The 24-dimensional Hamiltonian is discussed in detail in Ref. 202, and hence for simplicity here we only write the simpler three- and four-dimensional ones explicitly:

$$\mathbf{H} = \begin{bmatrix} H_0 - \Delta + \sum_i \kappa_i^1 Q_i & \lambda Q_{10a} \\ \lambda Q_{10a} & H_0 + \Delta + \sum_i \kappa_i^2 Q_i \end{bmatrix} \quad (3.1)$$

In Eq. (3.1), $H_0$ is a three- or four- dimensional harmonic oscillator Hamiltonian

$$H_0 = \sum_i \frac{\omega_i}{2} \left( -\frac{\partial^2}{\partial Q_i^2} + Q_i^2 \right) \quad (3.2)$$

$\Delta$ is the energy of the state (i.e., $2\Delta$ is the energy difference between the two electronic states), $\kappa_i^j$ are the gradients of the excitation energy with respect to the normal coordinate $Q_i$ at the reference geometry, and $\lambda$ is the vibronic-coupling constant. In the case of pyrazine, the two electronic states are $S_1[^1B_{3u}(n\pi^*)]$ and $S_2[^1B_{2u}(\pi\pi^*)]$, and only the totally symmetric $\nu_{10a}$ vibrational mode can couple these two states. The other, non-totally symmetric modes are $\nu_1$ and $\nu_{6a}$ for the three-dimensional model and $\nu_1$, $\nu_{6a}$, and $\nu_{9a}$ for the four-dimensional one. The parameters for the three- and four-dimensional models are listed in Tables I and II, and those for the 24-dimensional model are listed in Ref. 202. For the three- and four-dimensional models, numerically converged results were obtained using the Newton interpolation formula [45] and Fourier techniques, [44]; and for the 24-dimensional model, (nearly) numerically converged results were obtained using version 8 of the MCTDH program [210]. In all cases the

TABLE I

Parameters for the Three Dimensional $S_1$-$S_2$ Vibronic Coupling Model of Pyrazine [208]. All Quantities are in eV

|  | $\omega$ | $\kappa^1$ | $\kappa^2$ | $\lambda$ | $\Delta$ |
|---|---|---|---|---|---|
| $v_1$ | 0.126 | 0.037 | −0.254 |  |  |
| $v_{6a}$ | 0.074 | −0.105 | 0.149 |  |  |
| $v_{10a}$ | 0.118 |  |  | 0.262 |  |
|  |  |  |  |  | 0.450 |

calculations were performed in the diabatic representation. This representation is advantageous because all the required integrals can be evaluated analytically, but there is a penalty for using it. In the diabatic representation, the coupling between the two states is broad and in particular it is nonnegligible in the Franck–Condon region, implying that when we begin the simulation the system is already in a nonadiabatic region. As a consequence, a large number of basis functions are spawned and the spawning should begin before the actual simulation begins. To overcome the second problem we "pre-spawn" basis functions on the initially unpopulated electronic state ($S_1$ in this case). After selecting initial conditions for the basis functions on the $S_2$ state (these were chosen from the Wigner transform of the ground vibrational wavefunction on $S_0$), a mirror basis, with the same position and momentum parameters, was placed on the $S_1$ state. None of these ("virtual") basis functions on the $S_1$ state were populated at $t = 0$; however, once the simulation started, their amplitudes quickly changed. In order to control the number of basis functions spawned, we have (1) set an upper limit (of 900) on the size of the nuclear basis set (i.e., once the number of basis functions on both electronic states reaches 900, spawning is no longer allowed regardless of the magnitude of the nonadiabatic coupling) and (2) ramped the spawning threshold $\lambda_0$ during the propagation (i.e., the magnitude of $\lambda_0$ and $\lambda_f$ increases as a function of time). (Option 1 was always used, whereas 2

TABLE II

Parameters for the Four Dimensional $S_1$-$S_2$ Vibronic Coupling Model of Pyrazine [209]. All Quantities are in eV

|  | $\omega$ | $\kappa^1$ | $\kappa^2$ | $\lambda$ | $\Delta$ |
|---|---|---|---|---|---|
| $v_1$ | 0.12730 | 0.0470 | 0.2012 |  |  |
| $v_{6a}$ | 0.07400 | −0.0964 | 0.1194 |  |  |
| $v_{9a}$ | 0.15680 | 0.1594 | 0.0484 |  |  |
| $v_{10a}$ | 0.09357 |  |  | 0.1825 |  |
|  |  |  |  |  | 0.46165 |

was enforced in only some of the calculations.) In all the three-, four-, and 24-dimensional calculations, the size of the initial basis set (on each electronic state) was 20, 40, and 60, respectively, and only the $S_2$ state was populated at $t = 0$. The comparison to the exact methods (Fourier or MCTDH) focuses on the electronic absorption spectrum. The absorption spectrum obtained using MCTDH for the 24-dimensional model Hamiltonian has been shown to be in good agreement with the experimentally measured one. Certain aspects of this class of models are more challenging than our previous tests—high dimensionality and existence of conical intersections. Other aspects are not; the zeroth-order Hamiltonian is harmonic.

The three panels of Fig. 20 compare the FMS and numerically exact electronic absorption spectra. For the three- and four-dimensional models, the FMS method reproduces both the width of the spectrum and its finer features (location and intensity of the various peaks) very well. For the 24-dimensional model, the agreement in the width is good but the finer details are not reproduced as well as for the lower-dimensionality models. Presumably this is a consequence of the relatively small size of the nuclear basis set. It is quite possible that, in the case of a 24-dimensional model, the basis set is too small to correctly describe the spreading of the wavepacket. In Ref. 211, the convergence of the FMS method as a function of the initial basis set size and spawning threshold is discussed in detail, and here we only make a few comments. As the spawning threshold is increased, fewer basis functions are spawned and therefore the results gradually deteriorate. In the case of the three-dimensional model, the rate at which the results deteriorate is quite slow as demonstrated in Fig. 21. The exact results are compared to FMS results with very different spawning thresholds, from 2.5 to 20 atomic units. In all cases the envelopes of the FMS and exact absorption spectra are in good agreement. However, the details of the spectrum are not reproduced when the spawning threshold is increased. As might be expected, the short-time behavior of the wavefunction which gives rise to the spectral envelope is less sensitive to the spawning threshold.

## IV. OUTLOOK AND CONCLUDING REMARKS

In many chemical and even biological systems the use of an *ab initio* quantum dynamics method is either advantageous or mandatory. In particular, photochemical reactions may be most amenable to these methods because the dynamics of interest is often completed on a short (subpicosecond) timescale. The AIMS method has been developed to enable a realistic modeling of photochemical reactions, and in this review we have tried to provide a concise description of the method. We have highlighted (a) the obstacles that should be overcome whenever an *ab initio* quantum chemistry method is coupled to a quantum propagation method, (b) the wavefunction ansatz and fundamental

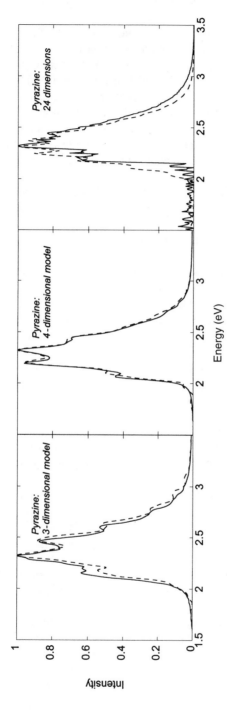

**Figure 20.** The $(S_0 \rightarrow S_2)$ absorption spectrum of pyrazine for reduced three- and four-dimensional models (left and middle panels) and for a complete 24-vibrational model (right panel). For the three- and four-dimensional models, the exact quantum mechanical results (full line) are obtained using the Fourier method [43,45]. For the 24-dimensional model (nearly converged), quantum mechanical results are obtained using version 8 of the MCTDH program [210]. For all three models, the calculations are done in the diabatic representation. In the multiple spawning calculations (dashed lines) the spawning threshold ($\lambda_0$) is set to 0.05, the initial size of the basis set for the three-, four-, and 24-dimensional models is 20, 40, and 60, and the total number of basis functions is limited to 900 (i.e., regardless of the magnitude of the effective nonadiabatic coupling, we do not spawn new basis functions once the total number of basis functions reaches 900).

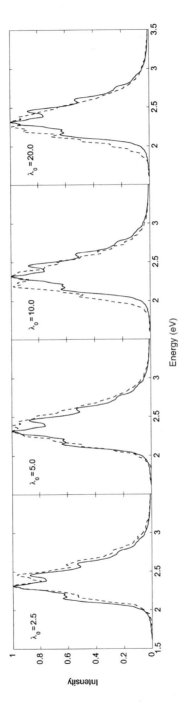

**Figure 21.** The $(S_0 \rightarrow S_2)$ absorption spectrum of pyrazine for the reduced three-dimensional model using different spawning thresholds. Full line: Exact quantum mechanical results. Dashed line: Multiple spawning results for $\lambda_0 = 2.5, 5.0, 10,$ and 20. (All other computational details are as in Fig. 20.) As the spawning threshold is increased, the number of spawned basis functions decreases, the numerical effort decreases, and the accuracy of the result deteriorates (slowly). In this case, the final size of the basis set (at $t = 0.5$ ps) varies from 860 for $\lambda_0 = 2.5$ to 285 for $\lambda_0 = 20$.

equations of the AIMS method, (c) required and optional approximations, (d) the analysis of the results, and (e) some advanced topics that are not required for a basic understanding of the method. The accuracy of the AIMS and FMS methods has been discussed. When the FMS method was first introduced, its accuracy (and computational efficiency) was tested using simple one- and two-dimensional problems. More recently, these tests have been extended to larger systems of three, four, and 24 dimensions. In this review we have discussed only the latter set of tests, because the older, and to some extent easier, tests have been extensively discussed in past publications. The applications section showed that it is indeed possible to model photochemical reactions from first principles for molecules of general chemical interest.

Throughout the review we have emphasized possible directions for improvements and extensions. There is considerable room for improvement in our treatment of both the electronic structure and nuclear dynamics—in certain cases with minimal increase in computational effort. Consider first the electronic structure and the case of ethylene. The electronic structure treatment that we have used does not do full justice to the Rydberg states of the molecule. We are currently exploring the role of the Rydberg states by carrying out calculations on ethylene with more extended basis sets; while computationally tractable, these calculations are quite challenging. Certainly, such studies on cyclobutene and butadiene, while of great interest to us, are not yet possible. While the ever-increasing speed of computers will likely make these possible in the near future, there is a clear need to implement analytical gradients in our calculations and also for new approaches to the electronic structure of excited states. Time-dependent density functional theory is one promising avenue. Others include combinations of interpolation and direct dynamics strategies, as well as hybrid quantum mechanical/classical electrostatic models of potential energy surfaces. The latter direction is currently being implemented and the hybrid interpolation/direct dynamics strategy has already been implemented (for a triatomic system) as discussed in Section III.A.

Similarly, improvement in the accuracy of the nuclear dynamics would be fruitful. While in this review we have shown that, in the absence of any approximations beyond the use of a finite basis set, the multiple spawning treatment of the nuclear dynamics can border on numerically exact for model systems with up to 24 degrees of freedom, we certainly do not claim this for the *ab initio* applications presented here. In principle, we can carry out sequences of calculations with larger and larger nuclear basis sets in order to demonstrate that experimentally observable quantities have converged. In the context of AIMS, the cost of the electronic structure calculations precludes systematic studies of this convergence behavior for molecules with more than a few atoms. A similar situation obtains in time-independent quantum chemistry—the only reliable way to determine the accuracy of a particular calculation is to perform a sequence of calculations in a hierarchy of increasing basis sets and electron correlation. What is critically different about time-independent quantum chemistry is that well-defined and extensively tested hierarchies exist—for example, the correlation consistent basis sets of Dunning and co-workers [194,196,212] and the increasing orders of perturbation theory, MPn [213]. Developing such hierarchies for the FMS method is an important goal that is prerequisite to the widespread use of AIMS. We are working toward this goal, but it is important to recognize that it will only be useful if it arises from an extensive set of applications. It is not fruitful to propose a computational hierarchy unless the incremental improvements going from one step to the next are similar throughout, and at the present stage it appears that this can only be determined empirically.

Practical use of the AIMS method requires certain approximations (see Section II.D) with the most severe being the use of a first-order saddle point approximation for the Hamiltonian (diagonal and off-diagonal) matrix elements. Even in the case of predetermined analytical potential energy surfaces, this approximation is required if the matrix elements of the Hamiltonian cannot be evaluated analytically. When the SP approximation was first introduced, its accuracy was tested [74]. Although the results were favorable, more rigorous tests using an extensive set of applications/models should be performed. Also, as discussed in Section II.D, the current implementation of the SP approximation is wasteful. Although we have information about the PESs for all possible pairs of nuclear basis functions (populated or not), individual matrix elements of the Hamiltonian (each involving one pair of basis functions) are evaluated using only a single data point (i.e., one relevant pair of basis functions). If the proper weighting scheme for all the other data points were developed, we could improve the accuracy of the SP approximation without any increase in computational expense. Additional increase in accuracy, again without increasing computational cost, could be obtained by incorporating information about the PESs from previous time-points in the calculation. Preliminary attempts in this direction have been discussed in Section II.F, but clearly much more can and should be done.

We hope that we have convinced the reader that, even though significant improvements can be expected in the future, the AIMS method is currently practical and useful for problems that are of chemical interest. In the near future we expect to be able to handle condensed phase and biological systems. This will open a new challenging and exciting area for applications of *ab initio* quantum dynamics.

## Acknowledgments

We began to work on the AIMS method under the guidance of Professor R. D. Levine. We are grateful for his insight, knowledge, and enthusiasm. We would like to thank Dr. G. Ashkenazi for providing us his efficient and elegant Fourier propagation and Wigner transform codes.

Dr. K. Thompson has taught us about interpolation methods. It was a pleasure to work with him on the hybrid interpolation/*ab initio* AIMS method. J. Quenneville diligently worked with us on the photochemistry of ethylene, and we greatly appreciate his contributions. The MCTDH program enabled us to extensively test our method. We thank Professor H.-D. Meyer for providing the program and promptly answering all our questions about it. Dr. L. Murga contributed to the initial coding of the vibronic coupling models. Professor D. G. Truhlar and Dr. M. Hack have presented us with new challenging tests for the FMS method (not presented here) and have engaged us in stimulating discussions about semiclassical and surface-hopping dynamics. Financial support for this work was provided by the U.S. Department of Energy through the University of California under Subcontract No. B341494, the National Science Foundation (CHE-97-33403), and the National Institutes of Health (PHS-5-P41-RR05969). TJM is the grateful recipient of CAREER, Research Innovation, Beckman Young Investigator, and Packard Fellow awards from NSF, Research Corporation, The Beckman Foundation, and the Packard Foundation, respectively.

# References

1. C. Leforestier, *J. Chem. Phys.* **68**, 4406 (1978).

2. R. Car and M. Parrinello, *Phys. Rev. Lett.* **55**, 2471 (1985).

3. M. Parrinello, *Solid State Comm.* **102**, 107 (1997).

4. R. Car and M. Parrinello, *NATO ASI Ser., Ser. B* **186**, 455–476 (1989).

5. D. K. Remler and P. A. Madden, *Mol. Phys.* **70**, 921 (1990).

6. M. E. Tuckerman, P. J. Ungar, T. Vonrosenvinge, and M. L. Klein, *J. Phys. Chem.* **100**, 12878–12887 (1996).

7. M. C. Payne, M. P. Teter, D. C. Allan, T. A. Arias, and J. D. Joannopoulos, *Rev. Mod. Phys.* **64**, 1045 (1992).

8. B. Hartke and E. A. Carter, *Chem. Phys. Lett.* **189**, 358 (1992).

9. B. Hartke and E. A. Carter, *J. Chem. Phys.* **97**, 6569–6578 (1992).

10. Z. Liu, L. E. Carter, and E. A. Carter, *J. Phys. Chem.* **99**, 4355–4359 (1995).

11. S. Hammes-Schiffer and H. C. Andersen, *J. Chem. Phys.* **99**, 523 (1993).

12. S. A. Maluendes and M. Dupuis, *Int. J. Quant. Chem.* **42**, 1327 (1992).

13. J. Jellinek, V. Bonacic-Koutecky, P. Fantucci, and M. Wiechert, *J. Chem. Phys.* **101**, 10092 (1994).

14. M. Klessinger and J. Michl, *Excited States and Photochemistry of Organic Molecules*, VCH Publishers, New York, 1995.

15. M. Garavelli, F. Bernardi, M. Olivucci, T. Vreven, S. Klein, P. Celani, and M. A. Robb, *Faraday Disc.* **110** (1998).

16. R. Ditchfield, J. E. Del Bene, and J. A. Pople, *J. Am. Chem. Soc.* **94**, 703 (1972).

17. J. B. Foresman, M. Head-Gordon, J. A. Pople, and M. J. Frisch, *J. Phys. Chem.* **96**, 135–149 (1992).

18. K. B. Wiberg, C. M. Hadad, J. B. Foresman, and W. A. Chupka, *J. Phys. Chem.* **96**, 10756 (1992).

19. R. Singh and B. M. Deb, *Phys. Rep.* **311**, 47–94 (1999).

20. I. Frank, J. Hutter, D. Marx, and M. Parrinello, *J. Chem. Phys.* **108**, 4060–4069 (1998).

21. K. Yabana and G. F. Bertsch, *Int. J. Quantum. Chem.* **75**, 55 (1999).

22. K. K. Docken and J. Hinze, *J. Chem. Phys.* **57**, 4928 (1972).

23. H.-J. Werner and W. Meyer, *J. Chem. Phys.* **74**, 5794 (1981).

24. H.-J. Werner and W. Meyer, *J. Chem. Phys.* **74**, 5802 (1981).

25. B. H. Lengsfield III and D. R. Yarkony, *Adv. Chem. Phys.* **82**, 1 (1992).

26. H. F. Schaefer and C. F. Bender, *J. Chem. Phys.* **55**, 1720 (1971).

27. P. J. Hay, T. H. Dunning, Jr., and W. A. Goddard III, *J. Chem. Phys.* **62**, 3912–3924 (1975).

28. M. J. S. Dewar, J. A. Hashmall, and C. G. Venier, *J. Am. Chem. Soc.* **90**, 1953–1957 (1968).

29. A. I. Krylov, *Chem. Phys. Lett.* **338**, 375–384 (2001).

30. E. R. Davidson, *J. Phys. Chem.* **100**, 6161 (1996).

31. W. A. Goddard III, T. H. Dunning, Jr., W. J. Hunt, and P. J. Hay, *Acc. Chem. Res.* **6**, 368–376 (1973).

32. M. Ben-Nun and T. J. Martínez, *Chem. Phys. Lett.* **298**, 57 (1998).

33. M. Ben-Nun and T. J. Martínez, *Chem. Phys.* **259**, 237 (2000).

34. K. Ruedenberg and G. J. Atchity, *J. Chem. Phys.* **99**, 3799 (1993).

35. W. Domcke and C. Woywod, *Chem. Phys. Lett.* **216**, 362 (1993).

36. C. Galloy and J. C. Lorquet, *J. Chem. Phys.* **67**, 4672 (1977).

37. R. Cimiraglia, M. Persico, and J. Tomasi, *Chem. Phys.* **53**, 357 (1980).

38. T. J. Martínez, M. Ben-Nun, and R. D. Levine, *J. Phys. Chem.* **100**, 7884 (1996).

39. T. J. Martínez, M. Ben-Nun, and R. D. Levine, *J. Phys. Chem.* **101A**, 6389 (1997).

40. M. Ben-Nun and T. J. Martínez, *J. Chem. Phys.* **108**, 7244 (1998).

41. M. Ben-Nun and T. J. Martínez, *J. Chem. Phys.* **110**, 4134 (1999).

42. E. J. Heller, *J. Chem. Phys.* **62**, 1544 (1975).

43. D. Kosloff and R. Kosloff, *J. Comp. Phys.* **52**, 35 (1983).

44. R. Kosloff, *Annu. Rev. Phys. Chem.* **45**, 145 (1994).

45. G. Ashkenazi, R. Kosloff, S. Ruhman, and H. Tal-Ezer, *J. Chem. Phys.* **103**, 10005 (1995).

46. E. J. Heller, *J. Chem. Phys.* **75**, 2923 (1981).

47. D. J. Tannor and E. J. Heller, *J. Chem. Phys.* **77**, 202 (1982).

48. E. J. Heller, R. L. Sundberg, and D. Tannor, *J. Phys. Chem.* **86**, 1822 (1982).

49. E. J. Heller, *Acc. Chem. Res.* **14**, 368 (1981).

50. A. R. Walton and D. E. Manolopoulos, *Chem. Phys. Lett.* **244**, 448 (1995).

51. S. M. Anderson, T. J. Park, and D. Neuhauser, *Phys. Chem. Chem. Phys.* **1**, 1343 (1998).

52. S. M. Anderson, J. I. Zink, and D. Neuhauser, *Chem. Phys. Lett.* **291**, 387 (1999).

53. A. E. Cardenas and R. D. Coalson, *Chem. Phys. Lett.* **265**, 71 (1997).

54. A. E. Cardenas, R. Krems, and R. D. Coalson, *J. Phys. Chem. A* **103**, 9569 (1999).

55. S.-I. Sawada, R. Heather, B. Jackson, and H. Metiu, *J. Chem. Phys.* **83**, 3009 (1985).

56. S. Sawada and H. Metiu, *J. Chem. Phys.* **84**, 227 (1986).

57. H.-D. Meyer and W. H. Miller, *J. Phys. Chem.* **70**, 3214 (1979).

58. H.-D. Meyer, *Chem. Phys.* **82**, 199–205 (1983).

59. G. D. Billing, *Int. Rev. Phys. Chem.* **13**, 309 (1994).

60. G. Stock, *J. Chem. Phys.* **103**, 1561–1573 (1995).

61. P. Kramer and M. Saraceno, *Lecture Notes in Physics*, Vol. 140, *Geometry of the Time-Dependent Variational Principle*, Springer, Berlin, 1981.

62. E. Deumens, A. Diz, R. Longo, and Y. Ohrn, *Rev. Mod. Phys.* **66**, 917 (1994).

63. E. J. Heller, *J. Chem. Phys.* **64**, 63 (1976).

64. R. T. Skodje and D. G. Truhlar, *J. Chem. Phys.* **80**, 3123 (1984).

65. R. C. Brown and E. J. Heller, *J. Chem. Phys.* **75**, 186 (1981).

66. M. Ben-Nun, T. J. Martínez, and R. D. Levine, *Chem. Phys. Lett.* **270**, 319–326 (1997).

67. M. Ben-Nun, T. J. Martínez, and R. D. Levine, *J. Phys. Chem.* **101A**, 7522 (1997).

68. G. C. Schatz and M. A. Ratner, *Quantum Mechanics in Chemistry*, Prentice-Hall, Englewood Cliffs, NJ, 1993.

69. R. K. Preston and J. C. Tully, *J. Chem. Phys.* **54**, 4297 (1971).

70. M. F. Herman and E. Kluk, *Chem. Phys.* **91**, 27 (1984).

71. J. Matthews and R. L. Walker, *Mathematical Methods of Physics*, W. A. Benjamin, New York, 1990.

72. K. Ruedenberg, *J. Chem. Phys.* **19**, 1433 (1951).

73. T. J. Martínez and R. D. Levine, *Chem. Phys. Lett.* **259**, 252 (1996).

74. T. J. Martínez and R. D. Levine, *J. Chem. Soc. Faraday Trans.* **93**, 940 (1997).

75. M. F. Trotter, *Proc. Am. Math. Soc.* **10**, 545 (1959).

76. E. C. G. Stueckelberg, *Helv. Phys. Acta.* **5**, 369 (1932).

77. E. E. Nikitin, *Theory of Elementary Atomic and Molecular Processes in Gases*, Clarendon, Oxford, 1974.

78. R. P. Feynman and A. R. Hibbs, *Quantum Mechanics and Path Integrals*, McGraw-Hill, New York, 1965.

79. R. P. Feynman, *Statistical Mechanics*, Addison-Wesley, Redwood City, CA, 1972.

80. L. S. Schulman, *Techniques and Applications of Path Integration*, Wiley, New York, 1981.

81. H. Kleinert, *Path Integrals in Quantum Mechanics*, World Scientific, Singapore, 1990.

82. B. J. Berne and D. Thirumalai, *Annu. Rev. Phys. Chem.* **37**, 401 (1986).

83. M. J. Gillan, in *Computer Modelling of Fluids, Polymers, and Solids*, C. R. A. Catlow, S. C. Parker, and M. P. Allen, eds., Kluwer, Dordrecht, 1990.

84. D. M. Ceperley, *Rev. Mod. Phys.* **67**, 279 (1995).

85. N. Makri, *Comp. Phys. Commun.* **63**, 389 (1991).

86. M. E. Tuckerman, B. J. Berne, G. J. Martyna, and M. L. Klein, *J. Chem. Phys.* **99**, 2796 (1993).

87. D. Chandler and P. G. Wolynes, *J. Chem. Phys.* **74**, 4078 (1981).

88. M. E. Tuckerman, D. Marx, M. L. Klein, and M. Parrinello, *J. Chem. Phys.* **104**, 5579 (1996).

89. D. Marx and M. Parrinello, *J. Chem. Phys.* **104**, 4077 (1996).

90. M. E. Tuckerman, D. Marx, M. L. Klein, and M. Parrinello, *J. Chem. Phys.* **104**, 5579–5588 (1996).

91. M. E. Tuckerman, D. Marx, M. L. Klein, and M. Parrinello, *Science* **275**, 817 (1997).

92. M. E. Tuckerman and G. J. Martyna, *J. Phys. Chem. B* **104**, 159–178 (2000).

93. M. Shiga, M. Tachikawa, and S. Miura, *Chem. Phys. Lett.* **332**, 396–402 (2000).

94. M. Pavese, D. R. Berard, and G. A. Voth, *Chem. Phys. Lett.* **300**, 93 (1999).

95. J. C. Tully and R. K. Preston, *J. Chem. Phys.* **55**, 562 (1971).

96. J. C. Tully, *J. Chem. Phys.* **93**, 1061 (1990).

97. N. Makri and W. H. Miller, *J. Chem. Phys.* **91**, 4026 (1989).

98. T. D. Sewell, Y. Guo, and D. L. Thompson, *J. Chem. Phys.* **103**, 8557 (1995).

99. T. Vreven, F. Bernardi, M. Garavelli, M. Olivucci, M. A. Robb, and H. B. Schlegel, *J. Am. Chem. Soc.* **119**, 12687–12688 (1997).

100. M. Hartmann, J. Pittner, and V. Bonacic-Koutecky, *J. Chem. Phys.* **114**, 2123–2136 (2001).

101. M. Ben-Nun and T. J. Martínez, *J. Phys. Chem.* **103**, 6055 (1999).

102. M. F. Herman, *Annu. Rev. Phys. Chem.* **45**, 83–111 (1994).

103. N. Corbin and K. Singer, *Mol. Phys.* **46**, 671 (1982).

104. D. Hsu and D. F. Coker, *J. Chem. Phys.* **96**, 4266 (1992).

105. F. Arickx, J. Broeckhove, E. Kesteloot, L. Lathouwers, and P. Van Leuven, *Chem. Phys. Lett.* **128**, 310 (1986).

106. K. Singer and W. Smith, *Mol. Phys.* **57**, 761 (1986).

107. R. Heather and H. Metiu, *Chem. Phys. Lett.* **118**, 558 (1985).

108. R. Heather and H. Metiu, *J. Chem. Phys.* **84**, 3250 (1986).

109. K. G. Kay, *J. Chem. Phys.* **101**, 2250 (1994).

110. D. J. Tannor and S. Garashchuk, *Annu. Rev. Phys. Chem.* **51**, 553–600 (2000).

111. M. F. Herman, *J. Chem. Phys.* **85**, 2069 (1986).

112. W. H. Miller, *J. Phys. Chem.* **105**, 2942–2955 (2001).

113. X. Sun and W. H. Miller, *J. Chem. Phys.* **106**, 6346 (1997).

114. B. Jackson and H. Metiu, *J. Chem. Phys.* **85**, 4129 (1986).

115. S.-I. Sawada and H. Metiu, *J. Chem. Phys.* **84**, 6293 (1986).

116. R. D. Coalson, *J. Phys. Chem.* **100**, 7896 (1996).

117. C. C. Martens and J. Y. Fang, *J. Chem. Phys.* **106**, 4918 (1997).

118. A. Donoso and C. C. Martens, *J. Phys. Chem.* **102**, 4291 (1998).

119. C.-C. Wan and J. Schofield, *J. Chem. Phys.* **112**, 4447 (2000).

120. R. Kapral and G. Ciccotti, *J. Chem. Phys.* **110**, 8919 (1999).

121. C.-C. Wan and J. Schofield, *J. Chem. Phys.* **113**, 7047 (2000).

122. M. Ben-Nun and T. J. Martínez, *J. Phys. Chem.* **103**, 10517 (1999).

123. O. V. Prezhdo and P. J. Rossky, *J. Chem. Phys.* **107**, 825 (1997).

124. M. D. Hack, A. M. Wensmann, D. G. Truhlar, M. Ben-Nun, and T. J. Martínez, *J. Chem. Phys.* **115**, 000 (2001).

125. K. Thompson and T. J. Martínez, *J. Chem. Phys.* **110**, 1376 (1998).

126. M. Ben-Nun and T. J. Martínez, *J. Chem. Phys.* **112**, 6113 (1999).

127. J. Ischtwan and M. A. Collins, *J. Chem. Phys.* **100**, 8080 (1994).

128. M. J. T. Jordan, K. C. Thompson, and M. A. Collins, *J. Chem. Phys.* **102**, 5647 (1995).

129. M. J. T. Jordan, K. C. Thompson, and M. A. Collins, *J. Chem. Phys.* **103**, 9669 (1995).

130. M. J. T. Jordan and M. A. Collins, *J. Chem. Phys.* **104**, 4600 (1996).

131. M. A. Collins, *Adv. Chem. Phys.* **93**, 389–453 (1996).

132. M. A. Collins and K. C. Thompson, in *Chemical Group Theory: Techniques and Applications,* Vol. 4, D. Bonchev and D. H. Rouvray, eds., Gordon and Breach, Reading, MA, 1995, p. 191.

133. K. C. Thompson and M. A. Collins, *J. Chem. Soc. Faraday Trans.* **93**, 871 (1997).

134. R. A. Marcus and M. E. Coltrin, *J. Chem. Phys.* **67**, 2609 (1977).

135. C. J. Cerjan, S. Shi, and W. H. Miller, *J. Phys. Chem.* **86**, 2244 (1982).

136. D. G. Truhlar, A. I. Isaacson, and B. C. Garrett, in *Theory of Chemical Reaction Dynamics,* Vol. 4, M. Baer, ed., CRC Press, Boca Raton, FL, 1985.

137. G. C. Lynch, D. G. Truhlar, and B. C. Garrett, *J. Chem. Phys.* **90**, 3102 (1989).

138. M. Ben-Nun and T. J. Martínez, *J. Phys. Chem.* **102A**, 9607 (1998).

139. M. Ben-Nun, F. Molnar, H. Lu, J. C. Phillips, T. J. Martínez, and K. Schulten, *Faraday Disc.* **110**, 447 (1998).

140. T. J. Martínez, *Chem. Phys. Lett.* **272**, 139 (1997).

141. C. F. Melius and W. A. Goddard, III, *Phys. Rev. A* **10**, 1528 (1974).

142. *Gaussian Basis Sets for Molecular Calculations*, Vol. 0, S. Huzinaga, ed., Elsevier, Amsterdam, 1984.

143. E. Teller, *Isr. J. Chem.* **7**, 227 (1969).

144. G. J. Atchity, S. S. Xantheas, and K. Ruedenberg, *J. Chem. Phys.* **95**, 1862 (1991).

145. U. Manthe and H. Koppel, *J. Chem. Phys.* **93**, 1658 (1990).

146. R. Baer, D. Charutz, R. Kosloff, and M. Baer, *J. Chem. Phys.* **105**, 9141 (1996).

147. D. R. Yarkony, *J. Chem. Phys.* **114**, 2601–2613 (2001).

148. G. Orlandi, F. Zerbetto, and M. Z. Zgierski, *Chem. Rev.* **91**, 867 (1991).

149. B. A. Balko, J. Zhang, and Y. T. Lee, *J. Phys. Chem.* **101**, 6611 (1997).

150. E. M. Evleth and A. Sevin, *J. Am. Chem. Phys.* **103**, 7414–7422 (1981).

151. J. R. McNesby and H. Okabe, *Adv. Photochem.* **3**, 228 (1964).

152. P. Farmanara, V. Stert, and W. Radloff, *Chem. Phys. Lett.* **288**, 518 (1998).

153. P. G. Wilkinson and R. S. Mulliken, *J. Chem. Phys.* **23**, 1895 (1955).

154. R. McDiarmid and E. Charney, *J. Chem. Phys.* **47**, 1517 (1967).

155. P. D. Foo and K. K. Innes, *J. Chem. Phys.* **60**, 4582 (1974).

156. R. J. Buenker and S. D. Peyerimhoff, *Chem. Phys.* **9**, 75–89 (1976).

157. A. J. Merer and R. S. Mulliken, *Chem. Rev.* **69**, 639 (1969).

158. R. S. Mulliken, *J. Chem. Phys* **66**, 2448–2451 (1977).

159. W. Siebrand, F. Zerbetto, and M. Z. Zgierski, *Chem. Phys. Lett.* **174**, 119 (1990).

160. J. Ryu and B. S. Hudson, *Chem. Phys. Lett.* **245**, 448 (1995).

161. V. Bonacic-Koutecky, P. Bruckmann, P. Hiberty, J. Koutecky, C. Leforestier, and L. Salem, *Angew. Chem. Int. Ed. Engl.* **14**, 575 (1975).

162. L. Salem and P. Bruckmann, *Nature* **258**, 526–528 (1975).

163. L. Salem, *Science* **191**, 822–830 (1976).

164. C. Longstaff, R. D. Calhoon, and R. R. Rando, *Proc. Natl. Acad. Sci. USA* **83**, 4209–4213 (1986).

165. B. R. Brooks and H. F. Schaefer III, *J. Am. Chem. Soc.* **101**, 307–311 (1979).

166. R. J. Buenker, V. Bonacic-Koutecky, and L. Pogliani, *J. Chem. Phys.* **73**, 1836–1849 (1980).

167. R. J. Sension and B. S. Hudson, *J. Chem. Phys.* **90**, 1377–1389 (1989).

168. A. H. H. Chang, A. M. Mebel, X. M. Yang, S. H. Lin, and Y. T. Lee, *Chem. Phys. Lett.* **287**, 301 (1998).

169. V. D. Knyazev and I. R. Slagle, *J. Phys. Chem.* **100**, 16899 (1996).

170. M. M. Gallo, T. P. Hamilton, and H. F. Schaefer III, *J. Am. Chem. Soc.* **112**, 8714–8719 (1990).

171. J. H. Jensen, K. Morokuma, and M. S. Gordon, *J. Chem. Phys.* **100**, 1981–1987 (1994).

172. E. F. Cromwell, A. Stolow, M. J. J. Vrakking, and Y. T. Lee, *J. Chem. Phys.* **97**, 4029 (1992).

173. B. A. Balko, J. Zhang, and Y. T. Lee, *J. Chem. Phys.* **97**, 935–942 (1992).

174. A. Stolow, B. A. Balko, E. F. Cromwell, J. Zhang, and Y. T. Lee, *J. Photochem. Photobiol.* **62**, 285 (1992).

175. S. Satyapal, G. W. Johnston, R. Bersohn, and I. Oref, *J. Chem. Phys.* **93**, 6398 (1990).

176. R. B. Woodward and R. Hoffmann, *Ang. Chem. Int. Ed. Eng.* **8**, 781 (1969).

177. W. A. Goddard III, *J. Am. Chem. Soc.* **92**, 7520–7521 (1970).

178. W. A. Goddard III, *J. Am. Chem. Soc.* **94**, 793–807 (1972).

179. H. E. Zimmerman, *Acc. Chem. Res.* **4**, 272 (1971).

180. J. E. Baldwin and S. M. Krueger, *J. Am. Chem. Soc.* **91**, 6444 (1969).

181. W. G. Dauben, J. Rabinowitz, N. D. Vietmeyer, and P. H. Wendschuh, *J. Am. Chem. Soc.* **94**, 4285 (1972).

182. C. W. Spangler and R. P. Hennis, *Chem. Commun.*, 24 (1972).

183. M. O. Trulson, G. D. Dollinger, and R. A. Mathies, *J. Am. Chem. Soc.* **109**, 586 (1987).

184. M. O. Trulson, G. D. Dollinger, and R. A. Mathies, *J. Chem. Phys.* **90**, 4274 (1989).

185. W. J. Leigh, *Can. J. Chem.* **71**, 147 (1993).

186. N. C. Craig, S. S. Borick, T. Tucker, and Y.-Z. Xia, *J. Phys. Chem.* **95**, 3549 (1991).

187. K. B. Wiberg and R. E. Rosenberg, *J. Phys. Chem.* **96**, 8282 (1992).

188. M. K. Lawless, S. D. Wickham, and R. A. Mathies, *J. Am. Chem. Soc.* **116**, 1593 (1994).

189. M. K. Lawless, S. D. Wickham, and R. A. Mathies, *Acc. Chem. Res.* **28**, 493 (1995).

190. R. C. Lord and D. G. Rea, *J. Am. Chem. Soc.* **79**, 2401 (1957).

191. F. Negri, G. Orlandi, F. Zerbetto, and M. Z. Zgierski, *J. Chem. Phys.* **103**, 5911 (1995).

192. W. Adam, T. Oppenlander, and G. Zang, *J. Am. Chem. Soc.* **107**, 3921 (1985).

193. H.-J. Werner, P. J. Knowles, J. Almlöf, R. D. Amos, A. Berning, D. L. Cooper, M. J. O. Deegan, A.-J. Dobbyn, F. Eckert, S. T. Elbert, C. Hampel, R. Lindh, A. W. Lloyd, W. Meyer, A. Nicklass, K. Peterson, R. Pitzer, A. J. Stone, P. R. Taylor, M. E. Mura, P. Pulay, M. Schütz, H. Stoll, and T. Thorsteinsson (1998).

194. T. H. J. Dunning, *J. Chem. Phys.* **90**, 1007 (1989).

195. T. H. Dunning Jr. and P. J. Hay, in *Modern Theoretical Chemistry: Methods of Modern Electronic Structure Theory*, Vol. 3, H. F. Schaefer III, ed., Plenum, New York, 1977, p. 1.

196. R. A. Kendall, T. H. Dunning, Jr., and R. J. Harrison, *J. Chem. Phys.* **96**, 6796 (1992).

197. H. J. Werner and P. J. Knowles, *J. Chem. Phys.* **89**, 5803–5814 (1988).

198. P. J. Knowles and H. J. Werner, *Theor. Chim. Acta* **84**, 95–103 (1992).

199. I. Ohmine, *J. Chem. Phys.* **83**, 2348 (1985).

200. H. Koppel, W. Domcke, and L. S. Cederbaum, *Adv. Chem. Phys.* **57**, 59 (1984).

201. H. Koppel, L. S. Cederbaum, and W. Domcke, *J. Chem. Phys.* **89**, 2023 (1988).

202. A. Raab, G. A. Worth, H.-D. Meyer, and L. S. Cederbaum, *J. Chem. Phys.* **110**, 936 (1999).

203. H.-D. Meyer, U. Manthe, and L. S. Cederbaum, *Chem. Phys. Lett.* **165**, 73 (1990).

204. A. D. Hammerich, U. Manthe, R. Kosloff, H.-D. Meyer, and L. S. Cederbaum, *J. Chem. Phys.* **101**, 5623–5646 (1994).

205. U. Manthe, H.-D. Meyer, and L. S. Cederbaum, *J. Chem. Phys.* **97**, 3199–3213 (1992).

206. U. Manthe, *J. Chem. Phys.* **105**, 6989 (1996).

207. M. H. Beck, A. Jackle, G. A. Worth, and H.-D. Meyer, *Phys. Rep.* **324**, 1 (2000).

208. R. Schneider and W. Domeck, *Chem. Phys. Lett.* **150**, 235 (1988).

209. C. Woywod, W. Domcke, A. L. Sobolewski, and H.-J. Werner, *J. Chem. Phys.* **100**, 1400 (1994).

210. G. A. Worth, M. H. Beck, A. Jackle, and H.-D. Meyer, 2nd edition, University of Heidelberg, Heidelberg, 2000.

211. M. Ben-Nun and T. J. Martínez (work in progress).

212. D. E. Woon and T. H. Dunning, *J. Chem. Phys.* **98**, 1358 (1993).

213. A. Szabo and N. S. Ostlund, *Modern Quantum Chemistry*, McGraw-Hill, New York, 1989.

214. M. Ben-Nun, J. Quenneville, and T. J. Martínez, *J. Phys. Chem.* **104**, 5161 (2000).

215. J. Quenneville, M. Ben-Nun, and T. J. Martínez, *J. Photochem. Photobiol.* **144**, 229 (2001).

216. M. Ben-Nun and T. J. Martínez, *J. Am. Chem. Soc.* **122**, 6299 (2000).

# AUTHOR INDEX

Numbers in parentheses are reference numbers and indicate that the author's work is referred to although his name is not mentioned in the text. Numbers in *italic* show the pages on which the complete references are listed.

# SUBJECT INDEX